Squeezing Minds From Stones

Squeezing Minds From Stones

Cognitive Archaeology and the Evolution of the Human Mind

Edited by Karenleigh A. Overmann

and

Frederick L. Coolidge

OXFORD
UNIVERSITY PRESS

OXFORD
UNIVERSITY PRESS

Oxford University Press is a department of the University of Oxford. It furthers the University's objective of excellence in research, scholarship, and education by publishing worldwide. Oxford is a registered trade mark of Oxford University Press in the UK and certain other countries.

Published in the United States of America by Oxford University Press
198 Madison Avenue, New York, NY 10016, United States of America.

CIP data is on file at the Library of Congress
ISBN 978–0–19–085461–4

1 3 5 7 9 8 6 4 2

Printed by Sheridan Books, Inc., United States of America

CONTENTS

CONTRIBUTORS

Philip J. Barnard
Honorary Member
Medical Research Council Cognition and Brain Sciences Unit
University of Cambridge, UK

Tony Berlant, artist
Santa Monica, USA

Adam Brumm
Associate Professor
Australian Research Centre for Human Evolution
Environmental Futures Research Institute
Griffith University, Australia

James Cole
Principal Lecturer
School of Environment and Technology
University of Brighton, UK

Frederick L. Coolidge
Professor
Department of Psychology
University of Colorado
Colorado Springs, USA

Iain Davidson
Emeritus Professor of Archaeology
University of New England, Australia

Sophie A. de Beaune
Professor
Jean Moulin Lyon 3 University
Faculté des Lettres et Civilisations, Lyon, and UMR 7041 "Archéologies et Sciences de l'Antiquité,"
Nanterre, France

Kate Emery
Institute of Archaeology
University College London
England, UK

Tiago Falótico
Postdoctoral Researcher
Institute of Psychology
University of São Paulo, Brazil

Clive Gamble
Emeritus Professor
Centre for the Archaeology of Human Origins
Department of Archaeology
University of Southampton, UK

Michael D. Gumert
Associate Professor
Division of Psychology
School of Social Sciences
Nanyang Technological University, Singapore

Miriam Noël Haidle
Scientific Coordinator
Research Center "The Role of Culture in Early Expansions of Humans—ROCEEH" of the Heidelberg Academy of Sciences and Humanities, Senckenberg Forschungsinstitut und Naturmuseum, and Institut für Ur- und Frühgeschichte und Archäologie des Mittelalters, Abt. Ältere Urgeschichte und Quartärökologie, Cognitive Archaeology Unit
Tübingen, Germany

Derek Hodgson
Adjunct Professor
Department of Archaeology
University of York, UK

Mathieu Leroyer
Département d'Histoire de l'art et archéologie (UFR03)
Université de Paris 1
Panthéon-Sorbonne
France

Marlize Lombard
Professor
Centre for Anthropological Research/ Palaeo-Research Institute
University of Johannesburg, South Africa

Stephen J. Lycett
Associate Professor
Department of Anthropology (Laboratory for Evolutionary Anthropology and Anthropological Archaeology)
The State University of New York (SUNY), USA

William C. McGrew
Honorary Professor
School of Psychology and Neuroscience
University of St. Andrews, Scotland, UK

Mark W. Moore
Associate Professor
Archaeology and Palaeoanthropology
University of New England, Armidale, Australia

Eduardo B. Ottoni
Professor
Institute of Psychology
University of São Paulo, Brazil

Karenleigh A. Overmann
MSCA Research Fellow
Department of Psychosocial Science
University of Bergen, Norway

Matt Pope
Principal Research Fellow
Institute of Archaeology
University College London
England, UK

Shelby S. Putt
Postdoctoral Researcher
The Stone Age Institute and The Center for Research into the Anthropological Foundations of Technology
Indiana University, USA

Lana Ruck
Doctoral Student
Cognitive Science Program; Department of Anthropology
Indiana University, USA

Gonen Sharon
Associate Professor
Multidisciplinary Studies, Tel Hai College
Upper Galilee, Israel

Ceri Shipton
Faculty Member
Centre of Excellence for Australian Biodiversity and Heritage
Australian National University, Australia

Cory Stade
Visiting Fellow
Centre for the Archaeology of Human Origins
University of Southampton, UK

Dietrich Stout
Associate Professor
Department of Anthropology, Emory University
Atlanta, USA

Natalie Uomini
Researcher
Department of Linguistic and Cultural Evolution
Max Planck Institute for the Science of Human History
Jena, Germany

Lyn Wadley
Honorary Professor
Evolutionary Studies Institute
University of the Witwatersrand,
South Africa

Rex Welshon
Professor
Department of Philosophy
University of Colorado
Colorado Springs, USA

Thomas Wynn
Distinguished Professor
Department of Anthropology
and UCCS Center for Cognitive
Archaeology
University of Colorado
Colorado Springs, USA

Squeezing Minds From Stones

INTRODUCTION

COGNITIVE ARCHAEOLOGY AT THE CROSSROADS

Karenleigh A. Overmann and Frederick L. Coolidge

Sometime in the fall of 1974, a graduate student named Thomas Wynn walked across the campus of the University of Illinois at Champaign-Urbana, deep in thought as he mulled over potential projects for his doctoral research in Early Stone Age/Lower Paleolithic archaeology. Suddenly, the research possibility that popped into his mind quite literally stopped him in his tracks, mid-quad. Months earlier, in the spring semester, his doctoral supervisor, anthropologist Charles Keller, had given him the name of psychologist Jean Piaget and tasked him with assessing whether Piaget's ideas about cognitive development in children were of any use to an archaeologist. At the time, Wynn had concluded that there was no way to apply psychological theory to the archaeological record. But now, he was powerfully struck by the thought that Piaget's ideas on the ontogenetic development of cognitive abilities might explain how human cognition had evolved, as discerned through change in the forms of stone tools.

The idea that human cognitive evolution could be understood from stone tools—the only things that remain over the millions of years separating us from our earliest tool-using ancestors, apart from a few bones, either theirs or ones they modified—was perhaps something of a Zeitgeist in the 1970s, a reaction to the rigid materialism that characterized the "processual archaeology" advocated by the "new archaeologists" a decade earlier. Certainly, floating around the University of Illinois campus were ideas by Lewis Binford (1962, 1972; Binford & Binford, 1968) and pro-structuralist concepts of culture and science from Claude Lévi-Strauss (1962) and Edmund Leach (1973). In fact, similar ideas were emerging on both sides of the Atlantic, not only in the work of Wynn but also that of his contemporaries, scholars like archaeologists Glynn Isaac, Colin Renfrew, John Gowlett, and Iain Davidson, psychologist William Noble, evolutionary primatologist William McGrew, biological anthropologist Sue Taylor Parker, and evolutionary neurobiologist Kathleen Gibson.

The central challenge then, as it remains today, for an archaeology interested in cognition was this: How might every last bit of data be squeezed from the archaeological record and used to understand evolutionary change in human cognition, with the daunting goal of resolving the ages-old mystery, how did we become human? The inquiry is particularly challenging, not merely because of the vagaries that perplex any archaeological interpretation, but because those interpretations necessarily involve theoretical frameworks, definitions, assumptions, methods, and models that come from outside archaeology. Cognitive abilities had to be understood in psychological terms in order to operationalize them as behaviors whose traces might be archaeologically detectible. As a result, cognitive archaeology was no relatively straightforward matter of discovery and description, but an interdisciplinary endeavor that increasingly needed to know and apply things about how psychological constructs might be operationalized so that they could be detected as change in material forms. Included in this endeavor was examination of the ways in which neuroanatomy and neurofunctionality change over ontogenetic and evolutionary spans of time and how this relates to material change, and philosophical speculations about the nature of cognition, concepts, and symbols and the possible role of materiality in such phenomena.

Cognitive archaeology might be called the not-so-simple art of squeezing minds from stones—not coincidentally the title we have chosen for this volume. To introduce it, we first look briefly at the work of some of the pioneers of cognitive archaeology, and use it to demonstrate how, some four decades later, many of the same concerns and challenges still remain, though new theoretical frameworks and methodological techniques have been adopted to address them. Then we outline the book's contributions, which we have divided into four parts: general cognitive evolution, stone tools and the mind, social cognition, and more recent cognition (things like numbers, weapons, and trapping). Our contributors include many of the scholars who helped establish cognitive archaeology as a research inquiry, as well as some emerging scholars. This range reflects change in the discipline itself, which is still pondering unresolved questions like language origins, incorporating approaches such as neuroimaging, and broadening demographics such as gender (more women) and specialty (primate archaeology and modern art). We conclude the volume with the thoughts of one of the pioneers, Thomas Wynn, on the future of cognitive archaeology.

THE BEGINNINGS OF COGNITIVE ARCHAEOLOGY: A REACTION TO PROCESSUALISM

Processual archaeology demanded that interpretation be guided by theory, something that remains a continuing mandate. Yet it also provided something to react against, its insistence that the past be interpreted strictly according to the material evidence. As Leach (1973) expressed it, archaeology was ill-equipped to study ancient *peoples*: ". . . all the ingenuity in the world will not replace the evidence that is lost and gone for ever," and ". . . you should recognize your guesses for what they are" (p. 768). Nonetheless, processualism had also opened up the possibility that the archaeological record could be informative not just about the artifacts themselves, the discover-and-describe mandate of earlier schools of archaeological thought, but also regarding the lifestyles of those who made and used them. As Renfrew (1982) would later note, "inferring intelligent behavior from its material relics" was something only archaeology could do (p. 4). Admittedly, these were matters requiring new approaches and theoretical frameworks, longer chains of inference, and, occasionally, bigger leaps of faith than might fit comfortably with the injunction for a no-nonsense scientism. Binford (2001), for example, had introduced the idea that the archaeological record could be understood by gaining insights from contemporary hunter-gatherer societies. Such ethnic explanations would quickly be criticized (if not rejected outright) because modern traditional societies are hardly prehistoric—as Smith (1955) expressed it, such comparisons demonstrated only "what an incredible variety of codes of behaviour in fact actuate human conduct" (p. 5)—and therefore do not provide valid comparisons to societies that were, either in form or dignity. One of Binford's contemporaries, anthropologist Ralph Holloway (1969), was grappling with the similarly difficult and still unresolved question of between-species comparability: How special is humanity in comparison to non-human species? How might concepts like culture and cultural transmission be defined? And should such definitions be broad enough to show continuities between humans and non-human species or so narrowly construed that discontinuities emphasized human uniqueness?

In responding to the whiff of revolution in the air while expanding on some of these processual ideas, Thomas Wynn focused his doctoral research on change in lithic technologies—the Oldowan, Acheulean, and Levallois industries—over a span of about two million years, published as "The Intelligence of Later Acheulean Hominids" in *Man* (1979), and later as *The Evolution of Spatial Competence* (1989). Wynn used Piagetian developmental theory as his framework for explaining how changes in material form revealed evolutionary change in hominin spatial cognition and intelligence. The evidentiary, theoretical, and investigative scopes were narrowly drawn: They included lithic change, Piaget's developmental framework, human capacities for spatiality and intelligence, and little else. Essentially, increasing complexity in material form was an index for the emergence of new and increasingly complex psychological capabilities, a spectrum of evolutionary change that originated in continuity (an endpoint that was decidedly ape-like) and reached discontinuity (an endpoint anchored in the unique cognitive complexity of modern humans). Wynn would later team up with evolutionary primatologist William McGrew (Wynn & McGrew, 1989) to explore further the idea that human discontinuity had originated in an ape-like continuity. They compared the Oldowan industry to tool use by modern apes (especially chimpanzees), arguing that the capabilities required for the former fell within the range demonstrated by the latter, with the possible exceptions of carrying objects for longer distances and competing for food with large carnivores. Here they addressed the *minimum necessary competence*, the idea that as tools reflect only the lower boundary of the capabilities required for their production and use, they are likely to underestimate either the actual or potential cognitive abilities of their makers. This concept would become a mantra for Wynn, one he has passed on to his collaborators and students: Make no more assumptions than necessary for explaining a phenomenon.

Also working in North America but independently from Wynn, Sue Taylor Parker and Kathleen Gibson (1979) were attacking the problem from a slightly different angle. Rather than operationalizing a cognitive ability like spatial cognition and then looking for its traces in material change, they examined the origins of intelligence and language through ontogenetic development in cognition and evolutionary change in brain size. While thoroughly Piagetian in their orientation toward ontogenetic development, they nonetheless held the phylogenetic part of Piaget's theory at arm's length, calling his evolutionary model "Lamarckian and vitalistic" (p. 400). Their attempted compromise was comparing and finding parallels between four stages in the evolution of intellectual abilities for human children (Piaget's ontogenetic theory) and various primate groups—prosimians, Old World monkeys, great apes, and hominids—species whose differentiation was apparently based on non-Piagetian evolutionary criteria.[1] Positing tool use as the *sine non qua* of hominid intelligence, they included a panoramic number of contributing factors in their model of how tool use developed, including cooperation, shared attention, diet, hunting, sociality and social structures, food sharing and transportation, shelter construction and use, climate, and available environmental resources. Individuals able to acquire more resources through cognitive advantages in domains such as planning were thought to have been favored by both natural and social selection: more likely to survive, acquire mates, and perpetuate their advantageous behavioral skills and cognitive traits through cultural and genetic mechanisms.

At about the same time but halfway across the globe in East Africa, Glynn Isaac (1976) was analyzing material complexity in the archaeological record and deriving conclusions about human evolution. He identified early tools and some dependence on their use for subsistence as falling outside the range of modern apes. Like Wynn, he stuck close to artifacts and their properties, but where Wynn had construed cognitive change from morphological change, Isaac focused on what morphological properties might indicate about the transport of raw materials, modification and use of finished tools, and cultural differences between groups. Isaac incorporated evidence on fossils and diet, types of interdisciplinary data whose inclusion has since become fairly common in cognitive archaeology. Also unlike Wynn (at least at this period), Isaac was willing to speculate about language, a topic as controversial and undecided then as it is today. He took the economic behaviors and adaptive patterns associated with tool use (e.g., modes of subsistence that included preying on local fauna and establishing a home base from which to scavenge, things that suggested groups were cooperating to obtain and share food resources like meat) to be windows on information exchange and the emergence of language (Isaac & Isaac, 1975).

In Australia, Iain Davidson and William Noble (1989) were also taking up the challenge of language, further helping to resurrect language origins as an archaeological inquiry. Recognizing the need to find ways of detecting traces of language origins and development in the archaeological record, they hypothesized that language had originated in depictive images that had themselves originated in gesture. During the Upper Paleolithic, gestures recreating the shapes of objects like bison had yielded parietal art that represented bison, which in turn had yielded reflexivity on the meaning of bison, and such reflexivity to words for them. Basically, language required understanding an object's meaning, a mental leap from depiction to reference that assumed significant discontinuity in the way humans and non-human species construct, understand, and share social meanings. This approach differed from that being followed

[1] Whether Parker and Gibson (1979) were completely successful in rejecting Piaget's phylogenetic theory is debatable, given that their compromise ended up comparing evolutionary change in primates with Piaget's ontogenetic stages of cognitive development. The parallels the authors inferred between Piaget's *sensorimotor period* (birth to 2 years), *preoperations period* (2–7 years), *concrete operations period* (7–12 years), and *formal operations period* (12 years and thereafter) and what they called a *prosimian stage, Old-World monkey stage, great ape stage,* and *hominid stage* still equated prosimians with human infants, monkeys with human toddlers, and so on, characterizations that, while perhaps not as disparaging as similar parallels drawn to human societies by nineteenth-century theorists, remain inaccurate nonetheless.

in two related avenues of language origins research: the idea that language originated in gesture (the gestural origins hypothesis; Corballis, 1999; Gentilucci & Corballis, 2006), not as a transition to reference, as Davidson and Noble had argued, but as a communicative modality in itself, and the idea that language originated in tool use (Bradshaw & Nettleton, 1982; Higuchi, Chaminade, Imamizu, & Kawato, 2009), which recognized the neural and combinatorial similarities of language and the motor movements associated with tool behaviors. Davidson and Noble also doubted that culture could exist without language, a question that in recent decades has been answered with a resounding *yes*, since socially learned behaviors have also been demonstrated by non-human primates and cetaceans, species that presumably lack communication capabilities on par with human speech (Laland & Janik, 2006). Davidson and Noble also claimed that handaxes were the unintended results of repetitive behaviors rather than the deliberate imposition of shape enabled by increasingly complex spatial abilities, as had been argued by Wynn.

Davidson and Noble's (1989) idea that inchoate concepts were given form as material substances were manipulated and were recognized perhaps only after the formative act complemented the philosophical turn developing in Britain, where Colin Renfrew (1982) was working to create a "cognitive processualism." Renfrew disagreed that hominid cognitive evolution could be understood through Piagetian theory, the "old and dangerous principle that ontogeny follows phylogeny" (p. 14) and its converse, that phylogeny follows ontogeny. He also rejected processual elements such as the insistence that "an explanation can only be decently 'scientific' if expressed in the form of a universal law" (p. 10), while continuing the processual tradition of prioritizing objectivity and elaborating the processual idea that material forms were informative of how their makers had lived, to examine *how*—and perhaps even *what*—they had thought. In an archaeology of the mind, intelligent behavior could be recognized in the material forms found archaeologically, since material culture had an active role in constituting cultural reality. Archaeologists were being set free to interpret artifacts like stone cubes from Mohenjo-daro, a site of the Indus Valley civilization, as weights that implied both a system of conceptual mapping and the social conditions associated with it (Renfrew, 1982). Renfrew also noted that a cognitive archaeology would necessarily overlap with other disciplines—human psychology and neurophysiology, social and physical anthropology, comparative studies with other species, and even artificial intelligence, and needed coherent theory and sound analytical procedures to guide its inferences. Further, he recognized that the Cartesian distinction between mind and matter was problematic, an idea that would later inspire the even more philosophically oriented material engagement theory of cognitive archaeologist Lambros Malafouris (2013), which incorporates concepts from philosophy of mind.

Also in Britain, and sticking very close to the artifacts themselves as Wynn was doing, John Gowlett (1979, 1984) was focusing on the evolution of design form in lithic technologies. He suggested that analyzing a skill such as stone knapping (as reflected in finished artifacts) might be a fruitful way to approach cognitive evolution. Accordingly, he outlined biface manufacture as a complex process that included not only the series of strike decisions and strikes, but the selection and transport of raw material and the use of the tool as well. Gowlett agreed that handaxes showed an increase in cognitive complexity but argued, against Wynn, that it was a matter of standardization, not one of coordinating multiple visuospatial perspectives. He proposed the end-to-end process to be so complex that it exceeded the capacity of modern non-human primates, even as early as the Oldowan (something Wynn and colleague William McGrew would argue against in 1989). Gowlett further suggested that handaxes and cleavers reflected concepts in the minds of their makers, some kind of mental template that guided the imposition of form, which was at odds with the idea of concepts becoming tangible and appreciable in new ways through material engagement (Malafouris, 2010b, 2013). Gowlett also noted that human cognition was likely not understood well enough in ways that would enable characterization of the continuities and discontinuities with other species, particularly the hominid adaptation to the intergenerational accumulation and transmission of experience (in other words, culture).

In this volume, the 22 chapters from our contributors develop many of the same themes raised in the formative decade of cognitive archaeology: the validity and use of ethnoarchaeological and experimental methods; the question of continuities and discontinuities between humans and non-human species; the selection and application of theoretical frameworks, including the displacement of Piagetian theory by contemporary psychological and neuroscientific approaches to brain function and form; the incorporation of interdisciplinary data; the origin of language; the ability of construing intentionality from artifactual form; the philosophical turn in cognitive archaeology; and the riddle of intergenerational accumulation and transmission. And, of course, still with us is the desire to wring every last bit of possible data out of stone tools, as well as more recent technologies like bows and arrows or traps and snares.

FORTY YEARS AFTER: CONTEMPORARY COGNITIVE ARCHAEOLOGY

Despite the rejection of Binford's hunter-gatherer comparisons, today ethnoarchaeological and experimental approaches remain vibrant modes of understanding traditional and ancient technologies and the cognitive processes they involve, as extrapolated from the behaviors and brains of modern humans and non-human species. For example, in Chapter 18 in this volume, Sophie de Beaune uses ethnographic data from contemporary hunter-gatherers in Australia, Africa, and the Arctic. She does not equate their gendered division of labor to that construed for prehistoric peoples but, rather, uses it to demonstrate that archaeological analyses of gender are often skewed by researchers' own cultural expectations. This has since been exemplified by the surprise greeting of last year's announcement that a Viking grave, identified as that of a warrior because it contained weapons and horse bones, belonged in fact to a female and not a man, whose sex was recently identified through DNA analysis of her remains (Hedenstierna-Jonson et al., 2017). And as an example of the experimental approach, Mark Moore applies his personal wealth of flintknapping experience to the problem of construing intent in lithic remains: What does a modified stone tell us—if anything—about what its ancient maker intended when he or she knocked rocks together to produce flakes and perhaps refine them? How much of a goal is involved in making an artifact like a handaxe: Does it require a mental template that guides production by visualizing the end product, or is it simply the unintended consequence of exhausting a core, much as Davidson and Noble argued in the 1980s? And what light, if any, can experiments performed by modern stone-knappers shed on ancient intent to realize artifactual form? Shelby Putt's contribution on language (discussed later) also falls into the experimental category.

Holloway's concern with between-species comparability foreshadowed what is now a substantial interest in comparative analyses, particularly with non-human primates and especially in regard to their tool use and the transmission and reproduction of learned behaviors. William McGrew, Tiago Falótico, Michael Gumert, and Eduardo Ottoni extend this line of inquiry by reviewing a wide range of studies in primate archaeology, now a distinct subdiscipline of archaeology (Haslam et al., 2009, 2017). Unlike experiments with captive apes (e.g., Kanzi; Savage-Rumbaugh, Toth, & Schick, 2007), primate archaeology focuses on tool use in natural situations, conditions more directly comparable to those of early hominids. These authors compare the tools produced by apes and Old and New World monkeys to early lithic technologies like the Oldowan industry, further updating work by Wynn and McGrew (Wynn, Hernandez-Aguilar, Marchant, & McGrew, 2011; Wynn & McGrew, 1989). One key difference in tool-use behaviors is the degree to which tools are reused. Non-human primates often make new tools whenever one is needed (e.g., chimpanzee termite-fishing probes), and when they reuse tools like anvils for cracking open foods like shellfish and nuts, they do so one at a time. These behaviors have several effects. First, tools are not used by more than one individual at a time, limiting potential opportunities for engaging cognitive processes like joint attention, something Rex Welshon explores in his chapter. Second, the tools do not accumulate modifications, either because they are created to fulfill a need and discarded once they have been used to satisfy it or because, like an anvil, their form does not become refined and hence does not act to accumulate social knowledge. The question of what it takes to accumulate and transmit culture between generations is addressed by Stephen Lycett and Miriam Noël Haidle in their respective chapters.

For its part, Piagetian theory has fallen out of fashion, particularly in the phylogenetic aspects presciently rejected by Parker and Gibson. If Piagetian phylogenesis was "Lamarckian and vitalistic" (Parker & Gibson, 1979, p. 400), his ontogenetic theory applied to archaic, traditional, and industrialized societies had the effect of dividing them into adults and children (Piaget, 1928), characterizations as unpalatable as they are inaccurate. More current theoretical frameworks on psychological capabilities have been adopted in cognitive archaeology. An example is the working memory model of Baddeley and Hitch (1974), brought to his decades-long collaboration with Thomas Wynn by Frederick Coolidge and recapped in his contribution to this volume. Consistent with the turn toward contemporary psychological theory, Thomas Wynn and Tony Berlant explore the neuroaesthetics of art, something that can be seen developing in the increasing incorporation of symmetric forms in lithic technologies. The chapter by Wynn and Berlant recounts "the great handaxe junket," their five-year mission to seek the world's most compelling handaxes, as judged by their respective archaeological and artistic sensibilities; the exhibit opened in January 2018 at the Nasher Sculpture Center in Dallas (Farago, 2018). Derek Hodgson examines developments in spatial cognition, as discerned from change in the form of stone tools, using insights from the neuroscience of visuospatial memory, visuomotor control, attention, and planning. James Cole offers a similar examination of change in lithic form, arguing that it illuminates developments in the construct known as theory of mind, the ability to know that others have minds whose content can differ from one's own.

Modeling the mind and its evolutionary development continue to pose a challenge, especially as interdisciplinary data are brought to bear on the question. Philip Barnard looks at behaviors like tool use and their effects on mental and neural systems, differentiated in humans and non-human primates by developments in abilities like propositional meaning that underlie mental states, perception, and bodily control. He reviews the interacting cognitive subsystems (ICS) model, which enables mental capabilities to be analyzed and compared across extant species and the evolutionary development of human cognition. Dietrich Stout focuses on the continued incorporation of theory and methods capable of analyzing how organisms interact with their environments over time: Darwinian and neo-Darwinian evolutionary theory, Dunbar's social brain hypothesis, technical intelligence hypothesis, niche construction theory, phenotypic accommodation, and gene–culture co-evolution. With a more personal retrospective, Iain Davidson looks back at how cognitive archaeologists approached stone tools during the formative decade, finding pre-processual influences from as early as the nineteenth century. He uses *chaîne opératoire* analysis, comparisons with non-human primate tool use, and Barnard's cognitive subsystem model to argue for newer, more productive methods. One such method is the Budapest model of evolving cognitive systems (ECS), which seeks to understand ancestral cognitive states through comparative studies; the goal is to avoid modeling ancestral cognition as incomplete or deficient (e.g., as lacking language, which modern cognition includes), much as extant species are assumed competent (though alinguistic).

Language origins remains a topic of significant interest. Shelby Putt applies another experimental approach, the use of neuroimaging to see what parts of the brain are activate when someone flintknaps. Her focus is not how much intent is needed to realize form but rather to understand whether and to what extent the motor movements involved in flintknapping might overlap that of language. She notes Wynn's (1991) reminder that attempts to find the origins of language in tool behaviors must be justified theoretically and, to the extent practicable, empirically if they are to be plausible and persuasive. Putt notes that in modern knappers, instruction by means of language rather than silent imitation affects which parts of the brain become active in replicating early stone tools. The idea is that specific overlaps between the neural activation patterns for language and stone knapping is taken as supporting the hypothesis of common origin. In comparison, Cory Stade and Clive Gamble approach the question of language origins through theory of mind. Their hypothesis is that somewhere during the long prehistory of stone tool production, silent imitation alone became insufficient for reproducing artifacts with the complexity of the prepared core techniques; language instruction would have been needed for its transmission. These authors analyze the tool as a "third partner" that enables joint attention and shared understanding on the part of both teacher and student, a critical development in cognition (see Rex Welshon's chapter) and cumulative culture (see the chapters by Stephen Lycett and Miriam Noël Haidle).

The philosophical turn is represented here by several scholars. Karenleigh Overmann analyzes prehistoric numerical cognition. After reviewing the evolution of contributing abilities like quantity perception, she examines the archaeological record of Mesopotamia for the first unambiguous numbers, the role of material forms for counting in them, and what change in such material forms might suggest about possible archaeological signs and timelines for even earlier periods. This discussion is underpinned by material engagement theory (MET), the theoretical framework of cognitive archaeologist Lambros Malafouris (2010a, 2013; Malafouris & Renfrew, 2010). MET views the mind as not just *embodied* and *embedded* (cognition is affected by being in a body and situated in an environment) but also *extended* (material forms are not just causally linked to cognition but a constitutive component of it), *enacted* (cognition is the interaction of brains, bodies, and behaviors), and *evolving* (humans have by no means reached a terminal cognitive state). Overmann takes Malafouris' idea that "tools make minds" to suggest that change in material forms indexes change in behaviors and psychological processes. Marlize Lombard similarly looks at how tools co-create and transform behaviors, bodies, and brains. Mark Moore emphasizes the embodied and enactive aspects of stone knapping: The "decision about where to place the next blow, and how much force to use, is not taken by the knapper in isolation; it is not even processed internally. The flaking intention is constituted, at least partially, by the stone itself . . . [which], like the knapper's body, is an integral and complementary part of the intention to knap" (Malafouris, 2010b, p. 17). And Rex Welshon brings analytic philosophy to bear on key attributes of joint attention, such as its reflexivity, symmetry, and transitivity.

While all the contributions in this volume use the archaeological record to illuminate the ancient mind, some adhere more closely to stone tools, the material form with the greatest longevity and hence the best potential for illuminating the deepest prehistory of human cognitive evolution. In this vein is the contribution by Adam Brumm, Matt Pope, Mathieu Leroyer, and Kate Emery. These authors analyze scavenging and reuse in Lower Paleolithic stone tools, as demonstrated through their analysis of weathering and patina effects. Natalie Uomini and Lana Ruck compare and contrast three hypotheses on why the human species is so strongly right-handed—matters of social learning, fighting, or task complexity that might intensify right-handedness or maintain a minority of left-handers. These authors then speculate on how the various hypotheses might be examined and substantiated through archaeological analyses. Gonen Sharon examines the independent development of giant core techniques by distinct Acheulean populations separated by most of the African continent. He proposes their similarities as an instance of convergent evolution, with different groups converging on similar solutions to questions of technology and design in response to identical social needs and purposes. Ceri Shipton looks at what stone tools reveal about over-imitation, the tendency to recreate actions when knapping stone, whether those actions are causally efficacious or not. He then analyzes this evidence for what it can tell us about social transmission and the necessity for some of the instruction to take the form of language.

Finally, what complex technologies can reveal about cognitive change concerns the last two chapters of the volume. In her contribution, Marlize Lombard focuses on compound technologies like the bow and arrow as an index for change in cognitive plasticity, which she defines as human abilities for teaching and learning from one another, innovating both behaviorally and technologically, and responding flexibly and creatively when novel or complex situations are encountered. Lyn Wadley offers a similar analysis of snares and traps, which require and therefore imply cognitive abilities like planning and inhibition (i.e., delayed gratification). This illuminates the need for inferential argumentation posed by the lack of direct archaeological evidence. Traps and snares are not archaeologically attested, presumably because they were made of perishable materials. Rather, they are suggested by demographic differences in faunal remains (e.g., more bones than expected from species difficult to hunt because of size, speed, and habits). Such indirect evidence requires the use of so-called bridging arguments, statements that link archaeological findings to the conditions producing them, especially as understood through ethnoarchaeological comparisons and experimental data.

ACKNOWLEDGMENTS

For acting as our external reviewers, we thank Stephen Chrisomalis, Wayne State University, USA; Michael Chazan, University of Toronto, Canada; Robert Clowes, Universidade Nova de Lisboa, Portugal; Agustin Fuentes, University of Notre Dame, USA; Mick Gantley, University of Exeter, UK; John Gowlett, University of Liverpool, UK; Michael Haslam, University of Oxford, UK; Antonis Iliopoulos, School of Archaeology, University of Oxford, UK; Alastair Key, University of Kent, UK; Marc Kissel, University of Notre Dame, USA; Danielle Macdonald, University of Tulsa, USA; Manuel Martin-Loeches, Universidad Complutense de Madrid, España; Michael O'Brien, University of Missouri, USA; Eric Reuland, Universiteit Utrecht, Nederland; Patrick Roberts, Max Planck Institut, Deutschland; Enza Spinapolice, Università di Roma, Italia; Alexandra Sumner, DePaul University, USA; and Matthew Walls, University of Calgary, Canada.

Work on this volume was accomplished with the kind support and infinite patience of our families, and of course, the inspiration of our colleague, friend, and mentor, Thomas Wynn.

REFERENCES

Baddeley, A. D., & Hitch, G. J. (1974). Working memory. In G. H. Bower (Ed.), *The psychology of learning and motivation: Advances in research and theory* (Vol. 8, pp. 47–89). New York: Academic Press.

Binford, L. R. (1962). Archaeology as anthropology. *American Antiquity, 28*(2), 217–225.

Binford, L. R. (1972). *An archaeological perspective.* New York: Seminar Press.

Binford, L. R. (2001). *Constructing frames of reference: An analytical method for archaeological theory building using ethnographic and environmental data sets.* Berkeley, CA: University of California Press.

Binford, S. R., & Binford, L. R. (Eds.). (1968). *New perspectives in archaeology.* Chicago, IL: Aldine Publishing Company.

Bradshaw, J. L., & Nettleton, N. C. (1982). Language lateralization to the dominant hemisphere: Tool use, gesture and language in hominid evolution. *Current Psychology, 2*, 171–192.

Corballis, M. C. (1999). The gestural origins of language: Human language may have evolved from manual gestures, which survive today as a "behavioral fossil" coupled to speech. *American Scientist, 87*(2), 138–145.

Davidson, I., & Noble, W. (1989). The archaeology of perception: Traces of depiction and language. *Current Anthropology, 30*(2), 125–155.

Farago, J. (2018, February 2). Stone Age tools, or art? Or both? *New York Times*, p. C17. New York.

Gentilucci, M., & Corballis, M. C. (2006). From manual gesture to speech: A gradual transition. *Neuroscience and Biobehavioral Reviews, 30*(7), 949–960.

Gowlett, J. A. J. (1979). Complexities of cultural evidence in the Lower and Middle Pleistocene. *Nature, 278*(5699), 14–17.

Gowlett, J. A. J. (1984). Mental abilities of early man: A look at some hard evidence. In R. Foley (Ed.), *Hominid evolution and community ecology: Prehistoric human adaptation in biological perspective* (pp. 167–192). London: Academic Press.

Haslam, M., Hernandez-Aguilar, R. A., Ling, V., Carvalho, S., De la Torre, I., DeStefano, A., . . . Warren, R. (2009). Primate archaeology. *Nature, 460*(7253), 339–344.

Haslam, M., Hernandez-Aguilar, R. A., Proffitt, T., Arroyo, A., Falótico, T., Fragaszy, D. M., . . . Luncz, L. V. (2017). Primate archaeology evolves. *Nature Ecology and Evolution, 1*, 1431–1437.

Hedenstierna-Jonson, C., Kjellström, A., Zachrisson, T., Krzewińska, M., Sobrado, V., Price, N., . . . Storå, J. (2017). A female Viking warrior confirmed by genomics. *American Journal of Physical Anthropology, 164*(4), 853–860.

Higuchi, S., Chaminade, T., Imamizu, H., & Kawato, M. (2009). Shared neural correlates for language and tool use in Broca's area. *Neuroreport, 20*(15), 1376–1381.

Holloway, R. L. (1969). Culture: A human domain. *Current Anthropology, 10*(4), 395–412.

Isaac, G. L. (1976). Stages of cultural elaboration in the Pleistocene: Possible archaeological indicators of the development of language capabilities. *Annals of the New York Academy of Sciences, 280*(1), 275–288.

Isaac, G. L., & Isaac, B. (1975). Africa. In R. L. Stigler (Ed.), *Varieties of culture in the Old World* (pp. 8–48). New York: St. Martins Press.

Laland, K. N., & Janik, V. M. (2006). The animal cultures debate. *Trends in Ecology and Evolution, 21*(10), 542–547.

Leach, E. R. (1973). Concluding address. In C. Renfrew (Ed.), *The explanation of culture change: Models in prehistory. Proceedings of a meeting of the Research Seminar in Archaeology and Related Subjects held at the University of Sheffield, December 14–16, 1971* (pp. 761–771). London: Gerald Duckworth.

Lévi-Strauss, C. (1962). *The savage mind* (G. Weidenfield, Trans.). Chicago, IL: University of Chicago Press.

Malafouris, L. (2010a). Grasping the concept of number: How did the sapient mind move beyond approximation? In C. Renfrew & I. Morley (Eds.), *The archaeology of measurement: Comprehending heaven, earth and time in ancient societies* (pp. 35–42). Cambridge, UK: Cambridge University Press.

Malafouris, L. (2010b). Knapping intentions and the marks of the mental. In L. Malafouris & C. Renfrew (Eds.), *The cognitive life of things: Recasting the boundaries of the mind* (pp. 13–27). Cambridge, UK: McDonald Institute for Archaeological Research.

Malafouris, L. (2013). *How things shape the mind: A theory of material engagement*. Cambridge, MA: MIT Press.

Malafouris, L., & Renfrew, C. (2010). Introduction: The cognitive life of things: Archaeology, material engagement and the extended mind. In L. Malafouris & C. Renfrew (Eds.), *The cognitive life of things: Recasting the boundaries of the mind* (pp. 1–12). Cambridge, UK: McDonald Institute for Archaeological Research.

Parker, S. T., & Gibson, K. R. (1979). A developmental model for the evolution of language and intelligence in early hominids. *Behavioral and Brain Sciences, 2*, 367–408.

Piaget, J. (1928). Logique génétique et sociologie. *Revue Philosophique de La France et de l'Étranger, 105*, 167–205.

Renfrew, C. (1982). *Towards an archaeology of mind: An inaugural lecture delivered before the University of Cambridge on 30th November 1982*. Cambridge, UK: Cambridge University Press.

Savage-Rumbaugh, E. S., Toth, N. P., & Schick, K. D. (2007). Kanzi learns to knap stone tools. In D. A. Washburn (Ed.), *Primate perspectives on behavior and cognition* (pp. 279–291). Washington, DC: American Psychological Association.

Smith, M. A. (1955). The limitations of inference in archaeology. *Archaeological News Letter, 6*, 3–7.

Wynn, T. (1979). The intelligence of later Acheulean hominids. *Man, 14*, 371–391.

Wynn, T. (1989). *The evolution of spatial competence.* Chicago, IL: University of Illinois Press.

Wynn, T. (1991). Tools, grammar and the archaeology of cognition. *Cambridge Archaeological Journal, 1*(2), 191–206.

Wynn, T., Hernandez-Aguilar, R. A., Marchant, L. F., & McGrew, W. C. (2011). "An ape's view of the Oldowan" revisited. *Evolutionary Anthropology, 20*(5), 181–197.

Wynn, T., & McGrew, W. C. (1989). An ape's view of the Oldowan. *Man, 24*(3), 383–398.

1

A SIMIAN VIEW OF THE OLDOWAN
RECONSTRUCTING THE EVOLUTIONARY ORIGINS OF HUMAN TECHNOLOGY

William C. McGrew, Tiago Falótico, Michael D. Gumert, and Eduardo B. Ottoni

INTRODUCTION

Question: Why model human origins? Answer: Because we have no choice. That is, until we devise a time machine that will take us back to the late Miocene, so that we can collect first-hand behavioral data from our earliest hominin ancestors, we must model them (Andrews, 2015). In the absence of such a tool, we must collect data on living species, either from living *Homo sapiens* or from other pertinent taxa, and use it to proxy an inferential reconstruction. If we seek referential models, then the first port of call should be other members of the order Primates, the more closely related to us the better. First choice on phylogenetic and genetic grounds should be those living creatures with whom we last shared a common ancestor, which are the two members of the extant genus *Pan*, bonobo (*P. paniscus*) and chimpanzee (*P. troglodytes*) (Gruber & Clay, 2016). If we are interested in the evolutionary origins of technology, then the starting point should be the latter, a prodigious tool-maker and tool-user, rather than the former, which shows paltry technology in nature (Haslam, 2014).

Tom Wynn realized this long ago. It is well worth the effort to reread his attempt to use Piagetian theory on the development of intelligence, applied to the Oldowan (Wynn, 1981). His daring assertion was that the Oldowan lithic industry required only preoperational intelligence, which is currently demonstrated in the elementary technology of living apes. He made clear his hypothesis: "The evolution of a uniquely hominid intelligence had not occurred by Oldowan times" (Wynn, 1981). Thus, comparative analyses of the stone artifacts of extinct, early hominins and the comparable products of the behavior of at least one African ape species should be informative and helpful.

On this basis, Wynn and McGrew (1989) collaborated on a systematic and detailed comparison of the tools of the Oldowan and the tools of living chimpanzees. They concluded that there was nothing exclusively human about the oldest known archaeological evidence, but acknowledged caveats about long-distance carriage of objects and competition with large carnivores for large prey. By 2010, we could not deny that the earlier effort was outdated, not just by abundant new knowledge of wild chimpanzees, but also by comparably impressive findings on wild orangutans,

Pongo species (spp.) (Meulman & van Schaik, 2013). (Equally notable was the continuing absence of such data on the habitual use of tools by the other two living taxa of African apes, the bonobo, and the gorilla, *Gorilla* spp.) Thus, an expanded update was needed, but not just among the hominoids; by then, impressive new data on elementary technology were emerging from monkeys. This was acknowledged but only minimally, in two brief, inserted boxes, one each for capuchin monkeys (*Sapajus* spp.) and long-tailed macaques (*Macaca fascicularis*) (Wynn, Hernandez-Aguilar, Marchant, & McGrew, 2011). Now, with findings from both Old (Malaivijitnond., Lekprayoon, Tandavanittj, Panha, Cheewatham, & Hamada, 2007) and New World monkeys (Ottoni, 2015) in natural settings, it seems likely that our early ideas about an "ape adaptive grade" are obsolete.

At the same time, when knowledge of the Oldowan expanded between what we knew in 1989 and 2011 (Toth & Schick, 2009; Whiten, Schick, & Toth, 2009), unexpected new Paleolithic phenomena emerged. The long-standing earliest evidence of flaked stone tools dating to 2.6 million years ago (Mya) and the search for the elusive Pre-Oldowan (Panger, Brooks, Richmond, & Wood, 2002) have been superseded by recent finds from West Turkana, Kenya. These were heralded by earlier finds of cut-marked bones at 3.39 Mya from Dikika, Ethiopia (McPherron et al., 2010), but those preliminary results have not yet been followed up. Meanwhile, Harmand and colleagues (2015) have presented lithic evidence for a new industry, the Lomekwian, from Lomekwi 3 at Lake Turkana at 3.3 Mya. It is too early to judge to what extent the Lomekwian relates to the Oldowan (Lewis & Harmand, 2016), either as precursor or otherwise, but such analyses are eagerly awaited.

Thus, the multiple aims of this chapter are to update the record on chimpanzee technology in nature, and to add exciting new data from both New World (mostly from the bearded capuchin, *Sapajus libidinosus*) and Old World (mostly from the Burmese long-tailed macaque, *Macaca fascicularis aurea*) monkeys. Finally, we seek to collate these various findings for comparison with the Oldowan, seeking a useful model for the evolutionary origins of human technology, in the spirit pioneered by Tom Wynn.

METHODS

Wynn and McGrew (1989) and Wynn and colleagues (2011) devised a framework for point-by-point comparison in two tables, one on tools and procedures (Table 1.1), and the other on tool uses and *chaînes opératoires* (Table 1.2). Table 1.1 had four topics: tool type, groupings of types, manufacturing, and associative tool use. Table 1.2 had six topics: target processed, mode of processing, site of processing, objects carried, distance carried, and *chaînes opératoires*. For purposes of direct comparison, we stick here to these established criteria pioneered by Wynn. We have retained 9 of the 10 topics; the last one, *chaînes opératoires*, was based on a single secondary reference (Haidle, 2010), seemed to be of limited utility, and was thus removed for our comparison here.

Both previous treatments (Wynn & McGrew, 1989, Wynn et al., 2011) employed simple binary (presence/absence) criteria for inclusion of evidence; readers could seek out primary data from references listed for most, but not all, topics. Thus, in Table 1.1, types of tools were referenced to source, but manufacturing processes were not. In Table 1.2, targets processed were referenced to source, but not modes of processing

Table 1.1. Tools and Their Characteristics for the Chimpanzee (*Pan troglodytes* [*Pt*]), Burmese Long-Tailed Macaque (*Macaca fascicularis aurea* [*Mfa*]), and Bearded Capuchin (*Sapajus libidinosus* [*Sl*])

Tool Characteristics	*Pt*	*Reference (Pt)*	*Mfa*	*Reference (Mfa)*	*Sl*	*Reference (Sl)*
What are types of tools?						
Stone hammer	C	Benito-Calvo, Carvalho, Arroyo, Matsuzawa, & de la Torre, 2015; Carvalho, 2011	C	Malaivijitnond, Lekprayoon., Tandavanittj, Panha., Cheewatham, & Hamada 2007; Gumert, Kluck, & Malaivijitnond, 2009; Gumert & Malaivijitnond, 2012, 2013; Haslam, Gumert, Biro, Carvalho, & Malaivijitnond, 2013; Tan, Tan, Vyas, Malaivijitnond, & Gumert, 2015, Tan 2017	C	Canale, Guidorizzi, Kierulff, & Gatto, 2009; Falótico & Ottoni, 2016; Fragaszy, Izar, Visalberghi, Ottoni, & de Oliveira, 2004; Mendes et al., 2015; Moura & Lee, 2004; Ottoni & Mannu, 2001
Stone anvil	C	Benito-Calvo et al., 2015; Carvalho, 2011	C	Malaivijitnond et al, 2007; Gumert et al., 2009; Gumert & Malaivijitnond, 2012, 2013; Haslam et al., 2013; Tan et al., 2015, Tan 2017	C	Canale et al., 2009; Falótico & Ottoni, 2016; Fragaszy et al., 2004; Mendes et al., 2015; Moura & Lee, 2004; Ottoni & Mannu, 2001
Wooden hammer	C	Luncz, Mundry, & Boesch, 2012; Yamakoshi, 2011			H	Cutrim, 2013; Ottoni & Mannu, 2001
Wooden anvil	C	Luncz et al., 2012			C	Falótico & Ottoni, 2016; Mendes et al., 2015; Ottoni & Mannu, 2001; Visalberghi, Haslam, Spagnoletti, & Fragaszy, 2013

(*continued*)

Table 1.1. Continued

Tool Characteristics	*Pt*	*Reference (Pt)*	*Mfa*	*Reference (Mfa)*	*Sl*	*Reference (Sl)*
Flexible probe	H	O'Malley, Wallauer, Murray, & Goodall, 2012; Sanz & Morgan, 2011				
Stiff probe	C	Boesch et al., 2016; Humle, Yamakoshi, & Matsuzawa, 2011; Koops, Schöning, Isaji, & Hashimoto, 2015			H	Falótico & Ottoni, 2014; Souto et al., 2011
Spear	C	Pruetz et al., 2015			H	Falótico & Ottoni, 2014; Mannu & Ottoni, 2009
Sponge	H	Hobaiter, Poisot, Zuberbühler, Hoppitt, & Gruber, 2014				
Penetrator, perforator	C	Sanz & Morgan, 2013				
Mollusc shell pick and hammer			C	Gumert, Kluck, & Malaivijitnond, 2009, Tan et al., 2015		
How do tools vary?						
Hammers reflect prey traits (e.g., nut hardness)	C	Boesch & Boesch, 1984	C	Gumert et al., 2009; Gumert & Malaivijitnond, 2013	C	Liu, Fragaszy, & Visalberghi, 2016; Visalberghi, Addessi, et al., 2009
Different raw materials used for same tool (e.g., bark, grass, vine, etc.)	H	Hobaiter et al., 2014: moss or leaf for sponge			C	Mannu & Ottoni, 2009: arthropod probe

Same raw material used for varied tools	C	McGrew, Tutin, & Baldwin, 1979: leaf for termite fishing probe and sponge	C	Gumert et al., 2009; Tan et al., 2015: stone for axe hammer and pound hammer	C	Mannu & Ottoni, 2009: stone for digging and pounding
Water access design varies with same raw material (e.g., leaves)	C	Sousa, 2011			C	Mannu & Ottoni, 2009
Raw materials selected by weight/ size for task	C	Boesch et al., 2016; Carvalho, Matsuzawa, & McGrew, 2013	C	Gumert & Malaivijitnond, 2013	C	Falótico & Ottoni, 2016; Falótico, Siqueira, & Ottoni, 2017
Sex differences in tool use	C	Lonsdorf, 2005	H	Gumert, Hoong, & Malaivijitnond, 2011	C	Falótico & Ottoni, 2014
How are tools made?						
Raw material selectivity	C	Luncz & Boesch, 2014; Luncz, Wittig, & Boesch, 2015			C	Falótico & Ottoni, 2014; Fragaszy et al., 2004; Mannu & Ottoni, 2009; Ottoni & Mannu, 2001; Visalberghi et al., 2007; Visalberghi, Addessi, et al., 2009
Transport (e.g., from source to site of use)	C	Pascual-Garrido, Buba, Nodza, & Sommer, 2012	C	Haslam, Pascual-Garrido, Malaivijitnond, & Gumert, 2016, Gumert & Malaivijitnond, 2013	C	Falótico & Ottoni, 2014; Fragaszy et al., 2004; Mannu & Ottoni, 2009; Ottoni & Mannu, 2001
Simple topological spatial concepts (e.g., proximity, boundary, order, symmetry)	C	Wynn, 1981			C	Proffitt et al., 2016: flake stone
Manual and oral modification	C	Sanz, Morgan, & Gulick, 2004; Yamakoshi, 2011			C	Falótico & Ottoni, 2014; Mannu & Ottoni, 2009

(continued)

Table 1.1. Continued

Tool Characteristics	*Pt*	*Reference (Pt)*	*Mfa*	*Reference (Mfa)*	*Sl*	*Reference (Sl)*
Retouch	C	Carvalho, Biro, McGrew, & Matsuzawa, 2009				
Reuse	C	Boesch et al., 2016; Carvalho et al., 2009				
Processing related to raw material type (e.g., detach, reduce, reshape, combine)	C	Sousa, Biro, & Matsuzawa, 2009			C	Liu et al., 2016; Proffitt et al., 2016
How are tools associated?						
Kit	C	Sanz & Morgan, 2013	C	Gumert et al., 2009	C	Falótico & Ottoni, 2016
Set	C	Sanz & Morgan, 2013			H	Mannu & Ottoni, 2009
Composite (e.g., hammer and anvil)	C	Carvalho et al., 2012, 2013	C	Malaivijitnond et al., 2007; Gumert et al., 2009; Gumert & Malaivijitnond, 2012, 2013; Haslam et al., 2013; Tan, et al., 2015, Tan 2017	C	Falótico & Ottoni, 2016; Fragaszy et al., 2004; Mannu & Ottoni, 2009; Mendes et al., 2015; Moura & Lee, 2004; Ottoni & Mannu, 2001
Crafted	C	Sanz & Morgan, 2013			H	Falótico & Ottoni, 2016
Metatool	P	Matsuzawa, 2011b				
Assemblage	C	Carvalho et al., 2012, 2013	C	Haslam, Luncz, et al., 2016	H	Falótico & Ottoni, 2016
Multifunction	C	O'Malley et al., 2012: ant fish, termite fish with flexible probe			C	Mannu & Ottoni, 2009: dig, hammer with pounding stone

Note: Table compiled by authors. C, Customary; H, Habitual; P, Present.

Table 1.2. Tool-Use Procedures for the Chimpanzee (*Pan troglodytes* [*Pt*]), Burmese Long-Tailed Macaque (*Macaca fascicularis aurea* [*Mfa*]), and Bearded Capuchin (*Sapajus libidinosus* [*Sl*])

Tool-Use Procedures	*Pt*	*Reference (Pt)*	*Mfa*	*Reference (Mfa)*	*Sl*	*Reference (Sl)*
What is processed?						
Termite mound, nest	C	Sanz, Call, & Morgan, 2009; Sanz et al., 2004; Sanz & Morgan, 2011; Stewart & Piel, 2014			P	Souto et al., 2011
Ant nest	C	Pascual-Garrido, Umaru, Allon, & Sommer, 2013			P	Falótico & Ottoni, 2014
Beehive (honey)	C	Sanz & Morgan, 2009; Sommer, Buba, Jesus, & Pascual-Garrido, 2012			C	Falótico & Ottoni, 2014; Mannu & Ottoni, 2009
Nut	C	Luncz et al., 2012	C	Gumert et al., 2009; Gumert & Malaivijitnond, 2012; Luncz et al., 2017; Falótico, Spagnoletti, Haslam, Luncz, Malaivijitnond, & Gumert, 2017, Proffitt, Luncz, Malaivijitnond, Gumert, Svensson, Haslam, 2018	C	De Moraes, Souto, & Schiel, 2014; de Resende, Ottoni, & Fragaszy, 2008; Falótico & Ottoni, 2016; Fragaszy et al., 2004; Mannu & Ottoni, 2009; Moura & Lee, 2004; Ottoni & Mannu, 2001; Visalberghi et al., 2013, 2016
Hard-shelled fruit, seed	H	McGrew, Marchant, Wrangham, & Klein, 1999	P	Gumert & Malaivijitnond, 2012	C	De Moraes et al., 2014; Falótico & Ottoni, 2016; Mannu & Ottoni, 2009
Vertebrate	C	Pruetz et al., 2015	P	Gumert & Malaivijitnond, 2012	C	Falótico & Ottoni, 2014; Mannu & Ottoni, 2009

(*continued*)

Table 1.2. Continued

Tool-Use Procedures	*Pt*	*Reference (Pt)*	*Mfa*	*Reference (Mfa)*	*Sl*	*Reference (Sl)*
Water	C	Hobaiter et al., 2014; McGrew, Marchant, Payne, Webster, & Hunt, 2013			H	Mannu & Ottoni, 2009
Invertebrate	C	Bessa et al., 2015; Sanz, Deblauwe, Tagg, & Morgan, 2014	C	Malaivijitnond et al., 2007; Gumert et al., 2009; Gumert & Malaivijitnond, 2012; Tan et al., 2015; Tan, 2017	C	Falótico & Ottoni, 2014; Mannu & Ottoni, 2009; Cutrim, 2013
Algae	C	Boesch et al., 2016; Humle et al., 2011				
Oil palm heart	C	Yamakoshi, 2011				
How is it processed?						
Absorb	C	Hobaiter et al., 2014; Sousa, 2011; Sousa et al., 2009			H	Mannu & Ottoni, 2009
Block	H	Alp, 1997; Suzuki, Kuroda, & Nishihara, 1995				
Contain	C	Sousa, 2011				
Dig	C	Dutton & Chapman, 2015; Sommer et al., 2012			C	Falótico, Siqueira, et al., 2017; Mannu & Ottoni, 2009
Drape	H	Boesch et al., 2016				

Hammer	C	Matsuzawa, 2011b; Sanz & Morgan, 2009; Yamakoshi, 2011	C	Malaivijitnond et al., 2007; Gumert et al., 2009; Gumert & Malaivijitnond, 2012, 2013; Haslam et al., 2013; Tan, et al., 2015, Tan 2017	C	Falótico & Ottoni, 2016; Fragaszy et al., 2004; Liu et al., 2016; Mannu & Ottoni, 2009; Mendes et al., 2015; Moura & Lee, 2004; Visalberghi et al., 2007
Penetrate	C	Sanz & Morgan, 2013				
Prise	C	Sanz & Morgan, 2009	C	fulcrum hammer: Gumert et al., 2009; Tan et al., 2015	C	Falótico & Ottoni, 2016; Mannu & Ottoni, 2009
Probe	C	O'Malley et al., 2012; Sanz et al., 2009			C	Falótico & Ottoni, 2014; Mannu & Ottoni, 2009; Souto et al., 2011
Stab, poke	C	Pruetz et al., 2015			H	Falótico & Ottoni, 2014
Where is processing done?						
Mound, nest, hive	C	Allon, Pascual-Garrido, & Sommer, 2012			C	Falótico & Ottoni, 2014; Mannu & Ottoni, 2009
Streambed, waterhole, pool	C	Boesch et al., 2016; Humle et al., 2011; McGrew et al., 2013			H	Mannu & Ottoni, 2009
Tree, shrub, liana, etc. bearing food	C	Marchant & McGrew, 2005; Yamakoshi, 2011	C	Falótico, et al., 2017; Gumert & Malaivijitnond, 2012	C	Falótico & Ottoni, 2014, 2016; Souto et al., 2011
Kill site		Plummer & Stanford, 2000			C	Falótico & Ottoni, 2014
Prey refuge (e.g., cavity, crevice)	C	Pruetz et al., 2015			C	Falótico & Ottoni, 2014; Mannu & Ottoni, 2009

(*continued*)

Table 1.2. Continued

Tool-Use Procedures	*Pt*	*Reference (Pt)*	*Mfa*	*Reference (Mfa)*	*Sl*	*Reference (Sl)*
Anthropogenic habitat (e.g., provisioning site)	C	Carvalho et al., 2012; Luncz, Proffitt, Kulik, Haslam, & Wittig, 2016	C	Tan, 2017	C	Liu et al., 2016; Luncz, Falótico, et al., 2016
Coastal rocky shore, mangrove, sandy beach			C	Malaivijitnond et al., 2007; Gumert et al., 2009; Gumert & Malaivijitnond, 2012, 2013; Tan, et al., 2015, Tan 2017	H	Cutrim, 2013
What is carried?						
Raw material, lithic and organic	C	Teleki, 1974			C	Falótico & Ottoni, 2014
Hammer	C	Carvalho & McGrew, 2012; Luncz, Proffitt, et al., 2016	C	Gumert & Malaivijitnond, 2012, 2013; Haslam et al., 2016; Malaivijitnond et al., 2007	C	Corat, Siqueira, & Ottoni, 2016; Fragaszy et al., 2004; Visalberghi, Spagnoletti, et al., 2009
Anvil	C	Carvalho & McGrew, 2012				
Probe	H	Boesch et al., 2016; Pascual-Garrido et al., 2012			C	Falótico & Ottoni, 2014; Mannu & Ottoni, 2009
Nut	C	Carvalho et al., 2012	C	Falótico et al., 2017; Gumert & Malaivijitnond, 2012; Luncz et al., 2017,	C	Visalberghi, Addessi, et al., 2009
Hard-shelled fruit, seed	H	Marchant & McGrew, 2005			C	Mannu & Ottoni, 2009

Carcass of mammal, bird	C	Teleki, 1975			C	de Resende, Greco, Ottoni, & Izar, 2003; Falótico (unpubl. data); Ferreira, de Resende, Mannu, Ottoni, & Izar, 2002
Invertebrate			C	Gumert & Malaivijitnond, 2012	H	Cutrim, 2013
How far are things carried?						
Nut	265m	Carvalho et al., 2012; Carvalho, Cunha, Sousa, & Matsuzawa, 2008	50 m	Falótico et al., 2017; Gumert & Malaivijitnond, 2012	>10m	Corat et al., 2016
Hammer	2265m	Boesch & Boesch, 1984; Carvalho et al., 2012; Carvalho & McGrew, 2012; Luncz, Proffitt, et al., 2016	<100 m	Gumert & Malaivijitnond, 2013; Haslam et al., 2016	>10m	Corat et al., 2016
Anvil	<10 m	Carvalho et al., 2012; Carvalho & McGrew, 2012				
Probe	800m	Goodall, 1968; Pascual-Garrido et al., 2012			>15m	Falótico & Ottoni, 2014; Mannu & Ottoni, 2009
Animal prey	1000s m	Teleki, 1975	<4m	Haslam et al., 2016	100s m	de Resende et al., 2003; Ferreira et al., 2002
Raw material	100m	Teleki, 1974			>15m	Falótico & Ottoni, 2014

Note: Table compiled by authors. C, Customary; H, Habitual; P, Present. Distances are maxima recorded.

(Wynn et al., 2011). Here we seek to be comprehensive in citing primary sources for the non-human primate models. Moreover, we distinguish between established/regular and preliminary/uncommon findings: Following Whiten and colleagues (1999), we seek to limit inclusion to phenomena that are *habitual* (observed repeatedly in several individuals) or *customary* (observed in all or most able-bodied members of at least one age-sex class). We exclude most present or rare occurrences from the tables, but some of these are discussed in the text as suggestive or potentially important, even if anecdotal. Finally, for chimpanzees, with their extensive published literature on tool use, we add mostly new references, published since 2010, leaving readers to pursue earlier sources from Wynn and colleagues' (2011) exhaustive list.

Ideally, all taxa compared should be of equal status (i.e., apples compared with apples, and oranges with oranges, but not apples with oranges). In reality, what is presently known does not allow this. Of the apes, *P. troglodytes* is a single species, although all known complex lithic technology comes from only one of its four subspecies, *P. t. verus*. Each of the monkey genera represents a diverse and wide-ranging evolutionary radiation of simians, the macaques (*Macaca* spp.) and the tufted capuchins (*Sapajus* spp., formerly, *Cebus apella*). Both taxa focus on lithics. In terms of technology, most of the former comes from a single subspecies, and most of the latter comes from a single species, as cited earlier. Further, each of these "model" taxa has been studied at multiple sites, for which results vary—that is, they show to differing extents interpopulational and even intrapopulational variation (Luncz & Boesch, 2015). We cannot tackle the fascinating but complicated issues of cross-population/cultural variation here, so the data should be considered only as samples from the major taxa for which a published record of elementary technology exists.

Finally, at another level, the Oldowan cannot be attributed to a single hominin taxon, as there are multiple candidates in that time frame in the fossil record. The organisms responsible for its production could have been either australopithecines or early *Homo*. So, while data on living primates derive from the behavior and artifacts of living, accessible populations, the data from extinct taxa come from artifacts only, with no access to behavior and no confident attribution as to its makers or users. We can do nothing about this handicapping heterogeneity other than to note it.

Equally confounding is the issue of context: Primates occupy a variety of venues that range from prison-style laboratories to relatively natural ecosystems. (The caveat in the latter is necessary, because *all* living primates are influenced to some extent by *H. sapiens*, if only indirectly, for example, via climate change; see Hockings, McLennan, et al., 2015). Where to draw the line along this continuum of contexts is debatable and arbitrary, but because we seek evolutionary adaptation to forces of natural (rather than artificial) selection, we exclude the following types of data: from laboratories, zoological gardens, refuges and sanctuaries, safari and wildlife parks, temples, rehabilitation projects, or ranging in villages/towns. All of these venues entail major human influence, such as provisioning or confinement that restrains free-range foraging and therefore limits or eliminates access to predators and prey (Haslam, 2013). We do include (albeit uneasily) populations of primates that are secondarily feral (Moscovice et al., 2007), commensal with local humans (Pruetz et al., 2015), irregularly provisioned (Tan, 2017), or crop-raiders (Hockings, Bryson-Morrison, et al., 2015).

Similarly, we include only spontaneous, emergent behavior and exclude artificially induced behavior, even if it occurs in natural surroundings. Thus, field experimentation employing unnatural stimuli or conditions is omitted if it asks unnatural behavior of its subjects of study. However, we include gray-area exceptions to this, such as the "outdoor laboratory" at Bossou (Matsuzawa, 2011a) or similar studies at Fazenda Boa Vista (Hanna et al., 2015), if they record natural behaviors in response to natural stimuli (Carvalho et al., 2012; Haslam, Cardoso, Visalberghi, & Fragaszy, 2014; Gumert & Malaivijitnond, 2013). Of course, many studies of many types of behavior occur in the proscribed settings listed in the preceding two paragraphs. Arguably, such studies are necessary if we ever are to understand causal mechanisms and ontogeny, but those are not the aims of this qualitative exercise.

We have sourced our data from findings in the public domain, thus excluding unpublished reports, personal communications, and work in progress, unless a vital point is made, with necessary caveats. Ideally, all data on elementary technology should be ethological, that is, based on first-hand observations of behavior. However, at least for chimpanzees, most study populations are unhabituated, that is, not amenable to close-range study throughout their daily lives (McGrew, 2017). In such cases, we rely on indirect evidence, such as tools, raw materials, *débitage*, and altered target items, plus signs indicating the presence of the user organism, such as foot or hand prints, hair, feces, and DNA. Such evidence is probabilistic, just as it must be in archaeology, whether obtained from humans or non-humans (Haslam et al., 2009, 2017). Finally, we note that there are huge disparities in research effort across the taxa highlighted here: Wild chimpanzees have been studied at more than 120 field sites, for as long as six decades (McGrew, 2017); no other primate taxon even approaches this extent of research. Thus, whatever the lesser-studied monkeys show us is doubly impressive (see Figures 1.1 through 1.6).

FINDINGS

Our tables here show a three-way comparison of chimpanzee, capuchin, and macaque tool behavior. We can see both similarity and difference and consider this first before juxtaposing them to the features of Oldowan tools. Here, we summarize key points from data reviewed and cited in Tables 1 & 2. At the most basic level, all are tool-users. The macaques use tools to process more prey taxa than the chimpanzees and capuchins combined. All three taxa use lithic tools, but only one (capuchin) makes stone artifacts that could be tools, but then does not use them. All three leave excavated archaeological records that reveal stone percussors used to process hard objects, but only one (macaque) customarily uses hammers to crack open hard shells of both animals and plants. All three taxa show sex differences in tool use, but while macaques and chimpanzees concur on female predominance (albeit in different tasks, percussion and probing, respectively), capuchins show male predominance in probing but not percussion. Overall, our comparison shows more similarities between chimpanzees and capuchins than for either of those taxa with macaques. More detailed and extensive comparisons, however, are needed.

Table 1.1 shows that for tool type, chimpanzees have the most varied, followed by capuchins and macaques. Chimpanzees' use of woody vegetation for hammer and anvil, probes, spear, penetrator, and perforator is notable; all of these types have their virtues, but flexible probes require precise manipulation as the tools are threaded into the winding passageways of the nests of ants and termites. The stiffer probes of

Figure 1.1. Chimpanzee at Bossou cracks oil palm nut (*Elaeis guineensis*) with stone hammer and anvil (foreground), while others use vegetation to sponge water from a tree hole (background). Photo taken by and published with the permission of Susana Carvalho.

capuchins winkle out both lizards and rodents from crevices, but the action of use depends more on power than precision. Macaques have also been reported to use mollusc shells to crack open oysters and other shellfish. Macaques use various shells to process other molluscs, such as linear, pointed auger shells to access the soft innards of oysters.

Figure 1.2. Bearded capuchin monkey at Serra da Capivara bimanually cracks cashew nut with stone hammer and anvil. Photo taken by and published with the permission of Alejandra Pascual-Garrido.

Figure 1.3. Burmese long-tailed macaque monkey unimanually uses stone axe hammer to crack open oysters attached to a boulder. Photo taken by and published with the permission of Michael D. Gumert. Previously published as Fig. 4 (p. 187) in Wynn et al. (2011), "An ape's view of the Oldowan" revisited, Evolutionary Anthropology.

Figure 1.4. Adult chimpanzee at Gombe fishes for termites (*Macrotermes* sp.) with flexible probe of vegetation. Photo taken by and published with the permission of Alejandra Pascual-Garrido.

Figure 1.5. Male capuchin monkey at Serra da Capivara uses stiff probe of vegetation to poke at iguanid lizard in crack of tree branch. Photo taken by and published with the permission of Tiago Falótico.

Figure 1.6. Burmese long-tailed macaque bimanually uses pounding hammer to crack open molluscs. Photo taken by and published with the permission of Michael D. Gumert.

Tool characteristics show more similarity than difference across the three taxa, at least on the parameters listed here. All use hammers that reflect the hardness of the prey-encasing outer shell, whether this be nut or mollusc. All select raw materials according to utilitarian variables such as size, mass, hardness, and rigidity. All use the same type of raw material for more than one form of extractive foraging. Macaques cannot be compared for material selectivity for probes or leaf-tools.

All three taxa prefer raw materials with appropriate features for the task at hand, as inferred from their non-random choices from the array of natural or sometimes unnatural objects available. All transport raw materials or tools from source to place of use. For tool-making, chimpanzee and capuchin are remarkably similar, with only retouching and reusing of tools apparently lacking in the latter, but macaque tool-making is minimal. For the key traits of modes of tool-making, when applied to lithics, only capuchins detach, reduce and reshape such raw materials, when they extract quartz cobbles from conglomerate matrix, reduce them to powder (for ingestion), and create flakes (unused) in the process.

For associated tool use, all three taxa have toolkits of differing tool types, and all leave behind assemblages amenable to successful archaeological recovery. All have composite tools, chiefly hammer and anvil, and all show multifunctionality, in that the same type of tool shows two or more uses. However, only chimpanzees and capuchins have tools sets and crafted tools, and only chimpanzees show metatools, in the form of wedge-stones that improve the function of stone anvils.

The items processed with tools and how tools are used have notable similarities and differences. Table 1.2 shows a remarkable congruence between chimpanzees and capuchins in the range of items processed by elementary technology, including both plant and animal prey. Items processed by macaques are similar, except that insects and drinking water appear to be absent from their subsistence technology. Of the 22 modes of processing action listed by McGrew (2013), chimpanzees show 10, capuchins 6, but macaques only 2. Most of the modes lacking in the macaques involve tools of plant material. All three use tools at vegetative sources, whether these are trees, shrubs, lianas, or other vegetation, such as grasses. Only macaques, however, specialize in coastal resources, harvesting many invertebrate taxa from intertidal zones, although capuchins living on mangrove areas have been reported to process crustacean and gastropods using wooden tools (Cutrim, 2013).

Transport varies too. All three taxa carry raw materials. Regarding stones, all three carry their hammers, but only capuchins and chimpanzees transport anvils as well; macaques use fixed anvils, typically boulders, rocky outcrops, or exposed roots. All carry nuts and hard-shelled fruits from source to processing sites. Transport of animal prey reflects preferred taxa: Chimpanzees and capuchins carry mammal or reptile prey, often as a "movable feast" after making a kill, with possessors being trailed by pressing cadgers. Macaques move mollusc and crustacean prey from acquisition site to anvil site, and like chimpanzees and capuchins, will also be followed when carrying crabs or other enticing prey items.

Distances for which raw materials, tools, or prey are carried vary by orders of magnitude. Most transport by macaques and capuchins is for only a few meters, except for vertebrate prey transported by capuchins. The latter sometimes carry killed lizards and rodents for hundreds of meters, usually seeking to avoid other monkeys trying to obtain a portion of the prey. Chimpanzees sometimes carry portable items of all types for hundreds of meters. They may do so with both raw materials and tools made in advance of their use, such as for termite fishing. While insect prey are eaten on the spot, vertebrate prey, which may take hours to consume fully, sometimes are carried for thousands of meters, even over more than a single day.

Next we describe features of the Oldowan industry and show that even with new updates little has changed in how *Pan* and Oldowan technology compare as first shown by Wynn and colleagues' (2011). In Table 1.1, we see the types of tools are much the same, except for the penetrators and perforators more recently added to the repertoire of chimpanzee termite fishing. Variation in tools largely reflects selectivity and task demands, although these are specified to a greater degree in our tables here.

For tool-making techniques, the same similarities and differences persist, with the latter starkly manifest in knapping. No living ape or monkey in nature has yet been seen to intentionally produce lithic flakes, except as a byproduct of their tool activities. Associated tools (i.e., complex and multicomponent) are more difficult to compare across the reports, as the taxonomy of such technology has sharpened up in recent times, and the typology has expanded (for definitions, see McGrew, 2013, supplementary information). For chimpanzees, the number of associated categories has gone from 4 to 10, and of these 10, 7 are seen in the apes, leaving only 3 (sequential, secondary, and construction) unknown at habitual or customary levels. If we were to extend chimpanzee tool use beyond subsistence, then construction would be demonstrated easily, in the apes' daily making of interwoven, spring-loaded sleeping platforms and nests; see Koops, McGrew, de Vries, & Matsuzawa, 2012; Samson, 2012; Stewart, Piel, & McGrew, 2011. Recategorizing the Oldowan into these same terms is not easy (see later discussion); the industry seems to reflect at least 6 of the 10 (kit, composite, crafted, assemblage, secondary, multifunction). The others (construction, metatool, sequential, set) are difficult to discern archaeologically in the absence of behavior—for example, sequential use of one tool to acquire another seems totally dependent on behavioral rather than artifactual data. Notably present in the Oldowan but absent in the apes is secondary tool use, that is, use of a tool to make another tool.

In Table 1.2, the list of what is processed in tool use by apes is largely unchanged, although specific items have been added to the prey list, such as land snails. For the Oldowan, the list is short for a simple reason: Most of the prey items listed for living apes are perishable and so do not persist in the fossil record. The list of processed items for the Oldowan is a function of taphonomic processes that yield fossil evidence, such as cut-marks on bone, but this has not changed notably since Wynn and colleagues' (2011) report.

Similarly and misleadingly limited is the number of modes used by Oldowan tool-users (Table 1.3). Only four can be shown from the products of behavior (as opposed to its direct observation): cut, hammer/pound, dig, and rub/scratch. The latter two would require valid measures of microwear on used surfaces, such as soil-abraded flakes or bones or scrapers used to work wood.

Oldowan processing sites are even more delimited by missing evidence. Organic indicators of localized subsistence activities are irretrievable unless they are assemblages of fossilized bones showing signs of processing by cut-marks or breakage patterns. Realistically, this means the remains of large-bodied vertebrates concentrated at kill sites or found elsewhere in isolation. This contrasts with both site-specific features of prey, such as beehives, or contextual variables, such as groves of nut-bearing trees, which are readily available to researchers following wild apes but which are usually archaeologically invisible.

Detectable carriage of items in the Oldowan is limited to stones and bones (Hayden, 2008), while for living apes, the range is much greater, as it includes a variety of plant and animal parts. However, distance carried seems to show a distinct difference between Oldowan hominin and living ape, with the former's transport being almost an order of magnitude longer in distance, that is, thousands rather than hundreds of meters.

DISCUSSION

Wynn and McGrew's (1989) and Wynn and colleagues' (2011) idea of an ape adaptive grade, based on the elementary technology of chimpanzees and orangutans, implicitly excluded other living non-human primates. In those former times, little was known to indicate otherwise, although at least two species of monkeys were waiting in the wings. In discussing the "apeness" of the Oldowan, Wynn and colleagues (2011) listed nine features shared by living apes and Oldowan hominins: (1) use of tools to access food; (2) use of tools to process food; (3) discrimination and selection of raw materials; (4) selection of tools in advance of use; (5) carrying tools and food; (6) reuse of activity areas; (7) hierarchically organized procedures; (8) use of flexible procedures; and (9) cultural differences. All of these also are now reported for both *Sapajus libidinosus* and *Macaca fascicularis aurea*. Thus, the ape adaptive grade has to be expanded to the simian adaptive grade. Of course, in principle, the grade could even be enlarged further, to include prosimians, other mammals, or even birds. We, however, do not assess that here.

Table 1.3. Oldowan Tools, Characteristics, and Use Procedures

Oldowan Tool Characteristics	*Oldowan Tool-Use Procedures*
What were types of tools? • Stone hammer • Stone anvil • Flaked core • Unmodified flake • Modified flake • Battered cobble • Unmodified cobble • Modified bone (rare) **How did tools vary?** • Lithic raw materials vary • Same raw material used for different tool types • Raw materials selected by weight, size, etc. **How were tools made?** • Raw material selectivity • Transport of raw materials • Simple topological spatial concepts • Manual modification (e.g., bipolar knapping) • Use of knappable angles • Contiguous placing of knapping strikes (e.g., core rotation) • Retouch (e.g., core rejuvenation, trimming) • Processing related to raw material type **How were tools associated?** • Kit • Composite • Crafted • Assemblage • Secondary (e.g., bipolar) • Multifunction	**What was processed?** • Bone for marrow • Animal soft tissue (e.g., flesh, tendon, etc.) • Plant part (e.g., fruit, root, stem, corm, etc.) **How was it processed?** • Cut (e.g., slice/cleave animal or plant tissue) • Dig • Hammer/pound (e.g., crush prey part, split bone) • Penetrate **Where was processing done?** • Death site of animal prey • Secure site for processing plant prey **What was carried?** • Raw material • Hammer • Flaked core • Large flake • Cobble • Animal body parts **How far were things carried?** • Hundreds or thousands of meters (up to 13 km)

But what of the blank cells in the columns for capuchin and macaque in Tables 1.1 and 1.2, which are correspondingly filled for chimpanzee? It would be rash to say that any such empty cells are unfillable. As studies build on capuchins and macaques, we may uncover other forms of tool behavior. Chimpanzees have far more study sites and time invested into searching for such evidence. This is not, however, comparatively matched in tool-using monkeys. The findings for the capuchins come mainly from only two study sites, Fazenda Boa Vista and Serra da Capivara, Brazil, while those for the macaques come mostly from only one site, Piak Nam Yai, Thailand. The subjects of study have been habituated to daylong, close-range observation for only a few years, if at all. Contrast this with 10 study sites of fully habituated chimpanzees, most of which have records compiled over decades. Furthermore, there are more than 110 other chimpanzees study sites of shorter duration (McGrew, 2017). Thus the total research effort for either of the monkey species pales beside that of the apes. Who knows what remains to be seen in these simians?

So, which of the three non-human primate taxa offers the best referential model for the Oldowan? Each has its strong points: Chimpanzees show the widest variety of modes of tool use and manufacture, including more types of associative tool use, but their lithic subsistence tools are limited to the simplest percussion of hammer and anvil. Capuchins exceed the others in actually flaking stone (although those flakes remain yet unused), but they fall short in some aspects of precision use of tools, such as flexible probes and their accessories. Macaques are clearly lithic specialists, with the widest range of stone tools and prey processed by them, but they fall short as varied tool-users and makers. Rather than assigning which is the better model, each has something important to offer for modeling the evolutionary origins of the Oldowan. The traits that are common to all three are arguably the ones that should form the basis of reconstructing the technology of (e.g.) a last common ancestor. The common denominator that is present in all three is the use of hammer stones to pound open encased food items on fixed anvils. Using this technique, the only food item processed across all three is hard-shelled nuts.

We must, however, also account for the two gaps asserted by Wynn and McGrew (1989) and Wynn and colleagues (2011) when we consider comparison to the Oldowan. First, is the distances that tools or their raw materials are carried by living apes compared to extinct hominins: Only for transport of vertebrate prey does ape transport exceed a kilometer, and this extension seems to be merely because of the greater time taken to consume a large prey. That is, a red colobus monkey carcass (e.g.) carried by a chimpanzee predator is not being taken to any specific place for processing or consumption, rather, it is being eaten on the go. All other non-human primate transports can reach hundreds of meters, but more often is only in tens, and usually only in ones. In contrast, hominins transported lithic raw materials or tools for up to 13 kilometers.

Is this an apples-versus-oranges comparison? Living primate transports typically are single *journeys* observed of an individual bout of carrying, at the end of which the item carried is consumed or used and then abandoned. This is very different from archaeologically measuring the *distance* from origin of raw materials (typically rocky outcrops) to the site of their modification and use, with no knowledge of how the stones got there and, if carried, over how many journeys over how much time by how many individuals. It is likely, therefore, that in the daily usage of Oldowan tools, they were transported similarly to how apes and monkeys are observed to carry stones. To determine how far these non-humans tools can be transported over their lifespan as a tool, like the uncovered Oldowan tools, would require focal sampling of tools and anvils. Happily, such research is now underway (Luncz, Proffitt, et al., 2016).

The second gap is competition between primates and carnivores over animal prey, where the former might be monkeys or apes or hominins and the latter might be felids, canids, or hyaenids, contesting access to mammalian carcasses—that is, *scavenging*.

Here, it matters not who made the kill or who usurped it, either sooner or later, but only who consumed it. There is apparently no contest: Living primates rarely scavenge meat (Watts, 2007, but cf. Morris & Goodall, 1977), whereas abundant evidence of cut-marks, percussive fracturing, and carnivore tooth-marks on fossil bones indicates some kind of interaction between hominins and carnivores (Stanford & Bunn, 2001). How and when this interaction occurred is less easy to discern, as usurping possession of a fresh kill does not equal splintering abandoned long bones. Just how important such competition was to Oldowan hominins is unknown, but the distinction between non-human primate and human remains sharp. Few data are available on chimpanzee–carnivore interactions. The only habituated group of wild, savanna-living chimpanzees lacks sympatric large carnivores (Pruetz et al., 2015), but other candidate study sites (McGrew, Baldwin, Marchant, Pruetz, & Tutin, 2014; Stewart & Piel, 2014) are available.

With the advent of the Pre-Oldowan, so far in the discovery of cut-marks at Dikika (McPherron et al., 2010) and flakes and cores at Lomekwi (Harmand et al., 2015), these issues are pushed notably further back in time. This chapter is not the place to assess these new and exciting sets of findings, but two points especially stand out, however tentatively, that have important implications: The only hominins known to exist at 3.3 million years ago had brains no larger than living African apes, and the two percussive techniques hypothesized to have been used to create the Lomekwian seem to resemble ape and monkey hammer and anvil extractive foraging (bipolar) and ape and monkey hard-shelled fruit smashing (passive). Perhaps it is time to call upon Tom Wynn for another of his acute assessments of the new evidence, given that his 1981 hypothesis has proven to be so useful.

ACKNOWLEDGMENTS

We thank the following researchers for aid: Susana Carvalho, Adriana Hernandez-Aguilar, Sonia Harmand, Michael Haslam, Kim Hockings, Kathelijne Koops, Elizabeth Lonsdorf, Alejandra Pascual-Garrido, and Craig Stanford. For photos, we thank Susana Carvalho, Alejandra Pascual-Garrido, and for useful comments, two anonymous reviewers. We are grateful to the editors for their patience!

REFERENCES

Allon, O., Pascual-Garrido, A., & Sommer, V. (2012). Army ant defensive behaviour and chimpanzee predation success: Field experiments in Nigeria. *Journal of Zoology, 288*(4), 237–244.

Alp, R. (1997). "Stepping-sticks" and "seat-sticks": New types of tools used by wild chimpanzees (*Pan troglodytes*) in Sierra Leone. *American Journal of Primatology, 41*(1), 45–52.

Andrews, P. (2015). *An ape's view of human evolution*. Cambridge, UK: Cambridge University Press.

Benito-Calvo, A., Carvalho, S., Arroyo, A., Matsuzawa, T., & de la Torre, I. (2015). First GIS analysis of modern stone tools used by wild chimpanzees (*Pan troglodytes verus*) in Bossou, Guinea, West Africa. *PloS One, 10*(3), e0121613.

Bessa, J., Sousa, C., & Hockings, K. J. (2015). Feeding ecology of chimpanzees (*Pan troglodytes verus*) inhabiting a forest-mangrove-savanna-agricultural matrix at Caiquene-Cadique, Cantanhez National Park, Guinea-Bissau. *American Journal of Primatology, 77*(6), 651–665.

Boesch, C., & Boesch, H. (1984). Mental map in wild chimpanzees: An analysis of hammer transports for nut cracking. *Primates, 25*(2), 160–170.

Boesch, C., Kalan, A. K., Agbor, A., Arandjelovic, M., Dieguez, P., Lapeyre, V., & Kühl, H. S. (2016). Chimpanzees routinely fish for algae with tools during the dry season in Bakoun, Guinea. *American Journal of Primatology, 79*(3), e22613.

Canale, G. R., Guidorizzi, C. E., Kierulff, M. C. M., & Gatto, C. A. F. (2009). First record of tool use by wild populations of the yellow-breasted capuchin monkey (*Cebus xanthosternos*) and new records for the bearded capuchin (*Cebus libidinosus*). *American Journal of Primatology, 71*(5), 366–372.

Carvalho, S. (2011). Extensive surveys of chimpanzee stone tools: From the telescope to the magnifying glass. In T. Matsuzawa, T. Humle, & Y. Sugiyama (Eds.), *The chimpanzees of Bossou and Nimba* (pp. 145–155). Tokyo: Springer.

Carvalho, S., Biro, D., Cunha, E., Hockings, K. J., McGrew, W. C., Richmond, B. G., & Matsuzawa, T. (2012). Chimpanzee carrying behaviour and the origins of human bipedality. *Current Biology, 22*(6), R180–R181.

Carvalho, S., Biro, D., McGrew, W. C., & Matsuzawa, T. (2009). Tool-composite reuse in wild chimpanzees (*Pan troglodytes*): Archaeologically invisible steps in the technological evolution of early hominins? *Animal Cognition, 12*(1), 103–114.

Carvalho, S., Cunha, E., Sousa, C., & Matsuzawa, T. (2008). *Chaînes opératoires* and resource-exploitation strategies in chimpanzee (*Pan troglodytes*) nut cracking. *Journal of Human Evolution, 55*(1), 148–163.

Carvalho, S., Matsuzawa, T., & McGrew, W. C. (2013). From pounding to knapping: How chimpanzees can help us to model hominin lithics. In C. M. Sanz, J. Call, & C. Boesch (Eds.), *Tool use in animals: Cognition and ecology* (pp. 225–241). Cambridge, UK: Cambridge University Press.

Carvalho, S., & McGrew, W. C. (2012). The origins of the Oldowan: Why chimpanzees (*Pan troglodytes*) still are good models for technological evolution in Africa. In M. Domínguez-Rodrigo (Ed.), *Stone tools and fossil bones: Debates in the archaeology of human origins* (pp. 201–221). Cambridge, UK: Cambridge University Press.

Corat, C., Siqueira, J. O., & Ottoni, E. B. (2016). Sequential organization and optimization of the nut-cracking behavior of semi-free tufted capuchin monkeys (*Sapajus* sp.). *Primates, 57*(1), 113–121.

Cutrim, F. H. R. (2013). *Padrão comportamental e uso de ferramentas em macacos-prego (*Sapajus libidinosus*) residentes em manguezal.* Doctoral thesis, University of São Paulo, Brazil.

De Moraes, B. L. C., Souto, A. da S., & Schiel, N. (2014). Adaptability in stone tool use by wild capuchin monkeys (*Sapajus libidinosus*). *American Journal of Primatology, 76*(10), 967–977.

de Resende, B. D., Greco, V. L. G., Ottoni, E. B., & Izar, P. (2003). Some observations on the predation of small mammals by tufted capuchin monkeys (*Cebus apella*). *Neotropical Primates, 11*(2), 103–104.

de Resende, B. D., Ottoni, E. B., & Fragaszy, D. M. (2008). Ontogeny of manipulative behaviour and nut-cracking in young tufted capuchin monkeys (*Cebus apella*): A perception-action perspective. *Developmental Science, 11*(6), 828–840.

Dutton, P., & Chapman, H. (2015). New tools suggest local variation in tool use by a montane community of the rare Nigeria–Cameroon chimpanzee, *Pan troglodytes ellioti*, in Nigeria. *Primates, 56*(1), 89–100.

Falótico, T., & Ottoni, E. B. (2014). Sexual bias in probe tool manufacture and use by wild bearded capuchin monkeys. *Behavioural Processes, 108*, 117–122.

Falótico, T., & Ottoni, E. B. (2016). The manifold use of pounding stone tools by wild capuchin monkeys of Serra da Capivara National Park, Brazil. *Behaviour, 153*(4), 421–442.

Falótico, T., Siqueira, J. O., & Ottoni, E. B. (2017). Digging up food: Excavation stone tool use by wild capuchin monkeys. *Scientific Reports, 7*(1), 6278.

Falótico, T., Spagnoletti, N., Haslam, M., Luncz, L. V., Malaivijitnond, S., & Gumert, M. D. (2017). Analysis of sea almond (*Terminalia catappa*) cracking sites used by wild Burmese long-tailed macaques (*Macaca fascicularis aurea*). *American Journal of Primatology, 79*, e22629.

Ferreira, R. G., de Resende, B. D., Mannu, M., Ottoni, E. B., & Izar, P. (2002). Bird predation and prey-transference in brown capuchin monkeys (*Cebus apella*). *Neotropical Primates, 10*, 84–89.

Fragaszy, D. M., Izar, P., Visalberghi, E., Ottoni, E. B., & de Oliveira, M. G. (2004). Wild capuchin monkeys (*Cebus libidinosus*) use anvils and stone pounding tools. *American Journal of Primatology, 64*(4), 359–366.

Goodall, J. van L. (1968). The behaviour of free-living chimpanzees in the Gombe Stream Reserve. *Animal Behaviour Monographs, 1*, 161–311.

Gruber, T., & Clay, Z. (2016). A comparison between bonobos and chimpanzees: A review and update. *Evolutionary Anthropology, 25*(5), 239–252.

Gumert, M. D., Hamada, Y., & Malaivijitnond, S. (2013). Human activity negatively affects stone tool-using Burmese long-tailed macaques *Macaca fascicularis aurea* in Laem Son National Park, Thailand. *Oryx, 47*(4), 535–543.

Gumert, M. D., Hoong, L. K., & Malaivijitnond, S. (2011). Sex differences in the stone tool-use behavior of a wild population of Burmese long-tailed macaques (*Macaca fascicularis aurea*). *American Journal of Primatology, 73*(12), 1239–1249.

Gumert, M. D., Kluck, M., & Malaivijitnond, S. (2009). The physical characteristics and usage patterns of stone axe and pounding hammers used by long-tailed macaques in the Andaman Sea region of Thailand. *American Journal of Primatology, 71*(7), 594–608.

Gumert, M. D., & Malaivijitnond, S. (2012). Marine prey processed with stone tools by Burmese long-tailed macaques (*Macaca fascicularis aurea*) in intertidal habitats. *American Journal of Physical Anthropology, 149*(3), 447–457.

Gumert, M. D., & Malaivijitnond, S. (2013). Long-tailed macaques select mass of stone tool according to food type. *Philosophical Transactions of the Royal Society of London. Series B, Biological Sciences, 368*(1630), 20120413.

Haidle, M. N. (2010). Working-memory capacity and the evolution of modern cognitive potential. *Current Anthropology, 51*(S1), S149–S166.

Hanna, J. B., Schmitt, D., Wright, K., Eshchar, Y., Visalberghi, E., & Fragaszy, D. M. (2015). Kinetics of bipedal locomotion during load carrying in capuchin monkeys. *Journal of Human Evolution, 85*, 149–156.

Harmand, S., Lewis, J. E., Feibel, C. S., Lepre, C. J., Prat, S., Lenoble, A., . . . Roche, H. (2015). 3.3-million-year-old stone tools from Lomekwi 3, West Turkana, Kenya. *Nature, 521*(7552), 310–315.

Haslam, M. (2013). "Captivity bias" in animal tool use and its implications for the evolution of hominin technology. *Philosophical Transactions of the Royal Society of London. Series B, Biological Sciences, 368*(1630), 20120421.

Haslam, M. (2014). On the tool use behavior of the bonobo-chimpanzee last common ancestor, and the origins of hominine stone tool use. *American Journal of Primatology, 76*(10), 910–918.

Haslam, M., Cardoso, R. M., Visalberghi, E., & Fragaszy, D. M. (2014). Stone anvil damage by wild bearded capuchins (*Sapajus libidinosus*) during pounding tool use: A field experiment. *PLoS One, 9*(11), e111273.

Haslam, M., Gumert, M. D., Biro, D., Carvalho, S., & Malaivijitnond, S. (2013). Use-wear patterns on wild macaque stone tools reveal their behavioural history. *PLoS One, 8*(8), e72872.

Haslam, M., Hernandez-Aguilar, R. A., Ling, V., Carvalho, S., de la Torre, I., DeStefano, A., . . . Warren, R. (2009). Primate archaeology. *Nature, 460*(7253), 339–344.

Haslam, M., Hernandez-Aguilar, R. A., Proffitt, T., Arroyo, A., Falótico, T., Fragaszy, D. M., . . . Luncz, L. V. (2017). Primate archaeology evolves. *Nature Ecology and Evolution, 1*, 1431–1437.

Haslam, M., Luncz, L. V., Pascual-Garrido, A., Falótico, T., Malaivijitnond, S., & Gumert, M. D. (2016). Archaeological excavation of wild macaque stone tools. *Journal of Human Evolution, 96*(13), 134–138.

Haslam, M., Pascual-Garrido, A., Malaivijitnond, S., & Gumert, M. D. (2016). Stone tool transport by wild Burmese long-tailed macaques (*Macaca fascicularis aurea*). *Journal of Archaeological Science: Reports, 7*, 408–413.

Hayden, B. (2008). What were they doing in the Oldowan? An ethnoarchaeological perspective on the origins of human behaviour. *Lithic Technology, 33*(2), 105–139.

Hobaiter, C., Poisot, T., Zuberbühler, K., Hoppitt, W. J. E., & Gruber, T. (2014). Social network analysis shows direct evidence for social transmission of tool use in wild chimpanzees. *PLoS Biology, 12*(9), e1001960.

Hockings, K. J., Bryson-Morrison, N., Carvalho, S., Fujisawa, M., Humle, T., McGrew, W. C., . . . Matsuzawa, T. (2015). Tools to tipple: Ethanol ingestion by wild chimpanzees using leaf-sponges. *Royal Society Open Science, 2*, 150150.

Hockings, K. J., McLennan, M. R., Carvalho, S., Ancrenaz, M., Bobe, R., Byrne, R. W., . . . Hill, C. M. (2015). Apes in the Anthropocene: Flexibility and survival. *Trends in Ecology and Evolution, 30*(4), 215–222.

Humle, T., Yamakoshi, G., & Matsuzawa, T. (2011). Algae scooping remains a puzzle. In T. Matsuzawa, T. Humle, & Y. Sugiyama (Eds.), *The chimpanzees of Bossou and Nimba* (pp. 117–122). Tokyo: Springer.

Koops, K., McGrew, W. C., de Vries, H., & Matsuzawa, T. (2012). Nest-building by chimpanzees (*Pan troglodytes verus*) at Seringbara, Nimba Mountains: Anti-predator, thermoregulatory, and anti-vector hypotheses. *International Journal of Primatology, 33*(2), 356–380.

Koops, K., Schöning, C., Isaji, M., & Hashimoto, C. (2015). Cultural differences in ant-dipping tool length between neighbouring chimpanzee communities in Kalinzu, Uganda. *Scientific Reports, 5*, 12456.

Lewis, J. E., & Harmand, S. (2016). An earlier origin for stone tool making: Implications for cognitive evolution and the transition to *Homo*. *Philosophical Transactions of the Royal Society of London. Series B, Biological Sciences, 371*(1698), 20150233.

Liu, Q., Fragaszy, D. M., & Visalberghi, E. (2016). Wild capuchin monkeys spontaneously adjust actions when using hammer stones of different mass to crack nuts of different resistance. *American Journal of Physical Anthropology, 161*(1), 53–61.

Lonsdorf, E. V. (2005). Sex differences in the development of termite-fishing skills in the wild chimpanzees, *Pan troglodytes schweinfurthii*, of Gombe National Park, Tanzania. *Animal Behaviour, 70*(3), 673–683.

Luncz, L. V., & Boesch, C. (2014). Tradition over trend: Neighboring chimpanzee communities maintain differences in cultural behavior despite frequent immigration of adult females. *American Journal of Primatology, 76*(7), 649–657.

Luncz, L. V., & Boesch, C. (2015). The extent of cultural variation between adjacent chimpanzee (*Pan troglodytes verus*) communities: A microecological approach. *American Journal of Physical Anthropology, 156*(1), 67–75.

Luncz, L. V., Falótico, T., Pascual-Garrido, A., Corat, C., Mosley, H., & Haslam, M. (2016). Wild capuchin monkeys adjust stone tools according to changing nut properties. *Scientific Reports, 6*, 33089.

Luncz, L. V., Mundry, R., & Boesch, C. (2012). Evidence for cultural differences between neighbouring chimpanzee communities. *Current Biology, 22*(10), 922–926.

Luncz, L. V., Proffitt, T., Kulik, L., Haslam, M., & Wittig, R. M. (2016). Distance-decay effect in stone tool transport by wild chimpanzees. *Proceedings of the Royal Society of London B: Biological Sciences, 283*, 20161607.

Luncz, L. V., Svensson, M. S., Haslam, M., Malaivijitnond, S., Proffitt, T., & Gumert, M. D. (2017). Technological response of wild macaques (*Macaca fascicularis*) to anthropogenic change. *International Journal of Primatology, 38*(5), 872–880.

Luncz, L. V., Wittig, R. M., & Boesch, C. (2015). Primate archaeology reveals cultural transmission in wild chimpanzees (*Pan troglodytes verus*). *Philosophical Transactions of the Royal Society of London. Series B, Biological Sciences, 370*, 20140348.

Malaivijitnond, S., Lekprayoon, C., Tandavanittj, N., Panha, S., Cheewatham, C., & Hamada, Y. (2007). Stone-tool usage by Thai long-tailed macaques (*Macaca fascicularis*). *American Journal of Primatology, 69*(2), 227–233.

Mannu, M., & Ottoni, E. B. (2009). The enhanced tool-kit of two groups of wild bearded capuchin monkeys in the Caatinga: Tool making, associative use, and secondary tools. *American Journal of Primatology, 71*(3), 242–251.

Marchant, L. F., & McGrew, W. C. (2005). Percussive technology: Chimpanzee baobab smashing and the evolutionary modelling of hominin knapping. In V. Roux & B. Bril (Eds.), *Stone knapping: The necessary conditions for a uniquely hominin behaviour* (pp. 341–350). Cambridge, UK: McDonald Institute for Archaeological Research.

Matsuzawa, T. (2011a). Field experiments of tool-use. In T. Matsuzawa, T. Humle, & Y. Sugiyama (Eds.), *The chimpanzees of Bossou and Nimba* (pp. 157–164). Tokyo: Springer.

Matsuzawa, T. (2011b). Stone tools for nut-cracking. In T. Matsuzawa, T. Humle, & Y. Sugiyama (Eds.), *The chimpanzees of Bossou and Nimba* (pp. 73–83). Tokyo: Springer.

McGrew, W. C. (2013). Is primate tool use special? Chimpanzee and New Caledonian crow compared. *Philosophical Transactions of the Royal Society of London. Series B, Biological Sciences, 368*(1630), 20120422.

McGrew, W. C. (2017). Field studies of *Pan troglodytes* reviewed and comprehensively mapped, focussing on Japan's contribution to cultural primatology. *Primates, 58*(1), 237–258.

McGrew, W. C., Baldwin, P. J., Marchant, L. F., Pruetz, J. D., & Tutin, C. E. G. (2014). Chimpanzees (*Pan troglodytes verus*) and their mammalian sympatriates: Mt. Assirik, Niokolo-Koba National Park, Senegal. *Primates, 55*(4), 525–532.

McGrew, W. C., Marchant, L. F., Payne, C. L. R., Webster, T. H., & Hunt, K. D. (2013). Well-digging by Semliki chimpanzees: New data on laterality and possible significance of hydrology. *Pan Africa News, 20*(1), 5–8.

McGrew, W. C., Marchant, L. F., Wrangham, R. W., & Klein, H. (1999). Manual laterality in anvil use: Wild chimpanzees cracking *Strychnos* fruits. *Laterality: Asymmetries of Body, Brain and Cognition*, *4*(1), 79–87.

McGrew, W. C., Tutin, C. E. G., & Baldwin, P. J. (1979). Chimpanzees, tools, and termites: Cross-cultural comparisons of Senegal, Tanzania, and Rio Muni. *Man*, *14*, 185–215.

McPherron, S. P., Alemseged, Z., Marean, C. W., Wynn, J. G., Reed, D., Geraads, D., . . . Béarat, H. A. (2010). Evidence for stone-tool-assisted consumption of animal tissues before 3.39 million years ago at Dikika, Ethiopia. *Nature*, *466*(7308), 857–860.

Mendes, F. D. C., Cardoso, R. M., Ottoni, E. B., Izar, P., Villar, D. N. A., & Marquezan, R. F. (2015). Diversity of nutcracking tool sites used by *Sapajus libidinosus* in Brazilian Cerrado. *American Journal of Primatology*, *77*, 535–546.

Meulman, E. J. M., & van Schaik, C. P. (2013). Orangutan tool use and the evolution of technology. In C. M. Sanz, J. Call, & C. Boesch (Eds.), *Tool use in animals: Cognition and ecology* (pp. 176–202). Cambridge, UK: Cambridge University Press.

Morris, K., & Goodall, J. van L. (1977). Competition for meat between chimpanzees and baboons of the Gombe National Park. *Folia Primatologica*, *28*(2), 109–121.

Moscovice, L. R., Issa, M. H., Petrzelkova, K. J., Keuler, N. S., Snowdon, C. T., & Huffman, M. A. (2007). Fruit availability, chimpanzee diet, and grouping patterns on Rubondo Island, Tanzania. *American Journal of Primatology*, *69*(5), 487–502.

Moura, A. C. de A., & Lee, P. C. (2004). Capuchin stone tool use in Caatinga dry forest. *Science*, *306*(5703), 1909.

O'Malley, R. C., Wallauer, W., Murray, C. M., & Goodall, J. van L. (2012). The appearance and spread of ant fishing among the Kasekela chimpanzees of Gombe: A possible case of intercommunity cultural transmission. *Current Anthropology*, *53*(5), 650–663.

Ottoni, E. B. (2015). Tool use traditions in nonhuman primates: The case of tufted capuchin monkeys. *Human Ethology Bulletin*, *30*(1), 22–40.

Ottoni, E. B., & Mannu, M. (2001). Semi-free-ranging tufted capuchins (*Cebus apella*) spontaneously use tools to crack open nuts. *International Journal of Primatology*, *22*(3), 347–358.

Panger, M. A., Brooks, A. S., Richmond, B. G., & Wood, B. A. (2002). Older than the Oldowan? Rethinking the emergence of hominin tool use. *Evolutionary Anthropology*, *11*(6), 235–245.

Pascual-Garrido, A., Buba, U., Nodza, G., & Sommer, V. (2012). Obtaining raw material: Plants as tool sources for Nigerian chimpanzees. *Folia Primatologica*, *83*(1), 24–44.

Pascual-Garrido, A., Umaru, B., Allon, O., & Sommer, V. (2013). Apes finding ants: Predator-prey dynamics in a chimpanzee habitat in Nigeria. *American Journal of Primatology*, *75*(12), 1231–1244.

Plummer, T. W., & Stanford, C. B. (2000). Analysis of a bone assemblage made by chimpanzees at Gombe National Park. *Journal of Human Evolution*, *39*(3), 345–365.

Proffitt, T., Luncz, L. V., Falótico, T., Ottoni, E. B., de la Torre, I., & Haslam, M. D. (2016). Wild monkeys flake stone tools. *Nature*, *539*(7627), 85–88.

Proffitt, T., Luncz, V.L., Malaivijitnond, S., Gumert, M., Svensson, M.S., Haslam, M. D. (2018). Analysis of wild macaque stone tools used to crack oil palm nuts. *Royal Society Open Science*, 5: 171904.

Pruetz, J. D., Bertolani, P., Ontl, K. B., Lindshield, S., Shelley, M., & Wessling, E. G. (2015). New evidence on the tool-assisted hunting exhibited by chimpanzees (*Pan troglodytes verus*) in a savannah habitat at Fongoli, Sénégal. *Royal Society Open Science*, *2*, 140507.

Samson, D. R. (2012). The chimpanzee nest quantified: Morphology and ecology of arboreal sleeping platforms within the dry habitat site of Toro-Semliki Wildlife Reserve, Uganda. *Primates, 53*(4), 357–364.

Sanz, C. M., Call, J., & Morgan, D. B. (2009). Design complexity in termite-fishing tools of chimpanzees (*Pan troglodytes*). *Biology Letters, 5*, 293–296.

Sanz, C. M., Deblauwe, I., Tagg, N., & Morgan, D. B. (2014). Insect prey characteristics affecting regional variation in chimpanzee tool use. *Journal of Human Evolution, 71*, 28–37.

Sanz, C. M., & Morgan, D. B. (2009). Flexible and persistent tool-using strategies in honey-gathering by wild chimpanzees. *International Journal of Primatology, 30*(3), 411–427.

Sanz, C. M., & Morgan, D. B. (2011). Elemental variation in the termite fishing of wild chimpanzees (*Pan troglodytes*). *Biology Letters, 7*, 634–637.

Sanz, C. M., & Morgan, D. B. (2013). Ecological and social correlates of chimpanzee tool use. *Philosophical Transactions of the Royal Society of London. Series B, Biological Sciences, 368*(1630), 20120416.

Sanz, C. M., Morgan, D. B., & Gulick, S. (2004). New insights into chimpanzees, tools, and termites from the Congo Basin. *American Naturalist, 164*(5), 567–581.

Sommer, V., Buba, U., Jesus, G., & Pascual-Garrido, A. (2012). Till the last drop: Honey gathering in Nigerian chimpanzees. *Ecotropica, 18*, 55–64.

Sousa, C. (2011). Use of leaves for drinking water. In T. Matsuzawa, T. Humle, & Y. Sugiyama (Eds.), *The chimpanzees of Bossou and Nimba* (pp. 85–96). Tokyo: Springer.

Sousa, C., Biro, D., & Matsuzawa, T. (2009). Leaf-tool use for drinking water by wild chimpanzees (*Pan troglodytes*): Acquisition patterns and handedness. *Animal Cognition, 12*(S1), 115–125.

Souto, A., Bione, C. B. C., Bastos, M., Bezerra, B. M., Fragaszy, D. M., & Schiel, N. (2011). Critically endangered blonde capuchins fish for termites and use new techniques to accomplish the task. *Biology Letters, 7*(4), 532–535.

Stanford, C. B., & Bunn, H. T. (Eds.). (2001). *Meat-eating and human evolution*. Oxford, UK: Oxford University Press.

Stewart, F. A., & Piel, A. K. (2014). Termite fishing by wild chimpanzees: New data from Ugalla, western Tanzania. *Primates, 55*(1), 35–40.

Stewart, F. A., Piel, A. K., & McGrew, W. C. (2011). Living archaeology: Artifacts of specific nest site fidelity in wild chimpanzees. *Journal of Human Evolution, 61*(4), 388–395.

Suzuki, S., Kuroda, S., & Nishihara, T. (1995). Tool-set for termite-fishing by chimpanzees in the Ndoki forest, Congo. *Behaviour, 132*(3), 219–235.

Tan, A. W. Y. (2017). From play to proficiency: The ontogeny of stone-tool use in coastal-foraging long-tailed macaques (*Macaca fascicularis*) from a comparative perception-action perspective. *Journal of Comparative Psychology, 131*(2), 89–115.

Tan, A. W. Y., Tan, S. H., Vyas, D. K., Malaivijitnond, S., & Gumert, M. D. (2015). There is more than one way to crack an oyster: Identifying variation in Burmese long-tailed macaque (*Macaca fascicularis aurea*) stone-tool use. *PLoS One, 10*(5), e0124733.

Teleki, G. (1974). Chimpanzee subsistence technology: Materials and skills. *Journal of Human Evolution, 3*(6), 575–594.

Teleki, G. (1975). Primate subsistence patterns: Collector-predators and gatherer-hunters. *Journal of Human Evolution, 4*(2), 125–184.

Toth, N. P., & Schick, K. D. (2009). The Oldowan: The tool making of early hominins and chimpanzees compared. *Annual Review of Anthropology, 38*, 289–305.

Visalberghi, E., Addessi, E., Truppa, V., Spagnoletti, N., Ottoni, E. B., Izar, P., & Fragaszy, D. M. (2009). Selection of effective stone tools by wild bearded capuchin monkeys. *Current Biology, 19*(3), 213–217.

Visalberghi, E., Albani, A., Ventricelli, M., Izar, P., Schino, G., & Fragaszy, D. M. (2016). Factors affecting cashew processing by wild bearded capuchin monkeys (*Sapajus libidinosus*, Kerr 1792). *American Journal of Primatology, 78*(8), 799–815.

Visalberghi, E., Fragaszy, D. M., Ottoni, E. B., Izar, P., De Oliveira, M. G., & Andrade, F. R. D. (2007). Characteristics of hammer stones and anvils used by wild bearded capuchin monkeys (*Cebus libidinosus*) to crack open palm nuts. *American Journal of Physical Anthropology, 132*(3), 426–444.

Visalberghi, E., Haslam, M., Spagnoletti, N., & Fragaszy, D. M. (2013). Use of stone hammer tools and anvils by bearded capuchin monkeys over time and space: Construction of an archeological record of tool use. *Journal of Archaeological Science, 40*(8), 3222–3232.

Visalberghi, E., Spagnoletti, N., Ramos da Silva, E. D., Andrade, F. R. D., Ottoni, E. B., Izar, P., & Fragaszy, D. M. (2009). Distribution of potential suitable hammers and transport of hammer tools and nuts by wild capuchin monkeys. *Primates, 50*(2), 95–104.

Watts, D. P. (2007). Scavenging by chimpanzees at Ngogo and the relevance of chimpanzee scavenging to early hominid behavioral ecology. *Journal of Human Evolution, 54*(1), 125–133.

Whiten, A., Goodall, J. van L., McGrew, W. C., Nishida, T., Reynolds, V., Sugiyama, Y., . . . Boesch, C. (1999). Cultures in chimpanzees. *Nature, 399*(6737), 682–685.

Whiten, A., Schick, K. D., & Toth, N. P. (2009). The evolution and cultural transmission of percussive technology: Integrating evidence from palaeoanthropology and primatology. *Journal of Human Evolution, 57*(4), 420–435.

Wynn, T. (1981). The intelligence of Oldowan hominids. *Journal of Human Evolution, 10*(7), 529–541.

Wynn, T., Hernandez-Aguilar, R. A., Marchant, L. F., & McGrew, W. C. (2011). "An ape's view of the Oldowan" revisited. *Evolutionary Anthropology, 20*(5), 181–197.

Wynn, T., & McGrew, W. C. (1989). An ape's view of the Oldowan. *Man, 24*(3), 383–398.

Yamakoshi, G. (2011). Pestle-pounding behavior: The key to the coexistence of humans and chimpanzees. In T. Matsuzawa, T. Humle, & Y. Sugiyama (Eds.), *The chimpanzees of Bossou and Nimba* (pp. 107–115). Tokyo: Springer.

2

HOMO ARTIFEX
AN EXTENDED EVOLUTIONARY PERSPECTIVE ON THE ORIGINS OF THE HUMAN MIND, BRAIN, AND CULTURE

Dietrich Stout

INTRODUCTION

For nineteenth-century evolutionists, it seemed clear that human tool-using abilities were both distinctive and radically transformative. Darwin (1871), for example, proposed a scenario in which bipedality arose to free the hands and arms "for prehension and other purposes" (p. 143), and the subsequent evolution of the "perfect hand" allowed invention of the "various weapons, tools, traps, &c. [including boats and fire], . . . by which man in the rudest state has become so preeminent" (p. 137). For Darwin, "these several inventions" were key examples of human intellectual exceptionalism, being "the direct result of the development of [human] powers of observation, memory, curiosity, imagination, and reason" (p. 137). He thus contended that "[man] manifestly owes [his] immense superiority to his intellectual faculties, his social habits, which lead him to aid and defend his fellows, and to his corporeal structure" (p. 137). Engels (2003 [1873]) extended this argument to theorize an explicit link between tools, sociality, and language evolution, proposing that

> mastery over nature began with the development of the hand, with labour, and widened man's horizon at every new advance. He was continually discovering new, hitherto unknown properties in natural objects. On the other hand, the development of labour necessarily helped to bring the members of society closer together by increasing cases of mutual support and joint activity, and by making clear the advantage of this joint activity to each individual. In short, men in the making arrived at the point where *they had something to say* to each other. (p. 73, emphasis original)

As is so often the case, aspects of this early work now appear remarkably prescient in anticipating contemporary emphasis on such topics as cooperation, sharing, imitation learning, and joint attention. To a certain degree, however, Darwin's and Engels' attempts to ground these social phenomena in human anatomical and technological evolution have received less attention in recent years.

Through much of the twentieth century, an emphasis on technology in human evolution was mainstream. The cultural neo-evolutionism and technological determinism of Leslie White (1949), with its important influences on Lewis Binford and processual archaeology, is one key example. Regarding human biological and cognitive evolution, there is Oakley's (1949) *Man: The Tool-Maker*, in which he echoed Darwin and Engels by contending that "man is a social animal, distinguished by 'culture': by the ability to make tools and communicate ideas. Employment of tools appears to be his chief biological characteristic," which drove the evolution of our advanced "powers of mental and bodily co-ordination" (pp. 1–2). On the subject of stone tools specifically, Oakley argued that "even the crudest Paleolithic artifacts indicate considerable forethought. . . . Using a hammerstone to make a hand-axe, and striking a stone flake to use in shaping a wooden spear, are activities which epitomize the mental characteristics of man" (p. 15). Washburn (1960) and Holloway (1967) similarly emphasized the presence of powerful evolutionary feedback relationships between bodies, brains, tools, and sociality.

By the late 1970s, however, a more exclusive emphasis on the cognitive demands of sociality was gathering steam. Humphrey (1976) was particularly influential in arguing that primates are "much cleverer than they need to be" to solve the "practical problems of living," such as finding food and avoiding predation (p. 306). This argument is curiously evocative of Wallace's (1870) earlier objection that natural selection could not explain advanced human intellect, and it is based on a similar contention that practical, ecological challenges (specifically including tool use) are simply not that demanding of intelligence. But whereas Wallace sought the answer to this conundrum in some form of supernatural intervention, Humphrey instead proposed that the chief selective pressure favoring increased intelligence was the "ability of an individual to outwit [one's] fellows." The work of Humphrey and others eventually developed into the Machiavellian intelligence hypothesis (Byrne & Whiten, 1988), supported by qualitative evidence of primate deception and social strategizing.

A few years later, Dunbar (1992) undertook to test Humphrey's (1976, p. 316) specific prediction of a "positive correlation across species between 'social complexity' and 'individual intelligence.'" By operationalizing intelligence ("information-processing capacity") as a ratio of neocortex to rest-of-brain volume and social complexity as average group size, Dunbar was able to show a strong correlation between the two over a wide range of primate species. This was taken as evidence that brain size constrains group size, so that evolutionary increases in group size (for whatever reason) would generate concomitant selective pressure for brain-size increase to handle the increase in social complexity. At the same time, Dunbar showed that neocortex ratio was *not* correlated with various measures of ecological complexity (e.g., percentage of fruit in diet; range size). This straightforward and decisive result convinced many skeptics that social complexity was not merely important, but was almost exclusively responsible for generating the selective pressures leading to primate brain expansion.

Dunbar's "social brain hypothesis" has been hugely influential, both within academia and without (e.g., Bennett, 2013). In archaeology, it has contributed to skepticism regarding the long-assumed importance of tool-making in human cognitive and brain evolution and led to the suggestion that technological change over human evolution was epiphenomenal to more fundamental developments in social cognition (Gowlett, Gamble, & Dunbar, 2012). With respect to stone tool-making in particular, so long a mainstay of cognitive archaeology (Isaac, 1976; Wynn, 1989), recent assessments have paralleled Humphrey's (1976) view that such "practical problems" are relatively undemanding of what he called "creative intellect" (p. 305). Thus, Coolidge and Wynn (2005) conclude that stone tools can provide evidence of spatial and procedural learning abilities but not of executive functions, and Mithen (1996, p. 76) suggests that Paleolithic tool-making was supported by a specialized cognitive domain lacking the "cognitive fluidity" characteristic of modern humans.

More recently, however, there has been renewed interest in ecological and technical cognition. For example, Navarrete, Reader, Street, Whalen, and Laland (2016) report comparative evidence in support of a "technical intelligence" hypothesis (cf. Byrne, 1997), linking technical innovation rates to encephalization in primates. Looking to neuroanatomy, there is the proposal of Genovesio, Wise, and Passingham (2014) that the frontoparietal "multiple-demand system" central to modern human general intelligence (Duncan, 2010) originally evolved as an adaptation for more efficient foraging decisions incorporating relational properties such as order, number, duration, length, distance, and proportion. More broadly, there is a growing realization that the dichotomy between "ecological" and "social" cognition is a false one owing to the underlying importance of social learning across so many aspects of primate (and other animal) behavior. Ironically, this was already explicitly recognized by Humphrey (1976, p. 310) when he wrote that "one of the chief functions of society is to act as a 'polytechnic school' for the teaching of subsistence technology," an insight that goes all the way back to the writings of Darwin and Engels cited earlier.

Many researchers now agree that humans occupy a "cultural niche" including complex adaptive technologies, practices, and beliefs that have been accumulated over generations and are beyond the ability of any one individual to reinvent in a single lifetime (Boyd, Richerson, & Henrich, 2011). The advent of such "cumulative culture" is thought to have underwritten the remarkable demographic success of modern humans (Hill, Barton, & Hurtado, 2009), to have been a key factor in hominin brain expansion (van Schaik, Isler, & Burkart, 2012), and to have produced many of the cognitive and behavioral characteristics that distinguish our species (Tomasello, 1999). At the same time, there is growing attention to the concrete materiality of culture and cognition, reflecting a theoretical (re)discovery of the importance of context (cf. Cole, 1996) in shaping human thought, action, and evolution. For example, Malafouris (2004) draws on Andy Clark's (e.g., Clark, 2008; Clark & Chalmers, 1998) concept of an extended mind to argue that artifacts help to actively constitute cognitive systems, rather than simply influencing internal cognition. Recent developments in evolutionary theory may help provide an integrative framework for these various threads. Particularly promising are calls for a new "extended evolutionary synthesis" that recognizes the importance of environmental and cultural, as well as genetic, inheritance (Laland et al., 2015). Psychological and evolutionary attention to context might be integrated through the concept of (cultural, cognitive, and developmental) niche construction (e.g., Flynn, Laland, Kendal, & Kendal, 2013), as illustrated in Fragaszy and colleagues' (2013) treatment of primate social learning. Work at developmental timescales holds particular promise as a bridge between behavioral and evolutionary levels of analysis. Byrge, Sporns, and Smith (2014) have made a key contribution by tracing avenues of causal interaction between dynamic processes over a range of spatiotemporal scales, from neural activity, plasticity, and development to somatic action and growth, physical niche construction, and social scaffolding. These recent advances make this an opportune time to revisit perennial ideas regarding cognitive, social, and somatic interactions in human evolution posited by Darwin, Engels, and so many others.

THE HUMAN TECHNOLOGICAL NICHE

These new insights and perspectives are particularly promising for the cognitive archaeology of human evolution, as they open novel opportunities for inference from material evidence to cognitive and evolutionary processes. Of central importance is the concept of an evolving human niche involving complex feedback relationships between social, behavioral, somatic, neural, and cognitive change, the proper study of which will require a transdisciplinary, integrative approach (Fuentes, 2015; Stout & Hecht, 2017). Early work in this direction has already produced various coevolutionary scenarios and timelines (Hill, Barton, & Hurtado, 2009; Isler & van Schaik, 2012; Sterelny, 2011). Further progress in testing such ideas and resolving issues of timing and causation will necessarily rely on direct evidence from the fossil, archaeological, and paleoenvironmental records (Antón, Potts, & Aiello, 2014), together with the development of robust middle-range theory for the interpretation of this evidence (Stout & Khreisheh, 2015).

Clearly, a properly evolutionary cognitive archaeology (Wynn & Coolidge, 2016) requires grounding in evolutionary theory, including the fundamental processes of variation, differential survival and reproduction, and inheritance that underpin natural selection. It is thus useful to begin, as did Darwin (1871, p. 136), with the observation that humans are a highly successful species. Even without agriculture, it has been estimated that *Homo sapiens* would have attained a global population of more than 70 million and a total biomass greater than any other large vertebrate (Hill et al., 2009). Such demographic potential seems paradoxical in a large-brained primate known for its slow and costly development. A growing consensus finds the solution to this paradox in a human strategy of alloparenting (Hrdy, 2009; Kramer, 2010) or "biocultural reproduction" (Bogin, Bragg, & Kuzawa, 2014), in which individuals other than the parents donate resources (e.g., time, effort, food) to help support offspring. Obviously, for these alloparents to have resources available for contribution, they must reliably produce a surplus beyond what they themselves require for survival. It is this surplus production by helpers that allows human mothers to produce large-brained children (Isler & van Schaik, 2012) with the shortest interbirth interval of any ape and a total fertility rate three times that of chimpanzees (Kramer, 2010).

How is it that Pleistocene human foragers, in contrast to other apes, were able to reliably produce the surpluses that fueled their demographic success? Embodied capital theory (Kaplan, Gurven, Winking, Hooper, & Stieglitz, 2010; Kaplan, Hill, Lancaster, & Hurtado, 2000) proposes that humans have evolved a tightly integrated strategy in which a focus on high-value, difficult-to-acquire food resources provides the surplus nutrition needed to fund growth, survival, and reproduction, and is in turn enabled by the increased longevity and brain size that allow learning of the requisite foraging skills. Cognitive and affective adaptations for prosociality (Hill et al., 2009), which are necessary for biocultural reproduction, also provide a venue for social learning and teaching, ultimately generating the human "cultural niche" described by Boyd and colleagues (2011). Integrating these various theoretical strands leads to a picture of a distinctly human way of life reliant on cognitive, affective, and life-history adaptations supporting the intergenerational reproduction and accumulation of foraging skills (Shennan & Steele, 1999), including the production and use of tools. We (Stout & Hecht, 2017; Stout & Khreisheh, 2015) have previously referred to this integration of embodied capital and cultural evolutionary theory as describing a specifically *technological* (as opposed to broadly "cultural") niche in order to highlight this critical interaction of material production and social organization.

It is important to stress that this concept of a technological niche is not meant to imply a narrowly material or utilitarian view of human nature and evolution—quite the opposite, in fact. Perhaps surprisingly, the term *technology*, so pervasive in our daily lives, did not come into widespread usage until well into the twentieth century. As argued by Leo Marx (1997), the new term emerged to fill a twofold conceptual void. First, it served to recognize (and assign higher social status to) the newly emerging and radical transformative power of technology, far beyond anything implied by preexisting concepts of the *mechanical* or *useful* (i.e., non-intellectual or non-creative) *arts*. Second, it provided a name for the new forms of social organization seen, for example, in the complex web of materials, social and economic relations, institutions, and regulations that constitute a unitary "thing" like "automotive technology." For Marx, *technology* is a "hazardous" concept precisely because the reification of this sprawling domain of human activity encourages a tacit focus on the material aspects of technology and an obfuscation of its broader social dimensions (cf. Ingold, 1996). This has important political and ideological implications that concern Marx. For students of human evolution, it serves as a warning against overly narrow conceptions of "technology" in human evolution. To speak of a human technological niche is to speak of something simultaneously material, economic, social, and cultural. Whereas this insight was obscured by dichotomous arguments pitting "social" versus "ecological" explanations of primate and human intelligence, it appears to have been more intuitive for nineteenth-century theorists and is now being rediscovered by contemporary accounts of brain evolution.

AN EXTENDED SYNTHESIS FOR HUMAN BRAIN EVOLUTION

The modern synthesis revolutionized evolutionary biology in the mid-twentieth century by integrating neo-Darwinism with population genetics, zoology, botany, paleontology, and natural history (Pigliucci, 2009). Now, a growing number of researchers are pursuing a further extension by explicitly addressing the effects of reciprocal causation, inclusive inheritance, and developmental bias on evolutionary processes (Laland et al., 2015). This new extended synthesis provides a novel framework for understanding complex feedback relationships in human brain evolution, in particular the potential interactions of behavioral and neural plasticity with cultural inheritance and developmental constraint.

Expensive Cultural Brains

The most comprehensive comparative account of brain-size evolution currently available is that of van Schaik and colleagues. This account grounds the *technological niche* concept just discussed by integrating economic and cultural elements under the headings of "expensive" (Isler & van Schaik, 2014) and "cultural" (van Schaik, Isler, & Burkart, 2012) brain hypotheses. The underlying assumption is that, all else being equal, bigger brains are generally advantageous. The expensive brain framework thus seeks to explain interspecific variation in brain size in terms of *net* fitness effects that also take energetic and life history constraints into account. Larger brains can evolve only if mortality is low enough to reward investments in such embodied capital and a sufficient energy budget can be found through increased intake and/or reallocation. Critically, these relationships are inherently bidirectional, as enlarging the brain may also lower mortality (e.g., through predation avoidance) and increase energy intake (foraging productivity). The "cultural brain" (van Schaik & Burkart, 2011; van Schaik et al., 2012) element then adds the possibility of gene–culture co-evolution. Modeling indicates that, if baseline conditions of frequency, learning ability, and skill complexity are met, social learning can increase the mean fitness of a population and lead to cumulative cultural evolution (Henrich & McElreath, 2003). This generates yet another potential feedback relationship, in which increasingly complex, socially learned skills both fund and require greater investment in neural tissue, as well as requiring/promoting social tolerance, slower life histories, and extensive resource transfers.

The broad generality of intelligence under this framework is consistent with comparative evidence of a correlation between brain size and behavioral flexibility or "general intelligence" (Reader, Hager, & Laland, 2011; Reader & Laland, 2002) across primates. Indeed, general intelligence appears to be highly evolvable due to conserved developmental mechanisms favoring the disproportionate expansion of flexible association networks (Buckner & Krienen, 2013; Finlay & Uchiyama, 2015), and might represent an "equifinal" response to many different selection pressures acting on brain size and cognitive function. As mentioned earlier, human general intelligence has been linked to a frontoparietal multiple demand (MD) system proposed to support "complex multi-component behavior" through sequential mental programming that divides complex problems into component sub-problems (Duncan, 2010). The MD system is one of several such distributed association networks that together account for much of human (Power et al., 2011) and monkey (Neubert, Mars, Thomas, Sallet, & Rushworth, 2014) neocortex and supports domain-general cognitive functions critical to flexible, intelligent behavior. The possibility that the emergence and expansion of such networks is a natural outgrowth of widely shared developmental mechanisms and constraints might help to explain the convergent evolution of general intelligence in taxa ranging from birds to cetaceans (van Horik, Clayton, & Emery, 2012).

Plasticity, Evolution, and Development

As Deacon (1997, p. 193) provocatively put it, "brain evolution should be impossible" according to the conventional gene-centered evolutionary theory of the modern synthesis. Indeed, it is difficult to see how random mutational changes to such a complex integrated system could be anything other than catastrophic. The resolution to this apparent paradox is, of course, that brain development is not the simple expression of an evolved genetic "blueprint," but rather is itself an evolutionary process of remarkable flexibility and adaptability. Briefly, cell division in the proliferative zones of the developing neural tube generates an overabundance of initially undifferentiated neurons that migrate along radial glial cells to assemble the various subdivisions of cortex (Rakic, 2009). During this process, neuron identity (destination and connectivity tendencies) is determined by signaling molecules secreted by local patterning centers located in the proliferative zones. Following arrival in the cortical plate, excess neurons that fail to establish viable connections providing sufficient external stimulation are selectively "pruned" through a process of programmed cell death (apoptosis). Events that increase or decrease neuron proliferation, alter patterning centers or the cellular mechanisms determining neuron identity, or change the pattern or extent of external stimulation will all tend to alter cortical organization in the adult. In this way, conserved developmental constraints and mechanisms (Finlay & Uchiyama, 2015) can interact with processes of developmental selection (Edelman, 1987) to produce functional systems even in the face of quite significant environmental or genetic perturbation. One notable human example of this adaptability is the recruitment of occipital visual cortex for language processing in congenitally blind adults (Bedny, Pascual-Leone, Dodell-Feder, Fedorenko, & Saxe, 2011).

On an evolutionary scale, the *tethering hypothesis* of Buckner and Krienen (2013) proposes that disproportionate expansion of the cortical mantle during brain enlargement (itself a result of the conserved order of neuronal proliferation; see Finlay and Darlington [1995]) tends to produces gaps between the chemical signaling gradients that pattern cortical differentiation. Neurons in these gaps are less likely to be incorporated into the highly structured "canonical" neural circuits for perception and action that have been built by natural selection and might normally be expected to be "pruned" during development. However, general cortical expansion also means that there will be multiple gaps, resulting in a distributed system of cortical regions that might survive apoptosis by forming viable connections *with each other*. The tethering hypothesis thus suggests that competition and developmental selection in developmental patterning gaps fosters the emergence of "non-canonical" association networks characterized by dense internal connections with each other rather than with more developmentally constrained peripheral sensorimotor systems. This is consistent with recent evidence of a topological connectivity gradient in human and macaque cortex, such that increasingly interconnected association cortices are located at increasing distances from primary sensorimotor cortex (Margulies et al., 2016).

A key implication of the tethering hypothesis is that putatively unconstrained association cortices should be relatively variable in anatomy and connectivity across individuals. In fact, association areas do display significantly greater variability in resting state functional connectivity when compared to unimodal sensorimotor cortex (Mueller et al., 2013). This is consistent with the fact that association areas are the latest developing portions of cortex (Hill et al., 2010) and are thus likely to be particularly sensitive to environmental and behavioral influences on developmental selection. The importance of this general trend in human brain evolution specifically is supported by evidence of low heritability for cortical morphology (sulcal dimensions) relative to overall brain size in humans, a pattern that contrasts with high heritability of both in chimpanzees (Gómez-Robles, Hopkins, Schapiro, & Sherwood, 2015). In other words, human cortical morphology appears to be more developmentally plastic and less genetically constrained, much as predicted by the tethering hypothesis. Gómez-Robles and colleagues (2015, p. 14802) further propose that this increased plasticity may have provided "a neurobiological basis for socially and culturally mediated behavioral evolution," a concept very much in line with the *cultural brain* and *technological niche* concepts discussed earlier.

The suggestion that organismal plasticity could interact with social heredity to facilitate adaptation and accelerate evolution is another idea that has been around since the nineteenth century (Morgan & Harris, 2015). In essence, plasticity allows organisms to explore new phenotypes (somatic and/or behavioral) in response to environmental variation. These new phenotypes reciprocally alter organisms' interactions with the world (potentially including environmental and/or cultural niche construction), leading to altered selection pressures and further evolution. Such plasticity is obviously advantageous to organisms, but it is also thought to come at some cost (e.g., temporary phenotype–environment mismatches, investments in learning). Thus, in cases where the new organism–environment interactions become stable over time, it is expected that selection will act to reduce costs by assimilating the previously plastic response as an automatic, "canalized" part of normal development (Ancel, 1999). This may be seen as a special case of the more general process of exaptation (evolutionary repurposing) followed by secondary adaptation envisioned by Gould and Vrba (1982). In fact, we have previously proposed a scenario of neuroanatomical plasticity, genetic assimilation, behavioral co-option, and secondary adaptation for the derivation of the modern human language capacity from technological precursors (Hecht, Gutman, Khreisheh, et al., 2015).

NICHE CONSTRUCTION AND ENVIRONMENTAL INHERITANCE

Many questions remain, however, about the relative importance of plasticity versus canalized adaptation in human brain evolution. As we have seen, there is substantial evidence that evolution has built increasing plasticity into brain development and function as a necessary corollary of increasing size. Insofar as plasticity is included in the costs of building a large, "expensive" brain (Isler & van Schaik, 2014; Murren et al., 2015) for any reason, there may have been little pressure or opportunity for the developmental canalization of particular circuits, including those supporting such prototypical human cognitive "specializations" as language (Bedny et al., 2011), theory of mind (Heyes & Frith, 2014), and imitation (Oostenbroek et al., 2016). This raises the possibility that much of the story behind the evolution of human cognition may be cultural rather than strictly biological (Tomasello, 1999), working through social processes such as developmental niche construction (Flynn et al., 2013) and cumulative culture evolution.

Niche construction (Odling-Smee, Laland, & Feldman, 2003) refers to a reciprocal evolutionary process whereby organisms modify their environment (including their social environment) and are in turn exposed to altered selection pressures. To the extent that such modifications are durable or consistently reproduced over time, they may be described as an environmental inheritance paralleling the genetic inheritance more conventionally emphasized by the modern synthesis. Developmental niche construction occurs when these inherited environmental alterations consistently affect the context of development (e.g., exposure to particular situations, artifacts, or behaviors), potentially leading to ontogenetic evolution (Heyes, 2003). A good nonhuman example is the nut-cracking behavior of capuchins, which attracts the attention of young monkeys and exposes them to the necessary materials even without any intentional teaching (Fragaszy et al., 2013). Human developmental niche construction obviously extends to a much wider range of skills, from object manipulation and walking (Byrge et al., 2014) to reading and thinking about the mental states of others (Heyes & Frith, 2014), and includes a wide range of influences ranging from the intentional to the incidental. As outlined earlier, the reproduction and elaboration of these skills across generations forms the foundation of the human way of life, our technological niche. It remains to be seen whether such processes led to the evolution of any specific and (relatively) innate cognitive adaptations in addition to a more generalized increase in brain size, plasticity, and behavioral flexibility. If such specializations exist, we would expect to find them first in relatively heritable, peripheral sensorimotor systems and with respect to behaviors and stimuli that have been relatively invariant over long periods of time.

Stone Tools and Human Brain Evolution

Out of the wide array of critical human skills that could be studied, stone tool-making is one of the most ancient and best represented in the archaeological record. This has been the core motivation behind an experimental research program into the "neuroarchaeology" of Paleolithic stone tool-making (Stout & Hecht, 2015), including simple Oldowan-style flake production and more elaborate Acheulean-style shaping. Results provide indications of likely foci for specific brain adaptations as well as evidence of more general cognitive demands that may have interacted with the evolution of other key human capacities.

Oldowan-style flake production is a simple technology, with relatively little requirement for planning and limited contingency between successive actions (Stout, 2011; Wynn, Hernandez-Aguilar, Marchant, & McGrew, 2011). Functional imaging with fluorodeoxyglucose positron emission tomography (FDG-PET) (Stout & Chaminade, 2007) indicates that the most salient metabolic demands occur in brain regions supporting visual perception, bodily awareness, and motor control. Interestingly, some of these foci of activity for Oldowan knapping correspond to regions of derived human functionality identified in comparative studies of macaques. These include regions of dorsal intraparietal sulcus that provide additional central visual field representations and increased sensitivity to the extraction of three-dimensional form from relative motion cues (Orban et al., 2006) and that may also have enhanced connectivity to frontal motor planning regions in humans (Hecht et al., 2013). Such perceptual abilities would likely be relevant to a wide array of tool and foraging behaviors other than stone knapping, and it is not known if they are unique to humans or shared with other apes. Notably, they are situated on relatively peripheral, sensorimotor portions of cortex (Margulies et al., 2016) where canalized adaptions might be considered more likely and pertain to perceptual phenomena that are likely to be of consistent relevance in an increasingly technological niche.

Taking a broader perspective, it is clear that acquisition of even simple flaking skill requires practice. Data are limited, but reasonable proficiency for a modern human probably requires on the order of 10–50 hours, with substantial individual variability (Stout, Hecht, Khreisheh, Bradley, & Chaminade, 2015; Stout & Khreisheh, 2015). This appears to reflect the development of embodied skills for the perceptual perception of core affordances and motor coordination of strike accuracy and appropriate force (Nonaka, Bril, & Rein, 2010). In fact, we found that even 100–200 hours' practice still did not result in reliable prediction of fracture patterns (Stout et al., 2015). Clearly, the demands of perceptual-motor skill acquisition should be taken into account when evaluating the cognitive implications of prehistoric technologies, and particularly the self-regulatory capacities and social scaffolding that may have been necessary for sustained, deliberate practice (Nonaka et al., 2010; Stout, 2010). Experimental evidence (Morgan et al., 2015) underlines the potential co-evolutionary relationships between tool-making, teaching, and language in this context.

Skill is also reflected in the brain, with expert knappers showing greater activation of inferior parietal lobe (IPL) involved in representation of the body-tool system and its capacities for action (Stout, Toth, Schick, & Chaminade, 2008). Again, there is some evidence suggesting tool-related human adaptations in this region, including functional evidence of a region of anterior left IPL specialized for the perception of handheld tools (Peeters et al., 2009; Peeters, Rizzolatti, & Orban, 2013) and anatomical evidence of increased connectivity with inferior frontal cortex (IFC) in humans compared to chimpanzees and macaques (Hecht, Gutman, Bradley, Preuss, & Stout, 2015). Perhaps most intriguingly, it has been shown that these connections to IFC are plastically enhanced by stone tool-making training in modern humans (Hecht, Gutman, Khreisheh, et al., 2015). These functional and structural traits would appear to be good candidates for adaptive canalization over human evolution, as they are located in relatively peripheral structures, and the perceptual-motor coordination they support is likely to have been of broad and continued relevance within an evolving technological niche. Nevertheless, experimental training results clearly show that substantial plasticity remains in this system even in adults, and the relative contributions and interactions of genetic, developmental, and cultural processes in generating modern human phenotypes are yet to be specified.

With respect to more central systems, it is only with more complex Late Acheulean–style shaping that we see increased functional involvement of classic, prefrontal association cortex (Stout et al., 2008, 2015). This effect is most consistent in the right inferior frontal gyrus (rIFG), which has been found to respond to both execution (Stout et al., 2008) and observation (Stout, Passingham, Frith, Apel, & Chaminade, 2011) of handaxe production, as well as to be a focus of white matter remodeling during tool-making training (Hecht, Gutman, Khreisheh, et al., 2015). Prefrontal cortex generally is associated with the higher-order cognitive or "executive" control of action, with the rIFG specifically supporting response inhibition and task switching (Levy & Wagner, 2011). Indeed, a recent study (Chavan, Mouthon, Draganski, van der Zwaag, & Spierer, 2015) found white matter changes associated with learning a classic "go/no-go" inhibitory control task in almost exactly the same rIFG location as Hecht and colleagues' (2015) tool-making effect, and transcranial magnetic stimulation (TMS) data demonstrate a causal role for rIFG inhibition in generating complex action sequences (Dippel & Beste, 2015). This general contribution to cognitive control processes may underpin rIFG involvement in superficially diverse activities ranging from manual sequence learning (Seitz & Roland, 1992) to language processing (Matchin & Hickok, 2016; Vigneau et al., 2011). Further, Duncan (2010) includes rIFG in his MD system supporting general intelligence, and the bilateral IFG (loosely, "Broca's area") has been consistently implicated in the control of hierarchically structured behavior (Koechlin & Jubault, 2006) across such modalities ranging from language to action sequencing, music, and mathematics (Fitch & Martins, 2014). Structurally, IFG is relatively distant from the sensorimotor periphery (Margulies et al., 2016) and thus might be expected to be particularly plastic and sensitive to and reliant on social and environmental inputs during development. For this reason, the cultural evolution and concrete reproduction (Roepstorff, Niewöhner, & Beck, 2010) of structured practices like language, music, and tool-making must be considered alongside more traditional evolutionary processes like natural selection in attempts to understand the modern human neurocognitive phenotype.

CONCLUSION

In keeping with the extended evolutionary perspective developed here, increasingly complex stone tool-making could have been one factor selecting for the enhanced behavioral flexibility afforded by an expansion of association cortex, perhaps mediated by nonspecific increases in overall brain size. Such evolution is an inherently complex and reciprocal process in which conventional dichotomies of organism versus environment, culture versus biology, cognition versus action, social versus ecological, evolutionary versus developmental, and proximate versus ultimate explanation may be misleading. The scope of interactions and relationships involved in this view of evolution may appear daunting, but it also offers new and exciting prospects for insight into aspects of human evolution long considered inaccessible to empirical inquiry (e.g., Lewontin, 1998). The real promise of "extended" conceptions of mind, culture, and evolution is that they are simultaneously grounded in the physical world and capable of seamlessly integrating dynamic interactions across multiple levels of spatiotemporal organization. Thus, neural activity drives behaviors that evoke further neural activity which, over time, will alter the patterns of functional and anatomical brain connectivity that help to shape behavior (Byrge et al., 2014). This constructive process unfolds over developmental time in a context of patterned practices and structured environments that shapes individual behavior (Flynn et al., 2013) and is itself reproduced on historical timescales by the accumulated action of individuals. At an even larger scale, these unfolding developmental, social, and environmental dynamics help to shape both the variation and the selective pressures that drive evolutionary change (Laland et al., 2015) to the biological systems that in turn constrain development and behavior. As we have argued elsewhere, investigating these complex and contingent interactions will require a dedicated, empirical research program that "evaluates comparative evidence of brain and behavioral variation in light of *i)* evolutionary and developmental process, *ii)* primary archaeological and paleontological evidence of evolutionary timing and context, and *iii)* the ethnographic, ethological, and experimental analogies needed to interpret this primary evidence" (Stout & Hecht, 2017, p. 7862).

ACKNOWLEDGMENTS

I would like to thank the editors for the invitation to contribute to this volume, and Tom Wynn for the occasion. The neuroarchaeological research reviewed here was made possible by the collaborative efforts and intellectual contributions of Thierry Chaminade and Erin Hecht.

REFERENCES

Ancel, L. W. (1999). A quantitative model of the Simpson–Baldwin effect. *Journal of Theoretical Biology, 196*(2), 197–209.

Antón, S. C., Potts, R., & Aiello, L. C. (2014). Evolution of early *Homo*: An integrated biological perspective. *Science, 345*(6192), 1236828–1–13.

Bedny, M., Pascual-Leone, A., Dodell-Feder, D., Fedorenko, E., & Saxe, R. (2011). Language processing in the occipital cortex of congenitally blind adults. *Proceedings of the National Academy of Sciences of the United States of America, 108*(11), 4429–4434.

Bennett, D. (2013). The Dunbar number, from the guru of social networks: How a technophobic Oxford primatologist became Silicon Valley's social networking guru. *Bloomberg Businessweek.* Retrieved from https://www.bloomberg.com/news/articles/2013-01-10/the-dunbar-number-from-the-guru-of-social-networks

Bogin, B., Bragg, J., & Kuzawa, C. (2014). Humans are not cooperative breeders but practice biocultural reproduction. *Annals of Human Biology, 41*(4), 368–380.

Boyd, R., Richerson, P. J., & Henrich, J. (2011). The cultural niche: Why social learning is essential for human adaptation. *Proceedings of the National Academy of Sciences of the United States of America, 108*(Suppl. 2), 10918–10925.

Buckner, R. L., & Krienen, F. M. (2013). The evolution of distributed association networks in the human brain. *Trends in Cognitive Sciences, 17*(12), 648–665.

Byrge, L., Sporns, O., & Smith, L. B. (2014). Developmental process emerges from extended brain–body–behavior networks. *Trends in Cognitive Sciences, 18*(8), 395–403.

Byrne, R. W. (1997). The technical intelligence hypothesis: An additional evolutionary stimulus to intelligence? In A. Whiten & R. W. Byrne (Eds.), *Machiavellian intelligence II: Extensions and evaluations* (pp. 289–311). Cambridge, UK: Cambridge University Press.

Byrne, R. W., & Whiten, A. (Eds.). (1988). *Machiavellian intelligence: Social expertise and the evolution of intellect in monkeys, apes, and humans.* Oxford, UK: Clarendon Press.

Chavan, C. F., Mouthon, M., Draganski, B., van der Zwaag, W., & Spierer, L. (2015). Differential patterns of functional and structural plasticity within and between inferior frontal gyri support training-induced improvements in inhibitory control proficiency. *Human Brain Mapping, 36*(7), 2527–2543.

Clark, A. (2008). *Supersizing the mind: Embodiment, action, and cognitive extension.* New York: Oxford University Press.

Clark, A., & Chalmers, D. J. (1998). The extended mind. *Analysis, 58*(1), 7–19.

Cole, M. (1996). *Cultural psychology: A once and future discipline.* Cambridge, MA: Harvard University Press.

Coolidge, F. L., & Wynn, T. (2005). Working memory, its executive functions, and the emergence of modern thinking. *Cambridge Archaeological Journal, 15*(1), 5–26.

Darwin, C. (1871). *The descent of man, and selection in relation to sex* (Vol. 1). London: John Murray.

Deacon, T. W. (1997). *The symbolic species: The co-evolution of language and the brain.* New York: W. W. Norton.

Dippel, G., & Beste, C. (2015). A causal role of the right inferior frontal cortex in implementing strategies for multi-component behaviour. *Nature Communications, 6*, 1–10.

Dunbar, R. I. M. (1992). Neocortex size as a constraint on group size in primates. *Journal of Human Evolution, 22*(6), 469–493.

Duncan, J. (2010). The multiple-demand (MD) system of the primate brain: Mental programs for intelligent behaviour. *Trends in Cognitive Sciences, 14*(4), 172–179.

Edelman, G. M. (1987). *Neural Darwinism: The theory of neuronal group selection*. New York: Basic Books.

Engels, F. (2003). The part played by labour in the transition from ape to man. In R. C. Scharff & V. Dusek (Eds.), *Philosophy of technology: The technological condition—An anthology* (pp. 71–77). London: Blackwell. [Originally published 1873]

Finlay, B. L., & Darlington, R. B. (1995). Linked regularities in the development and evolution of mammalian brains. *Science, 268*(5217), 1578–1584.

Finlay, B. L., & Uchiyama, R. (2015). Developmental mechanisms channeling cortical evolution. *Trends in Neurosciences, 38*(2), 69–76.

Fitch, W. T., & Martins, M. D. (2014). Hierarchical processing in music, language, and action: Lashley revisited. *Annals of the New York Academy of Sciences, 1316*(1), 87–104.

Flynn, E. G., Laland, K. N., Kendal, R. L., & Kendal, J. R. (2013). Target article with commentaries: Developmental niche construction. *Developmental Science, 16*(2), 296–313.

Fragaszy, D. M., Biro, D., Eshchar, Y., Humle, T., Izar, P., De Resende, B. D., & Visalberghi, E. (2013). The fourth dimension of tool use: Temporally enduring artefacts aid primates learning to use tools. *Philosophical Transactions of the Royal Society of London. Series B, Biological Sciences, 368*(1630), 1–10.

Fuentes, A. (2015). Integrative anthropology and the human niche: Toward a contemporary approach to human evolution. *American Anthropologist, 117*(2), 302–315.

Genovesio, A., Wise, S. P., & Passingham, R. E. (2014). Prefrontal–parietal function: From foraging to foresight. *Trends in Cognitive Sciences, 18*(2), 72–81.

Gómez-Robles, A., Hopkins, W. D., Schapiro, S. J., & Sherwood, C. C. (2015). Relaxed genetic control of cortical organization in human brains compared with chimpanzees. *Proceedings of the National Academy of Sciences of the United States of America, 112*(48), 14799–14804.

Gould, S. J., & Vrba, E. S. (1982). Exaptation—A missing term in the science of form. *Paleobiology, 8*(1), 4–15.

Gowlett, J. A. J., Gamble, C., & Dunbar, R. I. M. (2012). Human evolution and the archaeology of the social brain. *Current Anthropology, 53*(6), 693–722.

Hecht, E. E., Gutman, D. A., Bradley, B. A., Preuss, T. M., & Stout, D. (2015). Virtual dissection and comparative connectivity of the superior longitudinal fasciculus in chimpanzees and humans. *NeuroImage, 108*, 124–137.

Hecht, E. E., Gutman, D. A., Khreisheh, N., Taylor, S. V, Kilner, J., Faisal, A. A., . . . Stout, D. (2015). Acquisition of Paleolithic toolmaking abilities involves structural remodeling to inferior frontoparietal regions. *Brain Structure and Function, 220*(4), 2315–2331.

Hecht, E. E., Gutman, D. A., Preuss, T. M., Sanchez, M. M., Parr, L. A., & Rilling, J. K. (2013). Process versus product in social learning: Comparative diffusion tensor imaging of neural systems for action execution–observation matching in macaques, chimpanzees, and humans. *Cerebral Cortex, 23*(5), 1014–1024.

Henrich, J., & McElreath, R. (2003). The evolution of cultural evolution. *Evolutionary Anthropology: Issues, News, and Reviews, 12*(3), 123–135.

Heyes, C. M. (2003). Four routes of cognitive evolution. *Psychological Review, 110*(4), 713–727.

Heyes, C. M., & Frith, C. D. (2014). The cultural evolution of mind reading. *Science, 344*(6190), 1243091-1-6.

Hill, J., Inder, T., Neil, J., Dierker, D., Harwell, J., & van Essen, D. C. (2010). Similar patterns of cortical expansion during human development and evolution. *Proceedings of the National Academy of Sciences of the United States of America, 107*(29), 13135–13140.

Hill, K., Barton, M., & Hurtado, A. M. (2009). The emergence of human uniqueness: Characters underlying behavioral modernity. *Evolutionary Anthropology: Issues, News, and Reviews, 18*(5), 187–200.

Holloway, R. L. (1967). The evolution of the human brain: Some notes toward a synthesis between neural structure and the evolution of complex behavior. *General Systems, 12*, 3–19.

Hrdy, S. B. (2009). *Mothers and others: The evolutionary origins of mutual understanding*. Cambridge, MA: Harvard University Press.

Humphrey, N. K. (1976). The social function of intellect. In P. P. G. Bateson & R. A. Hinde (Eds.), *Growing points in ethology* (pp. 303–317). Cambridge, MA: Cambridge University Press.

Ingold, T. (1996). Social relations, human ecology, and the evolution of culture: An exploration of concepts and definitions. In A. Lock & C. R. Peters (Eds.), *Handbook of human symbolic evolution* (pp. 178–203). Oxford, UK: Clarendon Press.

Isaac, G. L. (1976). Stages of cultural elaboration in the Pleistocene: Possible archaeological indicators of the development of language capabilities. *Annals of the New York Academy of Sciences, 280*(1), 275–288.

Isler, K., & van Schaik, C. P. (2012). How our ancestors broke through the gray ceiling: Comparative evidence for cooperative breeding in early *Homo*. *Current Anthropology, 53*(S6), S453–S465.

Isler, K., & van Schaik, C. P. (2014). How humans evolved large brains: Comparative evidence. *Evolutionary Anthropology: Issues, News, and Reviews, 23*(2), 65–75.

Kaplan, H., Gurven, M., Winking, J., Hooper, P. L., & Stieglitz, J. (2010). Learning, menopause, and the human adaptive complex. *Annals of the New York Academy of Sciences, 1204*(1), 30–42.

Kaplan, H., Hill, K., Lancaster, J., & Hurtado, A. M. (2000). A theory of human life history evolution: Diet, intelligence, and longevity. *Evolutionary Anthropology: Issues, News, and Reviews, 9*(4), 156–185.

Koechlin, E., & Jubault, T. (2006). Broca's area and the hierarchical organization of human behavior. *Neuron, 50*(6), 963–974.

Kramer, K. L. (2010). Cooperative breeding and its significance to the demographic success of humans. *Annual Review of Anthropology, 39*, 417–436.

Laland, K. N., Uller, T., Feldman, M. W., Sterelny, K., Müller, G. B., Moczek, A., . . . Odling-Smee, F. J. (2015). The extended evolutionary synthesis: Its structure, assumptions and predictions. *Proceedings of the Royal Society of London B: Biological Sciences, 282*(1813), 1–14.

Levy, B. J., & Wagner, A. D. (2011). Cognitive control and right ventrolateral prefrontal cortex: Reflexive reorienting, motor inhibition, and action updating. *Annals of the New York Academy of Sciences, 1224*(1), 40–62.

Lewontin, R. C. (1998). The evolution of cognition: Questions we will never answer. In D. Scarborough & S. Sternberg (Eds.), *An invitation to cognitive science. Volume 4: Methods models, and conceptual issues* (pp. 107–132). Cambridge, MA: MIT Press.

Malafouris, L. (2004). The cognitive basis of material engagement: Where brain, body and culture conflate. In E. DeMarrais, C. Gosden, & C. Renfrew (Eds.), *Rethinking materiality: The engagement of mind with the material world* (pp. 53–61). Cambridge, UK: McDonald Institute for Archaeological Research.

Margulies, D. S., Ghosh, S. S., Goulas, A., Falkiewicz, M., Huntenburg, J. M., Langs, G., . . . Smallwood, J. (2016). Situating the default-mode network along a principal gradient of macroscale cortical organization. *Proceedings of the National Academy of Sciences of the United States of America, 113*(44), 12574–12579.

Marx, L. (1997). "Technology": The emergence of a hazardous concept. *Social Research, 64*(3), 965–988.

Matchin, W., & Hickok, G. (2016). "Syntactic perturbation" during production activates the right IFG, but not Broca's area or the ATL. *Frontiers in Psychology, 7*, 1–14.

Mithen, S. J. (1996). *The prehistory of mind: The cognitive origins of art, religion and science.* London: Thames & Hudson.

Morgan, T. J. H., & Harris, P. L. (2015). James Mark Baldwin and contemporary theories of culture and evolution. *European Journal of Developmental Psychology, 12*(6), 666–677.

Morgan, T. J. H., Uomini, N. T., Rendell, L. E., Chouinard-Thuly, L., Street, S. E., Lewis, H. M., . . . Laland, K. N. (2015). Experimental evidence for the co-evolution of hominin toolmaking teaching and language. *Nature Communications, 6*(6029), 1–18.

Mueller, S., Wang, D., Fox, M. D., Yeo, B. T. T., Sepulcre, J., Sabuncu, M. R., . . . Liu, H. (2013). Individual variability in functional connectivity architecture of the human brain. *Neuron, 77*(3), 586–595.

Murren, C. J., Auld, J. R., Callahan, H., Ghalambor, C. K., Handelsman, C. A., Heskel, M. A., . . . Schlichting, C. D. (2015). Constraints on the evolution of phenotypic plasticity: Limits and costs of phenotype and plasticity. *Heredity, 115*(4), 293–301.

Navarrete, A. F., Reader, S. M., Street, S. E., Whalen, A., & Laland, K. N. (2016). The coevolution of innovation and technical intelligence in primates. *Philosophical Transactions of the Royal Society of London. Series B, Biological Sciences, 371*(1690), 1–10.

Neubert, F.-X., Mars, R. B., Thomas, A. G., Sallet, J., & Rushworth, M. F. S. (2014). Comparison of human ventral frontal cortex areas for cognitive control and language with areas in monkey frontal cortex. *Neuron, 81*(3), 700–713.

Nonaka, T., Bril, B., & Rein, R. (2010). How do stone knappers predict and control the outcome of flaking? Implications for understanding early stone tool technology. *Journal of Human Evolution, 59*(2), 155–167.

Oakley, K. P. (1949). *Man: The tool-maker.* London: British Museum Press for the Trustees of the British Museum.

Odling-Smee, F. J., Laland, K. N., & Feldman, M. W. (2003). *Niche construction: The neglected process in evolution.* Princeton, NJ: Princeton University Press.

Oostenbroek, J., Suddendorf, T., Nielsen, M., Redshaw, J., Kennedy-Costantini, S., Davis, J., . . . Slaughter, V. (2016). Comprehensive longitudinal study challenges the existence of neonatal imitation in humans. *Current Biology, 26*(10), 1334–1338.

Orban, G. A., Claeys, K., Nelissen, K., Smans, R., Sunaert, S., Todd, J. T., . . . Vanduffel, W. (2006). Mapping the parietal cortex of human and non-human primates. *Neuropsychologia, 44*(13), 2647–2667.

Peeters, R. R., Rizzolatti, G., & Orban, G. A. (2013). Functional properties of the left parietal tool use region. *NeuroImage, 78*, 83–93.

Peeters, R. R., Simone, L., Nelissen, K., Fabbri-Destro, M., Vanduffel, W., Rizzolatti, G., & Orban, G. A. (2009). The representation of tool use in humans and monkeys: Common and uniquely human features. *Journal of Neuroscience, 29*(37), 11523–11539.

Pigliucci, M. (2009). An extended synthesis for evolutionary biology. *Annals of the New York Academy of Sciences, 1168*(1), 218–228.

Power, J. D., Cohen, A. L., Nelson, S. M., Wig, G. S., Barnes, K. A., Church, J. A., . . . Petersen, S. E. (2011). Functional network organization of the human brain. *Neuron, 72*(4), 665–678.

Rakic, P. (2009). Evolution of the neocortex: A perspective from developmental biology. *Nature Reviews: Neuroscience, 10*(10), 724–735.

Reader, S. M., Hager, Y., & Laland, K. N. (2011). The evolution of primate general and cultural intelligence. *Philosophical Transactions of the Royal Society of London. Series B, Biological Sciences, 366*(1567), 1017–1027.

Reader, S. M., & Laland, K. N. (2002). Social intelligence, innovation and enhanced brain size in primates. *Proceedings of the National Academy of Sciences of the United States of America, 99*(7), 4436–4441.

Roepstorff, A., Niewöhner, J., & Beck, S. (2010). Enculturing brains through patterned practices. *Neural Networks, 23*(8), 1051–1059.

Seitz, R. J., & Roland, P. E. (1992). Learning of sequential finger movements in man: A combined kinematic and positron emission tomography (PET) study. *European Journal of Neuroscience, 4*(2), 154–165.

Shennan, S., & Steele, J. (1999). Cultural learning in hominids: A behavioral ecological approach. In H. O. Box & K. R. Gibson (Eds.), *Mammalian social learning: Comparative and ecological perspectives* (pp. 367–388). Cambridge, UK: Cambridge University Press.

Sterelny, K. (2011). From hominins to humans: How *sapiens* became behaviourally modern. *Philosophical Transactions of the Royal Society of London. Series B, Biological Sciences, 366*(1566), 809–822.

Stout, D. (2010). The evolution of cognitive control. *Topics in Cognitive Science, 2*(4), 614–630.

Stout, D. (2011). Stone toolmaking and the evolution of human culture and cognition. *Philosophical Transactions of the Royal Society of London. Series B, Biological Sciences, 366*(1567), 1050–1059.

Stout, D., & Chaminade, T. (2007). The evolutionary neuroscience of tool making. *Neuropsychologia, 45*(5), 1091–1100.

Stout, D., & Hecht, E. E. (2015). Neuroarchaeology. In E. Bruner (Ed.), *Human paleoneurology* (pp. 145–175). Berlin: Springer.

Stout, D., & Hecht, E. E. (2017). Evolutionary neuroscience of cumulative culture. *Proceedings of the National Academy of Sciences of the United States of America, 114*(30), 7861–7868.

Stout, D., Hecht, E. E., Khreisheh, N., Bradley, B. A., & Chaminade, T. (2015). Cognitive demands of lower Paleolithic toolmaking. *PLoS One, 10*(4), 1–18.

Stout, D., & Khreisheh, N. (2015). Skill learning and human brain evolution: An experimental approach. *Cambridge Archaeological Journal, 25*(4), 867–875.

Stout, D., Passingham, R. E., Frith, C. D., Apel, J., & Chaminade, T. (2011). Technology, expertise and social cognition in human evolution. *European Journal of Neuroscience, 33*(7), 1328–1338.

Stout, D., Toth, N. P., Schick, K. D., & Chaminade, T. (2008). Neural correlates of early Stone Age tool-making: Technology, language and cognition in human evolution. *Philosophical Transactions of the Royal Society of London. Series B, Biological Sciences, 363*(1499), 1939–1949.

Tomasello, M. (1999). *The cultural origins of human cognition*. Cambridge, MA: Harvard University Press.

van Horik, J. O., Clayton, N. S., & Emery, N. J. (2012). Convergent evolution of cognition in corvids, apes and other animals. In J. Vonk & T. K. Shackelford (Eds.), *The Oxford handbook of comparative evolutionary psychology* (pp. 80–101). Oxford, UK: Oxford University Press.

van Schaik, C. P., & Burkart, J. M. (2011). Social learning and evolution: The cultural intelligence hypothesis. *Philosophical Transactions of the Royal Society of London. Series B, Biological Sciences, 366*(1567), 1008–1016.

van Schaik, C. P., Isler, K., & Burkart, J. M. (2012). Explaining brain size variation: From social to cultural brain. *Trends in Cognitive Sciences, 16*(5), 277–284.

Vigneau, M., Beaucousin, V., Hervé, P.-Y., Jobard, G., Petit, L., Crivello, F., . . . Tzourio-Mazoyer, N. (2011). What is right-hemisphere contribution to phonological, lexico-semantic, and sentence processing? Insights from a meta-analysis. *NeuroImage, 54*(1), 577–593.

Wallace, A. R. (1870). *Contributions to the theory of natural selection: A series of essays.* London: Macmillan and Co.

Washburn, S. L. (1960). Tools and human evolution. *Scientific American, 203*(3), 62–75.

White, L. (1949). *The science of culture: A study of man and civilization* (2nd ed.). New York: Grove Press.

Wynn, T. (1989). *The evolution of spatial competence.* Chicago: University of Illinois Press.

Wynn, T., & Coolidge, F. L. (2016). Archeological insights into hominin cognitive evolution. *Evolutionary Anthropology: Issues, News, and Reviews, 25*(4), 200–213.

Wynn, T., Hernandez-Aguilar, R. A., Marchant, L. F., & McGrew, W. C. (2011). "An ape's view of the Oldowan" revisited. *Evolutionary Anthropology, 20*(5), 181–197.

3

LOOKING AT ROCKS TOGETHER
TOOL PRODUCTION, JOINT ATTENTION, AND OFFLINE COGNITION

Rex Welshon

INTRODUCTION

Tom Wynn's cognitive archaeology of stone tool production has made a unique, influential, and lasting contribution to our knowledge surrounding the evolution of cognition and consciousness in early *Homo* species. In the 1970s and 1980s, he more or less invented the field of cognitive archaeology single-handedly (Wynn, 1989; Wynn [2017] offers an assessment that gives others credit). Since then, his ongoing reflections on using the production of stone tools as a window into early Paleolithic cognitive life have yielded a string of consistently remarkable articles and books (Coolidge & Wynn, 2005, 2009; Wynn, 1991, 2002; Wynn & Coolidge, 2004, 2010, 2012; Wynn, Haidle, Lombard, & Coolidge, 2017). In these publications, he explores the dimensions of the expert mind by investigating the expansion of and neural substrates for the cognitive abilities involved in making increasingly sophisticated stone tools. In doing so, he takes advantage of everything from Piagetian developmental psychology to cognitive scientific work on working memory to explain how the expert mind emerges, and he even suggests embedding the expert mind in a larger social context.

In this chapter, I discuss a small cluster of issues prompted by reflection on parts of Wynn's work. These issues surround the roles played in cognitive evolution by reflexive conscious experience, symmetric joint attention, transitive planning, and representational cognition. After defining these terms, I review archaeological findings regarding knapped tool production. The core of the chapter applies those findings and the phenomena of reflexive conscious experience, joint attention, and transitive planning to Wynn's claim that cognitive evolution in the Paleolithic period is a set of enhancements to working memory. I discuss how the deepening temporal, causal, and social dimensions required for stone tool production imply both working memory enhancements and the development of increasingly social and offline kinds of conscious cognition that go beyond augmenting working memory.

REFLEXIVITY, SYMMETRY, TRANSITIVITY, AND REPRESENTATION

Reflexivity and Conscious Experience

In general, a reflexive relation R is a relation that holds between any *x* and itself, as in the case of the relation "is identical with itself." Conscious experience is reflexive because it self-represents. In order to understand what self-representation amounts to, more needs to be said about what conscious experience is. *Conscious experience* is the set of psychological states that (1) are subjectively organized; (2) have a phenomenal or qualitative character, and (3) are widely accessible for subsequent cognitive and/or emotional responses. What makes conscious experience conscious at all is that it is subjectively organized; what makes conscious experience the conscious experience it is is its qualitative character; and what makes conscious experience psychologically useful is its wide accessibility at a particular time and over time. All three features of conscious experience are widely discussed in the philosophy and neuroscience of consciousness literature.[1] Our interest here lies immediately in the subjectivity of conscious experience.

The subjective character of conscious experience underwrites the distinction between an organism consciously experiencing exteroceptive and interoceptive inputs and an entity transducing and processing such inputs without consciously experiencing them as, for example, a thermostat does.[2] Alternatively, we may say that the difference between a thing transducing and processing inputs and an organism being conscious or aware of those exteroceptive and interoceptive inputs and their causes is the subjectivity of awareness. If this is the case, then it is only when an organism is aware that the things it is aware of cause its being aware of them that transduction and processing become conscious experience.

In the philosophical literature, two accounts of subjectivity dominate all others. The first is a *primitivist* account of subjectivity found, among other places, in the phenomenological tradition descending from Edmund Husserl. The second is a *reductionist* account of subjectivity; it comes from the representationalist tradition of understanding consciousness. According to the primitivist account, what makes subconscious transduction and processing conscious is an unanalyzable—hence, primitive—modality of experience, namely, the subjective modality. According to the reductionist account, what makes subconscious transduction and processing

[1] For contemporary discussions of subjectivity and reflexivity, see, among many others, Block, Flanagan, and Güzeldere (1998); Kriegel (2009); Metzinger (2003); Revonsuo (2006); Searle (2000); and Zahavi (2008). For contemporary discussions of qualitative character, see, among many others, Block (2001, 2005, 2007a, 2007b); Dennett (2006); Levine (2001); and Nagel (1974). For contemporary discussions of accessibility, see, among many others, Baars (1988) and Block (2001, 2007a, 2007b).

[2] Subjective character is more complex than the anodyne property of there being an embodied spatiotemporal point of projection in experience, while also being less complex than a property that entails possession of a self-concept as a condition of instantiation. For discussion and some relevant references, see Welshon (2013).

conscious may be analyzed as—and hence, reduced to—a feature of representational psychological states, namely, their subjective feature.

The difference between unconscious transduction and conscious experience occurs not only across thermostats and humans and across other species and humans but also across individual humans and even within an individual human. We also subconsciously transduce some exteroceptive and interoceptive inputs and consciously experience other inputs. The phenomenon of blindsight is an example.[3] In blindsight, exteroceptive sensory input is transduced and processed but is not consciously experienced (alternatively put, in blindsight, we are not aware of the exteroceptive input). For the primitivist and reductionist alike, conscious visual states are thus additionally complex in comparison to subconscious transduction/processing states by being subjectively organized. For the primitivist, the additional complexity of having subjective character consists in a state being experienced as "for-me" in an unanalyzable way. Hence, a subconscious visual state is, in the primitivist account, a visual state that is not for-me in the relevant sense. For the reductionist, likewise, a conscious visual state is conscious because it has subjective character. For the reductionist, the additional complexity of having subject character consists in that state having a structure such that the state represents itself. Hence, a subconscious visual state is, in the reductionist account, a visual state that fails to self-represent.

Both primitivists and reductionists thus claim that conscious states always come loaded with and reveal a unique reflexivity. For present purposes, either account is agreeable. However, to simplify discussion, we will assume the self-representationalist account of subjective character—that is, that the reflexivity of a conscious psychological state is a reflexive self-representation in which a psychological state not only represents its exteroceptive or interoceptive causes but also represents itself as a conscious state.

Two quick points are in order. First, the reflexivity of a conscious state is not to be confused with the kind of reflectivity that discloses the self to the subject. That is, the structure of a *reflexive* conscious state does not therefore also qualify it as a *self-reflective* state. Even if all conscious states are reflexive, only some reflexive conscious states are self-reflective: A self-reflective conscious state is already reflexive simply because it is a conscious state, but a self-reflective conscious state adds to that reflexive structure the additional facet of being about the self. Second, not all reflexively structured conscious states are *monitoring* conscious states. Monitoring conscious states are conscious states about other conscious states, as occurs, for example, when a person

[3] Blindsight is a disorder in which individuals have no qualitatively endowed visual perception but remain able to discriminate objects and events in the unseen field. Blindsight is most frequently associated with dysfunction in or destruction of the early stages of the cortical visual pathway in occipital cortex. Damage to this region causes a black spot (scotoma) in the visual field. While individuals deny seeing anything in the visual field contralateral to damaged cortical regions, they can if forced to identify an object's location in that field with considerable, albeit still abnormally low, reliability. Other features of blindsight are equally curious. For example, some blindsighted individuals report that they are aware of something they do not perceive and that they perceive visual afterimages of items they do not see. For discussion, see, among others, Cowey (2004); Holt (2003); and Milner and Goodale (1995).

realizes her bias against a certain political figure or when she reflects that her bias is unjustified. In general, self-reflective and monitoring conscious states are higher-order conscious states because they are conscious states about other conscious states. In this way, self-reflective states, monitoring states, and all of the other higher-order conscious states are more complex cognitive achievements than baseline subjectively structured conscious states. We return to this matter presently.

Symmetry and Joint Attention

In general, a symmetric relation R is one between two individuals, x and y, such that if x is R to y, then y is R to x. Shared, or as we shall refer to it, *joint attention*, is symmetric: If Marcia is jointly attending to something with Jan, then Jan is jointly attending to that thing with Marcia. Of course, either Marcia or Jan can be attentive without the other being attentive. However, that is not *joint* attention.

Not all conscious states are attentional states, much less joint attentional states. We often consciously experience an auditory landscape without attending to any of its particular facets, as, for example, when we are aware of the background drone of a motorcycle but do not focus our attention on it. Likewise, not all attentional states are conscious states. That is, some attentional states are subconscious states. Examples are noteworthy but not hard to find. Many parents, for instance, can recall an episode when they saved a child from a threat so quickly that the threat was not consciously processed, yet the reaction to save the child would not have occurred had they not been subconsciously attending to the environment.

Cognitive science has typed attention across various dimensions, two of which are germane. First, attention is either voluntary or involuntary. *Voluntary* or *goal-driven* or *endogenous attention* is prompted by internally initiated voluntary cognitive activity that is congruent with a person's goals or desires. *Involuntary* or *stimulus-driven* or *exogenous attention* is prompted by features of perceptual experience intruding on ongoing cognitive activity regardless of one's goals or desires. Second, attention can be overt or covert. *Overt attention* directs a sense organ or sense organs to what is attended; *covert attention* does not. We can attend to a feature in our visual field, for example, without turning our eyes to the feature. Similarly, we can attend to a particular sound or smell without turning our heads to its source.

There are then four general type of attention, each with subvarieties: endogenous overt, endogenous covert, exogenous overt, and exogenous covert. *Endogenous overt* attention happens when an intention to attend to some feature occurs and is accompanied by overt behavior to that feature (as when I decide to turn my head to see whether there is a coyote in the scrub oak 20 yards from my window). *Endogenous covert* attention happens when an intention to attend to some feature occurs without overtly turning to the attended object (as when I decide to listen to the coyote yipping while I am typing at the computer). *Exogenous overt* attention happens when some feature intrudes in a noteworthy manner and overt behavior occurs toward that feature (as when I turn my eyes away from the computer to look at the coyote). *Exogenous covert* attention happens when some feature intrudes in a noteworthy manner and no overt motor behavior occurs toward that feature (as when I catch the coyote in the corner of my visual field).

Joint attention is that type of endogenous overt attention that occurs whenever (1) there is an object two (or more) subjects are attending to, (2) there is a causal connection between the two subjects' acts of attending to the object, (3) the two subjects' experiences exploit their understanding of the cognitive features of attention, and (4) each subject is aware that the object is present to both subjects and that each subject's attending to it is mutually manifest to the other (Eilan, 2005; also see Metcalfe & Terrace, 2013; Moore & Dunham, 1995; Seeman, 2011). What is crucial in joint attention is that each subject's attention to the object is manifest to the other subject. That is, what makes joint attention joint is that attention is symmetric *across* subjects. The symmetry of joint attention distinguishes it from the similar but distinct case of two subjects individually attending to the same object but not sharing their attention.

Examples make the difference clear. It is one thing for each of two individuals, Calvin and Joe, to attend individually to balancing a ladder on sloping ground and another for them to attend jointly to balancing the ladder. In the latter case, unlike the former case, each is aware that the other is also attending to balancing the ladder and each is therefore better able to coordinate and cooperate with the other to balance the ladder. Similarly, it is one thing for each of two individuals to hunt after an animal and another for them to attend jointly to hunting that animal. For, again, if each is aware that the other is also hunting the animal, they are better prepared to cooperate with one another and so to coordinate their actions to achieve their shared goal.

Transitivity and Planning

In general, a transitive relation R is one between three individuals, *x*, *y*, and *z*, such that if *x* is R to *y*, and if *y* is R to *z*, then *x* is R to *z*.[4] Conscious planning is transitive. If Marcia understands that her presentation to the board requires setting a date for the presentation, and if she understands that fixing a date for the presentation requires looking at the calendar, then Marcia understands that her presentation to the board requires looking at the calendar. As with planning, so too with following a procedure: If Marty understands that tiling his bathroom requires setting the tiles in mastic, and if he understands that setting the tiles in mastic requires placing the mastic on the wall before setting the tiles, then Marty understands that setting the tiles in mastic requires placing the mastic on the wall before setting the tiles. Similarly, if an individual is preparing a hafting adhesive for attaching a spear point to a shaft, she understands that in order to achieve the proper viscosity and plasticity, the adhesive must heat and cool repeatedly, and she understands that heating requires putting the material over a fire and that cooling requires removing the material from the fire to let it rest.

[4] We are not interested here in transitive verbs. A transitive verb is simply a verb that takes an object. An example is "Don baked a cake." Here, the verb "bake" takes "cake" as its object. Other verbs are intransitive because they do not take an object. An example is "Over the last year, Tonya has certainly grown!" Many verbs can appear as either transitive or intransitive. Compare "When he's stressed out, George eats" with "When he's having a low blood sugar moment, George typically eats a Snickers bar."

The transitivity of planning and following a procedure reveals the accessibility of conscious psychological states to an individual and to more than one individual. For the transitivity of conscious states breaks the solipsistic loop of sequentially ordered reflexive conscious states and expands the symmetric loop of joint attentional states by supplementing them with an additive component that is, in principle, infinite.

Consider this last point in more detail. Transitivity is the basis of repetition, iteration, and recursion. If a note x of a western meadowlark's song at a particular time occurs later as note y, and if note y occurs later still as note z, then *z repeats x*. Likewise for *iteration*, which is a repetitive process that adds to repetition the condition that a process's prior outputs become the next inputs to the process. When Marcia paints the trim in her kitchen, she follows a transitive iterative process in which the outcome of each earlier step feeds into the next step as input. Since filling holes in the trim is required before sanding, and since sanding is required before priming, and since priming is required before applying finish paint, then filling holes in the trim is required before applying finish paint.

Recursion is an iterative repetitive process that adds to iteration the condition that the outputs of a process step that become the inputs of the next step are preserved in every subsequent step. Consider counting the number of handshakes at a party at which everyone shakes everyone else's hand exactly once. Suppose there are just two people, a and b, at the party. Since handshaking is symmetric, it is straightforward that the number of handshakes between a and b is 1, and the function for handshakes where there are two people is: $(2) = 2 - 1$, that is, 1. Suppose next that there are three people, a, b, and c, at the party. If so, then a and b shake hands, a and c shake hands, and b and c shake hands. Hence, there are $2 + 1 = 3$ total handshakes. And the function for handshakes where there are three people is: $(3) = (3 - 1) + (2 - 1)$, that is, $2 + 1$, that is, 3.

What if there are four people at the party? We can provide a recursive function to answer this question by focusing on the logic of the answer we gave for the case of two and three people and generalizing from that answer. In general, where the number of people at the party is n (where $n > 2$), the number of handshakes is determined by:

$$(n) = (n - 1) + (n - 1), n > 2$$

So, to compute the answer to our question about four people, there are:

$$(4) = (4 - 1) + (3 - 1) + (2 - 1), \text{ that is, } 3 + 2 + 1 = 6 \text{ handshakes.}$$

Likewise, if there are 10 people at the party, there are:

$$(10) = 9 + 8 + 7 + 6 + 5 + 4 + 3 + 2 + 1 = 45 \text{ handshakes.}$$

What is recursive about this iteration is that each previous step of the iteration plugs into every additional step of the iteration.

Putting this reflections about reflexivity, symmetry, and transitivity together, we may say that an episode of joint endogenous attention between two subjects, x and y, on an object o, utilizing a sequentially ordered procedure P, has the following structure. First, each of x and y is individually consciously attending to object o utilizing procedure *P*. That means that *x* and *y* each have reflexively structured (that is, self-representing) conscious and endogenously overt attentional states about *o* and procedure *P*. Second, since *x* and *y* are endogenously and overtly attending to object *o* utilizing procedure *P*, their conscious states have the additional vigilance that only attention brings to reflexively structured conscious states. Third, since *x* and *y* are jointly attending to object *o* utilizing procedure *P*, there is a causal connection between *x*'s and *y*'s attending to *o* and *P*, and each of *x* and *y* is aware that *o* and *P* are present to both *x* and *y* such that each subject's attending to *o* and *P* is mutually manifest to the other's attending to *o* and *P* and each subject's attending to *o* and *P* can be exploited for subsequent use of *o* and *P*. Fourth, since *P* is a sequentially ordered procedure, some of whose steps entail completing prior steps, each of *x* and *y* is transitively aware of the iterative and, perhaps, recursive steps of *P*.

One question, then, is this: Is knapping a stone tool a process that requires reflexive conscious states, episodes of symmetric joint attention, and transitive planning that incorporates iteration and recursion? I will argue that the answer is "yes."

Representation Hunger

In the previous section, representation was a basis for understanding certain features of conscious experience. But representation is remarkably fertile: Symptoms represent illnesses; limping gaits represent injuries; pheromones represent sexual availability; behaviors represent intentions; facial expressions represent moods; photos represent what they're photos of; maps represent geographical territories; charts, graphs, and diagrams represent relations, processes, and other organizational and structural features; and words and sentences of language represent, often with considerable success, everything under the sun. Despite this heterogeneity, a core set of features is common to the various species of representation. The representational relation is analyzable as follows: One item represents something else whenever the former *carries information about* or *stands in for* the latter.

Within the representation genus, there are at least three distinct species: icons, indexes, and symbols.[5] An *icon* is a sensory re-presentation, such as a drawing, a painting, or a photograph, of something else. An *index* is a stimulus-dependent representation that correlates particular perceptual information with, or points to (that is, indexes), something else. A limp indexes injury; an upheld hand at full arm extension indexes the direction to stop; olfactory detection of pheromones indexes sexual availability; particular kinds of vocal behavior index danger; other kinds of vocal behavior index contentment; still other kinds of vocal behaviors index amusement. Finally, a *symbol* is a stimulus-independent representation that has a conventional significance, acquired through association with other symbols, of representing something else. Symbols, unlike indexes and icons, bear no natural and no non-conventional relation to that which they represent. A sea hawk symbolically represents the professional football team from Seattle; an inscribed "s" is an ink squiggle that symbolically represents a particular phoneme; and an inscribed "sea hawk" is a collection of ink squiggles that

[5] The distinction between icons, indexes, and symbols goes back to the semiotic work of Charles Saunders Peirce (see Peirce, 1984 [1867], among others).

symbolically represents a concatenation of phonemes that symbolically represents a particular kind of bird that symbolically represents the football team from Seattle. Humans are, without question, the most enthusiastic and adventurous symbol users in the animal kingdom.

Different species of representations are at use in different species of cognition. Two species of cognition are germane for present purposes. Conscious cognition occurrently fed by sensory perception or interoception or engaged in a motor behavior task is different in certain ways from cognition that is relatively independent of sensory perception or interoception or is sustained by internal cognitive activity or not tasked with some motor behavior. Cognition that is in ongoing causal interaction with the external environment is *online* cognition, and cognition that is not in ongoing causal interaction with the external environment is *offline* cognition. Online cognition includes sensory perception (that is, exteroception) across the five modalities; joint attention; tasked motor behavior; conversation; and retrieval of procedural (implicit) memories. Offline cognition includes some kinds of covert endogenous attention; cognitive modeling and mental object rotation; abstract reflection; monitoring; engaging in certain kinds of problem-solving and rule-following; dreaming and daydreaming; retrieval of long-term declarative (semantic and episodic) memories; planning; and entertaining intentions.

Online cognition is online because it *causally couples* with the external environment; offline cognition is offline because it is *causally decoupled* from the external environment. Causally coupled cognition occurs whenever positive and negative feedback loops between cognition and the external environment occur, that is, whenever the results of some node in the causal process between cognition and environment are fed back into an earlier node in that causal process. Just so, cognition is causally decoupled from the external environment whenever it is not causally coupled with the external environment (Prinz, 2009).

Offline cognition is an evolutionarily significant development. Insects and reptiles are incapable of offline cognition and so "remain trapped in a (potentially very complex and context-variable) web of closed-loop interactions with the . . . reality upon which their survival depends" (Clark & Grush, 1999, p. 7). Humans and certain other mammals, on the other hand, "use models (internal and external) in place of directly operating upon the world" (Clark & Grush, 1999, p. 7). That is significant because decoupled offline cognition is often *representation-hungry,* which is to say that offline cognition often *entails* representations, whether they are icons, indexes, or symbols. Of course, not all cognition is representation-hungry. Perhaps no coupled cognitive processes are representational, and perhaps even some decoupled cognitive processes are non-representational (Gallagher, 2008). Still, at least some decoupled cognitive processes require representations because these kinds entail a causal node that replaces and stands in for the informational affordances of the perceptually presented external environment and the interoceptively presented internal environment. In short, where offline cognitive processes are decoupled and have representations as causal nodes, explanations of those offline processes *must* invoke representations (Clark & Toribio, 1994).[6]

[6] The category of representation is contentious in current cognitive neuroscience and philosophy of mind. Some philosophers argue that representational explanations are integral for most cognitive

Another question, then, is this: Is knapping a stone tool a process that requires representational offline cognition, and, if so, what kinds of representations are entailed? I will argue that the answer to the first part of the question is "yes" and that the answer to the second part of the question is "we don't yet really know."

TOOL PRODUCTION AND THE EMERGENCE OF OFFLINE COGNITION

Wynn has demonstrated that tool-making practices reveal evolving cognitive abilities over the period from the Lower Paleolithic, which started 2 million years ago, to the end of the Middle Paleolithic, about 30,000 years ago. One of his preferred ways to understand this cognitive evolution is as a series of enhancements to working memory. *Working memory* is, classically (that is, as found in the work of Baddeley), comprised of four components: a visuospatial sketchpad, a phonological loop, a central executive, and an episodic buffer. The phonological loop is composed of two subsystems, the phonological store and the articulatory loop, the first a temporary storehouse of sounds and the second responsible for sound production. The visuospatial sketchpad is likewise composed of two systems, the first devoted to processing visual pattern information such as color, texture, and shape, and the second devoted to processing spatial location and sequential movement information. The central executive is responsible for attention, behavior inhibition, decision-making, and planning. The episodic buffer integrates and temporarily stores information from different sources, including long-term memory, as a single, multimodally bound episode.[7]

We now use the reflexivity of conscious experience, the symmetry of joint attention, and the transitivity of iteration and recursion to identify and unpack what some of the dimensions of these enhancements to working memory may have been.

Many animals, from crows to sea otters to bottlenose dolphins, *use* tools of various kinds and for various purposes, but the vast majority of them use only a single kind of tool for a single purpose. An exception is chimpanzees. Chimps introduce novel cognitive developments in tool use: They not only use tools, they use several different kinds of tools, such as leaf-sponges, levers, probes, scoops, stick brushes, pestles, and stone hammers, and they use tools for a number of different reasons, such as termite, ant, and honey extraction; water retrieval; and nut cracking (Ambrose, 2001). These novel ways of using tools suggest both a nascent cognitive ability to dissociate acute from subacute needs and certain enhancements in working memory. For instance, their use of tools for termite and honey extraction and for nut cracking suggests incipient integration of ongoing conscious activity with long-term memory stores and

processes, others that representational explanations are irrelevant for all but a small set of arcane cognitive processes. For defenses of the former view, see, among many others, Clark and Toribio (1994); Clowes and Mendonça (2016); Fodor (1975, 1981); Gładziejewski (2016); Grush (2004); and Ash and Welshon (n.d.). For defenses of the latter view, see, among many others, Chemero (2000, 2009); Degenaar and Myin (2014); Haselager, van Dijk, and van Rooij (2008); Hutto and Myin (2013); and van Gelder (1995).

[7] There is no principled reason why there cannot also be olfactory, gustatory, and somatic stores. For further details, see, among many others, Baddeley (1986, 1996, 2000, 2002, 2003).

anticipated consequences, which implies an extension of cognitive time both back from the present into the past and forward from the present into the future.

More importantly, chimps *make* tools. For example, they choose from a number of sticks one that has the right diameter to extract termites from a termite mount, and they peel leaves from it so that it can be inserted without obstruction—both achievements again imply integrating the bubble of ongoing conscious experience (that is, working memory) with the past and the future. However, as interesting as chimp tool-making procedures may be, they are still simple in comparison with what even early members of the *Homo* line were capable of doing. Even early members of the *Homo* line were, since very near their point of emergence, employing more complex tool-making procedures than any used by any other species, chimps included. More than a million years ago, *H. erectus*, for instance, was already busy making tools using production procedures that were significantly more complicated than the most advanced tool-making procedures employed by chimps.

Paleolithic tool-making procedures revolve around the practice of *knapping*—striking a stone with another stone or something softer than a stone, such as bone or wood. Knapping is only one of any number of practical skills our ancestors probably deployed for survival. Many daily activities—cooking, waste disposal, social bonding, shelter construction, play, and reproduction—provided opportunities for cognitive growth across a diverse repertoire of motor skills, practices, plans, and technological improvements. However, whatever physical traces these technological improvements and the accompanying cognitive abilities and motor behavior skills required to implement them may once have left have vanished from the archaeological record. Luckily, stone tools have survived, and because we have them we can compare knapping procedures used by members of the various species of *Homo* with procedures used by pre- and non-*Homo* species so as to infer cross-genus working memory enhancements and intra-genus but cross-species working memory enhancements.[8]

Reflexivity and Tool-Making

Even the most basic kinds of stone tool production presuppose a distinction between acute and subacute needs, because the need to eat subordinates to undertaking and completing a tool production process. So, too, since planning of some kind is required for all kinds of tool production, extensions to cognitive time are also implied, where "extensions to cognitive time" means enhancements to working memory's capacity to integrate ongoing perceptual input, retrospection (the ability to represent the past through memory recall), and prospection (the ability to represent the future through counterfactual representations) into multidimensional conscious experience.

Planning to make a stone tool arguably entails certain representational conscious states, including reflexive self-representational states. The first representational state is that regarding the rock as occurrently presented in sensory perception. Second, there is

[8] For constraints on inferences from archaeological artifacts to cognitive abilities, see Botha (2006, 2008).

the prospective counterfactual representation of the rock after some knapping blows.[9] In order to plan, working memory must integrate both the occurrent and prospective representations. In addition, the knapper must be aware that the perceived rock and the prospective counterfactual representation of the rock are causally connected.[10] After all, being able to counterfactually represent a rock without also being aware that the rock as perceived is causally connected to that counterfactual representation would be nothing more than an epiphenomenal curiosity that is useless practically. Moreover, the knapper must also be aware that he or she is an agent who can, will, and does perform the motor behaviors required to make changes to the occurrently perceptually presented rock. Finally, the knapper must pay attention while engaged in the motor behavior that causes the rock to take shape.

These are not trivial cognitive achievements; they entail ongoing episodes of integrated retrospective representations of the past, occurrent sensory perceptual presentations and representations, and prospective counterfactual representations. None of these achievements is possible without psychological states that reflexively self-present. The reflexivity of offline conscious experience underwrites the cut between the way a rock is presented through sensory perception and (1) the way the rock is as counterfactually represented, (2) the prospective plan for knapping the shape presented in the counterfactual representation, (3) the attentively organized motor behavior that implements the plan in a particular procedure, and (4) the awareness of oneself as a causal agent implementing the plan through attentively organized motor behavior. Even if these cognitive achievements do not presuppose conceptual understanding of a particularly sophisticated kind, or reflective awareness of a self, or monitoring conscious states, or even declarative memory, all of them do presuppose reflexivity. Without reflexivity, it is not possible to recognize the difference between, on the one hand, entertaining the ongoing world as presented by sensory perception and, on the other, entertaining an offline counterfactual representation segregated from the ongoing world. Nor is it possible without reflexivity to recognize the awareness that attentively organized motor behavior achieves change in the world.

Transitivity and Tool-Making

The knapping techniques practiced by *H. erectus* and the stability of their lithic technology over more than a million years together suggest that their cognitive capacities to plan were limited to repetition and perhaps a few iterative steps. Even so, they were capable of transitively structured planning, and while they were not as adept as later

[9] It is not necessary to posit the existence of a prospective counterfactual representation of the finished tool in order to posit the existence of a prospective counterfactual representation of a rock having been shaped by a blow. Simple kinds of planning require only the latter. Hence, the argument here evades Davidson and Noble's "fallacy of the finished artefact." For details, see Davidson and Noble (1993).

[10] Space limitations preclude discussion of the implications of representing relations as causal relations or the implications of what causal cognition might consist in. For more on these matters, see the essays in McCormack, Hoerl, and Butterfill (2011).

species at extending that transitivity to more complex procedures with more than a couple of elements, that they were able to do it at all is significant.

An important cognitive development in transitive planning capacity occurs with the emergence of more complex knapping techniques. Consider the Levallois lithic technology of *H. heidelbergensis* and *H. neanderthalensis*. In Levallois technique, the knapper starts by hitting one stone with another, producing numerous edges around a core piece of stone. The knapper then hits the worked stone on the end, producing a pancake-shaped wafer. This wafer may be ready for use as is or it may be shaped again to create a more specialized tool. The core from which the wafer came can be used repeatedly until it becomes too reduced to produce more wafers.

Levallois technique is demonstrably iterative, for the output of one step in the procedure, namely, the pancake wafer, becomes input to subsequent steps. Once such iterative procedures are in place, more complex planning with one or more embedded component steps becomes available. The appearance of wooden spears 350,000 years ago at Neandertal sites and thereafter the appearance of hafted tools both evince the introduction of tool production procedures with embedded and even recursive component steps. Hafting requires production schedules at least days and sometimes weeks long as the various components are prepared. Adhesive production processes in particular are candidates for being recursive processes, for they require (1) cognitive multitasking; monitoring the progress of the adhesive's viscosity, texture, and plasticity; (2) recurrent attention to adhesive heating and cooling to ensure proper viscosity, texture, and plasticity; (3) multilevel thinking about the causal properties of various resources; (4) proper sequencing of adhesive production steps; and (5) proper sequencing of hafting once adhesives have been produced (Wadley, Hodgskiss, & Grant, 2009; also see Bradfield, Lombard, & Wadley, 2015; Wadley, 2010; Wynn, 2002).

These technical improvements entail constantly refreshed episodes of individual attention, retrospective memory recall, and prospective planning in order to get the sequencing of the procedure's steps right and to distinguish what is essential in the procedure from what is accidental. Consider the distinction between being essential and being accidental in more detail. Certain properties of a modern hammer—its head weight, arm length, and rigidity—are that in virtue of which the hammer drives nails, and just as other properties of a modern hammer—the material from which its arm is made, the color of its grip—can change without affecting the way the hammer works. Likewise, certain elements of hafting production procedures—such as using particular resources, completing steps in a particular order, combining certain outputs in a particular manner—cannot change without affecting the procedure. On the other hand, other elements—the diameter of firewood, time of day—can easily change without affecting the procedure.[11] Intelligent tool production not only relies on and exploits the difference between the procedure's essential elements and its accidental elements, it is also grounded in the knowledge of a procedure's essential

[11] For more on the distinction between essential and accidental properties of tools, see Campbell (2011).

properties, which suggests both episodes of iconic or indexical declarative memory and monitoring conscious states.[12]

Multicomponent production and construction techniques developed over the next several hundred thousand years, with the pioneering *H. neanderthalensis* eventually relinquishing technological innovation to *H. sapiens*. Thus do the transitivity of iterative and recursive tool production procedures in particular and the transitivity of planning in general deposit themselves ever more deeply into the conscious experience and daily cognitive lives of members of both *H. neanderthalensis* and early *H. sapiens*. By the Middle Paleolithic, they were probably also capable of specific declarative memory recall, sustained attention, iterative and recursive planning, and monitoring their own progress and even some of their own thoughts. Moreover, their accruing awareness of iterative and recursive multicomponent production procedures suggests a corresponding expansion and deepening of their awareness of their own causal agency. This expansion has two facets: their representation to themselves of various causal relations, and their deepening awareness of their status as causal agents.

Symmetry and Representation and Tool-Making

As iterative and recursive tool production procedures embedded themselves into the daily life of Paleolithic culture, representation-hungry offline cognition and indexical social representation probably also emerged for the first time. Consider each in turn.

It may not be possible to identify the exact point at which the complexity of a production procedure entails decoupled offline cognition that employs iconic or indexical representations (much less symbolic representations). Pre-Levallois knapping could have occurred without envisioning anything. The knapper, using only what was available through enhanced procedural and long-term memory, could have continued to strike the core until it acquired an edge without entertaining any counterfactual representations or invoking any declarative memories. However, the iterative elements in Levallois stone tool production pushes to the limit the capacity of procedural and long-term memory recall without prospective counterfactual representations to explain the technological improvements characteristic of this lithic technology (Coolidge & Wynn, 2018; Gamble, 1999). It is even more difficult to unpack the planning required to construct a hafted tool without introducing iconic and indexical declarative memories and iconic and indexical prospective counterfactual representations.

Other facets of multicomponent tool production deserve mention. Every additional step in a multicomponent construction technique further decouples acute needs from the subacute needs required to produce tools, and with that decoupling comes also the decoupling of satisfaction. A new kind of satisfaction that accompanies making tools appears for the first time, and with it, making tools becomes a goal in itself, which in turn leaves in its wake a new means of social differentiation (Coolidge & Wynn, 2009). The ever-increasing complexity of tool production procedures leads to the emergence of a class of experts with the various procedures. Moreover, the

[12] The distinction between *relying on* and *being grounded in* the knowledge of causal properties is due to Peacocke (2011).

products of those procedures—the points, axes, scrapers, cleavers, and other blades—become socially loaded with representational significance (Wynn & Coolidge, 2004). Tools become indexes of social stratification, means by which one's place in a social hierarchy is established, maintained, and lost. Stone tools, and the knowledge of how to construct and use them, come to represent intelligence, knowledge, utility, sexual attractiveness, and social power.

Again, wherever experts arise, there, too, are novices who learn from them. Hence, the increasingly complex stone tool production procedures lead inevitably to the need for experts and beginners, instructors and pupils, coming together in episodes of joint attention to nuances of the mastered procedures (Högberg, Gärdenfors, & Larsso, 2015). Thus, alongside the awareness of transitive causal relations and iterative and recursive knapping procedures, the symmetry of joint attention and its attendant cognitive benefits became fixed in Paleolithic consciousness. Not the least of these cognitive benefits is the instructor's and the pupil's mutual recognition that the other has a mind much like their own. That kind of recognition implies attributing to the other the same sort of experience as their own, experience that is rich with beliefs, desires, fears, hopes, plans, and bits of knowledge, all of which can be shared.

For example, joint attention between individuals on transitive knapping techniques is an obvious context in which monitoring conscious states and cooperative behavior can develop (Tomasello, 2008, Chapter 5). First, the symmetric circuitry of joint attention implies the occurrence of monitoring conscious states. As instructor and pupil jointly attend to the steps of the production procedure, they induce a feedback–feedforward cognitive loop that forms only by tacitly monitoring their own conscious states as states shared with the other. Second, the joint goal of teaching and learning how to knap implies coordination and cooperation between instructor and pupil. Instructor and pupil each share with the other both the goal that the pupil will develop knapping skills and the goal that the instructor will teach those skills. In addition, each acknowledges that the other shares these goals. It is because their understanding of these features and their awareness that each individual's joint attention with the other is mutually manifest to the other that successful instruction occurs.[13]

The increasingly sophisticated knapping procedures of *H. neanderthalensis* and early *H. sapiens* thus provide some of the fuel for the cognitive crucible in which our unusual suite of talents was forged.[14] Of course, one of our most unusual talents is our symbolic capacity, exemplified foremost by our linguistic abilities. We now conclude by suggesting that the conjunction of transitive planning and procedures and symmetric joint attention in episodes of instruction/learning inaugurate what we will call *reliable cognitive experience*, and that reliable cognitive experience in turn helped symbol use to emerge.

Chimps vocalize regularly, loudly, and with considerable variation. However, their vocal products are dominated by affect and emotion, are not recursively structured, and do not refer to anything outside occurrent experiential space. The famous

[13] Tomasello argues similarly that instruction is a crucial context for developing some of the unique facets of human cognition (see Tomasello, 2014, especially Chapter 3).

[14] Some of the fuel, but not all of the fuel: Tomasello (2014, Chapter 3) argues that collaborative foraging also played a constitutive role in the emergence of joint attention.

bonobos Kanzi and his sister Panbanisha are less vocal than most chimps, but their vocalizations are nevertheless as emotionally agitated and as context determined as those of other chimps. Still, these bonobos are interesting in a way that chimps are generally not: They are icon users. Kanzi and Panbanisha can carry on conversations (of a sort) with each other by pressing sequences of ideograms on a screen and sharing the ideogram sequences. Panbanisha is famous in part for using ideograms to request that Kanzi share some of his grapes by putting them on a table, and Kanzi is famous in part for satisfying her request (Savage-Rumbaugh et al., 1993; Savage-Rumbaugh, McDonald, Sevcik, Hopkins, & Rubert, 1986; Savage-Rumbaugh, Shanker, & Taylor, 1998; Savage-Rumbaugh & Lewin, 1994).

The cognitive preconditions for symbol use probably first appeared in a similarly quotidian setting and emerged from conscious experience already well stocked with iconic and indexical representational elements. We can picture a knapping tutorial in which an instructor demonstrates to a pupil how to prepare a core by rehearsing striking blows on a core and encouraging joint attention and imitation. However, we can also picture a tutorial in which the core is not present—the instructor can rehearse the striking blows in characteristic ways without the core being present.[15] When this occurs, the rehearsed arm motions indexically represent the blows on the core that the pupil is expected to mimic. The pupil's joint attention with the instructor on the latter's various motions and gestures provide the needed encouragement to do as the instructor does. This kind of jointly attentive activity, with the instructor leading and directing the pupil, cultivates cooperation between instructor and pupil. But it also and at the same time scaffolds reliable cognitive experience—that is, conscious experience whose most noteworthy aspect is its cognitive, as distinct from its perceptual, interoceptive, and affective, content. Given the reactants of joint attention and transitive planning, reliable cognitive experience distills out of conscious experience previously dominated by perception, interoception, affect, and acute need. This reliable cognitive experience in turn engenders improved endogenous attentional capacities for subacute problem-solving and offline cognition.

In such an increasingly reliable cognitive context, discovering instructional aids is predictable. We can imagine the instructor inscribing in the dirt or drawing an iconic representation of a step in the procedure and referring the pupil to it. Similarly, we can imagine the instructor vocalizing and gesturing in manners unique to the procedure. Both kinds of aid strengthen in the pupil the association of that learning aid with particular motor behaviors. Of course, none of the characteristic properties of language is yet manifest or presupposed in such a scenario. Neither grammar nor distal reference to anything outside the occurrent experiential field need be implied by the use of instructional aids, and there certainly need be nothing in instructional aids that approach the recursivity typical of embedded sub-sentential components. As far as that goes, there is no obvious symbol use in this scenario either. Icons and indexes are representational without being symbolic: icons represent what they represent by

[15] According to Tomasello, chimps do not mimic indexical motor behavior: "If an ape views someone hammering a nut, they know perfectly well what he is doing, but if they view him making a hammering motion in the absence of any stone or any nuts, they are simply perplexed" (Tomasello, 2014, p. 60).

being topographically similar to them, and indexes represent what they represent by sensory-dependent information transmission via correlation, whereas symbols represent what they represent by social convention when one set of phonemes or visual marks is agreed to represent something else.[16]

Since icons and indexes remain bound by the occurrent experiential field and the concreteness of practical concern jointly attended to, they are, unlike symbols, neither abstract nor conventionally arbitrary in any obvious way. However, given conscious experience outfitted with icons and indexes, symbols are not that far away, for both representation and social cooperation, two preconditions of symbol creation and use, are already present in reliable cognitive experience. Moreover, icons and indexes both provide the mind with a store of resources for offline cognition, resources that working and long-term memory can retrospectively recall and that can form parts of the counterfactual representations of future states of affairs. Finally, many of these decoupled representations come freighted with procedural memories and first-person episodic memories of the original experience's social facets.

CONCLUSION

Explaining the emergence of language requires more than the cognitive abilities and social dimensions discussed here. Among many other things, the identified cognitive and social dimensions of tool-making do not explain the development of distal reference, phonetic variation and chunking, the social conventions establishing fixed interpretations of phonetically chunked vocal behavior, shared semantic content, or the discovery of iteratively and recursively structured speech. Nor do any of the identified cognitive and social abilities individually entail the broader indexical, much less symbolic, cognitive abilities that are arguably prerequisites for language. Still, looking at stone tools, especially when informed by well-supported contemporary theories, reveals considerable cognitive life.

It is, of course, important not to overplay the cognitive implications of tool-making or overstate the implications those theories have for tool-making. For example, there is no reason to infer from the applicability of reflexivity and transitivity/recursion in explanations of the increasingly complex cognitive abilities that emerge from the dynamic social interactions involved in producing tools that conceptual understanding of reflexivity and transitivity/recursion needs to be attributed to Early or Middle Paleolithic individuals. Similarly, no individual involved in tutorials with a tool-making expert need have conceptually loaded beliefs about joint attention's symmetric structure if they are to benefit practically and cognitively from that symmetry. As Wynn (2017) points out regarding other contemporary theories, it is enough that stone tool-makers and the social interactions they participated in fall under the scope

[16] I therefore disagree with Tomasello (2014, Chapter 3), who argues that symbolic behavior becomes entrenched when icons, indexes, and mimicry are present. As noted, icons and indexes, while representational, are not yet symbolic. In the account offered here, symbol use is not a supervenient rider on iconic and indexical conscious experience, but an additionally complex and subsequent achievement that presupposes but is not reducible to iconic and indexical conscious experience.

of those theories for those theories' explanatory properties to find a home in our understanding of expanding cognitive and social powers. Acknowledging reflexivity of conscious experience, the transitivity of planning, and the symmetry of joint attention helps put what was happening hundreds of thousands, even millions, of years ago into a framework that can provide us insights into the evolving dimensions of the cognitively imbued environmental niche we created for ourselves.

It is a testament to his intellectual curiosity, tenacity, and rigor that Wynn appreciates the essential interplay between the archaeological record and contemporary cognitive science and neuroscience. From his early deployment of Piaget's ontogenetic developmental framework to his more recent recruitment of working memory and other resources from contemporary cognitive science and neuroscience, Wynn shows that he thinks long and hard about both the evidential and theoretical sides of an explanatory project. The result is an oeuvre that demonstrates the fertile benefit of laying down rigorous theoretical foundations for explaining, and so for understanding, how pregnant the archaeological record is.

ACKNOWLEDGMENTS

I would like to thank Leee Overmann for discussion of these and other issues and a referee for suggestions that improved the chapter.

REFERENCES

Ambrose, S. H. (2001). Paleolithic technology and human evolution. *Science, 291*(5509), 1748–1753.

Ash, M., & Welshon, R. (n.d.). Representations in a dynamic mind. (Under review).

Baars, B. B. (1988). *A cognitive theory of consciousness*. New York: Cambridge University Press.

Baddeley, A. D. (1986). *Working memory*. New York: Oxford University Press.

Baddeley, A. D. (1996). The fractionation of working memory. *Proceedings of the National Academy of Sciences of the United States of America, 93*(24), 13468–13472.

Baddeley, A. D. (2000). The episodic buffer: A new component of working memory? *Trends in Cognitive Sciences, 4*(11), 417–423.

Baddeley, A. D. (2002). Is working memory still working? *European Psychologist, 7*(2), 85–97.

Baddeley, A. D. (2003). Working memory: Looking back and looking forward. *Nature Reviews: Neuroscience, 4*(10), 829–839.

Block, N. (2001). Paradox and cross purposes in recent work on consciousness. *Cognition, 79*(1), 197–219.

Block, N. (2005). Two neural correlates of consciousness. *Trends in Cognitive Sciences, 9*(2), 46–52.

Block, N. (2007a). Consciousness, accessibility, and the mesh between psychology and neuroscience. *Behavioral and Brain Sciences, 30*(5/6), 481–499.

Block, N. (2007b). On a confusion about a function of consciousness. In *Consciousness, function, and representation: Collected papers* (Vol. 1, pp. 159–213). Cambridge, MA: MIT Press.

Block, N., Flanagan, O., & Güzeldere, G. (Eds.). (1998). *The nature of consciousness: Philosophical debates*. Cambridge, MA: MIT Press.

Botha, R. (2006). On the windows approach to language evolution. *Language & Communication, 26*(2), 129–143.

Botha, R. (2008). Prehistoric shell beads as a window on language evolution. *Language & Communication, 28*(3), 197–212.

Bradfield, J., Lombard, M., & Wadley, L. (2015). South African arrow poison recipes, their ingredients and implications for Stone Age archaeology. *Southern African Humanities, 27*(1), 29–64.

Campbell, J. (2011). Why do language use and tool use both count as manifestations of intelligence? In T. McCormack, C. Hoerl, & S. Butterfill (Eds.), *Tool use and causal cognition* (pp. 169–182). New York: Oxford University Press.

Chemero, A. (2000). Anti-representationalism and the dynamical stance. *Philosophy of Science, 67*(4), 625–647.

Chemero, A. (2009). *Radical embodied cognitive science*. Cambridge, MA: MIT Press.

Clark, A., & Grush, R. (1999). Towards a cognitive robotics. *Adaptive Behavior, 7*(1), 5–16.

Clark, A., & Toribio, J. (1994). Doing without representing? *Synthese, 101*, 401–431.

Clowes, R. W., & Mendonça, D. (2016). Representation redux: Is there still a useful role for representation to play in the context of embodied, dynamicist and situated theories of mind? *New Ideas in Psychology, 40*(Special Issue on Representation), 26–47.

Coolidge, F. L., & Wynn, T. (2005). Working memory, its executive functions, and the emergence of modern thinking. *Cambridge Archaeological Journal, 15*(1), 5–26.

Coolidge, F. L., & Wynn, T. (2009). *The rise of* Homo sapiens: *The evolution of modern thinking*. Oxford, UK: Wiley-Blackwell.

Coolidge, F. L., & Wynn, T. (2018). *The rise of* Homo sapiens: *The evolution of modern thinking* (2nd ed.). Oxford, UK: Oxford University Press.

Cowey, A. (2004). The 30th Sir Frederick Bartlett lecture: Fact, artefact, and myth about blindsight. *Quarterly Journal of Experimental Psychology Section A, 57*(4), 577–609.

Davidson, I., & Noble, W. (1993). Tools and language in human evolution. In K. R. Gibson & T. Ingold (Eds.), *Tools, language and cognition in human evolution* (pp. 363–388). Cambridge, UK: Cambridge University Press.

Degenaar, J., & Myin, E. (2014). Representation-hunger reconsidered. *Synthese, 191*(15), 3639–3648.

Dennett, D. C. (2006). *Sweet dreams: Philosophical obstacles to a science of consciousness*. Cambridge, MA: MIT Press.

Eilan, N. (2005). Joint attention, communication, and mind. In N. Eilan, C. Hoerl, T. McCormack, & J. Roessler (Eds.), *Joint attention: Communication and other minds: Issues in philosophy and psychology* (pp. 1–34). New York: Oxford University Press.

Fodor, J. A. (1975). *The language of thought*. Cambridge, MA: Harvard University Press.

Fodor, J. A. (1981). The mind-body problem. *Scientific American, 244*(1), 114–123.

Gallagher, S. (2008). Are minimal representations still representations? *International Journal of Philosophical Studies, 16*(3), 351–369.

Gamble, C. (1999). *The Palaeolithic societies of Europe*. Cambridge, MA: Cambridge University Press.

Gładziejewski, P. (2016). Action guidance is not enough: A plea for a two-factor theory of representation. *New Ideas in Psychology, 40*(Special Issue on Representation), 13–25.

Grush, R. (2004). The emulation theory of representation: Motor control, imagery, and perception. *Behavioral and Brain Sciences, 27*(3), 377–396.

Haselager, P., van Dijk, J., & van Rooij, I. (2008). A lazy brain? Embodied embedded cognition and cognitive neuroscience. In P. Calvo & T. Gomila (Eds.), *Handbook of cognitive science: An embodied approach* (pp. 273–290). Amsterdam: Elsevier.

Högberg, A., Gärdenfors, P., & Larsso, L. (2015). Knowing, learning and teaching—How homo became docens. *Cambridge Archaeological Journal, 25*(4), 847–858.

Holt, J. (2003). *Blindsight and the nature of consciousness*. New York: Broadview Press.

Hutto, D. D., & Myin, E. (2013). *Radicalizing enactivism: Basic minds without content*. Cambridge, MA: MIT Press.

Kriegel, U. (2009). *Subjective consciousness: A self-representational theory*. New York: Oxford University Press.

Levine, J. (2001). *Purple haze: The puzzle of consciousness*. New York: Oxford University Press.

McCormack, T., Hoerl, C., & Butterfill, S. (Eds.). (2011). *Tool use and causal cognition*. New York: Oxford University Press.

Metcalfe, J., & Terrace, H. S. (Eds.). (2013). *Agency and joint attention*. New York: Oxford University Press.

Metzinger, T. (2003). *Being no one: The self-model theory of subjectivity*. Cambridge, MA: MIT Press.

Milner, A. D., & Goodale, M. A. (1995). *The visual brain in action*. New York: Oxford University Press.

Moore, C., & Dunham, P. J. (Eds.). (1995). *Joint attention: Its origins and role in development*. New York: Psychology Press.

Nagel, T. (1974). What is it like to be a bat? *Philosophical Review, 83*(4), 435–450.

Peacocke, C. (2011). Representing causality. In T. McCormack, C. Hoerl, & S. Butterfill (Eds.), *Tool use and causal cognition* (pp. 148–168). New York: Oxford University Press.

Peirce, C. S. (1984). On a new list of categories. In E. C. Moore, M. H. Fisch, C. J. W. Kloesel, D. D. Roberts, & L. A. Ziegler (Eds.), *Writings of Charles S. Peirce: A chronological edition. Volume 2: 1867–1871* (pp. 49–59). Bloomington, IN: Indiana University Press.

Prinz, J. (2009). Is consciousness embodied? In P. Robbins & M. Aydede (Eds.), *The Cambridge handbook of situated cognition* (pp. 419–436). New York: Cambridge University Press.

Revonsuo, A. (2006). *Inner presence: Consciousness as a biological phenomenon*. Cambridge, MA: MIT Press.

Savage-Rumbaugh, E. S., & Lewin, R. (1994). *Kanzi: The ape at the brink of the human mind*. Toronto: John Wiley & Sons.

Savage-Rumbaugh, E. S., McDonald, K., Sevcik, R. A., Hopkins, W. D., & Rubert, E. (1986). Spontaneous symbol acquisition and communicative use by pygmy chimpanzee (*Pan paniscus*). *Journal of Experimental Psychology: General, 115*(3), 211–235.

Savage-Rumbaugh, E. S., Murphy, J., Sevcik, R. A., Brakke, K. E., Williams, S. L., Rumbaugh, D. M., & Bates, E. (1993). *Language comprehension in ape and child*. Chicago: University of Chicago Press.

Savage-Rumbaugh, E. S., Shanker, S., & Taylor, T. J. (1998). *Apes, language, and the human mind*. New York: Oxford University Press.

Searle, J. R. (2000). Consciousness. *Annual Review of Neuroscience, 23*, 557–578.

Seeman, A. (Ed.). (2011). *Joint attention: New developments in psychology, philosophy of mind, and social neuroscience*. Cambridge, MA: MIT Press.

Tomasello, M. (2008). *Origins of human communication*. Cambridge, MA: MIT Press.

Tomasello, M. (2014). *A natural history of human thinking*. Cambridge, MA: Harvard University Press.

van Gelder, T. (1995). What might cognition be, if not computation? *Journal of Philosophy, 92*(7), 345–381.

Wadley, L. (2010). Compound-adhesive manufacture as a behavioral proxy for complex cognition in the Middle Stone Age. *Current Anthropology, 51*(S1), S111–S119.

Wadley, L., Hodgskiss, T., & Grant, M. (2009). Implications for complex cognition from the hafting of tools with compound adhesives in the Middle Stone Age, South Africa. *Proceedings of the National Academy of Sciences of the United States of America, 106*(24), 9590–9594.

Welshon, R. (2013). Searching for the neural realizers of ownership unity. *Philosophical Psychology, 26*(6), 839–862.

Wynn, T. (1989). *The evolution of spatial competence.* Chicago: University of Illinois Press.

Wynn, T. (1991). Tools, grammar and the archaeology of cognition. *Cambridge Archaeological Journal, 1*(2), 191–206.

Wynn, T. (2002). Archaeology and cognitive evolution. *Behavioral and Brain Sciences, 25*(3), 389–402.

Wynn, T. (2017). Evolutionary cognitive archaeology. In T. Wynn & F. L. Coolidge (Eds.), *Cognitive models in Palaeolithic archaeology* (pp. 1–20). New York: Oxford University Press.

Wynn, T., & Coolidge, F. L. (2004). The expert Neandertal mind. *Journal of Human Evolution, 46*(4), 467–487.

Wynn, T., & Coolidge, F. L. (2010). Beyond symbolism and language: An introduction to Supplement 1, Working Memory. *Current Anthropology, 51*(S1), S5–S16.

Wynn, T., & Coolidge, F. L. (2012). *How to think like a Neandertal.* Oxford: Oxford University Press.

Wynn, T., Haidle, M. N., Lombard, M., & Coolidge, F. L. (2017). The expert cognition model in human evolutionary studies. In T. Wynn & F. L. Coolidge (Eds.), *Cognitive models in Palaeolithic archaeology* (pp. 21–43). Oxford: Oxford University Press.

Zahavi, D. (2008). *Subjectivity and selfhood: Investigating the first-person perspective.* Cambridge, MA: MIT Press.

4

EVOLUTION OF COGNITIVE ARCHAEOLOGY THROUGH EVOLVING COGNITIVE SYSTEMS
A CHAPTER FOR TOM WYNN

Iain Davidson

INTRODUCTION

I first met Tom Wynn in 1986 at the first World Archaeological Congress in Southampton in the company of Bill McGrew. Tom and Bill were working on their ground-breaking paper comparing chimpanzee material culture with that of early hominids at Olduvai (Wynn & McGrew, 1989). The take-home from their conference paper was that apes were the best model for early hominid behavior. Their conclusion tended toward the view that chimpanzees were more human than they had been thought to be. With my usual intuitive contrariness, I suggested that this implied the early hominids[1] were more ape-like than they had been thought to be.

On my return home to Australia, I began an intensive series of discussions with psychologist Bill Noble, which became an exploration of the evolutionary emergence of language. Several of our publications, beginning with "Archaeology of Perception," included comment on Wynn's early work on the cognitive implications of early stone tools (Davidson & Noble, 1989). Though we were not, at the time, generous enough to acknowledge just how bravely pioneering those studies were, I now marvel at the subtlety of Tom's recognition of the issue and his ingenuity in adapting Piaget's ontogenetic reasoning to the question of change in evolutionary time. In a later publication, Noble and I (Noble & Davidson, 1991) criticized Tom's work for its recapitulationist nature (Noble & Davidson, 1993) and constructed a new argument about how the emergence of symbols in the archaeological record indicated the emergence of language (Noble & Davidson, 1996). It is fair to say that Tom has never been fond of arguments about the evolution of language—which perhaps is not surprising, given the limited success of Holloway's (1969) search for syntax in stone tool-making. It is also fair to say that my work with Noble was dedicated to avoiding the concept of cognition, which, in my view, can now be seen as its major weakness.

One outcome of the publication of "Archaeology of Perception" was my participation in the 1989 Wenner-Gren conference in Portugal (Gibson & Ingold, 1993), where I encountered Tom again. On the first evening, one participant humorously

[1] This was before the shift from *hominid* to *hominin* that resulted from the cladistics assessment by Groves (1989).

observed that some people would not survive the conference. Tom and I have sometimes remarked to each other that the prediction was about one of us. But Tom had the flu, and in any case is one of the nicest people in archaeology, so the possibility of conflict over our published differences was minimal. We became good friends, and I have benefitted from his advice and his generous hospitality in Colorado Springs (see Figure 4.1). Though we may never agree about handaxes (Wynn, 1995, 2002), he has nevertheless always listened patiently when I have expressed my views about them (Davidson & Noble, 1993; Noble & Davidson, 1996).

Figure 4.1. Tom Wynn and his family in Colorado Springs, 1993. Photograph by the author.

Our common appreciation that these questions about cognition, language, and handaxes address really important issues in human evolution (though Tom did refer to them as "parochial" on one occasion) has allowed us to remain friends (I think), despite the continued divergence of our means of answering them. In 2008, we reconvened at the same location in Portugal for another Wenner-Gren conference (Wynn & Coolidge, 2010a), this time devoted to issues of Baddeley and Hitch's working memory model (Baddeley & Hitch, 1974) that have been the dominant feature of Tom's work since he teamed up with cognitive psychologist Fred Coolidge, in 2000 (Coolidge & Wynn, 2007, 2009, Wynn & Coolidge, 2007, 2012; also see Figure 4.2). In the interval, Bill Noble had returned to his more conventional research on the psychology of perception, and I had been involved in development of theory with cognitive scientist Phil Barnard and psychologist Richard Byrne (Barnard, Duke, Byrne, & Davidson, 2007; Byrne et al., 2004) during a research project based in Budapest in 2003.

There were two major products of the Budapest project. In the first, we explored the way in which "semantic roles, precursors of those expressed in language, are implicit in animal cognition" (Byrne et al., 2004) by comparing a set of such roles between humans and chimpanzees. I have explored this further with stone tools (Davidson, 2010a), mark-making (Davidson, 2014), and elsewhere in this chapter. This methodology forces us to emphasize the implications of these behaviors for understanding cognitive evolution, opening up new thinking in the archaeology.

The second product was an expansion (Barnard et al., 2007) of Barnard's long-standing exploration of a model of cognition (Barnard, 1985) that he calls Interacting Cognitive Systems (ICS). Barnard's approach (spelled out in Chapter 5 in this volume) has been to consider the human mind as nine interacting cognitive subsystems; the Budapest project began from Byrne's recognition that he could model the ape mind with six of those subsystems. The 2007 paper explored the mechanism by which new subsystems could evolve, through the identification, by hominins, of statistical regularities among actions and material products. This was a selective context for hominins to come to recognize the semantic value of such roles, and hence, to extract meaning in relation to them. Between an ape-like last common ancestor (LCA) and modern humans, there must have been creatures with seven and eight subsystems along the path to the emergence of the ninth subsystem. Most importantly, both ICS and the Budapest model of evolving cognitive systems (ECS) recognize the importance of the social and material connections to the world outside the hominin and human minds, rather than being restricted to the operation of brain or mental processes only (Davidson, 2010a).

During the course of the Budapest project, I wrote a paper with Bill McGrew exploring not the similarities between ape and hominin but the differences. We looked at the way in which differences might account for the natural selection of different evolutionary trajectories among the LCA populations, emphasizing the use of stone tools for cutting and the persistence of remains of knapping in the environment (Davidson & McGrew, 2005). For the 2008 conference, I looked at the conditions necessary for the colonization of Australia using watercraft constructed by the combination of many parts, classifying those conditions within the framework of the "mosaic emergence of different aspects" of Working Memory, and concluded that, far from being able to make a blanket determination of the cognitive requirements for colonization, any such

Figure 4.2. Tom Wynn at Cromlech dos Almendres, Evora, Portugal during Wenner-Gren Conference, 2008. Photograph by the author.

analysis would be very complicated (Davidson, 2010c). This was part of an ongoing methodology that, in this context, derives from Tom Wynn's initial exploration of Piaget's theories: breaking down the theory into elements that can be used as recognition criteria with empirical referents in the archaeological record.

Barnard, who had worked in Baddeley's research unit at the time he was developing the working memory model, was a participant at the 2008 Wenner-Gren conference. He explored the model of the evolution of the interactions between cognitive subsystems from the LCA to modern humans, and showed the formal "isomorphism" between the classic working memory model and the eight subsystem stage of that evolutionary sequence that characterized the period prior to the emergence of modern humans (Barnard, 2010b). The challenge of this approach is to break down the elements of the theoretical stages into elements that can be identified archaeologically, as was attempted at the 2013 Vienna conference of the European Society for the Study of Human Evolution (ESHE) during the focus session organized by Wynn and Coolidge (Barnard, Davidson, & Byrne, 2017; also see Figure 4.3).

The strands that emerge from this history are various: the role of knowledge about primate behavior in constructing arguments about hominin evolution; the relationship between ontogeny and phylogeny; the interpretation of stone tools; the importance (or otherwise) of language in the evolutionary emergence of modern human behavior; the importance of understanding theory about modern human cognition for understanding the cognitive requirements for different aspects of non-modern behavior; and the problems of converting models used to understand the cognition of living people into models that allow for dynamic change in evolutionary time. That is quite a list, and I am proud to have engaged with Tom Wynn in weaving related, but different, cloth over 30 years. It is worth noting that both warp and weft (perhaps seen as data and theory) are necessary to make the strands hold together and avoid holes in threadbare arguments.

In this chapter, I want to address questions about stone tools in the knowledge that Barnard has addressed more formal cognitive models and theory in Chapter 5 in

Figure 4.3. Tom Wynn with Fred Coolidge at the Freud Museum, Vienna, 2013. Photograph by the author.

this volume and Mark Moore has addressed questions about stone tools in Chapter 8. In doing so, I want to draw attention to difficulties of inference about cognition from the study of stone tools. My point is that neither Wynn, nor Piaget, nor Barnard, nor anyone else can use the material evidence from the past to assess models of cognitive evolution without carefully considering the hominin actions and intentions that produced what we (archaeologists) find. Solutions to the problems of the evolution of modern human cognition require theory development in cognitive science, as well as in archaeology. This is because both disciplines have developed their standard approaches to their more immediate problems without theory development in relation to evolution in the case of cognitive science, and to cognitive evolution in the case of archaeology. Archaeology, in particular, has developed means of creating narratives that ignore the theoretical background of the elements that contribute to the story.

As regards evolution in cognitive studies, I asked in 2010, "Is our understanding of the evolutionary emergence of modern human cognition inhibited by cognitive models that have been developed through attempts to understand the cognitive conditions of impaired modern humans rather than being based on an understanding of cognitive states of unimpaired ancestors?" (Davidson, 2010c). The Budapest team came up with a way around that problem by developing explicit models of alternative cognitive systems based on comparative studies. The challenge is repeatedly issued that people of classical Greece (or other favorite periods and places) would have been able to fly an aircraft, but their material culture prevented them from doing so. To argue this way tends to ignore the centrality of the fact that humans are cognitively different from apes, and, as Bill McGrew and Tom Wynn argued back when I first met them, modern apes are the best analogy for the cognitive abilities of the LCA. Arguably, the weak understanding of both the evolution of cognition and the cognitive abilities of early hominins were not so obvious when first Wynn, then Noble and I, set out on the path of developing our brands of cognitive archaeology. Approaches that will take us forward have evolved gradually by selection among the variations that we and others have produced. That is why this chapter is about the evolution of cognitive archaeology.

BIPEDALISM, CARRYING, JOINT ATTENTION, LEARNING

In criticizing the recapitulationist approach to cognitive evolution, Noble and I did not ignore ontogeny completely (Noble & Davidson, 1996). Indeed, we emphasized the important consequences of emergence of obligate bipedalism—that it became necessary for caregivers to carry their infants in front of them. This was even more necessary as selection operated against body hair (Dávid-Barrett & Dunbar, 2016). As a result of the consequent joint attention between carrier and carried, infants were enriched in the way they learned about the world and about the way the caregiver understood it. This phenomenon seems even more important as a result of the argument that "cooperative breeding" had high selective advantage for humans with helpless babies, and that human babies and infants seek attention from other humans through active monitoring of their faces (Hrdy, 2007). It is unknown how this essential feedback within the developmental sequence of human infants came about.

Given the universality of the experience that comprehension precedes production (Burling, 2005), children come to expect that the world is interpretable. In later experiments, it has been shown that when a human experimenter interacts with human infants, enculturated chimpanzee infants, and unenculturated chimpanzee infants, it is joint attention that enhances the development of human-like imitative learning for the first two groups, with implications for human-like cognition (Carpenter, Tomasello, & Savage-Rumbaugh, 1995). Again, such research takes for granted that infants have such joint attention, but there is another question about how humans came to have this behavior while African apes did not. The period of joint attention and enhanced learning was necessarily prolonged in early hominin infants as secondary altriciality emerged—in which babies were born at an earlier stage of development, particularly of the brain, probably as a result of metabolic constraints (Dunsworth, Warrener, Deacon, Ellison, & Pontzer, 2012). It remains controversial at what stage secondary altriciality emerged: Some argue that it was early in hominin evolution, being present in *Homo erectus*, others that it was later, around the time of the second leap in cranial capacity (Davidson, 1999).

There are further points about this research. First, we framed the question of the difference in the mode of communication in terms of the evolutionary emergence of its mechanism of ontogenetic acquisition. Then, we extended that argument to consider the cognitive consequences of the ontogenetic process and of its outcomes in enhanced communication.

None of these elements was a new observation about hominin and human evolution, but it was novel, at least to me, to combine them in this form to produce an argument about cognitive evolution. The combination set up the circumstances for a prolonged argument about the social construction of mind, almost entirely due to Noble (Noble & Davidson, 1996), and entailed an argument about the observation of semiotic signs that has proved useful in other circumstances (Davidson, 2014). One caricature of what we argued could be that we suggested that language emergence was a consequence of bipedalism, but unlike some other generalizations in such grand narratives, we offered a breakdown of the elements that could sustain the case. It seems reasonable to argue that the approach was valuable methodologically in breaking down arguments into their essential elements and recombining them in new ways. Importantly, the argument later fit into the Budapest model by emphasizing the behavioral context in which social and material connections to the world outside the hominin and human minds were enhanced.

SIGNS OF KNAPPING

Wynn and McGrew together conducted systematic comparisons between chimpanzee behavior and what was known of the earliest hominin behavior. Initially, they showed strong similarity between the two cases (Wynn & McGrew, 1989), but differences emerged when they revisited the comparison in light of new data (Wynn, Hernandez-Aguilar, Marchant, & McGrew, 2011). The apes seemed more like the human ancestors, but the hominins, in turn, seemed more ape-like. In the second review, the authors noted that among chimpanzees, tools had been observed in use in hunting (Pruetz & Bertolani, 2007), that sticks found next to holes in the ground could be interpreted as digging sticks for underground storage organs (Hernandez-Aguilar,

Moore, & Pickering, 2007), and that a case could be made for the use of cleavers to split large fruits into more manageable sizes (Koops, McGrew, & Matsuzawa, 2010). The enormous increase in the amount of data on all primates in the last 20 years has allowed more sophisticated analyses of the interrelations among them. Included in the recent raft of comparative studies is an analysis of the phylogeny of Hominoidea and Hominidae based on a selection of life history parameters (Duda & Zrzavý, 2013). This analysis was based entirely on characteristics of living species, and appears to allow the definition of characteristics of the LCA of chimpanzees and humans. In relation to the history described here, one important conclusion of such analysis is that the life histories and behavior of both modern species are the results of changes since the shared ancestor—something that is sometimes obscured by the use of modern species as models for early hominin behavior. Again, the use of variables with little or no archaeological signature makes it very difficult to test the veracity of the claims about LCA characteristics. They remain plausible speculations.

An alternative approach arises by looking at material culture. Using the criteria for identifying the cultural threshold of particular behaviors defined by McGrew and Tutin (1978), Davidson and McGrew (2005) argued that early hominin stone toolmaking probably should not be classified as cultural (Davidson, 2016). The question that emerged was this: Supposing chimpanzees are closer to the LCA than modern humans are, and given the similarities identified by Wynn and McGrew, what was it about Oldowan behavior that made a decisive difference in the long term? It is not a question of a small magic ingredient that suddenly transformed Oldowan tool-makers, but something that made a decisive difference to the environment of opportunity, despite the similarities identified in the comparisons. Such a difference might only have an effect in the long term, despite the pattern of behavior being consistent during all of that time. Davidson and McGrew (2005) made a couple of suggestions: (1) The production of stone tool debris in the environment was an example of niche construction, by establishing a new resource for the availability of tool stone, and (2) as discussed in the next section, flaked stone used for cutting made a decisive difference to the capacity to obtain food. In both cases, the behavior could be a product of six-subsystem cognition, but as more complex cognitive architectures evolved, new opportunities opened for hominins as a result of the initially simple behavioral change.

The fundamental observation for the niche construction argument is that hominins must have been carrying things with them, specifically hammers and the raw material from which to remove flakes. Chimpanzees also carry things, but, in the most famous example, they only carry hammers and not anvils (Boesch & Boesch, 1981). In the hominin case, as shown by the early knapping site Lokalalei (Roche et al., 1999), a series of flakes were removed from a core and left at the site. Other flakes were removed from the site. This is part of a pattern of activity where stone was flaked at a location different from the location of use of the products. Parallels to this can be found in chimpanzees, particularly through the carrying of hammers. Chimpanzees move hammers around the forest at the beginning of a new nutting season, and the statistics of the movements can be interpreted to suggest they remember the locations of hammers from one season to the next (Boesch & Boesch, 1984). Reuse is always for the same purpose. Given that stone flakes, from the earliest evidence, were used to cut different materials (Keeley & Toth, 1981), it seems likely that one of the opportunities that opened with the creation of new locations in the landscape where tools and tool stone could be acquired was that such materials could be recruited for different purposes. The crucial part of our argument was that "[w]hen hominins returned to the scene of earlier knapping events and repeated the actions of tool-making, possibly with different intentions, they set off on the path to reflective awareness and the addition of a symbolic component to their ape-like culture" (Davidson & McGrew, 2005)—the recognition of indexical signs of past activity being the essential prerequisite for the transformation of signs into symbols (Davidson, 2014).

A recent review of the evidence for flaking of long abandoned, and now patinated, stone tools cited published evidence of the process of scavenging and recycling from many different places. The widespread nature of this behavior can be seen from the fact that these locations were as far apart as Britain (Ashton, Cook, Lewis, & Rose, 1992), Africa (Leakey, 1971), and Flores, in Indonesia (Moore, Sutikna, Morwood, & Brumm, 2009; also see Chapter 7 in this volume). All of these examples seem to concern quite late episodes of scavenging. Of course, there is a bias here, as the reuse cannot be easily detected unless enough time has passed for a patina to form.

Repetition of actions that leave similar material products provides the circumstances for the identification of statistical regularities among these actions and products, not only by us as archaeologists but by the hominins themselves. This was a selective context for hominins to come to recognize the semantic value of such roles and, hence, to extract meaning in relation to them. The persistence of the products of the performance of roles in the production of stone flakes impacted hominin cognition.

CUTTING: SEVERING, SLICING, AND SHAVING . . . AND CONSTRUCTING FROM PARTS

By concentrating on the operations identified by Piaget, Tom Wynn was able to identify that aspects of stone artifact variation might provide insight into the cognitive abilities of early hominins (Wynn, 1979, 1981, 1993a, 1993b). The caricature of the position is that the use of stone tools was linked to the evolution of hominin cognition—a strawman that is often rejected. Insofar as Wynn offered a breakdown of that caricature, it was not in terms of the way stone tools required or facilitated cognitive changes, but more in terms of identifying the state of cognitive evolution that had already been reached according to the Piagetian indicators Wynn co-opted. Nevertheless, it was a significant innovation for theory of cognitive evolution and specified identification criteria in a quite different way from the way Parker and Gibson (1979) had used Piaget.

The problem that Noble and I identified is that all that could be done at the time was to look at stone tools as if the forms in which they were found and described were a product of the intention to produce such forms. In other words, if the archaeologist looks at the overall shape of a flaked stone object or the edge of the object and considers the flake removals that produced that shape (Wynn, 1979), then the argument is only as good as the assumption that that shape was desired by the knapper (Davidson, 2002). This is the inevitable precondition for interpreting archaeological evidence in typological terms, a weakness well demonstrated by Dibble (1989) and exposed by Hiscock and Attenbrow: "[H]ow can implements be designed for, and be efficient in, a specific use if their morphology is continuously changing?" (Hiscock & Attenbrow, 2005). The evidence is overwhelming that form is not a good guide to the way flaked stones were used (Beyries, 1987; Nowell et al., 2016), and recent experiments have shown that some of the typical forms said to have significance can actually arise from knapping that involves minimal assumptions about shaping (Moore & Perston, 2016). How can the grand narrative move away from this obsession with shape and frame the archaeological questions in ways that carry cognitive significance?

Bill McGrew, always on the lookout for ways to minimize the differences between chimpanzees and humans, remarked that chimpanzees do not need stone tools because they can obtain their food, be it plant or meat, by using their teeth. Early hominins could not use their teeth but did have stone tools, so perhaps the practice of cutting is one of the crucial steps in hominin evolution (Davidson & McGrew, 2005).

Chimpanzees in the wild produce stone flakes, accidentally, while cracking nuts, but they do not use them or appear to notice them (Davidson & McGrew, 2005). The fact that gorillas are not known to manipulate stones, but at least one monkey species (the South American capuchins) does (Proffitt et al., 2016), shows that stone use in monkeys, apes, and humans is a convergent behavior. The monkeys do not use the flakes that result from their activities (Proffitt et al., 2016), so the title of Proffitt et al.'s paper, "Wild Monkeys Flake Stone Tools," is misleading—the monkeys flake stone, they do not flake stone tools. Captive bonobos (Toth, Schick, Savage-Rumbaugh, Sevcik, & Rumbaugh, 1993) and an orangutan (Wright, 1972) have been taught to make flakes—and have also been taught to obtain rewards by cutting a string. So, cutting is not beyond the conceptual capacities of apes. But what is meant by "cutting?" Some aspects of this were addressed by Parker and Gibson (1979), citing Piaget, who emphasized how children cut off small parts off an object and ignore the rest.

In the case of bonobos or orangutans, cutting a string means separating a single thing into two parts, but the cutter has no interest in either part. This can be called *severing* (C1). This is conceptually different from use of the same sharp edge for *slicing* a piece of meat off a carcass (C2a), as Piaget's children did, and this is also conceptually different from cutting a *shaving* off a piece of wood in order to shape the shaved wood (C2b). Slicing and shaving have in common that the cutting involves separating the cut object into two parts, one of which is of interest and one of which is not; they differ in terms of whether the smaller or larger is the residue. The fact that, at Koobi Fora, flakes from Oldowan sites have signs of meat cutting and others of wood cutting (Keeley & Toth, 1981) suggests that both slicing and shaving were quite early in hominin stone tool production. Cut-marks on bones from Dikika (McPherron et al., 2010; but also see Sahle, El Zaatari, & White, 2017) suggest that slicing may be about the same age—at 3.3 million years ago (Mya)—as evidence of flaking stone to make sharp edges (Harmand et al., 2015).

There are two important extensions of this classification. First, chimpanzees could be said to do the equivalent of shaving when removing the shells of nuts by bashing the nut with a hammer or trimming the leaves off a grass stem to make a termiting probe. The trimming and sharpening of sticks prior to their use in hunting (Pruetz & Bertolani, 2007) is probably in this category too. But it is not the case that a Tasmanian digging stick made by shaving with a stone tool is the equivalent of a Tanzanian termiting probe (McGrew, 1987), because the digging stick cannot be made without the stone tool. As a result, the technology requires more than expedient combinations of tools. It is not clear that apes engage in an equivalent of slicing, though the suggestion that they use "cleaving" tools to reduce fruits to a manageable size could be considered as such (Koops et al., 2010).

The second extension is the removal of flakes from a core—behavior that is only present among hominins. This is the equivalent of slicing, as the desired product was the flake that was removed from the core and the other debris. It may be, therefore, that slicing was the big innovation in relation both to the production of stone tools and to their selective advantage. Zink and colleagues (Zink, Lieberman, & Lucas, 2014) have shown that slicing food—removing easily chewed fragments from a large mass—could have been critical in effective nutrition, and in the natural selection of aspects of masticatory anatomy. Such advantage could have provided the context for the selection that made slicing a prominent part of the behavioral repertoire.

Oakley (1952) defined humanity by our ability to make tools, but Goodall's (1964) observation of chimpanzee tool-making forced a refinement of the concept. A new definition of a tool was produced: a tool made with another tool. I have been unable to find the original source for this idea, which was taught to generations of beginning students from the mid-1960s: It is an idea repeated by Wynn and McGrew (1989),[2] who were misquoting Parker and Gibson (1979), who in turn were referring to a tool being used on an object.[3] The new definition was not only irresolvably circular but required refinement. This is a refinement that would specify conceptually different operations in the process of making a digging stick with a stone tool and would show how it is fundamentally more complex than a termiting probe (McGrew, 1987; Oswalt, 1976). There are two complementary but conceptually different actions: (1) slicing flakes off a core and ceasing to pay attention to the core, and (2) using the flakes to shave wooden flakes off a stick and ceasing to pay attention to the wooden "flakes."

The Budapest group pointed out how it was possible to identify that chimpanzees engage in semantic roles that are similar to those of humans (Fillmore, 1968) but are no more than "precursors of those expressed in language" (Byrne et al., 2004, p. 342). For example, the experiencer role of Fillmore's classification is identified when an animate being has a given experience or mental state: Among humans we identify it, for example, when "Daddy is cross"; among chimpanzees, piloerection and "waa" barks show that the chimpanzee Frodo is angry.

When such an analysis was extended to hominin uses of stone tools, the same range of semantic roles could be identified. Importantly, the enduring material product of six of the eight semantic roles in the form of stone artifacts and debris on the landscape provided a new environment of opportunity for hominins, as discussed in the previous section; in addition, such endurance facilitated the recognition of the patterns of the hominins' own behavior, through possibility of reflection on the relationships between what they produced and their roles in that production (Davidson, 2010a).

[2] On p. 389, Wynn and McGrew (1989) note, "The knappers also needed to use stone hammers to make the flaked stone tools; in other words they used tools to make other tools. This striking point has achieved some notoriety in discussions of the evolution of intelligence (e.g. Parker & Gibson 1979)."

[3] On p. 371, Parker and Gibson (1979) state, "True tool use (as opposed to simpler forms of prototool use) involves using one detached object (not a part of the animal's anatomy) to change the state of another object—that is, tool use requires a tool."

In the model of the evolving cognitive subsystems, it became possible to predict that even among late-stage hominins with eight subsystems, the "multimodal subsystem would be totally absorbed with managing its spatial-praxic, verbal, and bodily responses. There is no capability to reflect on meanings or, for example, to think about how to make a better tool while concurrently making one" (Barnard et al., 2017). There is inherent uncertainty about identifying whether hominins had late-stage eight subsystem minds or recruited all nine subsystems.

Once attention turned away from a single product, whether that be a flake or meat sliced off the parent material or the digging stick rather than the shaving removed to make it, to an interest in both parts (the flake and the core), the way was open to conceptualize the process of "partition" (Barnard et al., 2017). Only when the concept of "parts" was discovered would it become possible for hominins to conceptualize combining separate parts into a new single artifact, such as a tipped spear (Haidle, 2009) or a bow and arrow (Lombard & Haidle, 2012). This is to be distinguished from the composition of stereotyped and context-specific structures, such as nests (Parker & Gibson, 1979), using only components that were not themselves "made."

One approach to understanding the process of making stone tools has been to look at the *chaîne opératoire*, or operational sequence (OS), by which particular episodes of knapping were carried out (Bar-Yosef & Van Peer, 2009). We can illustrate the problems and strengths of such approaches using Moore's (2000a) analysis of a reduction sequence from Tasmania that showed what he has elsewhere (Moore, 2010) called the basic flake unit: Useable tools were made by the consequent, sequential application of simple principles of flake removal.

A consistent sequence of operations can occur without agency, as in the disarticulation of animal bodies in natural conditions without intervention from other animals (Hill & Behrensmeyer, 1984); similarly, consistent sequences can occur with agency but without necessarily involving intentionality, as in the repeated rote sequence of plucking and folding of stinging leaves practiced by gorillas (Byrne, 2003). The presence of a consistent sequence itself does not establish intentionality. Like the gorillas, who are capable of reproducing a sequence of actions every time they process stinging leaves, knapping hominins seem to have become able to chunk sequences of knapping actions into repeatable and repeated sequences, such as Moore's "basic flake unit." Wynn and Coolidge (2010b) explored the analogy with the processes of a game of chess, where remembered sequences of moves are an essential part of the way expert players compete.

Wynn and Coolidge applied their analogy to the argument that the Levallois technique of core preparation produced predetermined flakes, accepting that it involved sequences of routines with discrete sub-routines (Wynn & Coolidge, 2010b). Some caution about this interpretation is warranted (Davidson, 2010a), not only because of the evidence that the supposedly predetermined flakes were not actually the object of the knapping (Beyries, 1987; van Peer, 1992) but also because appropriate cores appeared much earlier than the appearance of the supposedly predetermined flakes (de la Torre, Mora, Domínguez-Rodrigo, de Luque, & Alcalá, 2003). Now, Moore and Perston (2016) have shown that flakes similar to those said to be predetermined arise during knapping when decisions about platform choices are made randomly (also see Chapter 8 in this volume). Interpreting the intentionality of the process of production is much more difficult than had been imagined.

In the Hunter Valley of mainland Australia (Moore, 2000b), the basic flake unit was the beginning of a more complex process where cores were prepared, then subjected to heat treatment (Brown et al., 2009; Delagnes et al., 2016), and then flaked once more to produce more specialist products. These, in turn, were ultimately hafted—a process that also required the production of both haft and gum. The combination of flaking modules indicates intentionality that was not necessarily established by sequences within the basic flake unit.

This example illustrates a feature of stone tool-making that resonates with some research in cognition, particularly in relation to working memory. Much research in working memory involves testing the ability to perform one task while distracted by the need to perform another. A typical task is to process lists of numbers, while attention is distracted by being presented with lists of words at the same time (Salamé & Baddeley, 1982). Such distractor tasks test the ability to store things in working memory. The initial ability to incorporate such tasks into a sequence when they completely alter the focus of attention represents a new cognitive ability. It can be recognized in the archaeological record through heat treatment of cores before the removal of flakes (Brown et al., 2009) or the use of crested blades in preparing cores to produce a sequence of blades (Soriano, Villa, & Wadley, 2007).

The ultimate tasks involving attention distraction were the construction of watercraft that brought people to Australia. This is true if either the craft had to be assembled from disparate materials present in different places or it had to be made by hollowing out a tree trunk with hafted stone tools; in either case, the watercraft was probably made for a purpose somewhat removed from the actions of making the craft. As a watercraft, it was probably used for fishing, probably with nets, as indicated by the remains from Timor (Kealy, Louys, & O'Connor, 2016; O'Connor, Ono, & Clarkson, 2011).

TOWARD A NARRATIVE FOR COGNITIVE EVOLUTION?

I tentatively define five stages (A through E) in the following narrative:

A. Cutting emerged about 3.5 Mya (McPherron et al., 2010) among hominins and probably represents the emergence of cognition beyond that of the LCA.
B. Knapping was present not long after stage A (Harmand et al., 2015) and represents a clear distinction from the abilities of the LCA in the capacity of hominins to divide the core into separate useable entities, recognize the usefulness of the part removed from the core, and use the part on third objects. Both removal of the flake from the core and the cutting made possible by the flakes suggest analogous functions of slicing.
C. These two novelties (stages A and B) are connected to meat acquisition and other enhanced food opportunities, particularly the improved nutrition yielded by eating sliced foodstuffs (Zink & Lieberman, 2016). They are in turn associated with the relaxation of selection against large brains and subsequent increases in body size. I argued previously (Davidson, 1999) that this stage of increase in brain size made the emergence of secondary altriciality advantageous.
D. Extension of knapping to stringing together repeated sequences of flake removals followed from the emergence of knapping. This required the capacity to recognize

that both the thing removed by knapping and the thing from which it was removed were useful objects. This represented the achievement of tasks involving relatively long chains of actions between the initiation of action and its consequent completion—but could have been part of a single sequence of actions.

E. The cognitive leap to constructing tasks that involved attention distraction (i.e., completion of a task composed of several sub-tasks that were different in nature from each other) was achieved by 150,000 years ago. This involved the recognition of the concept of "part" that could only follow from recognition of the thing removed and the thing it was removed from involved in stage D.

On the basis of evolution of ICS (Barnard et al., 2017), the seventh subsystem probably emerged between stages B and C, separating the effector subsystem into separate systems relating to the limbs on one hand and the articulators on the other. Vocal utterance under control separate from emotional states might have been possible at this stage, allowing the possibility of simple vocally guided instruction (Davidson, 2009). This, in turn, was probably part of the selective context for the emergence of the eighth cognitive subsystem between stages D and E. This involved the cognitive extension of such sequences to combinations of vocal utterances. Stage E led to the emergence of the ninth cognitive subsystem by which humans could imagine tools and tasks before they made them—processing them in the ninth, propositional, subsystem without any input from outside—and create new opportunities that did not arise from the contingencies of their current actions (Barnard et al., 2017). This cognitive system had the function of reflection that was central to the argument of Davidson and Noble (1989).

The extension of the analysis of human and chimpanzee behaviors to stone toolmaking has implications for picture-making (Davidson, 2014). In the production of images, four of the semantic roles can only be defined in terms of mental processes that are conceptually removed from the actions or roles. As with the stage E toolmaking, this was made possible by the differentiation in the central executive implied by the final, ninth subsystem of Barnard's cognitive scheme (Davidson, 2014). The achievement of mark-making involving repetitive marks capable of being used to carry meaning is one of the clearest indicators of the emergence of the modern human cognitive system.

WHAT DOES THIS MEAN FOR THE WAY IN WHICH COGNITIVE ARCHAEOLOGY HAS EVOLVED?

Almost 50 years ago, Holloway (1969) explored the interaction between arbitrary form and the imposition of form on artifacts. Despite the insights this approach undoubtedly brought to the discussion, the approach was unduly dependent on an understanding of the forms of early artifacts that derived primarily from archaeologists' need for typology. But the attempt opened the way for more productive approaches to cognitive evolution.

Parker and Gibson (1979), using Piaget's developmental scheme (at about the same time as Wynn) to speculate about the process of hominin evolution, emphasized first extractive foraging with tools, and then complex hunting. But given the state of the discipline at the time, they had little option but to accept the then-standard conclusions of archaeological analysis to identify such economic conditions. Wynn's (1979, 1981) breakthrough was to consider the details of one particular class of evidence from the archaeological record and carefully assess the operations necessary to make it. In other words, he went beyond both Holloway and Parker and Gibson in using the theory he espoused to produce new insights into the basic evidence of the record.

Davidson and Noble (Davidson & Noble, 1989; Noble & Davidson, 1991) considered not so much these details as the conceptual issues, pointing out that recapitulationism is not adequate as an explanation. In our view, the minds of early hominins should not be considered as incomplete modern minds but minds that required modeling in the same way that modern minds do but within a theoretical framework capable of making predictions about the mind in non-humans (including hominins) (Barnard, 2010a). Instead, we emphasized reflection and the impact language had on it. We did not, however, model either the modern mind or the way in which it might have evolved. We relied only on the definition of language as communication using symbols (avoiding the issue of syntax that Holloway had made central) and its role in the social construction of mind (Noble & Davidson, 1996). By identifying the importance of the early appearances of symbols and the accompanying greater information flow, planning depth, and conceptualization, we sought to identify the emergence of modern human behavior (Noble & Davidson, 1991). The weakness of our argument was the poor modeling of the mind and the fact that it came before an explosion of new, well-documented, and well-dated early candidates for symbol use in South Africa (d'Errico, Henshilwood, & Nilssen, 2001; d'Errico, Henshilwood, Vanhaeren, & van Niekerk, 2005; Henshilwood et al., 2011; Henshilwood, d'Errico, & Watts, 2009; Mackay & Welz, 2008; Texier et al., 2010) and Europe (Burdukiewicz, 2014; Caron, d'Errico, Del Moral, Santos, & Zilhão, 2011; Finlayson et al., 2012; Peresani, Fiore, Gala, Romandini, & Tagliacozzo, 2011; Peresani, Vanhaeren, Quaggiotto, Queffelec, & d'Errico, 2013; Radovčić, Sršen, Radovčić, & Frayer, 2015; Rodríguez-Vidal et al., 2014; Roebroeks et al., 2012; Soressi et al., 2013; Zilhão et al., 2010).

When Tom Wynn teamed up with Fred Coolidge, a neuropsychologist with a specialty in personality disorders, they shifted their attention to the way the mind works using a detailed examination of Baddeley's working memory model (Coolidge & Wynn, 2001, 2005; Wynn & Coolidge, 2004, 2006), concentrating principally on the interaction between Neandertals and modern humans. But as with the recapitulationists, they depended on a strong model suitable for modern people and considered how archaeological evidence did and did not fit into their framework. The use of the detail of working memory was tremendously productive about the specific interaction but much less productive about the evolution of hominin cognition generally.

Like working memory and partly in response to it, Barnard developed an ICS model with three features important for this discussion. First, it could be successfully applied to a variety of psychological conditions (Barnard & Teasdale, 1991); second, its relations with working memory could be specified (Barnard, 1999); and third, to a much greater extent than for working memory, it is possible to model cognition in relation to the inputs of materials and of other people through the senses (Davidson, 2010b). The distinctive contribution of the Budapest model of the evolutionary emergence of the nine-subsystem ICS from the six-subsystem LCA mind (Barnard et al., 2007, 2017) is that it generated a model of hominin cognitive evolution derived from the same theoretical assumptions underlying Barnard's ICS. Comparison with working memory showed that that was consistent with eight-subsystem cognition, associated with Neandertals (Barnard et al., 2007).

Finally, the Budapest model allows a solution to another paradox. In the 25 years since Noble and I sought to explore the emergence of language by appeal to the sharp distinction between Neandertals who did not have symbols and modern humans who did, the evidence has changed fundamentally. Most importantly, late Neandertals seem to have used ochres, feathers and other parts of birds, and shells in ways that are regarded as symbolic among later people. They appear to have had characteristics of nine-subsystem mental architecture, despite 350,000 or more years of separation. Either there was convergence on these common cognitive abilities or they were present but latent in the LCA of these species. Here, the important point is that the Budapest model provides the theoretical foundations for convergence. It does not specify the form in which the behavioral consequences of the ninth subsystem would manifest themselves, but the theory does allow that evolution can affect the system no matter how many subsystems it has or had. It is also the case that, because it is an evolutionary model specifying the conditions under which each new cognitive subsystem could emerge, it allows an understanding of the ambiguity of interpretation at the boundary at which a new subsystem emerges (Barnard et al., 2017). It is almost a prediction of the theory that the late Neandertals might manifest some of the behaviors thought to be typical of modern humans, despite their long separation.

So, the evolution of cognitive archaeology has been progressive. Threads I have drawn out here from my own previous work include the niche creation implicit in the products of knapping left in the environment after useful flakes have been removed, and the evolution of the processes of development. To this must be added the sophisticated modeling of cognition that recognizes both the inputs to the mind and its internal processes, and a mechanism by which the introduction of new variations can lead to the evolution of different configurations of mental processes. It may be that through these several threads a new fabric can be woven that will bring the study of cognitive evolution closer to recent developments in evolutionary theory (Laland et al., 2015).

Various attempts have made advances on the theoretical front or by specifying the methodological implications for the empirical record. As different attempts outlined here have specified more and more areas of interest, it has become obvious that new evidence alone (and there has been plenty of that) is not sufficient. The theory has had to keep pace, but the theory has had to be constructed appropriately for archaeology and not be merely borrowed from modern studies that are tackling different problems. No doubt there will be further iterations in the future, but we should always keep in mind the fundamental and original contribution Tom Wynn made in guiding us to this point.

ACKNOWLEDGMENTS

Changes were made in light of comments by two anonymous referees. For help in various ways over the years, I would like to thank and absolve from blame Helen

Arthurson, Phil Barnard, Adam Brumm, Dick Byrne, Elisabeth Culley, Mark Moore, April Nowell, Matt Pope, Philip R. Preston, J. Peter White, and, of course, Tom Wynn.

REFERENCES

Ashton, N. M., Cook, J., Lewis, S. G., & Rose, J. (Eds.). (1992). *High Lodge: Excavations by G. de G. Sieveking, 1962–8, and J. Cook, 1988*. London: British Museum Press for the Trustees of the British Museum.

Baddeley, A. D., & Hitch, G. J. (1974). Working memory. In G. H. Bower (Ed.), *The psychology of learning and motivation: Advances in research and theory* (Vol. 8, pp. 47–89). New York: Academic Press.

Barnard, P. J. (1985). Interacting cognitive subsystems: A psycholinguistic approach to short-term memory. In A. W. Ellis (Ed.), *Progress in the psychology of language* (Vol. 2, pp. 197–258). London: Lawrence Erlbaum Associates.

Barnard, P. J. (1999). Interacting cognitive subsystems: Modelling working memory phenomena within a multiprocessor architecture. In A. Miyake & P. Shah (Eds.), *Models of working memory: Mechanisms of active maintenance and executive control* (pp. 298–339). Cambridge, UK: Cambridge University Press.

Barnard, P. J. (2010a). Current developments in inferring cognitive capabilities from the archaeological traces left by stone tools: Caught between a rock and a hard inference. In A. Nowell & I. Davidson (Eds.), *Stone tools and the evolution of human cognition* (pp. 207–226). Boulder, CO: University Press of Colorado.

Barnard, P. J. (2010b). From executive mechanisms underlying perception and action to the parallel processing of meaning. *Current Anthropology, 51*(S1), S39–S54.

Barnard, P. J., Davidson, I., & Byrne, R. W. (2017). Toward a richer theoretical scaffolding for interpreting archaeological evidence concerning cognitive evolution. In T. Wynn & F. L. Coolidge (Eds.), *Cognitive models in Palaeolithic archaeology* (pp. 45–67). Oxford, UK: Oxford University Press.

Barnard, P. J., Duke, D. J., Byrne, R. W., & Davidson, I. (2007). Differentiation in cognitive and emotional meanings: An evolutionary analysis. *Cognition and Emotion, 21*(6), 1155–1183.

Barnard, P. J., & Teasdale, J. D. (1991). Interacting cognitive subsystems: A systemic approach to cognitive-affective interaction and change. *Cognition and Emotion, 5*(1), 1–39.

Bar-Yosef, O., & Van Peer, P. (2009). The *chaîne opératoire* approach in Middle Paleolithic archaeology. *Current Anthropology, 50*(1), 103–131.

Beyries, S. (1987). *Variabilité de l'industrie lithique au Moustérien: Approche fonctionnelle sur quelques gisements français*. Oxford, UK: British Archaeological Reports.

Boesch, C., & Boesch, H. (1981). Sex differences in the use of natural hammers by wild chimpanzees: A preliminary report. *Journal of Human Evolution, 10*(7), 585–593.

Boesch, C., & Boesch, H. (1984). Mental map in wild chimpanzees: An analysis of hammer transports for nut cracking. *Primates, 25*(2), 160–170.

Brown, K. S., Marean, C. W., Herries, A. I. R., Jacobs, Z., Tribolo, C., Braun, D. R., . . . Bernatchez, J. (2009). Fire as an engineering tool of early modern humans. *Science, 325*(5942), 859–862.

Burdukiewicz, J. M. (2014). The origin of symbolic behavior of Middle Palaeolithic humans: Recent controversies. *Quaternary International, 326*, 398–405.

Burling, R. (2005). *The talking ape: How language evolved*. Oxford, UK: Oxford University Press.

Byrne, R. W. (2003). Imitation as behaviour parsing. *Philosophical Transactions of the Royal Society of London. Series B, Biological Sciences, 358*(1431), 529–536.

Byrne, R. W., Barnard, P. J., Davidson, I., Janik, V. M., McGrew, W. C., Miklósi, Á., & Wiessner, P. W. (2004). Understanding culture across species. *Trends in Cognitive Sciences, 8*(8), 341–346.

Caron, F., d'Errico, F., Del Moral, P., Santos, F., & Zilhão, J. (2011). The reality of Neandertal symbolic behavior at the Grotte du Renne, Arcy-sur-Cure, France. *PLoS One, 6*(6), 1–11.

Carpenter, M., Tomasello, M., & Savage-Rumbaugh, E. S. (1995). Joint attention and imitative learning in children, chimpanzees, and enculturated chimpanzees. *Social Development, 4*(3), 217–237.

Coolidge, F. L., & Wynn, T. (2001). Executive functions of the frontal lobes and the evolutionary ascendancy of *Homo sapiens*. *Cambridge Archaeological Journal, 11*(3), 255–260.

Coolidge, F. L., & Wynn, T. (2005). Working memory, its executive functions, and the emergence of modern thinking. *Cambridge Archaeological Journal, 15*(1), 5–26.

Coolidge, F. L., & Wynn, T. (2007). The working memory account of Neandertal cognition—How phonological storage capacity may be related to recursion and the pragmatics of modern speech. *Journal of Human Evolution, 52*(6), 707–710.

Coolidge, F. L., & Wynn, T. (2009). *The rise of* Homo sapiens: *The evolution of modern thinking*. Oxford, UK: Wiley-Blackwell.

Dávid-Barrett, T., & Dunbar, R. I. M. (2016). Bipedality and hair loss in human evolution revisited: The impact of altitude and activity scheduling. *Journal of Human Evolution, 94*, 72–82.

Davidson, I. (1999). The game of the name: Continuity and discontinuity in language origins. In B. J. King (Ed.), *The origins of language: What nonhuman primates can tell us* (pp. 229–268). Santa Fe, NM: School of American Research Press.

Davidson, I. (2002). The finished artefact fallacy: Acheulean hand-axes and language origins. In A. Wray (Ed.), *The transition to language* (pp. 180–203). Oxford, UK: Oxford University Press.

Davidson, I. (2009). Comment on Bar-Yosef and Van Peer. *Current Anthropology, 50*(1), 119–120.

Davidson, I. (2010a). Stone tools and the evolution of hominin and human cognition. In A. Nowell & I. Davidson (Eds.), *Stone tools and the evolution of human cognition* (pp. 185–205). Boulder, CO: University Press of Colorado.

Davidson, I. (2010b). The archaeology of cognitive evolution. *Wiley Interdisciplinary Reviews: Cognitive Science, 1*(2), 214–229.

Davidson, I. (2010c). The colonization of Australia and its adjacent islands and the evolution of modern cognition. *Current Anthropology, 51*(S1), S177–S189.

Davidson, I. (2014). Origins of pictures: An argument for transformation of signs. In K. Sachs-Hombach & J. R. J. Schirra (Eds.), *Origins of pictures: Anthropological discourses in image science* (pp. 15–45). Cologne, Germany: Herbert von Halem Verlag.

Davidson, I. (2016). Stone tools: Evidence of something in between culture and cumulative culture? In M. N. Haidle, N. J. Conard, & M. Bolus (Eds.), *The nature of culture: Based on an interdisciplinary symposium "The nature of culture", Tübingen, Germany* (pp. 99–120). Dordrecht, The Netherlands: Springer Science + Business Media.

Davidson, I., & McGrew, W. C. (2005). Stone tools and the uniqueness of human culture. *Journal of the Royal Anthropological Institute of Great Britain and Ireland, 11*(4), 793–817.

Davidson, I., & Noble, W. (1989). The archaeology of perception: Traces of depiction and language. *Current Anthropology, 30*(2), 125–155.

Davidson, I., & Noble, W. (1993). Tools and language in human evolution. In K. R. Gibson & T. Ingold (Eds.), *Tools, language and cognition in human evolution* (pp. 363–388). Cambridge, UK: Cambridge University Press.

Delagnes, A., Schmidt, P., Douze, K., Wurz, S., Bellot-Gurlet, L., Conard, N. J., . . . Henshilwood, C. S. (2016). Early evidence for the extensive heat treatment of silcrete in the Howiesons Poort at Klipdrift Shelter (Layer PBD, 65 ka), South Africa. *PLoS One, 11*(10), 1–27.

de la Torre, I., Mora, R., Domínguez-Rodrigo, M., de Luque, L., & Alcalá, L. (2003). The Oldowan industry of Peninj and its bearing on the reconstruction of the technological skills of Lower Pleistocene hominids. *Journal of Human Evolution, 44*(2), 203–224.

d'Errico, F., Henshilwood, C. S., & Nilssen, P. (2001). An engraved bone fragment from c. 70,000-year-old Middle Stone Age levels at Blombos Cave, South Africa: Implications for the origin of symbolism and language. *Antiquity, 75*(288), 309–318.

d'Errico, F., Henshilwood, C. S., Vanhaeren, M., & Van Niekerk, K. L. (2005). *Nassarius kraussianus* shell beads from Blombos Cave: Evidence for symbolic behaviour in the Middle Stone Age. *Journal of Human Evolution, 48*(1), 3–24.

Dibble, H. L. (1989). The implications of stone tool types for the presence of language during the Lower and Middle Paleolithic. In P. A. Mellars & C. B. Stringer (Eds.), *The human revolution: Behavioural and biological perspectives in the origins of modern humans* (pp. 415–431). Princeton, NJ: Princeton University Press.

Duda, P., & Zrzavý, J. (2013). Evolution of life history and behavior in Hominidae: Towards phylogenetic reconstruction of the chimpanzee–human last common ancestor. *Journal of Human Evolution, 65*(4), 424–446.

Dunsworth, H. M., Warrener, A. G., Deacon, T. W., Ellison, P. T., & Pontzer, H. (2012). Metabolic hypothesis for human altriciality. *Proceedings of the National Academy of Sciences of the United States of America, 109*(38), 15212–15216.

Fillmore, C. J. (1968). The case for case. In E. Bach & R. T. Harms (Eds.), *Universals in linguistics theory* (pp. 1–88). New York: Holt, Rinehart and Winston.

Finlayson, C., Brown, K., Blasco, R., Rosell, J., Negro, J. J., Bortolotti, G. R., . . . Llanes, J. M. R. (2012). Birds of a feather: Neanderthal exploitation of raptors and corvids. *PLoS One, 7*(9), 1–9.

Gibson, K. R., & Ingold, T. (Eds.). (1993). *Tools, language and cognition in human evolution.* Cambridge, UK: Cambridge University Press.

Goodall, J. van L. (1964). Tool-use and aimed throwing in a community of free-living chimpanzees. *Nature, 201*(4926), 1264–1266.

Groves, C. P. (1989). *A theory of human and primate evolution.* Oxford, UK: Clarendon Press.

Haidle, M. N. (2009). How to think a simple spear. In S. A. de Beaune, F. L. Coolidge, & T. Wynn (Eds.), *Cognitive archaeology and human evolution* (pp. 57–74). Cambridge, MA: Cambridge University Press.

Harmand, S., Lewis, J. E., Feibel, C. S., Lepre, C. J., Prat, S., Lenoble, A., . . . Roche, H. (2015). 3.3-million-year-old stone tools from Lomekwi 3, West Turkana, Kenya. *Nature, 521*(7552), 310–315.

Henshilwood, C. S., D'Errico, F., Van Niekerk, K. L., Coquinot, Y., Jacobs, Z., Lauritzen, S.-E., . . . García-Moreno, R. (2011). A 100,000-year-old ochre-processing workshop at Blombos Cave, South Africa. *Science, 334*(6053), 219–222.

Henshilwood, C. S., D'Errico, F., & Watts, I. (2009). Engraved ochres from the Middle Stone Age levels at Blombos Cave, South Africa. *Journal of Human Evolution, 57*(1), 27–47.

Hernandez-Aguilar, R. A., Moore, J., & Pickering, T. R. (2007). Savanna chimpanzees use tools to harvest the underground storage organs of plants. *Proceedings of the National Academy of Sciences of the United States of America, 104*(49), 19210–19213.

Hill, A., & Behrensmeyer, A. K. (1984). Disarticulation patterns of some modern East African mammals. *Paleobiology, 10*(3), 366–376.

Hiscock, P., & Attenbrow, V. (2005). Reduction continuums and tool use. In C. Clarkson & L. Lamb (Eds.), *Lithics "Down Under": Australian perspectives on lithic reduction, use and classification* (pp. 43–55). Oxford, UK: Archaeopress.

Holloway, R. L. (1969). Culture: A human domain. *Current Anthropology, 10*(4), 395–412.

Hrdy, S. B. (2007). Evolutionary context of human development: The cooperative breeding model. In C. A. Salmon & T. K. Shackelford (Eds.), *Family relationships: An evolutionary perspective* (pp. 39–68). Oxford, UK: Oxford University Press.

Kealy, S., Louys, J., & O'Connor, S. (2016). Islands under the sea: A review of early modern human dispersal routes and migration hypotheses through Wallacea. *Journal of Island and Coastal Archaeology, 11*(3), 364–384.

Keeley, L. H., & Toth, N. P. (1981). Microwear polishes on early stone tools from Koobi Fora, Kenya. *Nature, 293*(5832), 464–465.

Koops, K., McGrew, W. C., & Matsuzawa, T. (2010). Do chimpanzees (*Pan troglodytes*) use cleavers and anvils to fracture *Treculia africana* fruits? Preliminary data on a new form of percussive technology. *Primates, 51*(2), 175–178.

Laland, K. N., Uller, T., Feldman, M. W., Sterelny, K., Müller, G. B., Moczek, A., . . . Odling-Smee, F. J. (2015). The extended evolutionary synthesis: Its structure, assumptions and predictions. *Proceedings of the Royal Society of London B: Biological Sciences, 282*(1813), 1–14.

Leakey, M. D. N. (1971). *Olduvai Gorge. Vol. 3, Excavations in beds I and II, 1960–63*. Cambridge, UK: Cambridge University Press.

Lombard, M., & Haidle, M. N. (2012). Thinking a bow-and-arrow set: Cognitive implications of Middle Stone Age bow and stone-tipped arrow technology. *Cambridge Archaeological Journal, 22*(2), 237–264.

Mackay, A., & Welz, A. (2008). Engraved ochre from a Middle Stone Age context at Klein Kliphuis in the Western Cape of South Africa. *Journal of Archaeological Science, 35*(6), 1521–1532.

McGrew, W. C. (1987). Tools to get food: The subsistants of the Tasmanian aborigines and Tanzanian chimpanzees compared. *Journal of Anthropological Research, 43*(3), 247–258.

McGrew, W. C., & Tutin, C. E. G. (1978). Evidence for a social custom in wild chimpanzees? *Man, 13*, 234–251.

McPherron, S. P., Alemseged, Z., Marean, C. W., Wynn, J. G., Reed, D., Geraads, D., . . . Béarat, H. A. (2010). Evidence for stone-tool-assisted consumption of animal tissues before 3.39 million years ago at Dikika, Ethiopia. *Nature, 466*(7308), 857–860.

Moore, M. W. (2000a). Lithic technology in Tasmania. *Archaeology in Oceania, 35*(2), 57–73.

Moore, M. W. (2000b). Technology of Hunter Valley microlith assemblages, New South Wales. *Australian Archaeology, 51*(1), 28–39.

Moore, M. W. (2010). "Grammars of action" and stone flaking design space. In A. Nowell & I. Davidson (Eds.), *Stone tools and the evolution of human cognition* (pp. 13–43). Boulder, CO: University Press of Colorado.

Moore, M. W., & Perston, Y. (2016). Experimental insights into the cognitive significance of early stone tools. *PLoS One, 11*(7), 1–37.

Moore, M. W., Sutikna, T., Morwood, M. J., & Brumm, A. (2009). Continuities in stone flaking technology at Liang Bua, Flores, Indonesia. *Journal of Human Evolution, 57*(5), 503–526.

Noble, W., & Davidson, I. (1991). The evolutionary emergence of modern human behaviour: Language and its archaeology. *Man, 26*(2), 223–253.

Noble, W., & Davidson, I. (1993). Tracing the emergence of modern human behavior: Methodological pitfalls, and a theoretical path. *Journal of Anthropological Archaeology, 12*(2), 121–149.

Noble, W., & Davidson, I. (1996). *Human evolution, language, and mind: A psychological and archaeological inquiry*. Cambridge, UK: Cambridge University Press.

Nowell, A., Walker, C., Cordova, C. E., Ames, C. J. H., Pokines, J. T., Stueber, D., . . . Al-Souliman, A. S. A. (2016). Middle Pleistocene subsistence in the Azraq Oasis, Jordan: Protein residue and other proxies. *Journal of Archaeological Science, 73*, 36–44.

Oakley, K. P. (1952). *Man: The tool-maker*. London: British Museum Press for the Trustees of the British Museum.

O'Connor, S., Ono, R., & Clarkson, C. (2011). Pelagic fishing at 42,000 years before the present and the maritime skills of modern humans. *Science, 334*(6059), 1117–1121.

Oswalt, W. H. (1976). *An anthropological analysis of food-getting technology*. New York: John Wiley & Sons.

Parker, S. T., & Gibson, K. R. (1979). A developmental model for the evolution of language and intelligence in early hominids. *Behavioral and Brain Sciences, 2*, 367–408.

Peresani, M., Fiore, I., Gala, M., Romandini, M., & Tagliacozzo, A. (2011). Late Neandertals and the intentional removal of feathers as evidenced from bird bone taphonomy at Fumane Cave 44 ky BP, Italy. *Proceedings of the National Academy of Sciences of the United States of America, 108*(10), 3888–3893.

Peresani, M., Vanhaeren, M., Quaggiotto, E., Queffelec, A., & d'Errico, F. (2013). An ochered fossil marine shell from the Mousterian of Fumane Cave, Italy. *PLoS One, 8*(7), 1–15.

Proffitt, T., Luncz, L. V, Falótico, T., Ottoni, E. B., De la Torre, I., & Haslam, M. (2016). Wild monkeys flake stone tools. *Nature, 539*(7627), 85–88.

Pruetz, J. D., & Bertolani, P. (2007). Savanna chimpanzees, *Pan troglodytes verus*, hunt with tools. *Current Biology, 17*(5), 412–417.

Radovčić, D., Sršen, A. O., Radovčić, J., & Frayer, D. W. (2015). Evidence for Neandertal jewelry: Modified white-tailed eagle claws at Krapina. *PLoS One, 10*(3), 1–14.

Roche, H., Delagnes, A., Brugal, J.-P., Feibel, C. S., Kibunjia, M., Mourre, V., & Texier, P.-J. (1999). Early hominid stone tool production and technical skill 2.34 Myr ago in West Turkana, Kenya. *Nature, 399*(6731), 57–60.

Rodríguez-Vidal, J., D'Errico, F., Pacheco, F. G., Blasco, R., Rosell, J., Jennings, R. P., . . . Finlayson, C. (2014). A rock engraving made by Neanderthals in Gibraltar. *Proceedings of the National Academy of Sciences of the United States of America, 111*(37), 13301–13306.

Roebroeks, W., Sier, M. J., Nielsen, T. K., Loecker, D. De, Parés, J. M., Arps, C. E. S., & Mücher, H. J. (2012). Use of red ochre by early Neandertals. *Proceedings of the National Academy of Sciences of the United States of America, 109*(6), 1889–1894.

Sahle, Y., El Zaatari, S., & White, T. D. (2017). Hominid butchers and biting crocodiles in the African Plio–Pleistocene. *Proceedings of the National Academy of Sciences, 114*(50), 13164–13169.

Salamé, P., & Baddeley, A. D. (1982). Disruption of short-term memory by unattended speech: Implications for the structure of working memory. *Journal of Verbal Learning and Verbal Behavior, 21*(2), 150–164.

Soressi, M., McPherron, S. P., Lenoir, M., Dogandžić, T., Goldberg, P., Jacobs, Z., . . . Texier, P.-J. (2013). Neandertals made the first specialized bone tools in Europe. *Proceedings of the National Academy of Sciences of the United States of America, 110*(35), 14186–14190.

Soriano, S., Villa, P., & Wadley, L. (2007). Blade technology and tool forms in the Middle Stone Age of South Africa: The Howiesons Poort and post-Howiesons Poort at Rose Cottage Cave. *Journal of Archaeological Science, 34*(5), 681–703.

Texier, P.-J., Porraz, G., Parkington, J. E., Rigaud, J.-P., Poggenpoel, C., Miller, C. E., . . . Verna, C. (2010). A Howiesons Poort tradition of engraving ostrich eggshell containers dated to 60,000 years ago at Diepkloof Rock Shelter, South Africa. *Proceedings of the National Academy of Sciences of the United States of America, 107*(14), 6180–6185.

Toth, N. P., Schick, K. D., Savage-Rumbaugh, E. S., Sevcik, R. A., & Rumbaugh, D. M. (1993). *Pan* the tool-maker: Investigations into the stone tool-making and tool-using capabilities of a bonobo (*Pan paniscus*). *Journal of Archaeological Science, 20*(1), 81–91.

van Peer, P. (1992). *The Levallois reduction strategy (monographs in world archaeology no. 13).* Madison, WI: Prehistory Press.

Wright, R. V. S. (1972). Imitative learning of a flaked-stone technology—The case of an orangutan. *Mankind, 8*(4), 296–306.

Wynn, T. (1979). The intelligence of later Acheulean hominids. *Man, 14*, 371–391.

Wynn, T. (1981). The intelligence of Oldowan hominids. *Journal of Human Evolution, 10*(7), 529–541.

Wynn, T. (1993a). Layers of thinking in tool behavior. In K. R. Gibson & T. Ingold (Eds.), *Tools, language and cognition in human evolution* (pp. 389–406). Cambridge, UK: Cambridge University Press.

Wynn, T. (1993b). Two developments in the mind of early *Homo*. *Journal of Anthropological Archaeology, 12*(3), 299–322.

Wynn, T. (1995). Handaxe enigmas. *World Archaeology, 27*(1), 10–24.

Wynn, T. (2002). Archaeology and cognitive evolution. *Behavioral and Brain Sciences, 25*(3), 389–402.

Wynn, T., & Coolidge, F. L. (2004). The expert Neandertal mind. *Journal of Human Evolution, 46*(4), 467–487.

Wynn, T., & Coolidge, F. L. (2006). The effect of enhanced working memory on language. *Journal of Human Evolution, 50*(2), 230–231.

Wynn, T., & Coolidge, F. L. (2007). A Stone-Age meeting of minds. *American Scientist, 96*(1), 44–51.

Wynn, T., & Coolidge, F. L. (2010a). Beyond symbolism and language: An introduction to Supplement 1, Working Memory. *Current Anthropology, 51*(S1), S5–S16.

Wynn, T., & Coolidge, F. L. (2010b). How Levallois reduction is similar to, and not similar to, playing chess. In A. Nowel & I. Davidson (Eds.), *Stone tools and the evolution of human cognition* (pp. 83–104). Boulder, CO: University of Colorado Press.

Wynn, T., & Coolidge, F. L. (2012). *How to think like a Neandertal.* Oxford, UK: Oxford University Press.

Wynn, T., Hernandez-Aguilar, R. A., Marchant, L. F., & McGrew, W. C. (2011). "An ape's view of the Oldowan" revisited. *Evolutionary Anthropology, 20*(5), 181–197.

Wynn, T., & McGrew, W. C. (1989). An ape's view of the Oldowan. *Man, 24*(3), 383–398.

Zilhão, J., Angelucci, D. E., Badal-García, E., d'Errico, F., Daniel, F., Dayet, L., . . . Zapata, J. (2010). Symbolic use of marine shells and mineral pigments by Iberian Neandertals. *Proceedings of the National Academy of Sciences of the United States of America, 107*(3), 1023–1028.

Zink, K. D., & Lieberman, D. E. (2016). Impact of meat and Lower Palaeolithic food processing techniques on chewing in humans. *Nature, 531*(7595), 500–503.

Zink, K. D., Lieberman, D. E., & Lucas, P. W. (2014). Food material properties and early hominin processing techniques. *Journal of Human Evolution, 77*, 155–166.

5

STICKS, STONES, AND THE ORIGINS OF SAPIENCE

Philip J. Barnard

INTRODUCTION

The emergence, use, and successive refinement of tools have been discussed within several forms of conventional evolutionary narrative. Hominins armed with sticks and stone tools had opportunities to exploit food resources more readily or otherwise outcompete species lacking them (e.g., Mithen, 1994, 1996). Some of the perhaps less functional attributes of, for example, handaxes have also been argued to have arisen from sexual selection (Kohn & Mithen, 1999). Although the latter represents just one account of the significance of handaxes, most would agree that hominin use of sticks and stones aided survival of the fittest. Cut-marks and residues left on edges show that stone tools were, without reasonable doubt, instrumental in butchering carcasses, breaking countless bones, and working wood. In doing so, they played a part in provisioning and protecting many thousands of generations of hominins. The nature and scope of mechanisms underlying changes in the architecture and capabilities of brains, minds, and social groups that came to support an increasing range of tool use are considerably less clear and the subject of much debate (see, e.g., the collection in Nowell & Davidson, 2010). It can be rather hard to establish with any clarity how a series of tiny adaptations of mental and neural capability can cumulate to yield qualitative differences in thinking patterns required to develop advanced multipart tools.

Unraveling the bases of such adaptations is a truly interdisciplinary and multilevel enterprise, particularly in a context where physical evidence is sparse and the possibilities for experiment limited because the hominin species under discussion are absent or even perhaps currently unknown to science. Ambiguities of interpretation abound. The landscape of mental labor within which those ambiguities have been vigorously debated is massive. Suffice it to note that the landscape has been populated by many kinds of intellectual artisans trained in archaeology, anthropology, biology, climatology, genetics, psychology, neuroscience, linguistics, semiotics, theology, philosophy, and more. The terminologies, methods, evidence, and inferential strategies of these different communities of practice are not always easy to harmonize. Many narratives have been and will continue to be formulated. Rather like the operation of the "spotlight" of human attention, specific accounts of how our minds evolved will tend to focus on a detailed part of the bigger picture, with high-resolution intellectual vision on some aspect of the advocates' primary discipline and some point in evolutionary time, while addressing other parts of the picture and other time periods with considerably less detail or perhaps even ignoring many issues that lie in their equivalent of intellectual peripheral vision (see Barnard, Davidson, & Byrne, 2017).

Our contributions to debates about cognitive evolution (Barnard, 2010b; Barnard et al., 2017; Barnard, Duke, Byrne, & Davidson, 2007) share, with several other recent approaches, a concern with broadening the focus of attention beyond specific properties of minds and tool use to encompass the role of embodiment and richer patterns of behavioral engagement with material culture (see e.g., Garofoli, 2015a, 2015b; Malafouris, 2013; Taylor, 2010). Our particular spotlight on the evolution of mental capabilities recruits a system-level approach to mental architecture called interacting cognitive subsystems (ICS). ICS originated as a candidate macro-theory of the mental architecture of *Homo sapiens* in the sense that it specifies the full range of mental resources that make up our minds and how those minds process sense data, think, and control actions, be they mediated by skeletal musculatures, vocal ones, or somatic effectors. It is a theory of broad scope and was developed over several decades of practical research in applied psychology to address, and account for, not only evidence associated with a wide range of experimental results from the psychological laboratory (e.g., Barnard, 1985, 1999; Su, Bowman, & Barnard, 2011) but also real-world phenomena that inherently required theories of broader scope than those confined to explaining laboratory phenomena. These have included analyses of how thought patterns and emotions become dysfunctional in various psychopathologies (e.g., Barnard, 2004; Barnard & Teasdale, 1991; Teasdale & Barnard, 1993); the human use of complex information technologies (Barnard, May, Duke, & Duce, 2000); and creative processes in the performing arts (see e.g., Barnard & deLahunta, 2018). As a macro-theory of broad scope, its prior application to cognition, meaning, affect, and creative thinking, as well as detailed analyses of the use of technologies, is important. It is at least indicative that the basic approach has a track record of broad descriptive and explanatory potential for addressing the use of tools by early hominins.

This chapter elaborates on an earlier conjecture by Barnard and colleagues (2007) that behaviors associated with tool use played a significant role, not just in survival of the fittest, but in laying the foundations of sapience. The basic idea is really quite simple. Over many millennia, tools enabled the gradual differentiation of a wide range of new behaviors and vocalizations associated with their manufacture, maintenance, and use, and that differentiation created necessary conditions for *deeper* semantic abstractions to emerge.

The example offered by Barnard and colleagues (2007) was that before the use of tools, most mammals could only break up foodstuffs by tearing them apart with paws, hands, or teeth. In this case, minds and the neural mechanisms underlying perception and the control of action really did not need to do that much to determine exactly how to act in any given set of circumstances when handling "stuff" of any particular kind. Their decision space was small. Following the emergence of tools, that decision space expanded. Foodstuffs of different types could be broken into pieces by pounding, and parcels of meat could be separated from bone or sinews or formed into different shapes and sizes by slicing. With tools, one could also cut or divide up all kinds of soft and hard materials in many different ways. The consequences of tool use could also create divisions of social roles—for example, when a band was moving around, those carrying babies and provisions might not also be able to look after the transport of weapons. With tool use, minds and brains needed to make *many more distinctions* in perception and the control of actions, be they skeletally enacted, vocally expressed, or just thought about. These extra distinctions necessitated, in one form or another, more mental and neural capacity. Such differentiation across behaviors was argued to be a prerequisite for specific concepts and deeper abstractions such as "partition" to emerge. Significant variation among related instances of actions was required before there would have been any benefit to grasping that a range of actions might have had more abstract properties in common or that might distinguish one type of action from another.

More capacity to make distinctions about what to do does not, in and of itself, allow us to develop a well-grounded theoretical account of those features of cognition and emotion that are associated with species such as great apes, hominins, and ourselves, where there is good evidence for the presence of enhanced cognitive abilities. Propositional meaning, wisdom, and advanced affective feelings such as grief or empathy, alongside creativity and multitasking, all contribute to our concept of sapience and, in doing so, require a more sophisticated theoretical picture. The next section introduces some of the basic concepts that will be called upon in elaborating why tools could have been such an important catalyst for cognitive evolution in general and our meaning systems in particular.

A THREE-PHASE TRAJECTORY FROM A BASIC MAMMALIAN MIND TO *HOMO SAPIENS*

Barnard and colleagues (2007) proposed that the mental capabilities of most mammals that lack advanced cognition can be modeled as a system composed of four subsystems in a very particular kind of architectural arrangement (see Figure 5.1, upper left quadrant). Three of these four subsystems specialized in processing sensory information. Two deal with distal perception—vision and audition—while the other, the body state subsystem, works with sensory information within and on the body envelope, including taste and smell. The fourth subsystem integrates over the products of all three sets of perceptual analyses and uses the "integrated" and more abstract multimodal pattern so generated to resolve what bodily responses and actions best suit the circumstances that hold in their physical setting at that point in time. In common with many other views proposing that emotion modules or programs originally evolved to support the adaptive control of action (see, e.g., Cosmides & Tooby, 2000), affective markers of gratification, safety, or discomfort are an integral, inherited part of the multimodal coding system and function to guide selection of what to do from within a repertoire of inherited fixed action patterns.

From this point of departure, it was argued that five further subsystems were added. The mechanism is akin to cell division in biology, and the same constraints apply across the trajectory. On each of the five steps a new daughter subsystem splits out of the parent multimodal subsystem to support narrower, more specialized functions that were, before separation, integral to multimodal processing but have now become statistically independent of the complementary functions left behind within the parent subsystem. Figure 5.1 is divided into four quadrants to illustrate our point of departure involving an architecture of four subsystems (upper left quadrant) and three of the five subsequent stages: six subsystems (upper right), eight subsystems (lower

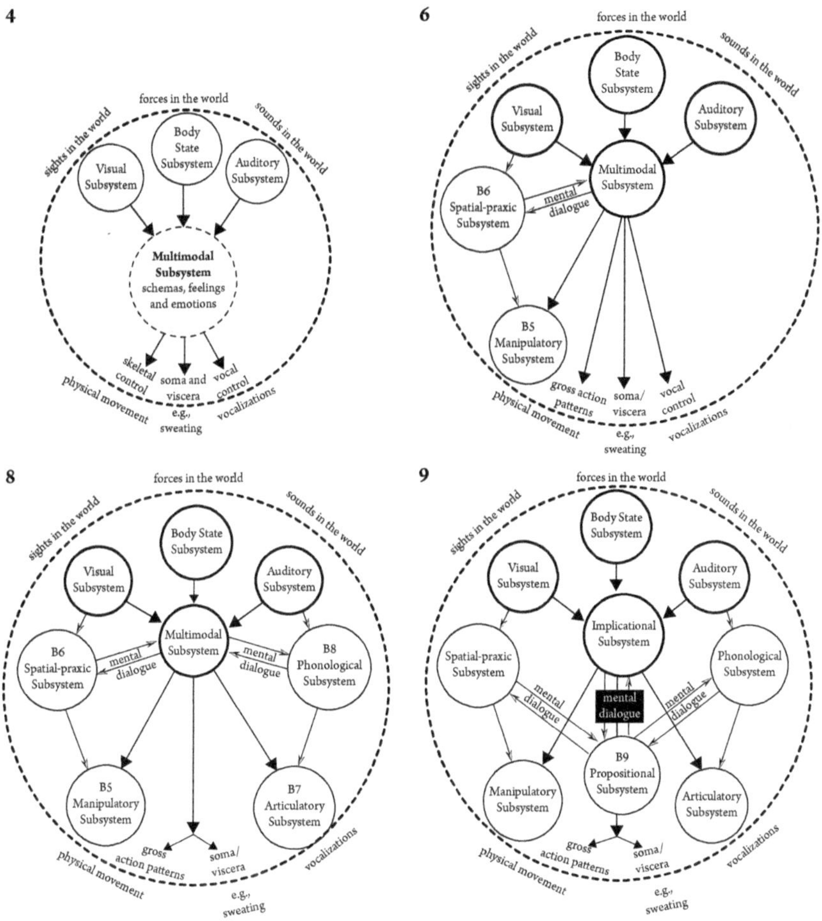

Figure 5.1. A family of mental architectures with four (top, left), six (top, right), eight (bottom, left), and nine (bottom, right) subsystems. In the nine-subsystem architecture, the bold arrow from the "Implicational" subsystem controlling gross action patterns and somatic/visceral responses passes behind the "Propositional" subsystem, not through or into it. The bidirectional arrows labeled "mental dialogue" not only index possibilities for inward and outward flows of information but also provide enabling conditions for the reordering and substitution of the elements of mental images. Image by the author.

left), and nine subsystems (lower right). The juxtaposition of four architectures, while informationally rich, enables key attributes of the overall trajectory to be summarized.

Mental mechanisms underlying tool use were argued to differentiate first in the visuospatial domain. This is a topic whose importance in archaeology was drawn to wider attention by Wynn (1989). Barnard and colleagues (2007) argued that some long-lost but early species of primate developed bimanual dexterity, and a subsystem specialized to control more intricate bimanual manipulation emerged (Figure 5.1 B5; upper right quadrant). This created conditions where the owner of a five-subsystem architecture was potentially manipulating a very wide range of materials—for example, twisting fruit, washing legumes, uncovering underground storage organs, grooming conspecifics, carrying stuff, moving stones, and so on. A collection of actions rather more complex than can be achieved with hoof or paw necessitated increased mental and neural capacity. This created conditions for a sixth subsystem to emerge that encodes and processes spatial-praxic information, a more abstract form that captures what is seen and what is enacted have in common (Figure 5.1 B6; upper right). This then put in place the foundations for "thinking" spatially, indexed by the reciprocal arrows marked "mental dialogue" between the multimodal and spatial-praxic subsystems. This internal dialogue enabled internal *mental imagery* for objects and actions on them in space—the first point where subsystems can *generate information internally, as well as substitute and reorder its elements,* rather than having the data streams running throughout the system being driven from external or bodily sources.

The second phase follows exactly the same pattern for the development of auditory vocal skills. As with manual dexterity, the next effector subsystem to be added (Figure 5.1 B7; lower left quadrant) dealt with vocal motor articulation and the kind of independent control of breath lacked by precursor hominin species. This enabled the quantitative differentiation of vocal forms in the speech of any owner of this seven-subsystem architecture. This, in turn, put in place the conditions for the eighth subsystem to emerge out of the multimodal subsystem that specialized in the abstract form capturing what heard sound and vocal articulation have in common. This introduces the phonological subsystem (Figure 5.1 B8; lower left quadrant). The owner of an eight-subsystem architecture has both visual and auditory verbal imagery as indexed by the two internal "mental dialogues" linked to the multimodal subsystem. It has a really quite sophisticated system of reorderable vocal components, and the resulting differentiation of communicative skills would be coordinated with spatial-praxic processing activity when engaging with material objects or social agents. The theory holds that the owner of an eight-subsystem architecture can "think" both with words and with spatial-praxic images.

The very confluence of talking about stuff being interacted with or acted upon means that the multimodal subsystem is now not just engaged with selecting physical actions on the basis of patterns of sensation. It is now selecting, generating, and coordinating visuospatial and reorderable verbal utterances at the same time as dealing with feedback from the sensory subsystems. As is evident from the architecture shown in the lower left quadrant of Figure 5.1, there are an awful lot of arrows going into and out of the multimodal component of an eight-subsystem architecture. Computationally, processes within the multimodal subsystem of this architecture will be differentiating information structures about properties underlying action (e.g., direction, force, location), abstract properties of agents, objects, and environs as attributes of the relationships among them, and of course, differentiating the means to talk about them. Such abstract properties and relationships are what we humans understand as the stuff of propositions, and hence, establishing similarities and differences among many actions of subdivision can come to frame the emergence of more abstract concepts like partition. As with all previous "cell divisions," this very differentiation created conditions for a ninth subsystem to emerge out of the multimodal subsystem that handled the properties that spatial-praxis and phonological material have in common. At this point, the original parent multimodal subsystem, in order to reflect the transition to sophisticated meanings, has been renamed as the "implicational" subsystem, and the final daughter subsystem is named the "propositional" subsystem (Figure 5.1 B9; lower right quadrant). Minds now have two qualitatively different forms of meaning.

Importantly, as a direct consequence of the principles governing subsystem division, the functionality of affect (or emotion) is retained in the original parent. The reconfiguration entails that the daughter propositional subsystem takes with it, and computationally assumes control of, the two internal dialogues with the spatial-praxic and phonological subsystems. This shift in "what controls what" can be examined by comparing the lower left with the lower right quadrant of Figure 5.1. Similarly, in addition to the parent retaining its direct flows from sensory subsystems, it also continues to hold responsibility for somatic and visceral patterns along with other gross action patterns such as withdrawal from pain. The presence of three internal dialogues, or *concurrent* processing loops, is what enables the owner of a nine-subsystem architecture to not only make tools and talk about what is happening but also think about how to make a better tool at the same time. This mental architecture has fully modern capabilities for thinking with and about meanings.

Clearly, there is much flesh to add to this skeletal account. Some of that flesh will be added in later sections, which will also refer to our other publications on this approach that provide even more substance. Throughout what follows, it is important to bear in mind that the specifications of mental architectures relate to the *computational* potential and constraints rather than those that apply in neural or behavioral systems. Relations among these three alternative levels of system characterization will be addressed at the end of this chapter. The approach holds that the cognitive capabilities of our last common ancestor with great apes, and, of course, great apes themselves, arise out of the potential and constraints of six-subsystem mental architectures. The sequence implies that only two reconfigurations (seven and eight) intermediate between the last common ancestor and *H. sapiens* with an architecture of nine subsystems. If we assume spatial cognition predates verbal cognition, then the order of emergence of subsystems is logically fixed. It is likely that many species existed across the era of *H. erectus,* where the available evidence appears to require architectures with no more than seven subsystems. Likewise, over the last 500,000 years and perhaps until the demise of the Neandertals/Denisovans, *many* archaic species may have existed with eight-subsystem architectures capable of making and using more enriched forms of material culture and transmitting it across generations. The juxtaposition of architectures in a single figure is needed here to illustrate key aspects of this macro-theoretic approach.

At each step, a single subsystem is added, but "system-level" properties that determine mental capabilities arise from not one but three attributes, *all of which are required to scaffold inferences when relating cognitive capabilities to behavior*—the domain specificity of the subsystems, the extent of concurrent processing, and the degree of abstraction achieved within the multimodal subsystem(s). As will be elaborated later, abstraction can progress from two levels deep in the four-subsystem architecture to a maximum of four levels deep in the nine subsystem one—and the basis of these abstractions will be referred to as second-order, third-order, and fourth-order invariants.[1] The more steps that can be traced through a processing sequence, the

[1] In mathematics, an *invariant* function, quantity, or property remains unchanged when a specified transformation is applied. Here the term is used to mean unchanging properties of patterns of information.

deeper abstractions that can be computed. For example, in a nine-subsystem architecture, "implicational" meanings can be contributed to via direct flows from sensory subsystems that yield second-order invariants and a lengthy indirect flow progressing from visual processing (first-order), through spatial-praxic (second-order), and propositional (third-order) processing, to yield fourth-order invariants for blending with the second-order invariants delivered via the more direct routings from the sensory subsystems.

Inspection of the full set of architectures in Figure 5.1 shows that the signature arrangement of the four-subsystem architecture *is retained across the full sequence*, with the inherited elements outlined in boldface. Later additions augment, rather than displace, the core mammalian mind. Of course, the inputs to the multimodal subsystem change as the architecture within which it is embedded morphs. The addition of each subsystem has very clear and inferable consequences for the nature of the system of significances or "meanings" that are computed within the multimodal subsystem as its inputs are augmented at differing levels of abstraction (summarized in Table 5.1 for later reference).

Modules, Architectures, and Methodological Implications for Cognitive Archaeology

Subsystems are, of course, a form of "domain-specific processing module" often called upon in evolutionary debates. Here they are defined by the nature of the mental code they use to intermediate between their inputs and outputs. They are quite different from the type of domain specificity invoked by, for example, Mithen (1996), who distinguishes technical, social, and natural history "intelligences" first as separate and then as interlinked by "cognitive fluidity." The family of processing architectures develops from four to nine subsystems. This is also very different from the modules often debated in evolutionary psychology (Cosmides & Tooby, 2000), where the number of modules can controversially and literally become "massive" (Sperber, 1994). The information-processing approach advocated here is most closely allied to the use of Baddeley and Hitch's (1974) model of human working memory (WM) by Coolidge and Wynn (2005). There are, however, some noteworthy differences.

ICS, as a systematically constrained macro-theory of the complete mental mechanism, is of broader scope than the WM model and can address a wide range of phenomena by refinement rather than adding new assumptions at will, warranted or otherwise, with each new subsystem addition. The WM model is, for example, relatively silent about the specification of sensory/perceptual mechanisms, bodily effectors, meaning, and emotion. They can only be addressed by adding assumptions that are not an integral part of the WM model itself. Further, each subsystem in ICS has a well-defined *internal* architecture that has three classes of "module internal" capabilities: an image, which supports phenomenological awareness of inputs to that subsystem; a memory record that models regularities of experience in that domain and can pattern complete when input is only partial; and processes that compute the invariants underlying their inputs and pass them to the next subsystem in line or

Table 5.1. Summary of Key Computational Features Underlying Cognitive Advance

No. of Subsystems	*Depth (Order) of Abstraction*	*Concurrent Processing Potential*
4 Core set	Two levels of invariants can be computed: • First-order sensory • Second-order multimodal	Concurrent processing occurs within each subsystem, but processing is sequential (or linear) across subsystems: • Sensation → Multimodal → Physical Effectors (PE)
6 Adds to the core: Manipulatory *and* Spatial-Praxic	Third-order spatial-praxic invariants can be computed and blended with other second-order ones in the multimodal subsystem.	Core sequential flow: • Sensation → Multimodal → Effector Subsystem → PE Augmented by one concurrent processing loop: • Spatial-Praxic ↔ Multimodal
8 Adds: Articulatory *and* Phonological	Third-order spatial-praxic *and* auditory-vocal invariants can be computed and blended with other second-order ones in the multimodal subsystem.	Core sequential flow: Sensation → Multimodal → Effector Subsystem → PE Augmented by two concurrent processing loops: • Spatial-Praxic ↔ Multimodal *and* • Phonological ↔ Multimodal
9 Adds: Propositional	Third-order invariants are now blended within the Propositional subsystem, while the Implicational subsystem blends second-order invariants with some fourth-order ones.	Core sequential flow: Sensation → Multimodal → Effector Subsystem → PE Augmented by three concurrent processing loops: • Spatial-Praxic ↔ Multimodal *and* • Phonological ↔ Multimodal *and* • Implicational ↔ Propositional Meanings

The bidirectional arrows index the three possible internal mental dialogues that can occur between subsystems shown in Figure 5.1.

Content adapted from Table 3.1 (pp. 58–59) in Barnard and colleagues (2017), Toward a richer theoretical scaffolding for interpreting archaeological evidence concerning cognitive evolution, *Cognitive Models in Palaeolithic Archaeology*, Oxford University Press.

to physical effectors (for detail, see Barnard, 1985; Teasdale & Barnard, 1993). It is the very specificity of the internal architecture that enables systematic task-specific refinements. It also supports analysis of mechanisms through which a daughter subsystem can subdivide the capability of the multimodal subsystem. In much the same way a molecule is composed of atomic components, a subsystem is itself composed of three types of components that can each subdivide both to yield a child with an identical internal architecture to its parent and to provide altered system-wide potential for information exchange, abstraction, and interconnectivity (Barnard et al., 2007; also see Table 5.1). Unlike many other forms of modular thinking, including the WM model, the ICS approach generates a systematic, sequential, and closely interrelated family of constrained mental architectures and allows these to "be married up" with species in the mammalian family and across the full span of hominin development. The process of linking properties of mental architectures to evidence in the archaeological record has been elaborated elsewhere for three classes of evidence relating to provisioning, tool use, and medication (Barnard et al., 2017).

Given the sparse distribution of that evidence, many would be happy to see more extensive and systematic use of well-grounded theoretical scaffolding to counteract a reliance on narratives that all too readily rely on some "single magic ingredient" and "just-so stories" to explain why a particular group of hominins are more mentally able than their precursors in the rest of the animal kingdom (see e.g., Byrne et al., 2004). Some investigators seeking improved arguments about the evolution of minds, such as those developing ideas associated with material engagement theory or radical embodied cognition, reject several fundamental assumptions associated with modular or representational approaches to cognition (e.g., Garofoli, 2015b; Malafouris, 2013). They seek to scaffold arguments about the evolution of minds in a qualitatively different way. In general, theoreticians schooled in Ockham's razor tend to shy away from complex, ambitious, and broad-reaching theoretical edifices of the kind so far introduced, and favor simpler explanations that are easier to challenge and disprove. Why consider a whole family of architectures rather than focus on interpreting a specific piece of evidence confining that interpretation to only those features that are strictly necessary? Is this not just more speculation that is unwarranted?

The counterargument is partly that the explananda[2] really are complicated and will not yield their secrets to accounts built around single-threaded lines of reasoning. More importantly, the weight of arguments can be assessed by a balanced assignment of effort between having ideas open to empirical test and relying on conjecture or thought experiments. It can be argued that the scientific community currently has access to, and can conduct field or laboratory studies on, owners of architectures with four subsystems (e.g., rats, dogs), five subsystems (e.g., monkeys and other dexterous animals), and six subsystems (definitely great apes, probably elephants, and some corvids). The vast literature on human psychology can also be used to assess the descriptive and explanatory adequacy of the nine-subsystem arrangement as a model of our own minds. Hence, empirical work is possible to test and validate whether the core assumptions of the theory can account for the mental capabilities of different members of the mammalian order on the basis of four, five, six, and nine subsystems—the larger

[2] *Explananda*: the stuff to be explained.

fraction of the complete trajectory. That leaves only the smaller fraction, those with seven and eight subsystems, to fill in as a part of a larger theoretical lattice. Where evidence is sparse, thought experiments, reconstructive work, and thinking about theoretical predictions that could be tested by new technologies applied to material in the archaeological record are crucial (see e.g., Wadley, 2013).

The next section will directly address the nature of cognitive-affective meanings across the trajectory from four to nine subsystems. The section that follows that is composed entirely of thought experiments about how differentiation of behaviors with tools may have played a very significant role in driving the transition from an eight-subsystem architecture to our own nine-subsystem architecture. Thought experiments in this context are all too often dismissed as unwarranted speculation in the behavioral and cognitive sciences, as well as in paleoanthropology. Not so in physics and cosmology, where thought experiments have played and do play a vital role in formulating complex ideas in quantum mechanics and relativity theory (see, e.g., Al-Khalili, 2011, p. 29). Thought experiments crystalize, and render thinkable, issues as a precondition for *refining theory* or indeed for *grasping the consequences of sometimes ineffable theoretical formulations* and to explore how they can be tested.

Multimodal Representation, Its Progression, and Augmentation in the Trajectory to Propositional and Implicational Meanings

The idea that the sequence of subsystem additions involved moving from two to four "levels of abstraction" was introduced earlier (Table 5.1, column 2). In this section, what is meant by levels of abstraction and how this connects to the concept of meaning will be illustrated and refined. The idea of levels of abstraction takes many well-known forms in discussions of evolution of advanced features of cognition and communication. In semiotics the distinction between iconic, indexical, and symbolic signs (Peirce, 1955) reflects three degrees of abstraction, while the literature on language evolution includes reference to the idea that animals can form proto-propositions (Hurford, 2007). It has also been argued that in certain forms of goal-based learning, animals must be using proposition-like structures rather than just simple associations (Dickinson, 2012). Of course, the Piagetian approach originally applied by Wynn (1979, 1985) in his analysis of the intelligence of hominids also invokes levels of abstraction for sensorimotor, concrete, and formal operations. Aspects of this particular approach have parallels in what follows. However, the present approach to abstraction is developed out of processing architectures and is best introduced with some concrete examples.

The individual subsystems within ICS each learn about what goes with what in the images they receive on the basis of rather simple statistical mechanisms—something like a principal components analysis with time as one of its dimensions (for detail, see Barnard et al., 2007). Sensory subsystems accomplish perceptual learning for visual patterns, patterns of bodily sensation, and auditory patterns, while the multimodal subsystem will take the products of all those first-order sensory analyses and do a second-order principal components analysis of deeper, more abstract relationships of the kind that have been extensively studied in the paradigms of classical and instrumental conditioning. This tunes animal mental mechanisms to react in similar ways to things or events with similar properties and in different ways to things or events

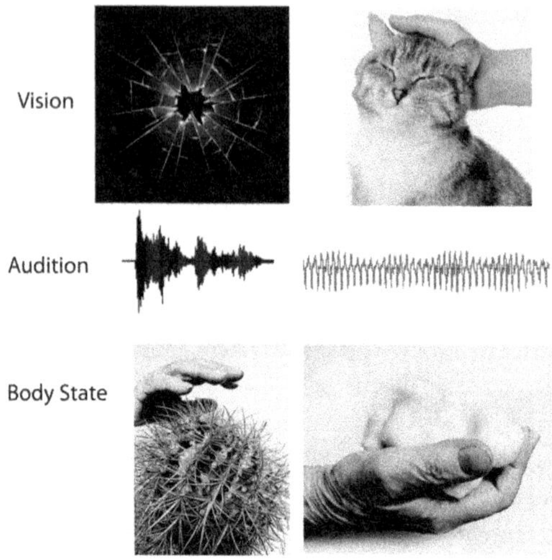

Figure 5.2. Illustrations of co-occurrences in the products of multimodal integration. Cat image by Ermolaev Alexander/Shutterstock; broken glass image by F. Vorobyov/Shutterstock; all other images by the author.

that differ from those prefigured earlier. The fundamental properties of sights and sounds tell us a great deal about objects, events, and places (e.g., Jenkins, 1985), as well as how to act in relation to them. Physics dictates that certain forms of multimodal correlations occur naturally: For example, large objects vibrate with lower frequencies than small objects, require more force to move them, and give greater resistive proprioceptive feedback.

The top row of Figure 5.2 shows the fracture pattern generated when a stone breaks a pane of glass alongside what would be "seen" when a cat is stroked, although here the motion needs to be imagined. The second row shows the sound patterns of breaking glass (left) and the cat purring in response to repetitive strokes (right). The third row shows a hand approaching the spines of a cactus and a hand holding balls of cotton wool. In everyday life, harsh sounds like glass breaking and pointed shapes will naturally co-occur, as will appearances of material that dynamically undulates with more periodic sounds. The multimodal subsystem is seen as modeling these co-occurrences, and when patterns include affective attributes linked to gratification, pain, frustration, or other forms of dissatisfaction, then these will be modeled too. There are clear contexts where such "spiky" patterns co-occur with discomfort, like pierced skin, and undulating, stroking, or fluffy patterns with comfort or pleasure. These models, as is the case with statistical semantics (see e.g., Landauer & Dumais, 1997), are high dimensional and inductive rather than simply associative. Mammals with four-subsystem architectures already build and rely on very intricate and effective "mental models" in guiding their action selection (e.g., think cats both big and small).

It was noted earlier that the core arrangement of four subsystems is inherited across the trajectory to nine, and so *experimental evidence* can be called upon to support the idea that implicational meanings in a nine-subsystem architecture blend multimodal inputs with propositional inputs. The upper part of Figure 5.3 shows two graphics

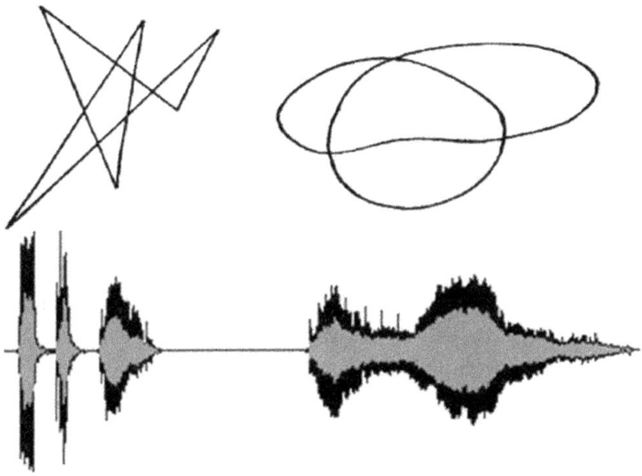

Figure 5.3. Abstract shapes of the type matched to non-words in the study by Davis (1961). Image courtesy of John Teasdale and used with his permission.

of the type originally developed by Köhler (1929). Davis (1961) showed that young children from markedly different cultures across the world reliably associate the novel non-word "Takete" with the novel jagged shape and a novel non-word like "Ulumoo" or "Maluma" with the equally novel more rounded and billowing shape. They have neither heard nor seen such material before but can make rapid and immediate use of their latent knowledge. Notice also that the auditory trace of the two spoken non-words bears generic family resemblances to those for breaking glass and purring. It has been shown that the same considerations apply to associations between tastes and the sounds of words (Gallace, Boschin, & Spence, 2011). We even describe the taste of lemons as having a degree of sharpness. Similar abstractions occur with *movements* involving purely abstract shapes (Heider & Simmel, 1944). Abstract shapes moving together in smooth and perhaps embracing patterns are interpreted as representing agents in contexts of social affiliation, while those moving with abrupt stabbing movements are interpreted as behaviors more typical of antagonist social contexts (Tavares, Lawrence, & Barnard, 2008).

These empirical studies all provide evidence for deep configural properties that enter into dendritic relationships rather than superficial properties of cleanly delineated scope. At best, we can oversimplify the essence of these patterns of invariants as involving something like "acute transition–discomfort/danger" or "smooth transition–comfort/safety." We would probably all agree that some such description is warranted. However, the detail inevitably remains rather ineffable, since the substrate for us humans, rather than a cat or rat, is a fourth-order abstraction that invokes many dimensions that are hard to capture fully in mere words or sentences.

A basic mammal like a cat is better thought of as abstracting and using such patterns but only implicitly. Meaning is nonetheless latent in their multimodal syntheses. In between the four- and nine-subsystem architectures, the proposed evolutionary trajectory includes three (6, 7, and 8) that have third-order abstraction as the deepest level at which invariant patterns within the multimodal subsystem can be extracted (Table 5.1). We can also at least consider hard evidence from great apes and other species as a checkpoint before considering architectures for those missing hominins hypothesized to have possessed seven- and eight-subsystem architectures via thought experiment. According to this form of theoretical logic, these would also have been restricted to the abstraction of third-order invariants. These would be more sophisticated than those second-order components available to mammals with just four subsystems, but the meanings would nonetheless remain latent rather than explicit. In theory, the sequence of six-, seven-, and eight-subsystem architectures should all have had forms of semantic distinctions implicit in the way their multimodal subsystems model the *particular* patterns of cross-modal inputs that subsystem receives. Of course, the literature on chimpanzee cognition has already considered possibilities for how non-human primates encode "categories" (e.g., Povinelli, 2000; Tomasello, Call, & Hare, 2003).

Barnard and colleagues (2007) briefly explored the consequences of the hypothesized dialogue between the spatial-praxic and multimodal subsystems. In this case, the spatial-praxic subsystem is already computing second-order invariants that were, pre-separation, part of the original core multimodal subsystem. The separation means that what is sent from the spatial-praxic subsystem to the multimodal subsystem is now a second-order abstraction, and the dialogue between the two means that the multimodal one can "see" and model deeper, third-order invariants alongside the second-order ones arriving direct from the sensory subsystems. It also means that affect can attach to more sophisticated organizations of both events and specifically visuospatial concepts. Barnard and colleagues point out that this particular dialogue can form the basis of a theoretical account of great apes' more advanced skills at visual memory, program-level imitation of action sequences, and some aspects of theory of mind, as well as certain instances where their emotional-linked behaviors are indicative of more intricate and abstract feelings about them.

Such inferences *follow from the computational constraints of the architecture,* rather than from chimpanzee culture per se. The key element of the argument is that the potential for a new subsystem to emerge occurs when two sources of complex and correlated feedback enter the multimodal subsystem. What is seen or heard will, for example, be correlated at a lag by proprioceptive feedback of the same skeletal or vocal action (Barnard et al., 2007). There is an emerging case that elephants have advanced spatial skills and display empathic-like feelings (e.g., Byrne, Bates, & Moss, 2009). As with chimpanzees but supported by more controversial evidence, some Asian elephants may also pass the mirror test of self-recognition (Plotnik, De Waal, & Reiss, 2006). Similarly, some corvids (i.e., crows) have also been referred to as "feathered apes" because they too show a range of advanced mental capabilities involving tool use and social cognition (Emery, 2004). Although these species are widely separated in the larger evolutionary tree, a case can be made that they could both have evolved a fifth and, plausibly, a sixth subsystem.

Elephants have flexible trunks whose tips are not only highly sensitive but also capable of quite remarkable "manipulative" dexterity when interacting with foodstuff as well as with other inedible objects and social agents. Few readers will be aware that birds have bones in their tongues, and many, including parrots and corvids, have very intricate muscle systems for lingual control. Curiously, Manegold (2009) found that a clade within the core group Corvidae (comprising Palearctic jays, crows, and ravens, as well as nutcrackers) all share a particular form of tongue morphology suggestive of having developed dexterous control analogous to the primate hand, amounting perhaps to an "opposable tongue." Arguably, then, both elephants and crows may have first developed a fifth effector subsystem followed by a sixth spatial-praxic subsystem allowing for the substitution and reordering of action elements and, hence, support some third-order abstraction and tool use. To reiterate a point made earlier, the particular forms of "semantic distinctions" implicit in the way multimodal models are formed by elephants or crows will naturally reflect the *particular* patterns of cross-modal inputs that their multimodal subsystem receives. Since we are dealing with noses and tongues (i.e., where smell or taste may directly contribute to action control), their spatial cognition may actually be qualitatively different from that of extant apes and extinct hominins. While all may be capable of substitution and reordering of action elements, the third-order invariants implicit in their systems of meaning will be species-specific. When it comes to the differentiation of tool use and, with that, the opening up of pathways to deeper abstraction, bimanual manipulative capabilities would seem to easily trump the possibilities offered by beaks and trunks.[3]

When the eighth subsystem is added in our hypothetical trajectory to full sapience, it is clearly necessary, at some point, to enter into the widespread debates that pervade the many volumes published on the evolution of language. The material that follows here will sidestep the vast bulk of those debates to enable a tight focus on the evolution of meaning systems in our particular hypothetical form of mental architecture. Here we will make heuristic use of ideas drawn from Fillmore's (1968) case grammar. This form of analysis considers the semantic roles required by a specific verb. Different verbs show different requirements for the kinds of cases that are either obligatory or optional, and these patterns reflect what goes on in the way stuff happens in our physical and mental worlds. Its use in the text that follows should in no way be taken to argue for the psychological or evolutionary reality of *this particular analysis*. In a context where we are exploring the origins of propositional meanings, it will simply be taken as a device to support thought experiments about the differentiation of meaning: what entities are involved, what properties they have, and what relationships are expressed.

Case grammar was also the form of categorization selected by Byrne and colleagues (2004) in discussing the many facets of what might be meant by culture. These authors point out that it is perfectly possible to *describe* the everyday activities in and surrounding chimpanzee culture, and many other species, using the case roles Fillmore proposed. These are all demonstrably implicit in the behavior of chimpanzees (see Table 5.2). Byrne and colleagues also note that such activities can only be *understood* in terms of the relationship between the action itself and the animate and inanimate entities that stand in different semantic roles to the action. Therefore, to the extent that a chimpanzee or later hominin does not just perform such actions but can actually *understand* their meaning at some more abstract level than a cat, it must possess

[3] Dolphins also meet the same criterion: With the capability to see with both vision and sonar, they too have the potential to grasp their more "referential" multimodal invariants. They also have intricate vocal articulation, whose properties are currently poorly understood.

Table 5.2. Case Roles Implicit in Behavior of Chimpanzees

Semantic Role	*Fillmore's Definition*	*Linguistic Example*	*Implicit Counterpart for Chimpanzees*
Agent	*Instigator* of an event	**John** opens the door	Chimpanzee, **Mike**, climbs a tree
Counter-agent	Force or resistance *against which* the action is carried out	John hit **the desk**	A chimpanzee strikes a *Strychnos* fruit against **a stone** to break it open
Object	Entity that *moves* or *changes* or whose position or existence is *under consideration*	**Mary** is 7 years old	Chimpanzee, **Figan**, is now alpha male
Result	Entity that comes *into existence* as a result of the action	Mary made **a cake**	Chimpanzee makes **a fishing probe** by stripping leaves from a grass stem
Instrument	Inanimate *stimulus* or immediate *physical cause* of an event	**The key** unlocked the door	**Spherical stone**, used as a hammer by a chimpanzee to crack nuts
Dative	Animate being *affected* by the action named by the verb	I gave my sweets to **Mary**	Female chimpanzee, **Flora,** is being groomed by another chimp
Experiencer	Animate being having a given *experience* or *mental state*	**Daddy** is cross	Piloerection and *waa* barks show that chimpanzee **Frodo** is angry
Locative	*Location* or *spatial orientation* of the state or action named by the verb	Toby sits by **the fire**	A group of male chimpanzees goes to the group's **periphery** and looks for intruders

Previously published as Table 1 (p. 343) in Byrne and colleagues (2004), Understanding culture across species, *Trends in Cognitive Sciences, 8*(8), 341–346.

some way of coding the deeper semantic relationships. It is the differentiation of such a system of semantic encoding that is explored next, when hominins come both to talk about and interact with their material and social cultures.

Semantic Differentiation in an Eight-Subsystem Architecture

In summarizing and comparing the ICS model with other theories, Welshon (2010, p. S195) eloquently describes the eight-subsystem architecture "as the crucible in which semantic reference and meaning are forged." Theoretically, within this architecture, the two internal mental dialogues involving spatial-praxic and phonological subsystems intersect in the multimodal subsystem of our eight-subsystem architecture (Figure 5.1, lower left quadrant). What speaking in words and the performance of actions in space have in common are reference to entities, properties, and relationships, and these must be induced at their multimodal intersection. This is the "crucible."

When viewed this way, meaning evolves in conjunction with differentiation in material culture and social engagement. While there are high-level similarities with material engagement theory (Malafouris, 2013), the details of the accounts are qualitatively different. The thought experiments presented here assume that over several hundreds of thousand years, eight-subsystem architectures roamed the earth. As that architecture was exploited, neural, mental, and behavioral capabilities all co-evolved in many tiny steps. Those species with more differentiated behaviors and the underlying mental and neural capacities to enact those behaviors were better able to survive and prosper. Further, as Barnard and colleagues (2017) note, the earliest eight-subsystem architectures would be little more capable than a fully exploited seven-subsystem architecture. After some 500,000 years of development and aided by the cultural transmission of skills, an eight-subsystem architecture may have had a behavioral repertoire not unlike those of an early, but underexploited, nine-subsystem architecture. The features summarized in Table 5.2 mean that the cognitive capabilities of an eight-subsystem architecture have the equivalent of an intellectual "glass ceiling." They can neither readily substitute and reorder the elements of ideas in the moment nor grasp fourth-order abstractions.

Since we can never know the full range of details, we must focus our thought experiments on patterns. Our thought experiments can take the long view and simply pose a few key questions about the bigger picture of differentiation. In what areas, and how, might actions have differentiated as a function of tool use? Each action is, by definition, different from other actions, and the underlying mental mechanism must discriminate among them, although of course, some different actions may achieve the same ends. Since each action is a response to different states of information in the world and in the mind, the kinds of case roles shown in Table 5.2 can help us organize how we address the differentiation of mental states associated with actions. We can therefore pose a second set of questions: How do properties of states of agents, objects, experiences, and so on that are relevant in determining action selection differentiate alongside the expansion of actions associated with tool use?

Before these questions are actually addressed, it is important to grasp that the individual answers are not as important as the pattern they imply. For some of the topics raised, there is already detailed evidence to consider and evaluate, but were that to be done here, with traditional reference to the many publications involved, it would detract from the thrust of the thought experiments. The point of the experiments is simple. Suppose we were to gather a jury of 12 of the types of intellectual artisans listed in the introduction who engage with cognitive archaeology. When faced with the evidence of a pattern, would they agree that there is a good case that tool use itself could have been instrumental in driving the developing cognitive capabilities across the evolutionary process, under selective pressure, rather than some happy consequence of, for example, genetic error or the arrival of a single new cognitive attribute that resolved the mystery of advanced cognition? Would the same jury consider that the degree of differentiation invoked through tool use to be smaller, similar, or larger than that required to support the foraging, social interaction, mating, infant care, and so on typical of pre–tool use mammals with four-subsystem architectures?[4]

[4] Given the existence of creationists, let us argue for a majority verdict rather than a unanimous one.

DIFFERENTIATION OF ACTIONS

Stone knapping requires carefully aimed blows of some force, while slicing meat or creating a point on the end of a stick requires timing strokes with an appropriate direction and degree of force for whatever is being separated by that particular stroke. Within each of these particular classes of action, there will be considerable and necessary *variation* in the nature of the discrete actions that can be performed and, hence, many state–action pairs to consider. Since we have evidence only of certain kinds of patterns (e.g., cut-marks on bone or the stone tools themselves, sometimes complemented by the *débitage* left from making them), the spotlight of academic attention is usually focused on a subset of those actions and the physical or mental conditions that need to be in place for them to be used effectively. The academic quest is more often than not to identify "what makes the difference" between the tools created by different generations of hominins that are actually observed in the record. Understandably, what remains often unattended in their intellectual peripheral vision is the bigger, and arguably rather more important, picture of the sheer range of actions on tools, with tools on something else, or just associated with tools that many of these species must actually have been engaged with.

In his discussion of chimpanzee culture, McGrew (1992) lists some 43 patterns of habitual tool use exhibited by a dozen groups in their natural habitats. In addition to the classic observations of cracking nuts with hammer and anvil or termite fishing they include use of clubs, leaves, missiles, and much more. If we were to not just consider identifying the different patterns of tool use but also do some more detailed analysis of the exact actions involved, not that unlike that undertaken by for the processing on foodstuffs by gorillas (Byrne & Byrne, 1993), then the differentiation of action patterns in great apes that are associated with their engagement with material culture could quite possibly come close to and maybe rank, *pari passu*, with differentiation of actions required to implement their foraging behaviors and social interactions not involving tools.

Were we now to switch our thought experiment to the likely material cultures of archaic species of *Homo*, then the output would potentially cover volumes rather than pages. The differentiation of actions and mental states involved in creating the types of spears known to have existed, or the control of fire, or the possible symbolic uses of ochre, feathers, or beads is usually at the focus of research attention. However, there were most likely not just these few but many other more mundane forms of material culture in play, and the set most likely increased substantially across the period in which archaic species existed. Evidence of survival following injury implies not just care but also the likely use of material to stem blood flow and protect from infection. It is not too hard to add to the list of really rather plausible actions. Tools with food debris may need to be wiped clean or washed to stop the development of bacteria, and certainly some tools appear to have been maintained by retouching. Archaic species may well have used flails of one sort or another to keep insects away, to dispatch snakes, or brush away mess. For all we know, species with eight-subsystem architectures could also have wrapped flakes in leaves during carriage to prevent cutting themselves.

There is clearly no point in pushing the generation of examples in thought experiments too far. If tool use can potentially be seen to approach or rank, *pari passu*, with other domains of great ape expertise, then we at least have to entertain the possibility that the differentiation of actions directly linked to tools was even greater with hominins equipped with our hypothetical seven-subsystem architecture, greater still with early species that were endowed with an eight-subsystem architecture, and even greater with eight-subsystem architectures after hundreds of thousands of years of behavioral developments in material engagement.

In addition, were we also to accept that owners of eight-subsystem architectures were speaking with some form of reorderable syntax, then adding a substantial range of vocalizations associated with tool use implies not just increased discrimination learning but a combinatorial explosion that would need to be both organized and managed within available neural or mental capacities. The general drift of this line of argument would hold that many of the vocal acts would involve some way of referencing actions, namely verbs or some form of "proto-verbs."

Differentiation of Real World and Mental States Involved in Triggering Actions

Alongside the differentiation of action patterns themselves, it clearly follows that the conditions under which specific actions are invoked would also undergo correlated differentiation. Since our social and physical worlds are obviously not random, there will be regularities to exploit in a manner directly like those considered in an earlier section that used the case of "Ulumoo and Takete" to illustrate key features of second-order but latent "semantic" abstraction. One obvious challenge is that our theoretically constrained sequence of architectures holds that an eight-subsystem architecture computes third-order regularities in its multimodal subsystem. These are going to be even more ineffable and difficult to explain in words that the second-order ones considered earlier. Although the case roles shown in Table 5.2 are far from fully fit for the purpose of generating a rich description of meaning spaces, they are quite adequate enough for some further thought experimentation on the topic of semantic role differentiation. Patterns underlying that differentiation, in this theoretical approach, must be modeled within, and by, multimodal processes and their memory records.

The literature on tool use by hominins has already identified much of importance about what kinds of distinctions early tool-users must have been exploiting when selecting one action rather than another. These have included new perceptual discriminations and categorizations of angles and shapes including abstract features, in the case of handaxes, such as their symmetry. Other points in focus have included physical properties such as the hardness and fracture properties of candidate stone materials for cores and soft and hard hammers. By extension, there would be both comparable and contrasting properties required when acting on wood, such as their rigidity, flexibility, length, girth, "sappiness," and perhaps much else, dealing with which tools were good for processing wood, digging up legumes, or meat processing. Again, we have to consider that only a fraction of the likely discriminations has already received the attention of researchers.

Were we to extend the thought experiment, then there are many other categories of action where the physical properties in focus vary. In sponging, it might be absorbency; in selecting bedding, it might be softness; for abrading surfaces, it might be roughness; and so on. The pertinent perceptual properties would extend to differentiation in effector muscular control of the actions, as well as to bodily feedback involving resistance to a strike or slice, the perturbing effects on movement of slicing through materials such as hide, flesh, and sinew, or missing the target trajectory for a blow, stroke of a flake, or use of a flail. The wider patterns in spatial-praxis and any vocalizations relating to them can be construed as elaborating the patterns within Fillmore's "object" and "counter-agent" roles.

The general point about the likely extent of differentiation should be reasonably clear, and potential differentiation of the other case roles in Table 5.2 can be illustrated, but the full scope of possible thought experiments are left open for readers to pursue. For differentiation in the result role, we can reuse the example of dividing foodstuffs mentioned in the introduction to yield smaller or larger portions with different physical or nutritional properties. A tool, of course, is itself both an instrument and a result, and the success of that result may include having one of the "right kind" of shape, sharpness, length, girth, or resilience to be fit for purpose. For the case role of location, elaborations might include distance, directions, and risks involved in navigating to locations with source materials for tool manufacture that would not have needed to be encoded before tool use was widespread.

Conspecific animate *agents* already differentiated by, for example, age, gender, rank, band membership, genetic relationship, or specific social affiliations within a group may now undergo expansion on tool-using dimensions. Status distinctions might include tool maker or user, and the range of different tools over which some competence is in place; they might also include expert or novice status, and which agents to watch or talk to in order to learn. Animals from other species may now, as a consequence of tool use, become behaviorally more relevant than before as possible sources of food and additionally differentiated in terms of the difficulty of killing them, butchering them, or the quality of nutrition gained for the effort expended.

In Fillmore's particular scheme, the semantic dative and experiencer roles are restricted to animate agents. Thought experiments on dative roles might naturally include topics like observing an agent affiliated with one's own group being in receipt of a recently cut food parcel, being poked by a stick, or even being cut or bruised in the course of using a tool of some sort. If the agent affected by the action is the self, then patterning in the experiencer case role comes into focus. Some simple observations on the likely effect of tools and tool use on the affective states have already been discussed by Barnard and colleagues (2007), including the likely frustration or annoyance if a hammer blow misses its mark and a core fracture goes wrong, or positive affect that might come with using a tool that was particularly "good" for its intended functions. Much else can safely be left at this point to the reader's imagination.

Differentiation of the Coordination and Control of Mental Processing

So far, the thought experiments have dealt with action differentiation arising in the physical world. They have focused on material engagement with tools. An eight-subsystem architecture can also be employed to think about those engagements, real and imagined, using the two internal mental dialogues (spatial-praxic and verbal) that intersect in the multimodal subsystem. The control and coordination of this eight-subsystem architecture, like all the others proposed for the mammalian order, are data driven. There is no need for any kind of "central executive" or "homunculus" to plan what to do next or arbitrate conflicts about which of several candidate actions to select (see Barnard, 1999, 2010b); modules with collective actions are, like a committee of equals or the Internet, controlled by the to and fro of exchanges among the internal components of subsystems and between subsystems. Events in the mind provide their own control conditions and are modeled in exactly the same way as events in the body or world. If those events get more complicated, as would occur in an eight-subsystem architecture with expanding use of material culture, then the control and coordination of processing activity "in mind" would increase in complexity in a manner that is correlated with the expansion of the behavioral repertoire. This would clearly compound the influence of tools on the functioning of mind and brain.

From these theoretically driven thought experiments, the case to be laid before our hypothetical jury of intellectual artisans is as follows. The scale of differentiation in multimodal patterns that arise from considering the states of entities, locations, agents, and actions that surround tool use is definitely significant as it expands across generations with our hypothetical eight-subsystem architecture. It must approach, rival, or perhaps exceed the demands on cognitive capacity of other engagements with animate agents and other stuff not mediated by tools. Precursor species to an eight-subsystem architecture who made no, marginal, or less extensive use of tools were perfectly able to survive, provision, protect themselves, and propagate their species. The *additional* richly diverse space of actions and conditions underlying the selection of tool-based actions, alongside thinking and talking about them, is catalytic in the sense that it has to be supported not just in the wetware of the brain but also in "mindware."

Faced with the challenges of differentiation, neural networks are really good at inductively finding what correlated patterns have in common. They will find an economic set of patterns of connections, or dimensions, to map their "inputs" onto their "outputs." In contrast, the use of something like a case grammar formulation enables us to get some thinkable intellectual traction at the level of mindware on how those patterns are constituted, even if the real deep analysis is more ineffable. Here, our ICS specification of an eight-subsystem architecture identifies the many types of inputs and outputs that need to be considered, and supports comparison with those of earlier and later arrangements of multimodal processing (Figure 5.1). The theoretical schema holds that the kinds of semantic roles linked to actions will be third-order abstractions, and so owners of eight subsystems that act on and have extensive verbal communication about tools would have a multimodal meaning system in which their thoughts and behaviors are governed by proto-propositional organizations.

The final transition in the sequence proposed by Barnard and colleagues (2007) is when the processing of propositional meanings becomes statistically independent of all other intersections in the multimodal subsystem. This separation can also be accomplished in a series of minute steps. At the point at which separation occurs, a new dialogue is set up between the now two sorts of meaning, processed in parallel in propositional and implicational subsystems. As with what is argued to have occurred when a sixth subsystem emerges and facilitates third-order abstraction, that third-order mental dialogue now enables the processing of implicational meaning to "see over" the regularities of meaning that were only latent in the predecessor architecture. This leads to both the ability to understand and make use of fourth-order abstractions such as "partition," "cause," "belief," or "grief" in both reasoning and in the direct selection and guidance of actions in world and mind. In the domain of cognition, the owner of an architecture with nine subsystems can be wise. However, where emotions are entwined within fourth-order abstractions, an owner of a nine-subsystem architecture can also be prepared to fight not just for their survival but for the propagation of an abstract belief.

The evolutionary trajectory in Figure 5.1 originated with a four-subsystem architecture. A standard mammal with this architecture has multimodal "feelings" derived from a state of its body in a state of the world in a moment in time. Our senses and feelings about self in a state of our own world and in our own bodies are several orders of magnitude more sophisticated. They are augmented by the products of processing meaningful abstractions about the world and states of mind and the capability to reflect on them. To the extent that these arguments hang together and offer themselves up for evaluation and test, without tool use and its gradual expansion over an evolutionary timescale, *Homo sapiens* might well not have come into existence.

CONCLUSION

This chapter has added just a little more substance to the case that the last stage of cognitive evolution involved the expansion and the subdivision of a single multimodal system of meaning to yield two types of meaning. An evolutionary transition was argued to be instrumental in the massively increased potential for differentiation and expansion of material and social cultures. The entailments of this approach for interpretation of archaeological evidence, and how these theoretical ideas might be more thoroughly tested, were extensively explored in work by Barnard and colleagues (2017).

The first half of this chapter focused on and summarized ICS, a theory of the composition and configuration of mental architecture. The second half of the chapter examined issues associated with behavior and the relationship between what happens in *behavioral architectures* when hominins interact with tools and how the relationship between neural capability, mental capability, and behavioral capability all co-evolve in lockstep with each other. The engagement with material culture is not in itself a theory about minds or a theory about behavior. Rather it is a theory about *the relationship* between mental architecture and behavioral architecture and the reciprocal influences that can hold between the two.

In the wider context of current debates in the area of cognitive archaeology, there is currently much discussion about the explanatory power of embodied or situated cognition (e.g., Garofoli, 2015a, 2015b) or theories of material engagement (see, e.g., Malafouris, 2013; Taylor, 2010), in which the emphasis is on developing explicit ideas about how behaviors with things shape the mind. In moving the attentional spotlight of research onto these issues, the technicalities of understanding how mechanisms of cognition work and evolve, brought into focus by Thomas Wynn some 40 years ago, are under some threat of being moved into intellectual peripheral vision and perhaps even toward academic extinction.

It was noted at the outset of this chapter that the views presented here directly parallel much of the discussion about material engagement theory while preserving a powerful role for specifically cognitive theory. Figure 5.4 provides a means of discussing the differences among approaches at different system levels. In this figure, introduced originally to discuss how best to model the use of modern information technologies

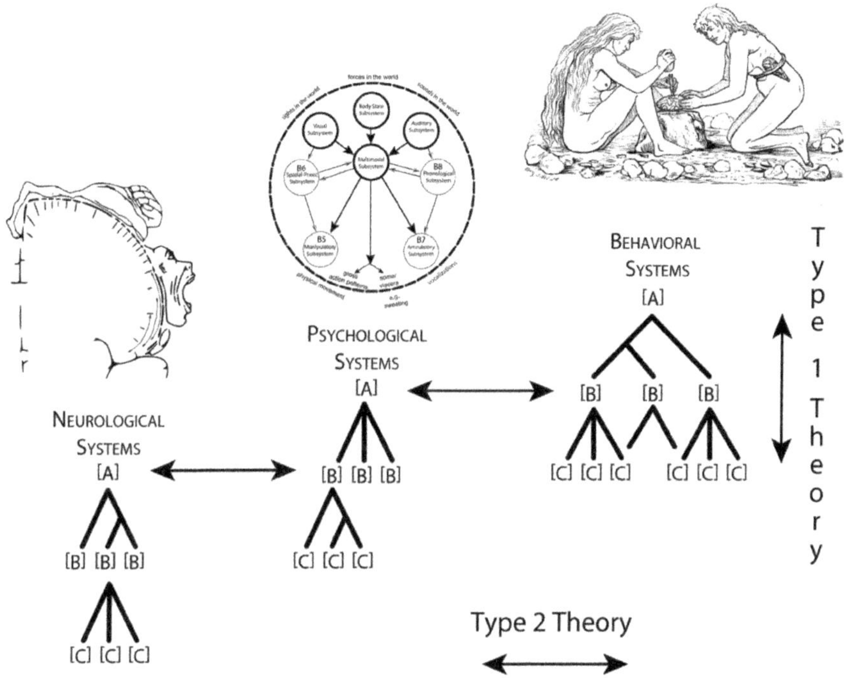

Figure 5.4. Relating theories of brains, minds, and behavior (type 1 theories) with theories of the relationships between neurological system theories and psychological theories and between psychological theories and behavioral ones (type 2 theories). Homunculus image by ralf@ark.in-berlin.de/Shutterstock. The image of two unclothed women co-operating in tool making is from an illustration generated by Victorian paleoanthropologist W. G. Smith (1894), *Man the Primeval Savage*, Edward Stanford (material in the public domain).

(Barnard et al., 2000) but also applied to clinical psychology (Barnard, 2009) and cognitive archaeology (Barnard, 2010a; Barnard et al., 2017), system theory is explored in the vertical dimension. In this vertical dimension, *A* indexes a whole system, *Bs* are the main components of that system, and *Cs* are the constituents of the components. This is referred to as a "type 1" theory and needs elaborating for the full specification of such theories (again, see Barnard et al., 2000). ICS is a type 1 theory that addresses the behavior of information in the mind.

The two other forms of type 1 theory in Figure 5.4 address the organization of actions in a behavioral system and the electrochemical behavior within a neurological system. When an agent engages with other physical entities or forces, it becomes a basic component of a behavioral system, and it is at this level that it is appropriate to develop theories of material engagement. Addressing how material engagement shapes the mind requires a *different* kind of theory that maps between levels of explanation, sometimes known as a bridging theory but referred to here as a type 2 theory. The arguments advanced in the second part of this chapter considered how the gradual differentiation in a behavioral system impacted the functioning and evolution of psychological systems. These arguments lie in the domain of a type 2 theory. Note that a type 2 theory can only be approached if both sides of the mapping are adequately specified. The ICS architecture and its detailed specification of components and their constituents directly address that requirement for a psychological system. The specification also guides and adds value to the analysis. A good example of added value explored earlier is that the theoretical specification raises issues not just in differentiating attributes for the control of overt action but also in differentiating the coordination and control of mental processing activity.

The arguments presented here have touched only briefly on mapping psychological systems to neurological systems. The three levels are included because we have argued elsewhere (Barnard, 2010a) that an analysis of psychological systems is required to mediate any discussion of relationships between brain architecture and archaeological evidence created by hominin behaviors. Nevertheless, in Figure 5.4 the "homunculus" depicted above the neurological system reminds us that neural subsystems specialized for dealing with bimanual manipulation and facial/vocal musculatures are heavily overrepresented. This relationship for hands and voice control could be argued to reflect how the emergence of the fifth and seventh subsystems not discussed here would, via a type 2 mapping theory, be directly associated with correlated increases in their underlying neural circuitry. The essence of the type 2 theoretical claims of this chapter concerned how differentiation of tool-using behaviors and the specific subsystems of ICS reciprocally influenced each other, and that without the differentiation of tool use, a nine-subsystem architecture might never have happened at all.

We have come a long way since Kenneth P. Oakley (1944) wrote on the topic of "man the tool-maker,"[5] and not only to the point of acknowledging "woman the tool-maker" (Bird, 1993). In concluding this chapter, there is a simple way to summarize the force of the arguments presented here. There is a very real sense in which we now might want to pay equivalent attention to the flip side of Oakley's coin and entertain the possibility that tools were the "man-maker," or at the very least, that their use played a very substantial catalytic role in enabling us to add the qualifier *sapiens* to the genus *Homo*.

ACKNOWLEDGMENTS

I am grateful to my collaborators on this specific project, Richard W. Byrne, David Duke, and Iain Davidson, for their support over many years and for the many insights they have contributed. The application of ideas from basic theory in the psychological sciences to our understandings of tool use by hominins owes much to four decades of work by Thomas Wynn. The substance of the arguments presented in this chapter have many attributes that reflect his enormous influence on the field.

REFERENCES

Al-Khalili, J. (2011). *Black holes, wormholes and time machines* (2nd ed.). Boca Raton, FL: CRC Press.

[5] With apologies on the marking of gender here; 1944 was a much more gender-marked era than the present.

Baddeley, A. D., & Hitch, G. J. (1974). Working memory. In G. H. Bower (Ed.), *The psychology of learning and motivation: Advances in research and theory* (Vol. 8, pp. 47–89). New York: Academic Press.

Barnard, P. J. (1985). Interacting cognitive subsystems: A psycholinguistic approach to short-term memory. In A. W. Ellis (Ed.), *Progress in the psychology of language* (Vol. 2, pp. 197–258). London: Lawrence Erlbaum Associates.

Barnard, P. J. (1999). Interacting cognitive subsystems: Modelling working memory phenomena within a multiprocessor architecture. In A. Miyake & P. Shah (Eds.), *Models of working memory: Mechanisms of active maintenance and executive control* (pp. 298–339). Cambridge, UK: Cambridge University Press.

Barnard, P. J. (2004). Bridging between basic theory and clinical practice. *Behaviour Research and Therapy, 42*(9), 977–1000.

Barnard, P. J. (2009). Depression and attention to two kinds of meaning: A cognitive perspective. *Psychoanalytic Psychotherapy, 23*(3), 248–262.

Barnard, P. J. (2010a). Current developments in inferring cognitive capabilities from the archaeological traces left by stone tools: Caught between a rock and a hard inference. In A. Nowell & I. Davidson (Eds.), *Stone tools and the evolution of human cognition* (pp. 207–226). Boulder, CO: University Press of Colorado.

Barnard, P. J. (2010b). From executive mechanisms underlying perception and action to the parallel processing of meaning. *Current Anthropology, 51*(S1), S39–S54.

Barnard, P. J., Davidson, I., & Byrne, R. W. (2017). Toward a richer theoretical scaffolding for interpreting archaeological evidence concerning cognitive evolution. In T. Wynn & F. L. Coolidge (Eds.), *Cognitive models in Palaeolithic archaeology* (pp. 45–67). Oxford, UK: Oxford University Press.

Barnard, P. J., & DeLahunta, S. (2018). Intersecting shapes in music and in dance. In D. Leech-Wilkinson & Helen M. Prior (Eds.), *Music and shape* (pp. 328–349). New York: Oxford University Press.

Barnard, P. J., Duke, D. J., Byrne, R. W., & Davidson, I. (2007). Differentiation in cognitive and emotional meanings: An evolutionary analysis. *Cognition and Emotion, 21*(6), 1155–1183.

Barnard, P. J., May, J., Duke, D., & Duce, D. (2000). Systems, interactions, and macrotheory. *ACM Transactions on Computer-Human Interaction (TOCHI), 7*(2), 222–262.

Barnard, P. J., & Teasdale, J. D. (1991). Interacting cognitive subsystems: A systemic approach to cognitive-affective interaction and change. *Cognition and Emotion, 5*(1), 1–39.

Bird, C. F. M. (1993). Woman the tool maker: Evidence for women's use and manufacture of flaked stone-tools in Australia and New Guinea. In H. du Cros & L. Smith (Eds.), *Women in archaeology: A feminist critique* (pp. 22–30). Canberra: Australian National University.

Byrne, R. W., Barnard, P. J., Davidson, I., Janik, V. M., McGrew, W. C., Miklósi, Á., & Wiessner, P. W. (2004). Understanding culture across species. *Trends in Cognitive Sciences, 8*(8), 341–346.

Byrne, R. W., Bates, L., & Moss, C. J. (2009). Elephant cognition in primate perspective. *Comparative Cognition & Behavior Reviews, 4*, 65–79.

Byrne, R. W., & Byrne, J. M. E. (1993). Complex leaf-gathering skills of mountain gorillas (*Gorilla g. beringei*): Variability and standardization. *American Journal of Primatology, 31*(4), 241–261.

Coolidge, F. L., & Wynn, T. (2005). Working memory, its executive functions, and the emergence of modern thinking. *Cambridge Archaeological Journal, 15*(1), 5–26.

Cosmides, L., & Tooby, J. (2000). Evolutionary psychology and the emotions. In M. Lewis & J. M. Haviland-Jones (Eds.), *Handbook of emotions* (2nd ed., pp. 91–115). New York: Guilford Press.

Davis, R. (1961). The fitness of names to drawings: A cross-cultural study in Tanganyika. *British Journal of Psychology, 52*(3), 259–268.

Dickinson, A. (2012). Associative learning and animal cognition. *Philosophical Transactions of the Royal Society of London. Series B, Biological Sciences, 367*(1603), 2733–2742.

Emery, N. J. (2004). Are corvids "feathered apes"? *Comparative Analysis of Minds, 181*, 1–33.

Fillmore, C. J. (1968). The case for case. In E. Bach & R. T. Harms (Eds.), *Universals in linguistics theory* (pp. 1–88). New York: Holt, Rinehart and Winston.

Gallace, A., Boschin, E., & Spence, C. (2011). On the taste of "Bouba" and "Kiki": An exploration of word–food associations in neurologically normal participants. *Cognitive Neuroscience, 2*(1), 34–46.

Garofoli, D. (2015a). A radical embodied approach to Lower Palaeolithic spear-making. *Journal of Mind and Behavior, 36*(1/2), 1–26.

Garofoli, D. (2015b). Do early body ornaments prove cognitive modernity? A critical analysis from situated cognition. *Phenomenology and the Cognitive Sciences, 14*(4), 803–825.

Heider, F., & Simmel, M. (1944). An experimental study of apparent behavior. *American Journal of Psychology, 57*(2), 243–259.

Hurford, J. R. (2007). *The origins of meaning: Language in the light of evolution.* Oxford, UK: Oxford University Press.

Jenkins, J. J. (1985). Acoustic information for objects, places, and events. In W. H. J. Warren & R. E. Shaw (Eds.), *Persistence and change: Proceedings of the first international conference on event perception* (pp. 115–138). Hillsdale, NJ: Lawrence Erlbaum Associates.

Köhler, W. (1929). *Gestalt psychology*. New York: Liveright Kovic.

Kohn, M., & Mithen, S. J. (1999). Handaxes: Products of sexual selection? *Antiquity, 73*(281), 518–526.

Landauer, T. K., & Dumais, S. T. (1997). A solution to Plato's problem: The latent semantic analysis theory of acquisition, induction, and representation of knowledge. *Psychological Review, 104*(2), 211–240.

Malafouris, L. (2013). *How things shape the mind: A theory of material engagement.* Cambridge, MA: MIT Press.

Manegold, A. (2009). Morphological characters of the tongue skeleton reveal phylogenetic relationships within the Corvidae (Oscines, Passeriformes). *Emu, 108*(4), 321–330.

McGrew, W. C. (1992). *Chimpanzee material culture: Implications for human evolution.* Cambridge, UK: Cambridge University Press.

Mithen, S. J. (1994). From domain specific to generalized intelligence: A cognitive interpretation of the Middle/Upper Palaeolithic transition. In C. Renfrew & E. B. W. Zubrow (Eds.), *The ancient mind: Elements of cognitive archaeology* (pp. 29–39). Cambridge, UK: Cambridge University Press.

Mithen, S. J. (1996). *The prehistory of mind: The cognitive origins of art, religion and science.* London: Thames & Hudson.

Nowell, A., & Davidson, I. (Eds.). (2010). *Stone tools and the evolution of human cognition.* Boulder, CO: University Press of Colorado.

Oakley, K. P. (1944). Man the tool-maker. *Proceedings of the Geologists' Association, 55*(2), 115–118.

Peirce, C. S. (1955). Logic as semiotic: The theory of signs. In J. Buchler (Ed.), *Philosophical writings of Peirce* (pp. 98–119). New York: Dover Publications.

Plotnik, J. M., De Waal, F. B. M., & Reiss, D. (2006). Self-recognition in an Asian elephant. *Proceedings of the National Academy of Sciences of the United States of America, 103*(45), 17053–17057.

Povinelli, D. J. (2000). *Folk physics for apes: The chimpanzee's theory of how the world works.* Oxford, UK: Oxford University Press.

Sperber, D. (1994). The modularity of thought and the epidemiology of representations. In L. A. Hirschfeld & S. A. Gelman (Eds.), *Mapping the mind: Domain specificity in cognition and culture* (pp. 39–67). Cambridge, UK: Cambridge University Press.

Su, L., Bowman, H., & Barnard, P. J. (2011). Glancing and then looking: On the role of body, affect, and meaning in cognitive control. *Frontiers in Psychology, 2*, 1–23.

Tavares, P., Lawrence, A. D., & Barnard, P. J. (2008). Paying attention to social meaning: An fMRI study. *Cerebral Cortex, 18*(8), 1876–1885.

Taylor, T. (2010). *The artificial ape: How technology changed the course of human evolution.* New York: Palgrave Macmillan.

Teasdale, J. D., & Barnard, P. J. (1993). *Affect, cognition and change: Re-modelling depressive thought.* Hove, UK: Lawrence Erlbaum Associates.

Tomasello, M., Call, J., & Hare, B. (2003). Chimpanzees understand psychological states—The question is which ones and to what extent. *Trends in Cognitive Sciences, 7*(4), 153–156.

Wadley, L. (2013). Recognizing complex cognition through innovative technology in Stone Age and Palaeolithic sites. *Cambridge Archaeological Journal, 23*(2), 163–183.

Welshon, R. (2010). Working memory, neuroanatomy, and archaeology. *Current Anthropology, 51*(S1), S191–S199.

Wynn, T. (1979). The intelligence of later Acheulean hominids. *Man, 14*, 371–391.

Wynn, T. (1985). Piaget, stone tools and the evolution of human intelligence. *World Archaeology, 17*(1), 32–43.

Wynn, T. (1989). *The evolution of spatial competence.* Chicago: University of Illinois Press.

6

THE ORIGIN OF CUMULATIVE CULTURE
NOT A SINGLE-TRAIT EVENT BUT MULTIFACTORIAL PROCESSES

Miriam Noël Haidle

INTRODUCTION

Over the last 20 years, cumulative culture has become one of the major elements in the study of cultural evolution, and it is often assumed to be a uniquely human trait (Boyd & Richerson, 1996; Dean, Vale, Laland, Flynn, & Kendal, 2014; Tomasello, 1999). In contrast to other animals, human groups have reached and permanently colonized nearly every land area of our home planet; they are exploring the deep sea as well as outer space. They have expanded the range of resources used by a single species to an unprecedented extent, even creating new materials. In large, communal efforts, they have cultivated plants and domesticated animals; they have built megacities and created political and religious organizations. Today they communicate in thousands of languages, hundreds of writing systems, and different codes, across continents and even in virtual worlds. Their multifaceted technology is modularized, allowing for the diverse recombination of tools in complex processes and chains that encompass composite and complementary elements, automata, and notional systems like mythical worlds, measurement units, and currency. All these things are high-end expressions of a cumulative culture that (1) has been acquired and maintained by social learning over generations and (2) is generating ongoing variations of performances[1] beyond the inventive possibilities of any single individual (cf. Tennie, Call, & Tomasello, 2009). The very definition of cumulative culture, however, poses the question of how it might be empirically identified in species other than modern humans. Indeed, groups of chimpanzees (McGrew, 2015; Whiten et al., 1999) and orangutans (van Schaik et al., 2003), New Caledonian crows (Hunt & Gray, 2003; St Clair, Klump, van der Wal, Sugasawa, & Rutz, 2016), and cetaceans (Whitehead & Rendell, 2014) demonstrate multiple traditions and basic cultural patterns (Whiten, 2016, 2017; Whiten, Caldwell, & Mesoudi, 2016), but they are often assumed to lack the cumulative aspect

[1] According to Caldwell, Atkinson, and Renner (2016, p. 191), "Cumulative cultural evolution is a process by which a series of social transmission events results in successive improvements in performance, arising due to an accumulation of modifications to the transmitted behaviours." Thus, the term *performances* as used herein refers to any behavior informed by cultural processes (e.g., reproduction, modification, abandonment, and so on).

of human culture, "where cultural traits are preserved and modified over successive generations resulting in a 'ratcheting up' of the complexity or efficiency of those traits" (Kempe, Lycett, & Mesoudi, 2014, p. 29).

Seen from a distance, there are clear differences in the performances of humans, great apes, and other animal species in their current form, and these are often attributed to cumulative culture being a specifically human trait. But when and how did cumulative culture emerge or, rather, develop? Ethological and archaeological data yield a blurred picture. Excavations at chimpanzee nut-cracking sites in the Taï Forest reveal traditions that have persisted over thousands of years (Mercader et al., 2007); in Brazil, capuchin sites contain stone hammers and anvils used to pound open hard food items that date back at least 600 years (Haslam et al., 2016). In experiments, chimpanzees demonstrate a "vital prerequisite for cumulative culture"—relinquishing ineffective behaviors, adopting more effective strategies they have witnessed conspecifics performing, and combining alternatives to realize even greater effectiveness—thereby demonstrating an ability to increase behavioral complexity through social learning (Davis, Vale, Schapiro, Lambeth, & Whiten, 2016; Vale, Davis, Lambeth, Schapiro, & Whiten, 2017). Similarly, homing pigeons have demonstrated the ability to accumulate progressive modifications across multiple generations, not through individual cognitive complexity but by learning and combining their collective intelligence (Sasaki & Biro, 2017). Thus, the acquisition and maintenance of traditions over generations and their modification over successive generations do not seem to be exclusively human phenomena. The finding that other species accumulate culture raises the question of whether human cultural accumulation involves distinct cognitive or behavioral performances.

For the most part, archaeological data capable of providing insight into the early cultural development of technology in the hominin lineage are limited to tools made of stone as the raw material. This is because the organic materials (e.g., wood, bark, leaves, blades of grass) typically employed by extant tool-using animals are highly perishable; while we can speculate that early hominins used similar materials, these have left no archaeological trace. As a result, we can examine only a fraction of the probable technological spectrum of early hominins. As the behavioral and cognitive context of the tools, their manufacture and use, and the acquisition of associated skills and know-how cannot be observed, they must instead be derived from the material record through several inferential steps (Haidle, 2014). The first evidence of the production of stone tools with cutting edges (i.e., secondary tools produced with other tools) comes from Lomekwi 3, Kenya (Lewis & Harmand, 2016); these tools have been dated to around 3.3 million years ago (Mya), though the species responsible for their manufacture currently remains unknown. The archaeological remains have been interpreted as demonstrating the use of at least two different manufacture techniques, the passive hammer and bipolar techniques.[2] "Several distinctively different modes" have been reconstructed, including the intentional knapping of flakes with successive use (Harmand et al., 2015, p. 312). Is the deliberate manufacture of stone tools an

[2] In the passive hammer technique, the core is directly hit against the anvil to produce flakes; as a proto-tool or aid, it works as a hammer itself. In the bipolar technique, the core is supported on the anvil and then struck with a hammer as a tool to produce flakes.

example for a ratchet effect in the complexity or efficiency of percussive techniques (Lombard, Högberg, & Haidle, 2018)? Did cumulative culture start then?

From 2.6 Mya onward, the signals become much more frequent and detailed. Different technological approaches were used at Gona (Semaw et al., 2003; Stout, Semaw, Rogers, & Cauche, 2010) and Omo, sites in Ethiopia (de la Torre, 2004), as well as Lokalalei 2C, in Kenya (Delagnes & Roche, 2005). Despite their differences, these are often simplistically lumped together in the category of Oldowan or mode I technology. Different stone raw materials were used not just because of variable resource availability, but because stone knappers were actively selecting raw materials for their flaking qualities (Braun, Plummer, Ferraro, Ditchfield, & Bishop, 2009; Goldman-Neuman & Hovers, 2012; Harmand, 2009; Stout, Quade, Semaw, Rogers, & Levin, 2005). The spectrum of technological knowledge broadens further with the use of different knapping techniques, the selection of raw materials for specific qualities, and geographic information about where desired raw materials could be found. Had this knowledge been socially transmitted and modified over generations? Did cumulative culture start then? And are the human capacities suggested by these processes, strategies, and technologies any different from the raw material selectivity and geographic knowledge seen, for example, in extant chimpanzees (Luncz, Proffitt, Kulik, Haslam, & Wittig, 2016; also see Boesch, Bombjaková, Boyette, & Meier, 2017; Toth & Schick, 2009) or capuchins (Visalberghi et al., 2009)?

The emergence of Acheulean bifacial technology around 1.8–1.7 Mya in Kenya (Lepre et al., 2011), Tanzania (Diez-Martín et al., 2015), and Ethiopia (Beyene et al., 2013) was attended by new requirements for understanding preparative actions in productive processes, something that is often perceived as a major transition in human cultural evolution. When teaching and the advanced communicative abilities needed to maintain cultural achievements are discussed, they are generally considered to have been necessary for, and thus to have emerged with, handaxes and cleavers (Davidson, 2016; de la Torre, 2016; Gärdenfors & Högberg, 2017; Morgan et al., 2015; Pradhan, Tennie, & van Schaik, 2012; Shipton & Nielsen, 2015; Tennie, Braun, Premo, & McPherron, 2016). But handaxes, the processes used to produce them, and probably also the concepts behind them show marked differences over time (Gallotti & Mussi, 2017). So, did cumulative culture start in the early Acheulean with an alteration in some capacity of social cognition? Or in a later Acheulean phase, as reflected by the more advanced bifacial technology? Or did it just result from change in a structural factor like demography, to occur much later with the rise of the Upper Paleolithic, when culture is presumed to have exploded (Caldwell, 2015; Powell, Shennan, & Thomas, 2009; Shennan, 2001)?

In this chapter, I review possible factors supporting the emergence of cumulative culture, coming to the general conclusion that several are necessary. I then discuss the developmental dimensions of these factors to model a developmental process with self-enhancing effects. Finally, I introduce a scenario dealing with the interplay of various basic factors and different dimensions of their development. As a consequence, I conclude that the onset of cumulative culture was not a single-trait event that took place in a relatively short period of time but, rather, the result of multifactorial and gradual processes that unfolded over millions of years.

IS CUMULATIVE CULTURE BASED ON A SINGLE TRAIT?

Several basic factors of cumulative culture have been proposed. In an early publication, Boyd and Richerson (1996) suggested observational learning or imitation as a central factor. Michael Tomasello (1999) identified joint intentionality as the key adaptation that enabled ancestral humans to develop new forms of cultural learning. In an experimental study of sequential problem-solving equated with cumulative culture, Dean and colleagues (Dean, Kendal, Schapiro, Thierry, & Laland, 2012) compared the ability to reach higher-level solutions in human children, chimpanzees, and capuchin monkeys. The authors showed that the success of human children in these experiments was significantly linked to teaching, communication and language, prosociality, and observational learning and imitation. While recognizing the importance of high-fidelity copying of cognitively opaque behavior, Caldwell (2015, p. 152) questioned the importance of imitation, which demands "crossmodal mapping . . . between observation of another's actions and one's own performance of the same action." As potentially crucial factors for developing cumulative culture, Dean and colleagues (2014) listed differences in (1) cognitive abilities like innovation, conservatism, imitation, adaptive filtering, teaching, complex communication, and prosociality; (2) social learning strategies like conformity and selective copying; (3) social structures such as monopolization and scrounging; and (4) demographic factors. Querbes and colleagues (Querbes, Vaesen, & Houkes, 2014), however, have used the definition of complexity to question whether demography is a factor in cultural change. Defining complexity as the density of interaction between the elements of a trait (Simon, 1962), they noted that "large populations tend to lose their advantage in sustaining cumulative cultural change" (Querbes et al., 2014, p. 7; for a general critique of population size to explain change in cultural complexity, also see Vaesen, Collard, Cosgrove, & Roebroeks, 2016).

Assessing the basic preconditions and sufficient factor(s) for generating cumulative culture is not an easy task. Several of the aspects just listed are themselves multifactorial (and cultural) performances that are not independent from each other, as for example, different forms of teaching and increasingly advanced forms of communication (Gärdenfors & Högberg, 2017). As a consequence, there are several problems with identifying necessary and sufficient traits. The assessment is founded on differences in performances of living humans and other animals. To depict the evolution of cumulative culture, the evolution of the associated performances must be tracked, along with their interdependencies. Additionally, the story of how cumulative culture evolved requires accepting the notion that a fully developed cumulative culture occurred at some point in early prehistory. Yet, the expression of the cumulative aspect of culture in the material record is a matter of some debate (as has been discussed), and many of the different associated performances can hardly be detected in the archaeological record. Assumptions may lead to assumptions may lead to assumptions.

Another perspective may help trace the relevant factors: Let's go back to the definition of cumulative culture given by Dean and colleagues (2014, p. 287) as the "modification, over multiple transmission episodes, of cultural traits (behavioral patterns transmitted through social learning) resulting in an increase in the complexity or efficiency of those traits." What does this mean? A group of individuals displays a certain performance; this is part of the environmental sphere of the individuals and does not involve any social interaction (→ environmental sphere). Individuals learn this performance in social context (→ individual and social sphere). Other individuals learn from them and so on, until one individual modifies the behavior (→ individual sphere). Besides the invention of something new from existing cultural patterns—be it a solution, a problem, or a link between the two—the transition from an individual invention to an innovation on the level of the social group is a crucial phase in the cumulative process (cf. Haidle, Garofoli, Scheiffele, & Stolarczyk, 2017; Renfrew, 1978): An old behavioral practice must be abandoned, against individual habits and social conformity, and replaced by a new one (→ individual and social sphere). Finally, if the new performance is widely accepted in the group, it becomes part of the general learning environment (→ environmental sphere). Succeeding generations of learners no longer recognize it as an innovation or a cumulative element but, rather, as a standard performance. Critical aspects in the process of cumulative culture can thus be identified: (1) copying a performance (by whatever means) over several generations by social learning (tradition); (2) modifying the behavior on the individual level (invention); and (3) adopting the new problem-solution against individual habits and group conformity (innovation), at least partially contradicting (1). In summary, the bases of cumulative culture cannot be sought in changes of only the social aspects but must also be sought in the individual and environmental ones.

In the different social, individual, and environmental spheres of development of cumulative culture, the totality of performances plays a major role. The non-genetic transmission of performances (e.g., playing soccer, forms of greeting, opening nuts with a hammerstone, but also forms of learning and transmitting knowledge and skills via different forms of teaching) from one individual to another is mainly influenced by the social sphere. Part of this sphere is the group's social structure, the social tolerance toward different kinds of members (e.g., young and old, male and female, kin or not), conforming or deviant behavior, the frequency and form of social interactions with different kinds of members, the capacity for joint intentionality, communicative abilities, and prosociality, to name only a few. Various performances in the social sphere support or hamper the transmission of well-established performances, as well as new or modified ones. In developing cumulative culture, the accession and intensification of interpersonal relationships (e.g., joint intentionality) as well as supportive, directed transmission (e.g., different forms of teaching) have become increasingly important to passing down complex information and new elements of performances against group conformity.

While the social transmission of performances is a necessary prerequisite of cumulative culture, it is not sufficient. Development of modifications and invention of new solutions take place in the individual sphere. The invention either (1) generates a new solution to an existing problem, which may be more efficient, more secure, more comfortable, more prestigious, fancier, or cheaper and easier to access—in other words, the new solution is somehow more appealing; (2) applies an existing solution to a new problem; (3) makes a new link between a known solution and a known problem; or (4) solves a new problem with a new solution. Typically, neither problems nor solutions are ever completely new but, rather, only modified in some respect (Haidle & Bräuer, 2011). Cumulative modifications are not restricted to more complex or efficient solutions in one progressive line, but each modification building upon an earlier version may alter a different aspect of a performance. Modifications can be beneficial in some respects but disadvantageous in others. The cumulative aspect of culture is also based on individual performances transferring known elements into new contexts or combining them with new elements. High-fidelity social copying hampers modifications. Instead, individual performances of curiosity and extended play (Riede, Johannsen, Högberg, Nowell, & Lombard, 2018), along with the abilities to decouple problems and solutions and to chunk and chain together parts of performances, support innovation via transfer and combination.

Chunking and chaining, or modularization, have become very important in human cultural evolution (Wynn, Haidle, Lombard, & Coolidge, 2017; Lombard, Högberg, & Haidle, 2018). In animal tool behavior, actions are generally within the scope of one behavioral unit and directed toward satisfying a specific need (e.g., hunger, thirst, defense, stimulation, etc.). In comparison, in human tool behavior, behavioral units have increasingly diverged from basic needs. Whereas, for example, chimpanzees deal with hammerstones only when cracking nuts, humans have gradually separated actions into modules with intermediate aims (e.g., matters like the acquisition of raw materials, the manufacture and use of primary and secondary tools, tool maintenance, and different steps of processing an end product, such as a simple wooden spear). These aims are decoupled from a current need, for example, to eat. If directly linked to the perception of prey and its use in hunting the prey down within one behavioral sequence, as chimpanzees use sticks to hunt for bush babies (Pruetz et al., 2015; Pruetz & Bertolani, 2007), the making of a simple wooden spear like those found at Schöningen, Germany from around 300 Ka (Schoch, Bigga, Böhner, Richter, & Terberger, 2015; Thieme, 1997) would be cognitively highly demanding. Though a single behavioral sequence, it would nonetheless be relatively highly demanding of cognitive resources (Haidle, 2010). Chunking and chaining, however, would reduce the complexity (in this sense, the number of elements within an action unit and of their interactions) of such tool production and use. Decreasing cognitive and behavioral complexity by fragmenting lengthy thought-and-action sequences into shorter, more manageable tasks becomes progressively important in applying different technologies toward end products like composite tools (e.g., hafted spears) and symbiotic and complementary tools (e.g., bow and arrow) (Lombard & Haidle, 2012). Chunking and chaining, or modularization, have a positive effect not only on the execution of complex performances but also on their social acquisition through learning and high-fidelity reproduction of complex solutions. Besides lowering the cultural acquisition costs that can constrain cumulative cultural evolution (Mesoudi, 2011), they simplify the modification (and preservation) of segments within a process and stimulate the transfer and combination of some elements of a performance to another. However, chunking and chaining are not sufficient to generate cumulative culture but, rather, act as a catalyst that boosts the range of individual and social performances.

The environmental sphere of development of cumulative culture grows with each new performance or modification. Each performance executed in a group setting represents a direct resource that can be modified in the individual sphere and transmitted in the social sphere. Additionally, each performance can serve as a prototype for chunking, modification, transfer, combination, and transmission. With the extension of resources and models, the potential scaffolds for new performances expand. The opportunities for inventions and innovations in technology also increase by the frequency of material engagement (Malafouris, 2013): The more one handles different aspects of the environment and becomes adept at a technological application, the more one may change, adjust, improve, and enhance dealing with the environment—in this specific interaction or in others. As a consequence, one may perceive small differences in performances, as well as in their effects, and even reflect on them, enabling an active transmission of the alteration. Accumulation of cultural performances can foster cumulative culture in aspects of modification and transmission.

EVENT OR PROCESS?

All of these individual and social performances that support cumulative culture can be traced to three interdependent dimensions of developmental processes in interaction with a specific environment or resource space (Haidle et al., 2015) (Figure 6.1). In the *evolutionary-biological dimension* (Figure 6.1(A)), the general anatomical and physiological range for the physical, mental, and behavioral aspects of

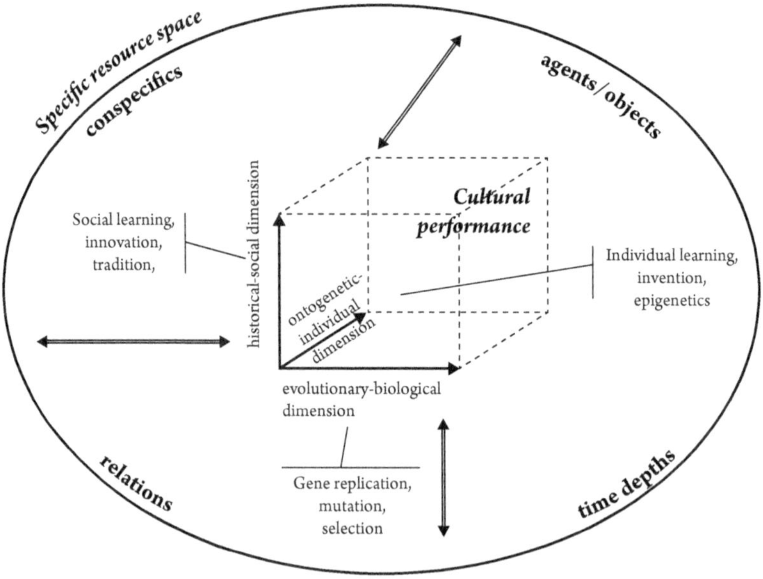

Figure 6.1. Development of cultural performance in three interdependent dimensions that interact (⇔) with each other as well as their context. The evolutionary-biological dimension (A) encompasses the physical, mental, and behavioral aspects of performance (i.e., the anatomical and physiological range determined by genetic factors). The ontogenetic-individual dimension (B) includes the experiential factors that influence physiological, mental, and behavioral development. The historical-social dimension (C) determines the availability of resources and standards that affect performance (e.g., precursors, models, habit, traditions, norms). The specific functional environment (resource space) (D) is the context in which performances develop, occur, and change; it includes the resources that influence cultural performance (e.g., social others, other species, cultural and natural artifacts, relations, and time). Adapted from Fig. 1 (p. 46) in Haidle et al. (2015), The nature of culture: An eight-grade model for the evolution and expansion of cultural capacities in hominins and other animals, *Journal of Anthropological Sciences*, and published under a Creative Commons License.

performance have been formed by genetic factors (e.g., replication, mutation, natural and sexual selection, genetic drift). Through the *ontogenetic-individual dimension* (Figure 6.1(B)), the general range of physical, mental, and behavioral aspects of performance is shaped by individual engagement and positive and negative experiences. As individuals learn to navigate and interact with their environments (e.g., avoid harmful settings, seek beneficial situations), they gain unique sets of experiences through their individual encounters with elements of the specific resource space, with conspecifics as well as other agents and objects. It is through the ontogenetic-individual dimension that individual performances impact the development of (cumulative) culture. The *historical-social dimension* (Figure 6.1(C)) marks the cultural range of a performance set in the historical-social context; it determines how performances are focused. The historical-social dimension narrows the range of variability given by the evolutionary-biological dimension but can broaden an individual's potential by making available the experiences of other individuals within the same framework of a specific time and cultural group. It is via the historical-social dimension that (cumulative) culture affects the development of individual performances. In human evolution, the historical-social dimension has expanded immensely to become important for increasingly larger portions of individual performances. Finally, the *specific functional environment* (or *resource space*) (Figure 6.1(D)) is where performances develop, take place, and change. It contains conspecifics as well as other biotic and abiotic agents and objects in specific relationships and in certain time depths. Agents and objects can be naturally occurring or artificially made, material or immaterial, and they encompass nourishment, prey, competitors, predators, parasites, symbionts, raw materials, artifacts, and performances. Conspecifics perform as parents, mates, allies, and competitors, as members of the community, kin group, peer group, and any culturally defined group (e.g., employment, interests, religious and political affiliations, etc.). The specific environment is constantly shaped by the performances of organisms and gives rise to material and social interactions.

The evolutionary-biological dimension forms the basis of the ontogenetic-individual and historical-social dimensions, but the specifications of the ontogenetic-individual and historical-social dimensions within a group contribute to shaping the selective and developmental environment of further development in the evolutionary-biological, ontogenetic-individual, and historical-social dimensions, constructing in the process a cultural niche (Laland & O'Brien, 2011). Intensified material engagement (Malafouris, 2013) generates an environment in which new performances like chunking and chaining are developed in the individual and transmitted and shaped in the historical-social dimension. Finally, changes in the evolutionary-biological dimension of cognitive elements like working memory systems become adaptive (Garofoli, 2016). The seed of joint intentionality and teaching performance, as another example, most probably was not a spontaneous genetic mutation; rather, enhanced social engagement between naïve and experienced individuals in the ontogenetic-individual and historic-social dimension generated a selective environment for evolutionary-biological adaptations (Haidle, 2017). Natural pedagogy (Csibra & Gergely, 2011) and theory of mind (Frith & Frith, 2005) are undoubtedly beneficial to the teaching performance yet are most likely outcomes rather than prerequisites of the developmental process.

Performances that foster cumulative culture via individual creativity, social interactions, and individual and social learning create at the same time a beneficial environment for the further development and maintenance of such performances. Modularization, or chunking and chaining, as one aspect of cumulative culture developed in the field of individual learning, has positive effects in all three different spheres. In the individual sphere, it fosters transfer/combination; in the social sphere, it facilitates social learning, and in the environmental sphere, it amplifies the sum of resources. The development and increasing application of aspects of cumulative culture can have self-enhancing effects. For example, through increased social engagement, the social sphere becomes self-enhancing. Cumulative culture and language support the development of each other (Dean et al., 2012; Sterelny, 2016b). Active forms of teaching facilitate the learning of language, and language can make teaching much more efficient (Laland, 2017). Language and teaching profit from prosociality and vice versa; prosociality supports imitation, and imitation can foster prosociality. The environmental sphere is self-enhancing via increased material engagement (Malafouris, 2013), creating an enriched material environment that forms a scaffold for increased material engagement. Additionally, developments in one sphere can receive positive feedback from other spheres. An example is the invention (individual sphere) of social tools (social sphere) supporting individual invention (individual sphere) (Sterelny, 2016a).

The development of cumulative culture relies on a complex network of interactive processes in the individual, social, and environmental spheres, each with several aspects. Accordingly, development cannot be reduced to a biological adaptation, as Tomasello (1999) suggested. Rather, it is based on ongoing processes in the ontogenetic-individual, historical-social, and evolutionary-biological dimensions in interaction with the specific environment, in both phylogeny and ontogeny. The performance of cumulative culture is not fixed in a specific genetic code; instead, genetic basics formed through evolutionary processes are individually developed in historical-social contexts, and they change throughout the phases of life history. Tennie and colleagues (Tennie, Walter, Gampe, Carpenter, & Tomasello, 2014), for instance, document limitations of the cultural ratchet effect in young children, who do not innovate if relatively efficient solutions are already known to them. As a lot of things must be learned in early childhood, the focus of acquiring new performances, if any, may lie in yet-unsolved problems. On the other hand, increased individual material and social engagement is not sufficient to explain the course of development in human evolution but must instead be combined with mutational enhancements of the nervous system within social and material environments that have been altered by human engagement (Garofoli, 2016). Still, there is no direction toward a final target that drives the three dimensions of the developmental model; direction is an illusion originating in analytical retrospective. When the development of human cumulative culture is considered in retrospect, however, an expansion of cumulative cultural performances can be observed.

EVER MORE COMPLEX AND EFFICIENT?

The *ratchet effect* is a widely used metaphor for cumulative cultural development (Tennie et al., 2009; Tomasello, 1999). A cultural trait is reproduced via social learning over generations, generating a platform for modifications that alter behavioral patterns, creating an advanced platform for further modifications. The ratchet metaphor implies continued progress toward complexity, with each further step impeding not only simple variation on the same level but also devolution to any earlier, less complex levels. In contrast to the progressive ratchet, Marlize Lombard (2016) developed the *mountaineering* metaphor. Within this concept, each platform reached functions as a place where a hook is fixed. The platform allows the climber to continue her ascent, traverse on the same level, proceed at different angles and directions, whether up or down, and even slide straight down to a much lower level. Regardless of the direction of movement, the climber arrives at a new platform where the next hook can be fixed. The mountaineering metaphor does not entail reaching a summit or any teleological goal but, rather, paths and potentialities that develop from earlier decisions, with new perspectives, with possibilities or constraints for proceeding in different directions being constantly offered. The mountaineering metaphor much more realistically conveys how culture develops, since efficiency and complexity (or better: more appealing conditions) are relative values that may concern only parts of a cultural trait and which may be assessed differently from different perspectives.

Whether it is envisioned as a ratchet or mountaineering, the cumulative aspect of culture includes inventor modification processes and user adoption processes. Using individuals are likely to have different backgrounds, interests, and influence on other users. Modern examples of innovative processes (and failures) reveal that what seems to be a more efficient and obviously "better" solution can be rejected from another perspective. In particular, the culture previously accumulated—in mountaineering terms, the path selected and the necessity to continue from the point reached—can hamper the adoption of innovations. This is illustrated by the failure of the improved Dvorak keyboard to supplant the traditional QWERTY layout[3] (Rogers, 1995). Cumulative culture is path dependent—modifications build on existing practices—yet the path does not necessarily lead "up" to greater efficiency and increasing complexity. The economic paradigm characterizing cumulative culture as a matter of increasing efficiency and greater complexity provides a restrictive perspective on its development and, accordingly, should be abandoned.

[3] The QWERTY keyboard is the familiar layout for Latin alphabets, designed in 1873 and named after the first six keys on the top left row of letters. The most frequently used letters were intentionally spaced to impede typing speed so the mechanical movements of the typewriter keys were less likely to jam. In comparison, the Dvorak keyboard was designed for electronic (non-mechanical) components; it differs from the QWERTY in several respects, including its reorganization of the vowel keys (left side) and most frequently used consonants (right side) to the home row, a feature that can significantly increase typing speed. Despite its advantages, most typists are disinclined to adopt the Dvorak keyboard, as the QWERTY layout works well enough and avoids the need to relearn the new design.

SCENARIO: FROM ACCUMULATION TO DONATED CULTURE

When the different factors that contribute to cumulative culture are taken into account (e.g., the three developmental dimensions, their interdependency with the social and material environment, and the path dependency of development), it is unrealistic to look for the onset of cumulative culture at a specific moment in animal or human evolution. A gradual developmental model is more appropriate. The scenario suggested next offers only a rough sketch, leaving aside the smaller ups and downs, branchings, and dead ends that generally characterize such developments. Although retrospective, the scenario follows the historical storyline from early to later stages and shows an expansion of cumulative cultural potential; doing so, it seems to be progressive. But it can also be compared to the development of velocity in cars: In a broad trend, the potential maximum speed increases, yet the average cruising speed is much lower. Real individual speed changes, up or down, and ranges from a walking pace to the highway maximum, owing to individual needs and decisions and environmental constraints such as speed limits, red lights, and traffic jams.

The *accumulation of traditions* can be set as the walking pace. Although not cumulative in the sense of modifying previously acquired cultural practices, "the addition of knowledge or behavior patterns to the behavioral repertoire of an individual or population" (Dean et al., 2014, p. 287) represents an initial branch of cumulative culture. The basis of accumulation is the social transmission of a variety of traditions. Individual competence to modify an existing practice might be still weak, as will be the ability to relinquish old habits (at the beginning of an innovation against group conformity) and replace them with modified behaviors. However, even if every single practice could have been invented by one individual alone, the whole set of practices is unlikely to have been invented by a single individual (but see later discussion for details). This early and basic branch of cumulative culture can be equated to the developmental grade of "basic cultural capacities" as proposed in the Evolution and Expansion of Cultural Capacities (EECC) model (Haidle et al., 2015). Accordingly, it would be expressed at least in chimpanzees (McGrew, 2015; Whiten et al., 1999) and orangutans (van Schaik et al., 2003), and likely in New Caledonian crows (Hunt & Gray, 2003; St Clair et al., 2016) and some cetaceans (Whitehead & Rendell, 2014) as well. The cumulative effect in this stage is weak, and its developmental speed is quite low. The learning environment is slightly enriched; social transmission may be limited to context-specific, goal-oriented learning without the precise copying of actions (see Logan, Breen, Taylor, Gray, & Hoppitt, 2016).

Cultural modifications mark a second branch in the development of cumulative culture. Socially transmitted practices are modified by individual inventions; in a social innovation process, these modifications replace older, customary practices. In experiments, homing pigeons showed that they are able to modify behavioral patterns learned in social contexts (Sasaki & Biro, 2017). Such performances may be limited to specific behaviors like finding an easy way home (as in the pigeons), but they may also be applied to a variety of behaviors. Chimpanzees were able to improve socially learned patterns (Davis et al., 2016; Vale et al., 2017), and, as discussed earlier, they are able to accumulate these altered behaviors.

As proposed in the EECC model, the cultural modification branch of cumulative culture has been strengthened by the modular cultural capacities associated with secondary tool use. Secondary tool use (e.g., used in stone tool production) expands the problem-solution distance, thus fostering the decoupling of problem and solution, as well as the development of chunking and chaining. This, in turn, supports social learning and simple, probably unintentional assistance. For example, the transition from pounding with hammerstones (common among chimpanzees and other primates) (Boesch et al., 2017; Proffitt et al., 2016) to cutting with stone flakes may have originated in an exaptation of an accidental byproduct applied as a new solution (a cutting edge) to a new problem (something to be cut open, out, or off). The cutting process became an existing problem requiring the modification of an existing solution: from pounding, for example, nuts with a hammerstone resulting in stone fragments as accidental byproducts, to the intentional production and use of flakes with cutting edges. The battering and core-reduction in the passive hammer and bipolar techniques at Lomekwi 3 around 3.3 Mya (Harmand et al., 2015; Lewis & Harmand, 2016) may represent an initial modification. In the Oldowan techno-complex, another modification was introduced with direct freehand percussion, suggested for several East African sites at 2.6 Mya, based on the technical characteristics of the stone tools found there (Delagnes & Roche, 2005; de la Torre, 2004; Semaw et al., 2003; Stout et al., 2010). Technological modifications in Developed Oldowan subsequently gave way to bifacial technology on slabs and nodules in the early Acheulean (Beyene et al., 2013; de la Torre, 2016; Diez-Martín et al., 2015; Lepre et al., 2011), later modified to bifacial technology on large flakes detached from slabs, and so on (Galotti & Mussi, 2017). Accompanying modifications in how stone tool were produced were modifications in how they were used. The resolution of the archaeological record in these early phases of human evolution is, of course, far too low to follow the process in detail and decide to what extend the individuals invented modifications on their own. Again, however, it is unlikely that the whole set of modifications was reinvented by single individuals again and again. The cumulative effect in the stage of cultural modifications is still weak, developmental speed still low. Chunking and chaining can facilitate modifications by supporting transfer and (re)combination of parts of solutions and problems. The learning environment is slightly enriched. Social learning is gaining importance and is probably extended beyond emulation by imitation and simple, possibly unintentional assistance (Gärdenfors & Högberg, 2017).

With the accumulation of traditions and the cultural modifications, the main factors of cumulative culture are in place. With *simple donated culture*, however, the cumulative effect becomes extended, and the developmental speed increases. Cumulative cultural development based on emulation, imitation, and simple, possibly unintentional assistance is limited not only by rising complexity but also by the sheer number of accumulating practices that must be acquired by naïve individuals without the purposeful help of an expert. Simple donation of cultural practices from experts to naïve individuals via motivation, intentional evaluative feedback, drawing attention, assistance, or demonstrating expands the possibilities of social learning (Gärdenfors & Högberg, 2017). Thus, the initiative for transmitting a practice is no longer limited to the naïve individual but can also be taken by the expert. Additionally, the expert can intensify the transmission process by focusing on important elements of the practice, reducing trial-and-error learning and introducing possible shortcuts. Active support from experts extends the amount and complexity of performances that can be learned. It fosters a deeper understanding of the practices, allowing transfer and recombination to become increasingly complex. The learning environment is significantly enriched. The introduction of simple donated culture has probably been a slow process linked to the development of joint intentionality (see earlier discussion). A marked expansion of the complexity and combination of different practices can be observed in composite technology (Barham, 2013) as, for example, in hafted tools, which combine stone tools with wooden hafts and some form of adhesive and/or binding material (all from different raw materials and manufactured and processed with different technologies). Although simple donated culture may occur much earlier than composite cultural capacities as described in the EECC model (Haidle et al., 2015)—for example, in intermediate evaluations of observable traits and characteristics in developed bifacial technology—composites are unlikely to have been maintained as a tradition without an active participation by experts in the transmission process. The cumulative effect of simple donated culture is stronger, developmental speed increasing. Transfer and recombination of cultural practice elements gradually becomes more complex. The learning environment is enriched and extended by new elements such as composite tools.

Complementary and notional cultural capacities (Haidle et al., 2015) dealing with non-transparent circumstances require *advanced donated culture*. In complementary tool sets (e.g., the bow and arrow; needle and thread), the primary action is executed on the controlling elements (bow; needle), but the major action is realized by the controlling elements that are effective on the target (arrow; thread) (Lombard & Haidle, 2012). The interactions between subject, goal, and different components of the tools become increasingly opaque. As an extension of social learning, formal teaching enables the learning and comprehension of non-transparent relationships by transmitting knowledge beyond specific examples. Such opaque learning situations exist in creating compound materials, constructing snares and traps, and applying subject-initiated agents (e.g., fire in the heat treatment of stone as a raw material) (Wadley, 2013). Furthermore, advanced donated culture is indispensable for developing and transmitting notional tools with socially negotiated elements (Haidle et al., 2015). Here, discourse between individuals about the meaning of notions (e.g., measurement units, values, ethics, supernatural beings) is central for the creative and maintenance processes. Yet, advanced mechanisms of donating culture can also facilitate and boost the transmission of practices that could have been achieved with one of the simpler transmission techniques (as described earlier). The cumulative effect of advanced donated culture is strong, developmental speed high. Transfer and recombination of cultural practice elements can become increasingly complex. The learning environment is significantly enriched and extended through new elements such as complementary and notional tools.

LATENT SOLUTIONS

Scenarios of the gradual development of cumulative culture, such as the one presented here, are challenged by a thought experiment called the "island test," proposed by Tomasello (1999) and developed further by Tennie and colleagues (Tennie, Premo, Braun, & McPherron, 2016, 2017). The "island test" is based on the idea that latent solutions can be distinguished from cultural behaviors transmitted by high-fidelity copying. Latent solutions are

> latently present in the individual and [are] expressed in the context of specific stimuli or when one recognizes the behavior (or its effects on the environment) expressed by others. . . . While cultural transmission allows for the accumulation of modifications through time—the so-called ratcheting effect of cumulative culture—latent solutions are more tightly bounded, or canalized, by each individual's cognitive and/or motor abilities, which are ultimately underwritten by genes. (Tennie et al., 2017, p. 653)

The concept of latent solutions poses several problems. Whether a solution is latent depends on the specific environment, including specific stimuli, the behaviors of other individuals, or their effects; these elements of the specific learning environment can be culturally modified (i.e., historically-socially) by the group over several generations. Latent behaviors are canalized by individual cognitive or motor abilities, or both, which have a genetic foundation but are individually trained, more or less, in the historical-social dimension (i.e., the cultural framework reproduced over generations in interaction with the specific learning environment). Thus, the zone of latent solutions expands with any form of social learning and any expansions of the specific learning environment that result. Orangutans, for example, "acquire virtually all their learned skills through exploration that is socially induced" (van Schaik et al., 2016, p. 1). Today, human children are raised in a completely culturally modified learning environment and are even actively trained in and by material and social engagement from birth on. Both orangutans and human children are, therefore, at no time completely naïve but rather rely on behaviors modeled by conspecifics and on environments modified by those behaviors, even if they lack a social example for a specific task. Expanding zones of latent solutions are a sign of cumulative culture.

When chimpanzee juveniles learn to crack nuts, they must acquire multiple elements of knowledge and skills (e.g., that nuts can be eaten; which nuts can be eaten; that nuts can be opened; that nuts can be opened with a hammer; what makes a good hammer; how a hammer is efficiently used; that an anvil is helpful; where to find anvils, hammers, and nuts). Individually, none of these is a big element whose learning entails high-fidelity copying, but all together they are many—and probably too many—for one individual to discover, particularly in view of the dozens of other behaviors and parts that must be learned. This leads us back to the island test. The thought experiment involves an individual (of whatever species) growing up alone on an island. Lacking all society, the island-isolated individual has never seen how to perform certain behaviors (e.g., making and using a stone tool), nor is there any evidence of the products of such behaviors on the island. Will this individual, naïve to a behavior of interest, prove able to invent it (e.g., producing and using an Oldowan flake)? If so, the behavior (here: Oldowan technology) fails the island test for cumulative culture, as cultural transmission is not required to develop the behavior. In other words, "it is consistent . . . with the expectations of a latent solution rather than a culturally transmitted technology" (Tennie et al., 2017, p. 653).

Testing the inventive possibilities of a single individual is, however, impossible for individuals raised in historical-socially developed cultural contexts. The juvenile chimpanzee learning to crack nuts is not socially isolated but instead observes conspecifics performing behaviors and can inspect the tools and products of the performance. It is not naïve, and it learns the skill over years in a social context with constant stimulation and occasional facilitation (Biro, Sousa, & Matsuzawa, 2006; Boesch, 1991). Hominins that discovered—probably bit by bit, generation by generation—how to use, manufacture, and improve flake tools with sharp cutting edges most likely relied on a comparable environment with social and material engagement from the beginning of their individual learning history. As is the case with chimpanzee juveniles and human children today, they discovered and invented within a specific zone of latent solutions. Learning to crack nuts and knap stone are not processes that can be learned by emulation (goal-oriented, low-fidelity copying) or imitation (process-oriented, high-fidelity copying) alone. There are several aspects that must be discovered, mastery requires significant time, and the risk of frustration is high. To spread within a group, the social stimulation for the behavior should be high and associated with a huge amount of play behavior (subject- or group-oriented, low-fidelity copying), at least initially. With experience, the emulative and imitative aspects gain importance. A distinction between latent solutions and high-fidelity copying cannot be drawn; there is no contradiction, but high-fidelity copying is expressed within a specific zone of latent solutions. As the island test does not consider latent solutions, it cannot be used to assess cumulative culture (for more discussion on latent solutions and the island test, see Tennie et al., 2017).

NOT A SINGLE-TRAIT EVENT BUT MULTIFACTORIAL PROCESSES

So, when and how did cumulative culture emerge? It is not a single-trait phenomenon but, rather, one composed of sets of performances. Some supporting the invention-and-modification aspects of cumulative culture operate in the individual sphere. Others supporting the innovation-and-accumulative aspects operate in the social sphere. A crucial point in cumulative culture is the interplay between the individual and social spheres in the introduction of novel practices against socially maintained traditions. The more and the longer cultural performances and their results are present in the environmental sphere, the more they can be a stimulus and affordance for sustaining and modifying practices. For cultural beings, as hominins were in prehistory and are today, performances develop in three dimensions—an evolutionary-biological, a historical-social, and an ontogenetic-individual dimension—that interact with each other and the specific environment. The development of cumulative culture is a gradual, interactive process with some self-enhancing elements. Cumulative culture is path dependent, as modifications are built on existing practices, yet the path is not necessarily one that leads "up" to greater efficiency and increasing complexity. The picture of a progressive ratchet falls short of capturing the variety in complexity, efficiency, and other factors that can reinforce but also contradict each other. Modifications must be appealing in some way in order to be accepted as an innovation; evaluation of the economic, social, practical, emotional, and environmental impacts is left to later generations. A rough sketch of how cumulative culture develops from the accumulation of traditions with cultural modifications via simple to advanced donated culture shows a multifactorial process, something that continues to develop and change.

ACKNOWLEDGMENTS

I am grateful to an anonymous reviewer who asked very detailed questions; I hope my thoughts are expressed more precisely now. The research presented here is part of the work of the center "The Role of Culture in Early Expansions of Humans," funded within the program of the Union of the Academies by the German Ministry (BMBF) and the states of Baden-Wurttemberg and Hesse. It is also supported by the German Research Foundation (DFG) in the context of project HA 2744–9, "Qualitative and quantitative differences in innovative behaviour in the Paleolithic—the example of Middle Stone Age techno-complexes of Southern Africa."

REFERENCES

Barham, L. S. (2013). *From hand to handle: The first industrial revolution.* Oxford, UK: Oxford University Press.

Beyene, Y., Katoh, S., WoldeGabriel, G., Hart, W. K., Uto, K., Sudo, M., . . . Asfaw, B. (2013). The characteristics and chronology of the earliest Acheulean at Konso, Ethiopia. *Proceedings of the National Academy of Sciences of the United States of America, 110*(5), 1584–1591.

Biro, D., Sousa, C., & Matsuzawa, T. (2006). Ontogeny and cultural propagation of tool use by wild chimpanzees at Bossou, Guinea: Case studies in nut cracking and leaf folding. In T. Matsuzawa, M. Tomonaga, & M. Tanaka (Eds.), *Cognitive development in chimpanzees* (pp. 476–508). Tokyo: Springer.

Boesch, C. (1991). Teaching among wild chimpanzees. *Animal Behaviour, 41*(3), 530–532.

Boesch, C., Bombjaková, D., Boyette, A., & Meier, A. (2017). Technical intelligence and culture: Nut cracking in humans and chimpanzees. *American Journal of Physical Anthropology, 163*(2), 339–355.

Boyd, R., & Richerson, P. J. (1996). Why culture is common, but cultural evolution is rare. *Proceedings of the British Academy, 88*, 77–93.

Braun, D. R., Plummer, T. W., Ferraro, J. V, Ditchfield, P. W., & Bishop, L. C. (2009). Raw material quality and Oldowan hominin toolstone preferences: Evidence from Kanjera South, Kenya. *Journal of Archaeological Science, 36*(7), 1605–1614.

Caldwell, C. A. (2015). Experimental studies of cumulative culture in modern humans: What are the requirements of the ratchet? In A. Mesoudi & K. Aoki (Eds.), *Learning strategies and cultural evolution during the Palaeolithic* (pp. 145–154). Tokyo: Springer.

Caldwell, C. A., Atkinson, M., & Renner, E. (2016). Experimental approaches to studying cumulative cultural evolution. *Current Directions in Psychological Science, 25*(3), 191–195.

Csibra, G., & Gergely, G. (2011). Natural pedagogy as evolutionary adaptation. *Philosophical Transactions of the Royal Society of London. Series B, Biological Sciences, 366*(1567), 1149–1157.

Davidson, I. (2016). Stone tools: Evidence of something in between culture and cumulative culture? In M. N. Haidle, N. J. Conard, & M. Bolus (Eds.), *The nature of culture: Based on an interdisciplinary symposium "The nature of culture", Tübingen, Germany* (pp. 99–120). Dordrecht, The Netherlands: Springer Science + Business Media.

Davis, S. J., Vale, G. L., Schapiro, S. J., Lambeth, S. P., & Whiten, A. (2016). Foundations of cumulative culture in apes: Improved foraging efficiency through relinquishing and combining witnessed behaviours in chimpanzees (*Pan troglodytes*). *Scientific Reports, 6*, 35953.

Dean, L. G., Kendal, R. L., Schapiro, S. J., Thierry, B., & Laland, K. N. (2012). Identification of the social and cognitive processes underlying human cumulative culture. *Science, 335*(6072), 1114–1118.

Dean, L. G., Vale, G. L., Laland, K. N., Flynn, E. G., & Kendal, R. L. (2014). Human cumulative culture: A comparative perspective. *Biological Reviews, 89*(2), 284–301.

Delagnes, A., & Roche, H. (2005). Late Pliocene hominid knapping skills: The case of Lokalalei 2C, West Turkana, Kenya. *Journal of Human Evolution, 48*(5), 435–472.

de la Torre, I. (2004). Omo revisited: Evaluating the technological skills of Pliocene hominids. *Current Anthropology, 45*(4), 439–465.

de la Torre, I. (2016). The origins of the Acheulean: Past and present perspectives on a major transition in human evolution. *Philosophical Transactions of the Royal Society of London. Series B, Biological Sciences, 371*(1698), 20150245.

Diez-Martín, F., Yustos, P. S., Uribelarrea, D., Baquedano, E., Mark, D. F., Mabulla, A. Z. P., . . . Domínguez-Rodrigo, M. (2015). The origin of the Acheulean: The 1.7 million-year-old site of FLK West, Olduvai Gorge (Tanzania). *Scientific Reports, 5*(17839), 1–9.

Frith, C. D., & Frith, U. (2005). Theory of mind. *Current Biology, 15*(17), R644–R645.

Gallotti, R., & Mussi, M. (2017). Two Acheuleans, two humankinds: From 1.5 to 0.85 Ma at Melka Kunture (Upper Awash, Ethiopian highlands). *Journal of Anthropological Sciences, 95*, 1–46.

Gärdenfors, P., & Högberg, A. (2017). The archaeology of teaching and the evolution of *Homo docens*. *Current Anthropology, 58*(2), 188–208.

Garofoli, D. (2016). Metaplasticit-*ies*: Material engagement meets mutational enhancement. In G. Etzelmüller & C. Tewes (Eds.), *Embodiment in evolution and culture* (pp. 307–336). Tübingen, Germany: Mohr Siebeck.

Goldman-Neuman, T., & Hovers, E. (2012). Raw material selectivity in late Pliocene Oldowan sites in the Makaamitalu Basin, Hadar, Ethiopia. *Journal of Human Evolution, 62*(3), 353–366.

Haidle, M. N. (2010). Working-memory capacity and the evolution of modern cognitive potential. *Current Anthropology, 51*(S1), S149–S166.

Haidle, M. N. (2014). Building a bridge—An archeologist's perspective on the evolution of causal cognition. *Frontiers in Psychology, 5*(1472), 1–15.

Haidle, M. N. (2017). Development of teaching performance: Comment on P. Gärdenfors and A. Högberg, The archaeology of teaching and the evolution of *Homo docens*. *Current Anthropology, 58*(2), 202–204.

Haidle, M. N., Bolus, M., Collard, M., Conard, N. J., Garofoli, D., Lombard, M., . . . Whiten, A. (2015). The nature of culture: An eight-grade model for the evolution and expansion of cultural capacities in hominins and other animals. *Journal of Anthropological Sciences, 93*, 43–70.

Haidle, M. N., & Bräuer, J. (2011). Special issue: Innovation and the evolution of human behavior. From brainwave to tradition—How to detect innovations in tool behaviour. *PaleoAnthropology, 2011*, 144–153.

Haidle, M. N., Garofoli, D., Scheiffele, S., & Stolarczyk, R. E. (2017). Die Entstehung einer Figurine? Material Engagement und verkörperte Kognition als Ausgangspunkt einer Entwicklungsgeschichte symbolischen Verhaltens. In G. Etzelmüller, T. Fuchs, & C. Tewes (Eds.), *Verkörperung—Eine neue interdisziplinäre Anthropologie* (pp. 251–279). Berlin: de Gruyter.

Harmand, S. (2009). Variability in raw material selectivity at the late Pliocene sites of Lokalalei, West Turkana, Kenya. In E. Hovers & D. R. Braun (Eds.), *Interdisciplinary approaches to the Oldowan* (pp. 85–97). Dordrecht, The Netherlands: Springer.

Harmand, S., Lewis, J. E., Feibel, C. S., Lepre, C. J., Prat, S., Lenoble, A., . . . Roche, H. (2015). 3.3-million-year-old stone tools from Lomekwi 3, West Turkana, Kenya. *Nature, 521*(7552), 310–315.

Haslam, M., Luncz, L. V, Staff, R. A., Bradshaw, F., Ottoni, E. B., & Falótico, T. (2016). Pre-Columbian monkey tools. *Current Biology, 26*(13), R521–R522.

Hunt, G. R., & Gray, R. D. (2003). Diversification and cumulative evolution in New Caledonian crow tool manufacture. *Proceedings of the Royal Society of London B: Biological Sciences, 270*(1517), 867–874.

Kempe, M., Lycett, S. J., & Mesoudi, A. (2014). From cultural traditions to cumulative culture: Parameterizing the differences between human and nonhuman culture. *Journal of Theoretical Biology, 359*, 29–36.

Laland, K. N. (2017). The origins of language in teaching. *Psychonomic Bulletin & Review, 24*(1), 225–231.

Laland, K. N., & O'Brien, M. J. (2011). Cultural niche construction: An introduction. *Biological Theory, 6*(3), 191–202.

Lepre, C. J., Roche, H., Kent, D. V, Harmand, S., Quinn, R. L., Brugal, J.-P., . . . Feibel, C. S. (2011). An earlier origin for the Acheulian. *Nature, 477*(7362), 82–85.

Lewis, J. E., & Harmand, S. (2016). An earlier origin for stone tool making: Implications for cognitive evolution and the transition to *Homo*. *Philosophical Transactions of the Royal Society of London. Series B, Biological Sciences, 371*(1698), 1–8.

Logan, C. J., Breen, A. J., Taylor, A. H., Gray, R. D., & Hoppitt, W. J. E. (2016). How New Caledonian crows solve novel foraging problems and what it means for cumulative culture. *Learning & Behavior, 44*(1), 18–28.

Lombard, M. (2016). Mountaineering or ratcheting? Stone Age hunting weapons as proxy for the evolution of human technological, behavioral and cognitive flexibility. In M. N. Haidle, N. J. Conard, & M. Bolus (Eds.), *The nature of culture: Based on an interdisciplinary symposium "The nature of culture", Tübingen, Germany* (pp. 135–146). Dordrecht, The Netherlands: Springer Science + Business Media.

Lombard, M., & Haidle, M. N. (2012). Thinking a bow-and-arrow set: Cognitive implications of Middle Stone Age bow and stone-tipped arrow technology. *Cambridge Archaeological Journal, 22*(2), 237–264.

Lombard, M., Högberg, A., & Haidle, M. N. (2018). Cognition: From capuchin rock pounding to Lomekwian flake production. *Cambridge Archaeological Journal*, (online first), doi:10.1017/S0959774318000550

Luncz, L. V, Proffitt, T., Kulik, L., Haslam, M., & Wittig, R. M. (2016). Distance-decay effect in stone tool transport by wild chimpanzees. *Proceedings of the Royal Society of London B: Biological Sciences, 283*(1845), 20161607.

Malafouris, L. (2013). *How things shape the mind: A theory of material engagement*. Cambridge, MA: MIT Press.

McGrew, W. C. (2015). The cultured chimpanzee: Nonsense or breakthrough? *Human Ethology Bulletin, 30*(1), 41–52.

Mercader, J., Barton, H., Gillespie, J., Harris, J., Kuhn, S. L., Tyler, R., & Boesch, C. (2007). 4,300–year-old chimpanzee sites and the origins of percussive stone technology. *Proceedings of the National Academy of Sciences of the United States of America, 104*(9), 3043–3048.

Mesoudi, A. (2011). Variable cultural acquisition costs constrain cumulative cultural evolution. *PLoS One, 6*(3), e18239.

Morgan, T. J. H., Uomini, N. T., Rendell, L. E., Chouinard-Thuly, L., Street, S. E., Lewis, H. M., . . . Laland, K. N. (2015). Experimental evidence for the co-evolution of hominin tool-making teaching and language. *Nature Communications*, *6*(6029), 1–18.

Powell, A., Shennan, S., & Thomas, M. G. (2009). Late Pleistocene demography and the appearance of modern human behavior. *Science*, *324*(5932), 1298–1301.

Pradhan, G. R., Tennie, C., & van Schaik, C. P. (2012). Social organization and the evolution of cumulative technology in apes and hominins. *Journal of Human Evolution*, *63*(1), 180–190.

Proffitt, T., Luncz, L. V, Falótico, T., Ottoni, E. B., De la Torre, I., & Haslam, M. (2016). Wild monkeys flake stone tools. *Nature*, *539*(7627), 85–88.

Pruetz, J. D., & Bertolani, P. (2007). Savanna chimpanzees, *Pan troglodytes verus*, hunt with tools. *Current Biology*, *17*(5), 412–417.

Pruetz, J. D., Bertolani, P., Ontl, K. B., Lindshield, S., Shelley, M., & Wessling, E. G. (2015). New evidence on the tool-assisted hunting exhibited by chimpanzees (*Pan troglodytes verus*) in a savannah habitat at Fongoli, Sénégal. *Royal Society Open Science*, *2*(4), 1–11.

Querbes, A., Vaesen, K., & Houkes, W. (2014). Complexity and demographic explanations of cumulative culture. *PLoS One*, *9*(7), e102543.

Renfrew, C. (1978). The anatomy of innovation. In D. R. Green, C. Haselgrove, & M. Spriggs (Eds.), *Social organization and settlement: Contributions from anthropology, archaeology and geography* (pp. 89–117). Oxford, UK: British Archaeological Reports.

Riede, F., Johannsen, N. N., Högberg, A., Nowell, A., & Lombard, M. (2018). The role of play objects and object play in human cognitive evolution and innovation. *Evolutionary Anthropology*, *26*.

Rogers, E. M. (1995). *Diffusion of innovations*. New York: Free Press.

Sasaki, T., & Biro, D. (2017). Cumulative culture can emerge from collective intelligence in animal groups. *Nature Communications*, *8*, 15049.

Schoch, W. H., Bigga, G., Böhner, U., Richter, P., & Terberger, T. (2015). New insights on the wooden weapons from the Paleolithic site of Schöningen. *Journal of Human Evolution*, *89*, 214–225.

Semaw, S., Rogers, M. J., Quade, J., Renne, P. R., Butler, R. F., Domínguez-Rodrigo, M., . . . Simpson, S. W. (2003). 2.6-million-year-old stone tools and associated bones from OGS-6 and OGS-7, Gona, Afar, Ethiopia. *Journal of Human Evolution*, *45*(2), 169–177.

Shennan, S. (2001). Demography and cultural innovation: A model and its implications for the emergence of modern human culture. *Cambridge Archaeological Journal*, *11*(1), 5–16.

Shipton, C., & Nielsen, M. (2015). Before cumulative culture. *Human Nature*, *26*(3), 331–345.

Simon, H. A. (1962). The architecture of complexity. *Proceedings of the American Philosophical Society*, *106*(6), 467–482.

St Clair, J. J. H., Klump, B. C., van der Wal, J. E. M., Sugasawa, S., & Rutz, C. (2016). Strong between-site variation in New Caledonian crows' use of hook-tool-making materials. *Biological Journal of the Linnean Society*, *118*(2), 226–232.

Sterelny, K. (2016a). Adaptable individuals and innovative lineages. *Philosophical Transactions of the Royal Society of London. Series B, Biological Sciences*, *371*(1690), 20150196.

Sterelny, K. (2016b). Cumulative cultural evolution and the origins of language. *Biological Theory*, *11*(3), 173–186.

Stout, D., Quade, J., Semaw, S., Rogers, M. J., & Levin, N. E. (2005). Raw material selectivity of the earliest stone toolmakers at Gona, Afar, Ethiopia. *Journal of Human Evolution*, *48*(4), 365–380.

Stout, D., Semaw, S., Rogers, M. J., & Cauche, D. (2010). Technological variation in the earliest Oldowan from Gona, Afar, Ethiopia. *Journal of Human Evolution, 58*(6), 474–491.

Tennie, C., Braun, D. R., Premo, L. S., & McPherron, S. P. (2016). The island test for cumulative culture in the Paleolithic. In M. N. Haidle, N. J. Conard, & M. Bolus (Eds.), *The nature of culture: Based on an interdisciplinary symposium "The nature of culture", Tübingen, Germany* (pp. 121–133). Dordrecht, The Netherlands: Springer Science + Business Media.

Tennie, C., Call, J., & Tomasello, M. (2009). Ratcheting up the ratchet: On the evolution of cumulative culture. *Philosophical Transactions of the Royal Society of London. Series B, Biological Sciences, 364*(1528), 2405–2415.

Tennie, C., Premo, L. S., Braun, D. R., & McPherron, S. P. (2017). Early stone tools and cultural transmission: Resetting the null hypothesis. *Current Anthropology, 58*(5), 652–672.

Tennie, C., Walter, V., Gampe, A., Carpenter, M., & Tomasello, M. (2014). Limitations to the cultural ratchet effect in young children. *Journal of Experimental Child Psychology, 126*, 152–160.

Thieme, H. (1997). Lower Palaeolithic hunting spears from Germany. *Nature, 385*(6619), 807–810.

Tomasello, M. (1999). The human adaptation for culture. *Annual Review of Anthropology, 28*(1), 509–529.

Toth, N. P., & Schick, K. D. (2009). The Oldowan: The tool making of early hominins and chimpanzees compared. *Annual Review of Anthropology, 38*, 289–305.

Vaesen, K., Collard, M., Cosgrove, R., & Roebroeks, W. (2016). Population size does not explain past changes in cultural complexity. *Proceedings of the National Academy of Sciences of the United States of America, 113*(16), E2241–E2247.

Vale, G. L., Davis, S. J., Lambeth, S. P., Schapiro, S. J., & Whiten, A. (2017). Acquisition of a socially learned tool use sequence in chimpanzees: Implications for cumulative culture. *Evolution and Human Behavior, 38*(5), 635–644.

van Schaik, C. P., Ancrenaz, M., Borgen, G., Galdikas, B., Knott, C. D., Singleton, I., . . . Merrill, M. (2003). Orangutan cultures and the evolution of material culture. *Science, 299*(5603), 102–105.

van Schaik, C. P., Burkart, J. M., Damerius, L., Forss, S. I. F., Koops, K., van Noordwijk, M. A., & Schuppli, C. (2016). The reluctant innovator: Orangutans and the phylogeny of creativity. *Philosophical Transactions of the Royal Society of London. Series B, Biological Sciences, 371*(1690), 20150183.

Visalberghi, E., Addessi, E., Truppa, V., Spagnoletti, N., Ottoni, E. B., Izar, P., & Fragaszy, D. M. (2009). Selection of effective stone tools by wild bearded capuchin monkeys. *Current Biology, 19*(3), 213–217.

Wadley, L. (2013). Recognizing complex cognition through innovative technology in Stone Age and Palaeolithic sites. *Cambridge Archaeological Journal, 23*(2), 163–183.

Whitehead, H., & Rendell, L. (2014). *The cultural lives of whales and dolphins*. Chicago, IL: University of Chicago Press.

Whiten, A. (2016). The evolution of hominin culture and its ancient pre-hominin foundations. In M. N. Haidle, N. J. Conard, & M. Bolus (Eds.), *The nature of culture: Based on an interdisciplinary symposium "The nature of culture", Tübingen, Germany* (pp. 27–39). Dordrecht, The Netherlands: Springer Science + Business Media.

Whiten, A. (2017). Culture extends the scope of evolutionary biology in the great apes. *Proceedings of the National Academy of Sciences of the United States of America, 114*(30), 7790–7797.

Whiten, A., Caldwell, C. A., & Mesoudi, A. (2016). Cultural diffusion in humans and other animals. *Current Opinion in Psychology, 8*, 15–21.

Whiten, A., Goodall, J. van L., McGrew, W. C., Nishida, T., Reynolds, V., Sugiyama, Y., . . . Boesch, C. (1999). Cultures in chimpanzees. *Nature, 399*(6737), 682–685.

Wynn, T., Haidle, M. N., Lombard, M., & Coolidge, F. L. (2017). The expert cognition model in human evolutionary studies. In T. Wynn & F. L. Coolidge (Eds.), *Cognitive models in Palaeolithic archaeology* (pp. 21–43). Oxford, UK: Oxford University Press.

7

HOMININ EVOLUTION AND STONE TOOL SCAVENGING AND REUSE IN THE LOWER PALEOLITHIC

Adam Brumm, Matt Pope, Mathieu Leroyer, and Kate Emery

The archaeological record is a fascinating chronicle of what may be mankind's greatest asset—the ability to recycle what others have left behind. (Ebert, 1992, p. 11)

INTRODUCTION

Thomas Wynn's research into the origin and evolution of human cognition revolves around a basic but fundamental problem: When and why did human cognition change from being ape-like to human-like in nature? In his classic paper with the primatologist William McGrew, Wynn argued that Oldowan stone technology is little different in essential form from the tool-using "cultures" of chimpanzees, our closest living evolutionary kin (Wynn & McGrew, 1989). More recently, however, the Australian archaeologist Iain Davidson, together with McGrew, has proposed a view in which Oldowan tool behavior, while ape-like in many respects, has at least one distinctly human-like characteristic: that of "niche creation" (Davidson & McGrew, 2005). *Niche creation*, or *niche construction*, refers to the processes by which organisms modify important components of their environment through their behavior, altering the conditions and selection pressures to which they and other organisms are exposed (Laland & Brown, 2006; Laland & O'Brien, 2010; Laland, Odling-Smee, & Feldman, 2001; Odling-Smee, Laland, & Feldman, 1996). Davidson and McGrew (2005) argue that chimpanzee tool behavior does not display this trait. They point out that chimpanzees are like hominins in the sense that they return to previously visited localities and perform similar activities, resulting in analogous patterns of debris accumulation and, thus, "site" formation (Haslam et al., 2009). However, the most complex forms of chimpanzee tool-use do not involve niche creation because the physical evidence for tool-using activities does not survive in the environment long enough to influence chimpanzee behavior over the long term. Most of the tools used by chimpanzees are composed of organic (i.e., plant-based) materials that deteriorate rapidly in the humid tropics. Moreover, the stone anvils and hammers used by chimpanzees for cracking nuts are always reused for the same purpose; thus far, there is no documented evidence for the suite of behaviors associated with one set of chimpanzee nut-cracking tools being recruited for use in different functional contexts.

Hominins were different, Davidson and McGrew (2005) propose. They contend that the very physical act of stone knapping would have modified the environment of opportunity for Oldowan hominins (their phrase), in the sense that the cores, flakes, debris, and other abandoned products of prior tool-making events provided a source of lithic material for later generations of tool users (also see Potts, 1988). The effect of the hominin habit of carrying and knapping stones, Davidson and McGrew (2005) argue, was the reduction of variability in the environmental distribution of stone through the creation of tool-provisioning opportunities away from fixed sources of raw lithic material. Stone knapping, according to this conception, would have created new niches for Oldowan hominins. This complex and unique behavior modified natural hominin habitats and created circumstances in which the enduring remnants of prior hominin behavior—older stone artifacts still visible in the landscape—could influence ongoing behavior.

On these grounds, Davidson and McGrew (2005) propose that from even the earliest record of stone knapping in East Africa, there is evidence for exclusive features found only in hominin tool-making cultures and that are not, so far as we are aware, shared with modern chimpanzee cultures. In fact, the model proposed by these scholars suggests that the ability of early hominins to scavenge and reuse stone artifacts discarded by other hominins may provide evidence for a crucial stage in the emergence of a uniquely human cognitive trait. As the authors note, "When hominins returned to the scene of earlier knapping events and repeated the actions of tool-making, possibly with different intentions, they set off on the path to reflective awareness and the addition of a symbolic component to their ape-like culture" (Davidson & McGrew, 2005, p. 813).

In this chapter, we examine the empirical evidence for the scavenging and reuse of stone artifacts by Lower Paleolithic hominins. We focus our attention on the existence of ancient stone tools that display indications of at least two distinct states of weathering, providing evidence for discrete phases of working separated by a time lapse. Artifacts of this kind, including handaxes, are present, though uncommon, in Lower Paleolithic assemblages. An examination of the circumstances under which differential patination occurs allows us to infer that the intensity of tool scavenging and reuse in the Lower Paleolithic may be underestimated. We consider the implications of this for our understanding of the evolution of hominin cognition.

IDENTIFYING SCAVENGING, REUSE, AND RECYCLING

There is much variation, and many inconsistencies, in how archaeologists use the terms *reuse* and *recycling* to refer to aspects of past human behavior, especially with regard to stone technology (Amick, 2007; Bamforth, 1986; Camilli & Ebert, 1992; Ebert, 1992; Schiffer, 1972, 1987; Schiffer, Downing, & McCarthy, 1981). Generally, an artifact is classified as reused when its journey through a technological system is rerouted to processes or stages of its life history through which it has already passed (Schiffer, 1972, 1987). Preston (2000) defines three forms of reuse: (1) *recycling* is when a used object is manufactured into a new item (i.e., reworked) and employed for a different purpose, such that both form and function are new; (2) *secondary use* is when an unmodified item is employed in a different activity, such that function but not form is new; and (3) *lateral cycling* is when neither form nor function changes, but the object is transferred from one user to another. The recycling and secondary use of items may or may not involve a change in the user, but lateral cycling always does (Schiffer, 1987; Schiffer et al., 1981). An important reuse mechanism is the reclamation and reuse of a previously deposited object, whether by the same or a different user (Medina, 2007)—behavior known as *scavenging*. Concerning the latter, as Schiffer (1987, p. 114) remarks, "virtually any object—regardless of mass—can be scavenged." Reuse is differentiated from *maintenance*, which is the repair or refurbishment of an item to prolong its use-life, resulting in a new form, but not necessarily a new function, for the object (Andrefsky, 2009; Preston, 2000; Schiffer, 1987).

Traces of the earlier forms of recycled artifacts will often still be recognizable after the process of modifying these implements into different forms, and hence recycling is the reuse process that is generally the most frequently identified from archaeological materials (Schiffer, 1987). Despite this, in most if not all cases, inferring the level of intent behind typological transformations in stone tools is complicated by a potentially wide range of variables that could equally account for these processes (Odell, 2001). Foremost among these variables are changes in tool form that may occur as a result of routine maintenance during tool use (Frison, 1968). It follows from this that the best evidence for the reuse of a previously used stone tool is the demonstration that it experienced more than one temporally distinct phase of reduction.

Artifact Repatination

Scholars have long recognized that certain prehistoric stone artifacts exhibit negative flake scars that can be ordered into a chronologically distinct series based on their degree of surface weathering (e.g., Barkai, Gopher, & Shimelmitz, 2006; Boucher de Perthes, 1999a, 1999b; Breuil, 1954; Clarkson, 2007; de Mortillet, 1885; de Pradenne, 1935; Henri-Martin, 1906, 1923; Hiscock & Attenbrow, 2005; Mora, de la Torre, & Martínez-Moreno, 2004; Rudner & Rudner, 1954; Smith, 1894; Sturge, 1911; Warren, 1902). With specimens such as these, some scars will appear more heavily weathered than others, and in instances where the less weathered scars and the more weathered ones overlap, the former will intrude into the latter, unambiguously showing that they were produced at a later point in time. This phenomenon is known variously as *differential patination, repatination*, or, not always accurately, *double patination* (Amick, 2007; Goodwin, 1960; Sturge, 1911). Stone artifacts displaying this feature can be unequivocally interpreted as the result of two (or more) phases of knapping separated by an intervening period of deposition and weathering—a time lapse. The formation of differentially patinated flake scars always involves a time lapse: Even though we cannot usually know the duration of this time lapse—it may be months, years, or millennia—repatinated artifacts provide clear evidence of it. Differentially patinated stone artifacts, therefore, provide the most clear-cut proxy evidence for tool scavenging and reuse.

It is usually possible to distinguish evidence for differential patination resulting from tool reuse from scars and edge-wear caused by post-depositional mechanical weathering processes, such as hydraulic tumbling or trample damage. A phenomenon similar to differential patination may also occur on stone tools owing to varying rates of chemical alteration and/or mechanical weathering or staining of artifact surfaces—such as happens, for example, to elevated faces or protruding parts of partially buried tools (e.g., Boucher de Perthes, 1999a; de Mortillet, 1885, p. 65; Evans, 1872; Goodwin, 1960; Howard, 1999; Prestwich, 1860; Sturge, 1911; Woodcock, 1981). In most cases, it is possible to distinguish these features from scavenging and reuse scars.

Repatinated Artifacts in Lower Paleolithic Contexts

A survey of the archaeological literature reveals the presence of stone artifacts with differentially weathered surfaces at a number of Lower Paleolithic sites. For instance, at the Oldowan site of Chesowanja in Kenya, dated to before 1.42 ± 0.07 million years ago (Mya), Gowlett and colleagues (Gowlett, Harris, Walton, & Wood, 1981, p. 125) identified one artifact that had been "retrimmed freshly on one surface and [that] is strong evidence for recurrent use of a site." Repatinated implements are also reported in ~1.6–1.5 Mya Developed Oldowan assemblages from Koobi Fora (Ludwig & Harris, 1998) and HWK East in the Olduvai Gorge (Leakey, 1971). In Koobi Fora, large transported cores were deposited at the site and reworked at a later stage, resulting in artifacts with differentially weathered scars (Ludwig & Harris, 1998, p. 99). Repatinated handaxes and other differentially weathered stone artifacts have also been identified in Acheulean assemblages from across Europe, such as in Britain and France (de Mortillet, 1906; Soriano, 2000; Villa, 1983, p. 185) and Belgium (Rutot, 1906). Outside Europe, differentially patinated implements, most often handaxes, are known from assemblages attributed to Acheulean industries in Africa (Aumassip, 2004; Kuman, 2001), the Near East (Bar-Yosef & Goren-Inbar, 1993; Gopher et al., 2005, p. 82; Le Tensorer, 1997; Marder, Milevski, & Matskevich, 2006, p. 240), and subcontinental India (Pappu & Akhilesh, 2006).

We have tabulated evidence for differentially patinated stone tools in Lower Paleolithic British assemblages, based on a review of the literature and our first-hand knowledge of key collections from the region (see Table 7.1). It has long been known that repatinated artifacts are present in a variety of British Acheulean and core-and-flake (i.e., Clactonian) assemblages. Over a century ago, for instance, the British antiquarian Worthington Smith described two repatinated handaxes from southeastern England (Figure 7.1), noting that "the re-pointing of [Acheulean handaxes] reminds one of the hammering of old plate armour into new forms, as commonly practised in mediaeval times" (Smith, 1894, p. 117). Notable examples of reworked handaxes were also identified at the Acheulean localities of Hoxne (Figure 7.2), Warren Hill (Figure 7.3), and Boxgrove (Figures 7.4–7.8). A Warren Hill handaxe exhibits *three* distinct states of patination (Figure 7.9), indicating an initial production and abandonment phase followed by two subsequent reworking episodes, each separated by a time lapse. A similar so-named triple-patinated handaxe was recovered from Boxgrove (Woodcock, 1981). The High Lodge assemblage yielded three additional examples of repatinated artifacts, a "flaked flake," core, and two scrapers (Figure 7.10).

Table 7.1. Finds of Differentially Patinated Artifacts in Lower Paleolithic Contexts in Britain

Site	*N*	*Type*	*Source*	*Notes*
East Dean	1	Handaxe	Brown (1893, p. 98, plate 1.4)	—
Kempston	1	Handaxe	Smith (1894, pp. 116–117)	See Figure 7.1
Caddington	1	Handaxe	Smith (1894, pp. 116–117)	—
Santon Downham	1	Handaxe	AB, pers. obs., February 2010	Warren Collection, Franks House, British Museum
Warren Hill (MIS 12/13)	6	Handaxe	Sturge (1911, p. 68)	In the extensive collection of handaxes from Warren Hill, Sturge (1911, p. 65) noted, "there are some implements which show re-working at a later stage—double patination." However, a count of these implements was not provided. Our own informal inspection of a small sample (~80 handaxes) of Sturge's Warren Hill material held in the British Museum revealed six repatinated specimens, including one handaxe showing evidence for three temporally distinct periods of working (see Figure 7.3).
Warsash	1	Handaxe	Burkitt, Paterson, and Mogridge (1939)	Several examples of handaxes with opposed notches patinated differently to the rest of the artifacts have been collected from the Warsash beachfront in more recent times (Nick Ashton, pers. comm., June 2008). It has been suggested that the additional scars may be due to historically recent reuse of the implements as fishing weights (Nick Ashton, pers. comm., June 2008). However, as Burkitt and colleagues (1939) pointed out, some handaxes recovered directly from the Warsash gravel pits were similarly notched, and these scars exhibit identical patinas to the rest of the artifacts (AB, pers. obs., July 2009).

(*continued*)

Table 7.1. Continued

Site	*N*	*Type*	*Source*	*Notes*
Witham	2	Handaxe	Turner and Wymer (1987)	Two differentially patinated handaxes were recovered from some large gravel-filled depressions at a Roman religious site at Essex.
Priory Bay (~MIS 11)	1	Handaxe	Francis Wenban-Smith, pers. comm., August 2009	A large repatinated handaxe was recovered during excavations at Priory Bay, on the eastern coast of the Isle of Wight in 2001. This artifact appears to have been scavenged from an active beachfront and, judging from the extent of weathering of the original scars, was heavily rolled prior to reworking.
Boxgrove (MIS 13)	15	Handaxe	Woodcock (1981); Pope (2002); Emery (2009)	Woodcock (1981, p. 111) reports the discovery in buried channel deposits at Amey's Eartham Pit of a reworked handaxe with differential patination indicating at least three distinct phases of reduction. Three handaxes showing evidence for two periods of working were also recovered from the junction of the lower chalky Coombe-Rock and the lower brickearth/buried channel deposits (Woodcock, 1981). In terms of the extensive lithic assemblage recovered from Quarry 1/B (Roberts & Parfitt, 1999), situated around 74 m from the relict beach cliff (the only natural raw material source at Boxgrove), we have identified evidence for differential patination on at least 10 handaxes from a total sample of 416. Of these, five were clearly made on scavenged handaxes, while the rest comprise handaxes manufactured from recycled flakes or other artifacts. Baker (2007) claims to have identified two examples of differentially patinated Boxgrove handaxes during his brief inspection of a sample of 16, giving a total of some 12.5% of recycled/scavenged artifacts in the assemblage. We have not been able to independently verify this claim. However, we noted many other examples of Boxgrove handaxes with potential evidence for differentially patinated surfaces, and in each case, it was unclear whether these specimens had been intentionally reworked by hominins.

Manor Farm	1	Handaxe	Woodcock (1981, p. 160)	—
Bognor Regis	1	Handaxe	Smith (1929)	The reworking episode probably relates to later human activity, possibly in an early Middle Paleolithic context.
High Lodge (MIS 13)	5	Handaxe ($n = 1$); scraper ($n = 2$); core ($n = 1$); flaked flake ($n = 1$)	AB, pers. obs., June 2009	See Figure 7.10
Hoxne (MIS 11)	5	Scraper ($n = 2$); flake ($n = 3$); flake blank core ($n = 2$); core ($n = 1$)	AB, pers. obs., June 2009	Two scrapers made on recycled flakes and a core made on an older, patinated core/flaked piece have been identified (AB, pers. obs., June 2009) in the "Upper Industry" at Hoxne (MIS 11) (Ashton, Lewis, Parfitt, Penkman, & Coope, 2008; Wymer & Singer, 1993). A flake and a distal flake fragment with differential weathering of flake scars, indicating that they were struck from patinated cores, and a core made on a weathered flake, were also noted in the same assemblage, providing further evidence for tool recycling. Similarly, in the Hoxne "Lower Industry," which is also assigned to MIS 11 (Ashton et al., 2008), a blade-like flake was removed from a patinated core, and a stained or patinated flake blank was worked bifacially and centripetally into a core. Wymer and Singer (1993, Table 4.16) note that a denticulate from the "Upper Industry" at Hoxne was made on an older, weathered flake. However, in the British Museum registration catalogue this specimen (P1980.4-1.2963) is reclassified as an unretouched flake. Following our own inspection of the piece, we conclude that the fresh scars on the ventral face are probably the result of post-depositional trampling.

(*continued*)

Table 7.1. Continued

Site	*N*	*Type*	*Source*	*Notes*
Barnham (MIS 12)	3	Flaked flake ($n = 1$); flake blank core ($n = 1$); core ($n = 1$)	AB, pers. obs., January 2010	A bifacially worked flaked flake excavated from the Area I light gray silts (Ashton, 1998b) was made on a blank struck from an older, patinated artifact of unidentified form, but probably a core. A flake blank core from the same context was also made on a weathered blank. A core made on a recycled flaked piece of unidentified morphology was recovered from the cobble band in Area I.
Swanscombe (MIS 12)	1	Core ($n = 1$)	AB, pers. obs., August 2009	A flake struck from a previously worked, patinated core is present in the assemblage recovered during Waechter's excavations of the Lower Gravel (~OIS 12) at Barnfield Pit, Swanscombe.
Elveden (MIS 12)	2	Flake blank core ($n = 1$); core ($n = 1$)	AB, pers. obs., January 2010	A patinated flake from the Area I brickearth (1995 British Museum excavations) was reworked as a core. A bifacial core made on an older patinated core or tested nodule was also recovered from Area III (top of the paleosol) (Ashton et al., 2005).

Note: Data compiled by the authors. pers. comm., personal communication; pers. obs., personal observation.

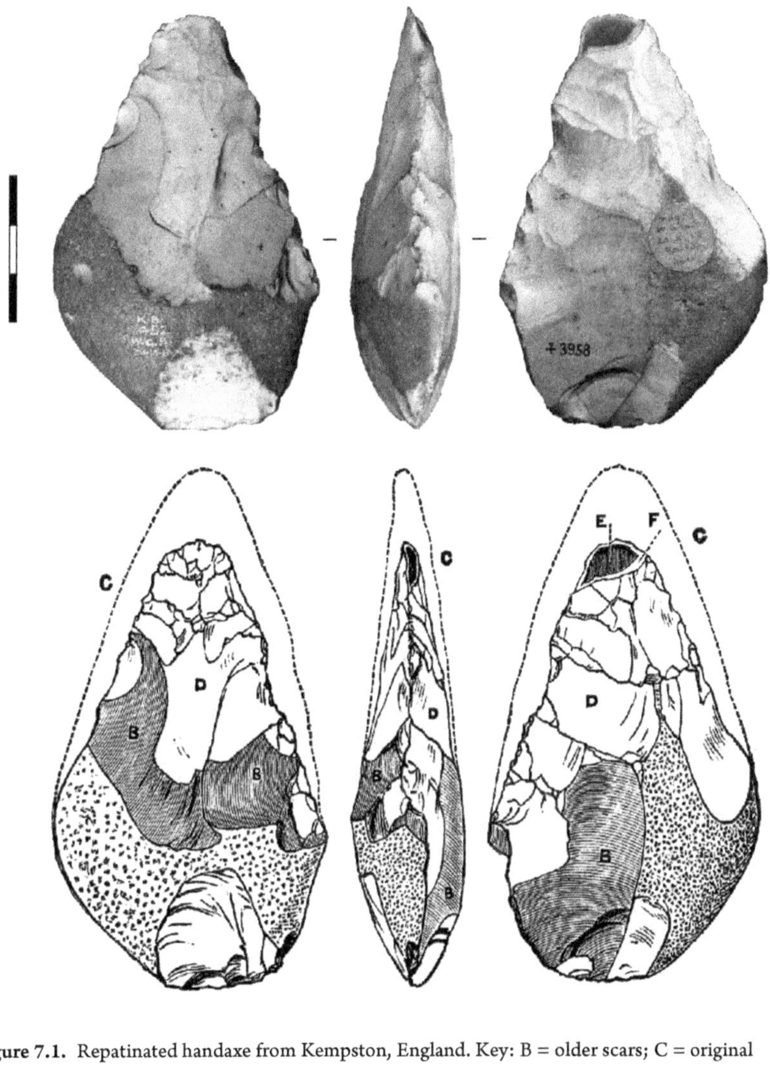

Figure 7.1. Repatinated handaxe from Kempston, England. Key: B = older scars; C = original dimensions of handaxe; D = reworking scars; E = taphonomic damage; F = thickness of patina. Scale is 30 mm. Images in upper row by Adam Brumm; permission to photograph specimen courtesy of the British Museum. Drawings in bottom row are from Smith (1894), *Man the Primeval Savage*, Edward Stanford (material in the public domain).

WAS TOOL SCAVENGING AND REUSE EXCEPTIONAL OR COMMONPLACE?

The identification of repatinated implements in Lower Paleolithic assemblages demonstrates that early hominins were scavenging and reusing previously discarded stone artifacts by at least ~1.6 Mya, and perhaps earlier. Moreover, reuse of older lithic artifacts was a component of Acheulean and other Early and Middle Pleistocene stone technologies. A more comprehensive review of the literature would doubtlessly reveal

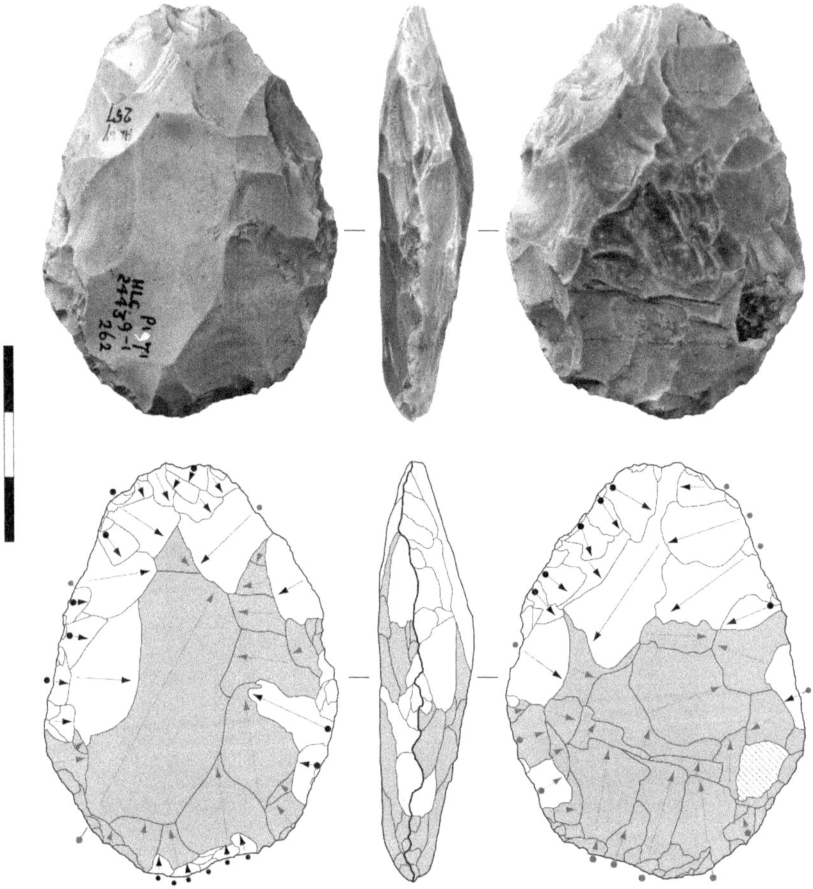

Figure 7.2. Repatinated handaxe from Hoxne, England. The older scars are gray and the later reworking scars are white. Key: arrows = orientation of flake removals; black circles = complete negative striking points; gray circles = missing negative striking points. Scale is 30 mm. Image by Adam Brumm; permission to photograph specimen courtesy of the British Museum.

further published examples; however, even the most exhaustive search is likely to remain incomplete, since it is probable that differentially patinated artifacts—commonly seen as little more than curiosities—have often gone unreported. In any case, it is clear from even a cursory review that repatinated artifacts are uncommon finds in the vast majority of Lower Paleolithic sites. Does the rarity of these objects suggest that tool scavenging and reuse was not a significant component of early hominin technology, or could it mean that this behavior is simply underrecognized?

The potential economic benefits of tool scavenging and reuse suggest that this behavior may have been a ubiquitous component of Lower Paleolithic stone technology. Theoretical modeling of cost–benefit factors associated with lithic procurement suggest that, all things being equal, foragers will attempt to behave in an economically efficient manner by minimizing extractive effort while also maximizing output of

Figure 7.3. Repatinated handaxe from Warren Hill, England. Patinated scars from the original phase of working are evident at the butt of the piece and contrast clearly with the darker, fresher scars at the tip and mid-section. Scale is 30 mm. Image by Adam Brumm; permission to photograph specimen courtesy of the British Museum.

tool-making stone (Elston, 1992; also see Andrefsky, 1994; Beck et al., 2002; Binford, 1979; Close, 1996; Kelly, 1988; Metcalfe & Barlow, 1992). This involves increasing the benefit–cost ratio of resource procurement by reducing as much as possible the indirect costs of securing stone, such as expenditures of food, labor, travel time, and transport. The fundamental benefit of this is a concomitant reduction in contingency risk, the probability of having insufficient tool-making stone to meet subsistence needs (Elston, 1992). Actualistic studies have also shown that increasing the time spent in processing stone for tool manufacture does not necessarily result in a higher return in terms of the number of useable tools (i.e., bifaces) (Elston, 1992). One implication of is that it will always make greater economic sense for foragers to focus on local production of tools, and to work whatever material is available on-site, than to make special trips to procure fresh material (cf. Binford, 1979; Gould, 1980; Hayden, 1977). For this reason, stone artifacts discarded by earlier knappers—which, in the case of certain kinds of rock (e.g., flint), can endure for thousands of years on the landscape without deleterious effects on their knapping properties—would have been a useful, and possibly very important, source of material for hominin tool-makers. The obvious benefit of scavenging older tools found on the landscape is that the most costly time and energy investments involved in lithic procurement and curation have already been incurred by others (Amick, 2007). Even when high-quality raw materials are located nearby to, or actually at, hominin activity areas or "bases," costs are still required to procure them. In the case of a primary chalk outcrop, for example, such costs involve traveling to the source; searching for and extracting blocks or nodules of flint; inspecting and assaying them for knapping quality; working suitable stone packages into flake blanks, bifaces, or cores; and finally, carrying large and heavy artifacts away from the source (Elston, 1992). These activities all require minimum investments of time and energy that can become costly when they detract from other economic activities (e.g., Stout, Semaw, Rogers, & Cauche, 2010).

Large tools like handaxes in particular may have been the continued focus of scavenging and reuse, owing to their potential for further reduction and greater

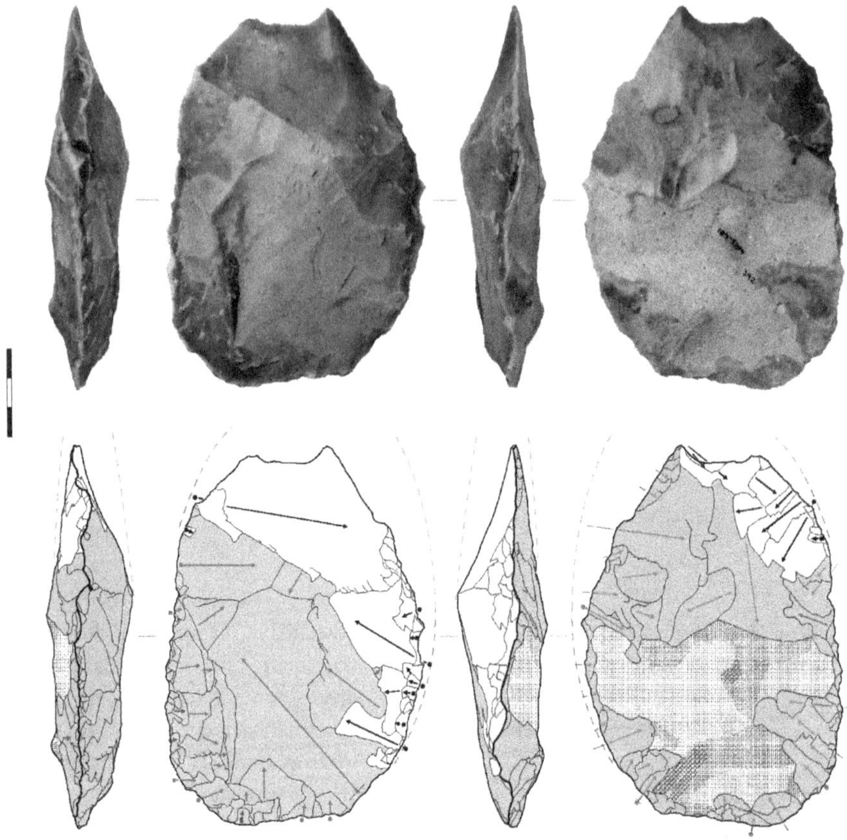

Figure 7.4. Repatinated handaxe from Boxgrove, England. The older scars are gray and the later reworking scars are white; broken lines indicate the approximate dimensions of the original piece, and arrows indicate orientation of tranchet removals The secondary working consists primarily of large tranchet removals affecting the tips of the original handaxes. The tranchet blow is associated with a less invasive transformation of the opposite face. The non-invasive removals extend along one margin, from the tranchet removal to the butt. Scale is 30 mm. Data and image by Mathieu Leroyer; permission to photograph specimen courtesy of the British Museum.

propensity to resist sedimentation processes when compared with smaller artifacts (e.g., Baker, 1978; Camilli & Ebert, 1992). As noted, there are several examples of individual handaxes in early British assemblages that have been scavenged and reused on at least two separate occasions. There is also evidence at Boxgrove and at other Middle Pleistocene sites in England of repeated discard of handaxes and other stone tools at points in the landscape where static resources were available, such as at permanent or semi-permanent water bodies with abundant flora and fauna and natural sources of lithic raw material (e.g., lag-gravels and cobble-bars) (Ashton, 1998a). Hominins also transported stone tools from these locations to other fixed resources in the landscape, leading to dense accumulations of lithic material at particular sites where group activities were focused. In addition, stone tools were carried further out on the landscape where fixed resources were unavailable and hominins were

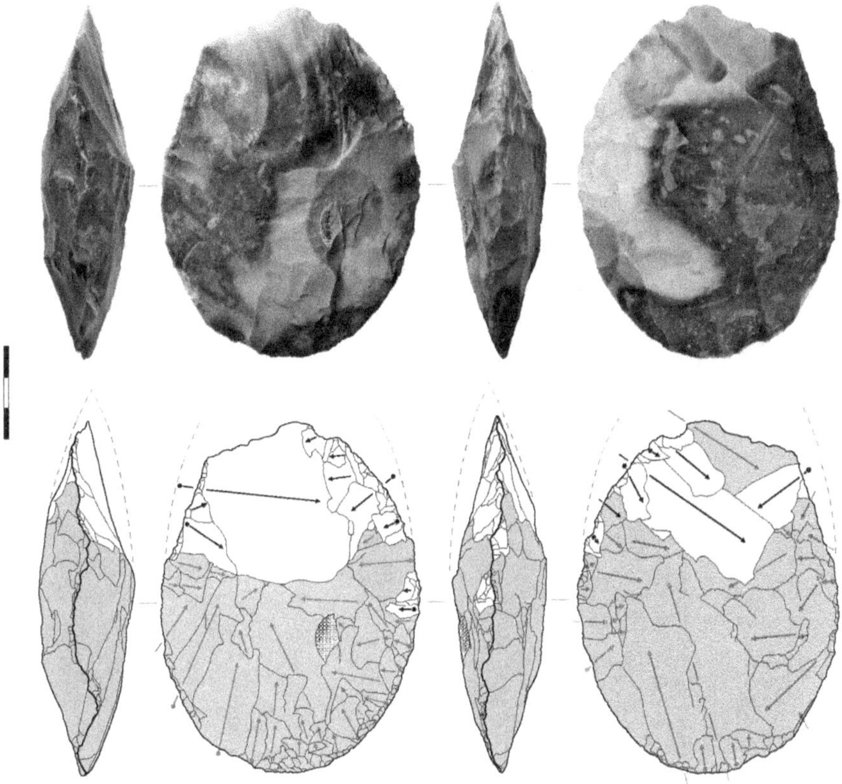

Figure 7.5. Repatinated handaxe from Boxgrove, England. The older scars are gray and the later reworking scars are white; broken lines indicate the approximate dimensions of the original piece, and arrows indicate orientation of tranchet removals. The secondary working consists primarily of a large tranchet removal affecting the tip of the original handaxe. Reduction seems to be limited to the upper part of the handaxe but affects the two opposite margins and is preceded by semi-invasive reduction of the tranchet removal scar. Scale is 30 mm. Data and image by Mathieu Leroyer; permission to photograph specimen courtesy of the British Museum.

more likely to encounter food procurement opportunities in random locations, such as the carcasses of large herbivores. The pattern of lithic deposition associated with mobile resources was characterized by single-event discards over much wider areas. In terms of opportunities for scavenging artifacts, the repeated signature of handaxe concentrations at fixed points in the environment is likely to have provided sources of reusable material at places where it was most likely to be reused, such as at favored sleeping sites and locales for group re-aggregation. However, we may also postulate that the repetitive patterns of carrying and discarding stone tools in parts of the landscape where mobile resources were situated would have led to the greater availability of reusable stone implements in these areas.

It is envisaged that as these structured patterns of artifact discard developed over time into distinct cultural landscapes (Pope, 2002; Pope & Roberts, 2005), opportunities for hominins to discover the enduring remains of earlier knapping

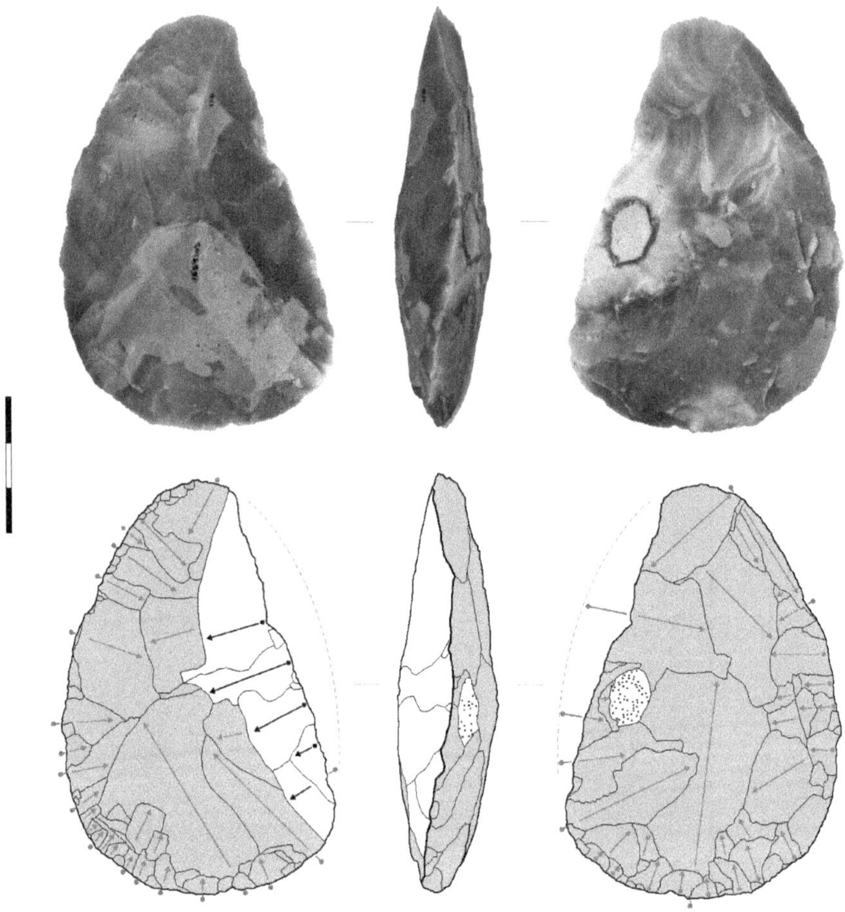

Figure 7.6. Repatinated handaxe from Boxgrove, England. The older series of scars are shaded gray and the later reworking scars are white; broken lines indicate the approximate dimensions of the original piece, and arrows indicate orientation of blows. The handaxe is a finely worked, patinated specimen with a series of five deep parallel blows along one margin, probably produced by hard-hammer percussion, and resulting in a denticulate scraper-like edge. Scale is 30 mm. Image by Adam Brumm; permission to photograph specimen courtesy of the British Museum.

activities would have become more frequent. The potential to scavenge previously flaked items would have increased in precisely those areas where they were most needed: at random points in the landscape where natural sources of raw material were unavailable. Long-term processes of tool reduction and land use thus would have led to the formation of culturally generated, and ostensibly static, sources of stone in areas where mobile resources were encountered, allowing increasingly minor investments in the procurement, reduction, and curation of lithic materials. These accumulations may also have acted to stimulate or "cue" other behavior at particular points in the landscape, such as the discard of curated tools and raw material. The potential for feed-back dynamics and patterns of self-organization in the record are certainly suggested by the exceptionally rich and well-documented archaeological evidence available from

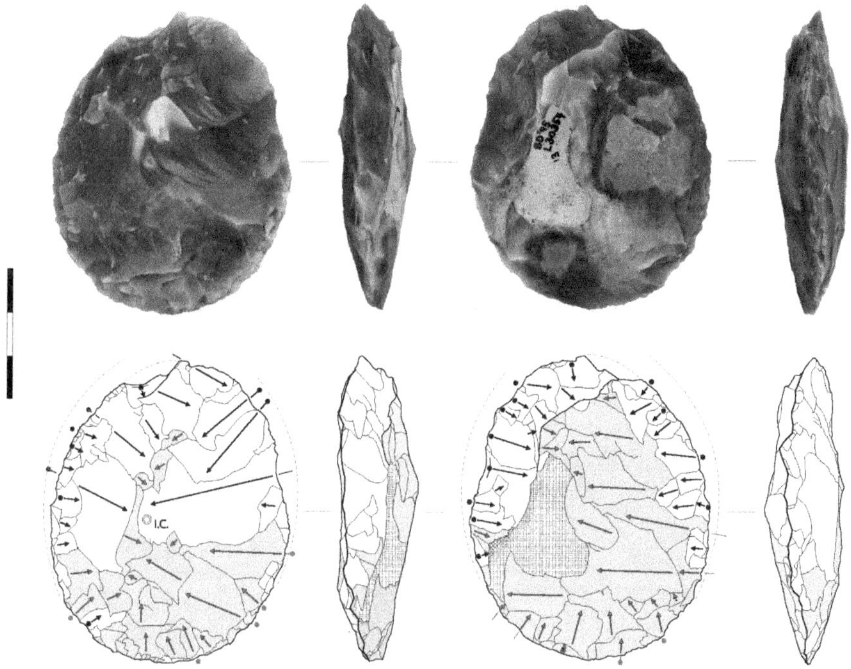

Figure 7.7. Repatinated handaxe from Boxgrove, England. The older series of scars are shaded gray and the later reworking scars are white; broken lines indicate the approximate dimensions of the original piece, and arrows indicate orientation of reworking. The handaxe is an extensively reworked ovate. On one margin the secondary reworking is carefully executed, except at the tip where a small bending fracture can be observed. By contrast, the other margin is modified by less regular removals. There is an incipient cone (I.C.) on the center of one face, but its temporal relationship with the recycling episode is difficult to establish. Scale is 30 mm. Data and image by Mathieu Leroyer; permission to photograph specimen courtesy of the British Museum.

Boxgrove (Pope, Russel, & Watson, 2006). These possibilities should be explored and considered alongside artifact scavenging in investigating the role that stone artifact accumulations in the landscape played in hominin ecology.

In sum, it can be anticipated on theoretical grounds that the scavenging and reuse of older stone artifacts was a pervasive aspect of hominin tool-making cultures during the Lower Paleolithic. As also noted, this prediction is not borne out by empirical evidence: Differentially patinated artifacts are almost always an uncommon feature of Lower Paleolithic assemblages. In the following section, however, we show that there are many circumstances in which previously flaked stone tools may be retrieved from the landscape and reused by hominins without resulting in discernable differences between the original set of scars and any subsequent additions.

COMPLEXITIES OF THE PATINATION PROCESS

Archaeologists use the term *patina* to refer to changes that occur to the surface of a lithic artifact as a result of exposure to chemical and/or mechanical weathering

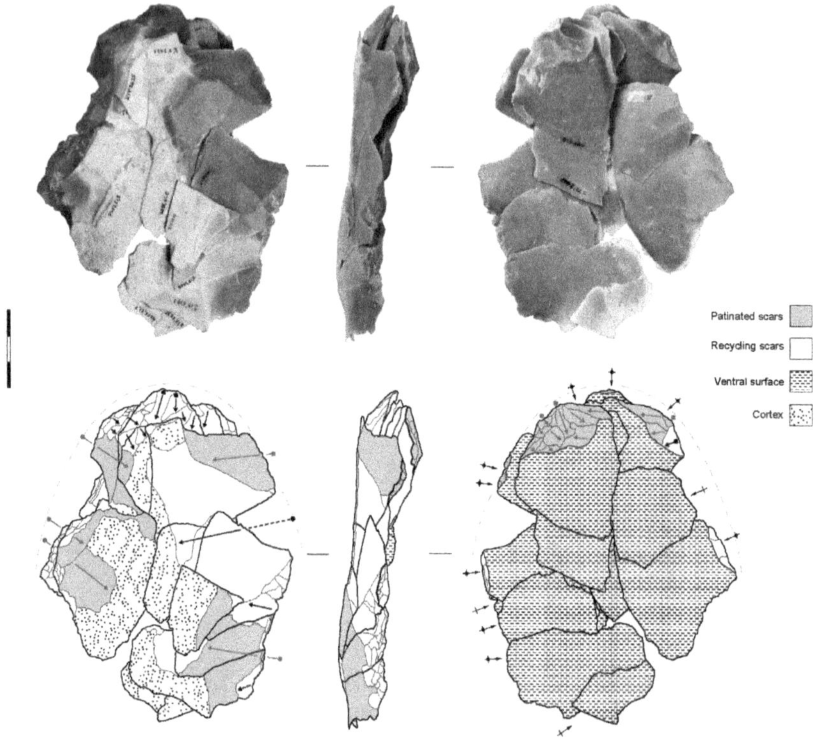

Figure 7.8. Refitted flakes from the reduction of a scavenged handaxe at Boxgrove. The refit set (Group 50) is from GTP17, the horse-butchery site, and consists of 17 flakes from the late-stage reduction of a well-developed rough-out or partially finished biface. Refitting shows that a partially complete biface/rough-out was introduced with at least six large flake removals from one face. The faceted platforms on the flakes comprising this refit group suggest that, in addition to this, extensive thinning of the opposite face had taken place. All of the previous removals have a blue-white patina, indicating that the introduced implement was scavenged from an exposed archaeological site. The resultant recycled tool was removed from the area. The older series of scars are shaded gray and the later reworking scars are white; broken lines indicate the approximate dimensions of the original piece. Key: arrows on scars = orientation of flake removals; black circles = complete negative striking points; gray circles = missing negative striking points. Small arrows = flake percussion axes; small arrow with black circle at butt = intact bulb of force; small arrow with cross at butt = missing bulb of force. Scale is 30 mm. Image by Adam Brumm; permission to photograph specimen courtesy of the British Museum.

processes (Andersen & Whitlow, 1983; Goodwin, 1960; Hogg, 1905; Howard, 1999, 2002; Hurst & Kelly, 1961; Purdy & Clark, 1987, p. 211; Rottländer, 1975; Schmalz, 1960). The term itself, however, encompasses a wide range of often quite different phenomena. Selective leaching and chemical dissolution are the most common processes of patina formation in the stone tool record. Surface alterations resulting from the accretion of mineral deposits or other substances (e.g., manganese staining) on artifacts are also sometimes referred to as patinas. This is not strictly accurate, however, as these processes involve additions to, rather than chemical alterations of, stone

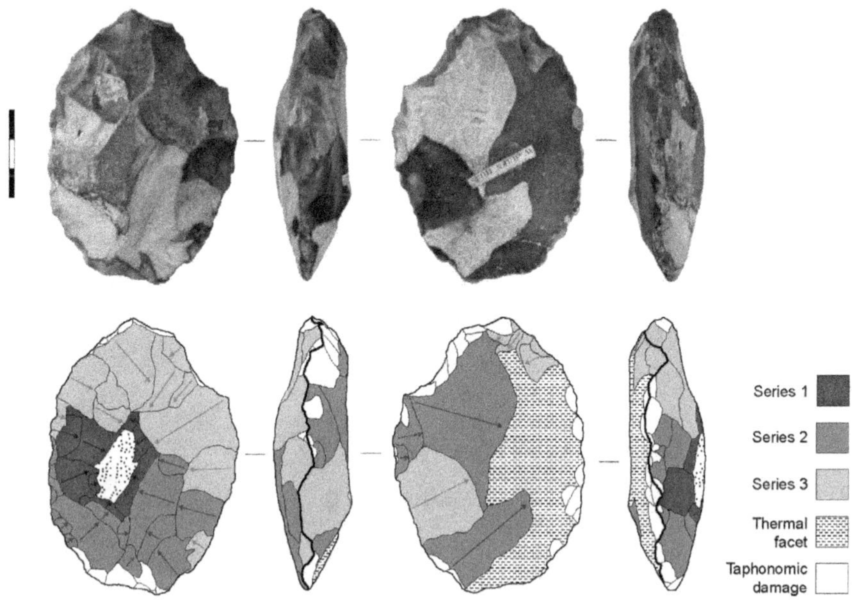

Figure 7.9. Repatinated handaxe from Warren Hill, England. The scars exhibit three different states of patination, indicating three temporally discrete periods of use. This handaxe was originally made on a thermal pot-lid or other frost-fractured piece. Knapping was concentrated predominately on one face throughout the different stages of reduction and recycling. Key: arrows = orientation of flake removals; black circles = complete negative striking points. Scale is 30 mm. Image by Adam Brumm; permission to photograph specimen courtesy of the British Museum.

tool surfaces (e.g., Cackler, Glascock, Neff, & Chiarulli, 1999; Meeks, Sieveking, Tite, & Cook, 1982; Shackley, 1989). Given the wide use of the term, it is sufficient for our purposes to define a patina as any alteration to a stone artifact that produces a visually perceptible difference between the exterior surface and the unaltered interior material—with the exception of changes caused by deliberate heat treatment of rocks.

Stone artifacts deposited into the archaeological record acquire an outer weathered layer as a result of interaction with the sediment matrix or through exposure to light and atmospheric conditions in surface contexts (Rottländer, 1975). A general rule of patina formation is that the thickness of weathered layers on artifacts depends on the length of exposure to weathering processes (Purdy & Clark, 1987). It has long been recognized, however, that the depth of patina formation on stone artifacts does not provide an absolute measure of their age (Burroni, Donahue, Pollard, & Mussi, 2002; Goodwin, 1960; Prestwich, 1860, p. 296; Viereck, 1964). This is owing to the fact that patina formation is contingent on complex environmental and material factors (Goodwin, 1960; Howard, 1999, 2002; Purdy & Clark, 1987; Rottländer, 1975; Schmalz, 1960) and is not always characterized by linear rates of weathering (Pope, Meierding, & Paradise, 2002; Wells, Hancock, & Fryer, 2008).

The specific nature of the atmospheric, aquatic, or soil environment in which artifacts are deposited (e.g., solution pH, humidity, temperature, ground and soil water composition) is critical to the process of patina formation (Polikreti, 2007; Purdy &

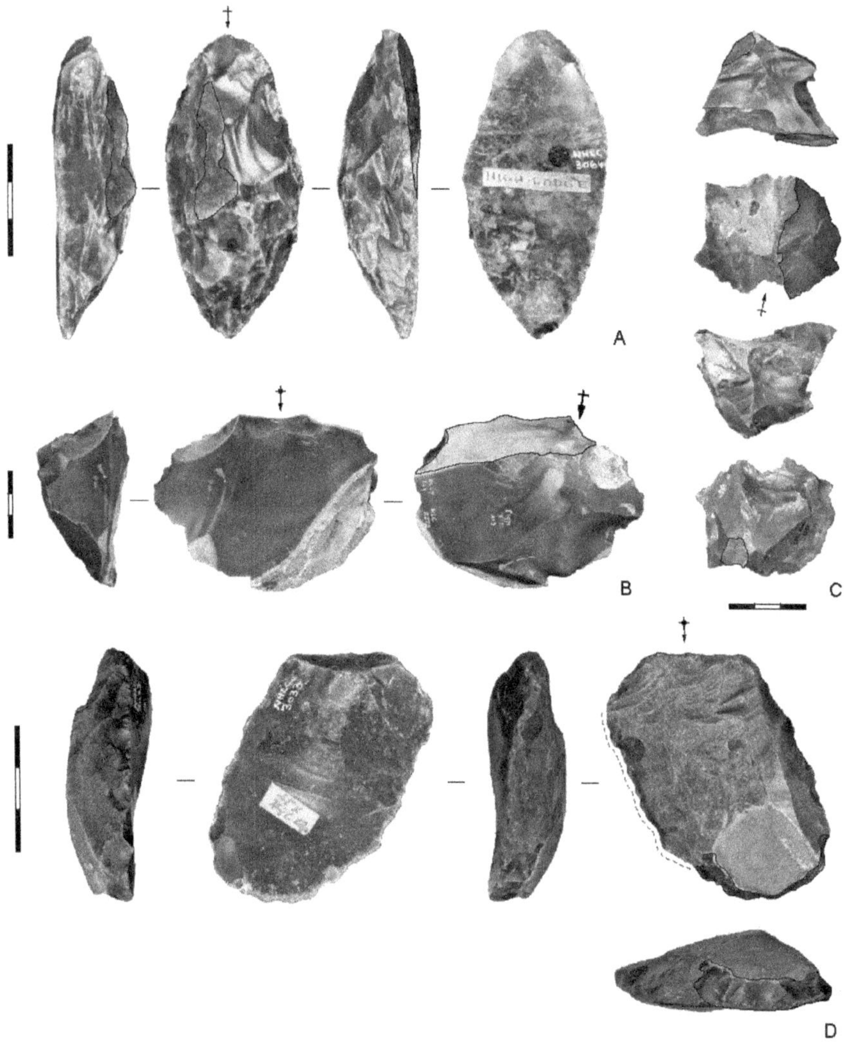

Figure 7.10. Repatinated implements from High Lodge, England. (A) Convergent convex sidescraper ("Old collection"); (B) flaked flake (de Sieveking collection); (C) flake blank core (de Sieveking collection); (D) end scraper made on recycled side scraper ("Old collection"). The patinated scars are shown in outline in A–C; the dashed line in D indicates the original retouch scars on the older scraper. It is made on a heavily patinated flake blank, as evidenced by the pronounced yellowish-brown "toad-belly" patina (Sturge, 1911) on the ventral face and on a remnant dorsal scar—although, it should be noted, the resulting recycled scraper was itself subsequently affected by chemical alteration and has a distinctly less developed toad-belly surface patina. The flaked flake (B) is from the Bed E sand and gravel and was struck from the ventral surface of a patinated flake blank. The resulting flake was bifacially reduced, and retouched. The heavily worked asymmetrical bifacial core (C) is from Bed C2 and was made on a thick, weathered flake. One face of the core also exhibits battering marks indicating a subsequent recycling phase as a hammerstone. Key: arrows = flake percussion axes; arrow with black circle at butt = intact bulb of force; arrow with cross at butt = missing bulb of force. Scale is 30 mm. Image by Adam Brumm; permission to photograph specimen courtesy of the British Museum.

Clark, 1987). However, examples of refitted artifacts with differing rates of chemical alteration (not due to scavenging) show that surface weathering of items often depends on microenvironmental factors and the nature and intensity, rather than the simple duration, of exposure to weathering processes (Ashton et al., 2005; Kelley, 1965; Ophoven, 1938). There are many microenvironmental factors that may contribute to differing rates of patina formation on stone tools in the same depositional context. For instance, biotic weathering agents (e.g., bacteria, lichen, algae, and fungi) can have a significant effect on patina formation, contributing to acid dissolution, oxidation, and various mechanical weathering mechanisms, and are not always distributed uniformly in the environment (Pope et al., 2002; also see Ackerman, 1964; Goodwin, 1960; Rottländer, 1975). Moreover, faster rates of patination may be promoted in artifacts deposited in association with decaying organic matter, owing to the release of acids (Burroni et al., 2002). Other variables, such as the elevation of artifacts in the landscape, also appear to play a role (e.g., Thompson, 2009), while taphonomic and use-related factors may result in differential weathering of implements found in close proximity to each another (Clemente-Conte, 1997; Collins, 1993; Rottländer, 1975).

The specific properties of the raw material, such as composition, mineralogy, and kinds and proportions of pigmenting impurities, are equally important determinants of patina formation (Purdy & Clark, 1987). Some raw material types are relatively resistant to weathering due to their hardness, while others (e.g., volcanic materials) may patinate rapidly under the same conditions (Rottländer, 1975). There is also variation within single rock types that can affect the rate at which individual artifacts made from the same materials weather. With fine-grained Cretaceous flint, for example, the specific proportions of its main silica components—namely, silicified skeletal fragments, spherical quartz lepispheres, and microfibrous quartz chalcedony "cement" infilling the voids between the preceding structural elements—vary both between sources and within nodules from the same geological outcrop (Bradley & Clayton, 1987). This variability affects the degree of porosity of materials, which in turn determines the rate at which water and other weathering agents diffuse into them (Howard, 2002; Hurst & Kelly, 1961; Purdy & Clark, 1987). Leaching is a selective extraction and starts with the most soluble components (Burroni et al., 2002). The presence of fine cracks, striations, pores, and fissures accelerates this process, and thus the surface roughness of artifacts influences the extent and magnitude of patina formation (Purdy & Clark, 1987). When unaltered flint is knapped, the propagating crack preferentially fractures around the lepispheres and through the weaker interstitial chalcedony, resulting in hemispherical domes or craters spaced according to the distribution and density of lepispheres (Bradley & Clayton, 1987). These features increase the surface roughness of knapped artifacts, providing weak points through which chemical agents may penetrate. However, if recrystallization of the silica "cement" occurs, its strength is increased and the fracture front consequently cuts across lepispheres as well as chalcedony. This results in artifacts with a more undulating surface microtopography, and, correspondingly, a reduced susceptibility to chemical attack. Importantly, flint nodules from the same sources often exhibit varying degrees of recrystallization, while the density of packing of lepispheres within the flint structure may also differ (Bradley & Clayton, 1987). We may therefore expect to find variation in the extent to which chemical weathering processes affect individual flint artifacts, even those made from material procured at the same geological outcrop.

There are other practical problems involved in identifying differential patination on stone tools. For instance, diagnostic evidence for the phenomenon can be particularly difficult to identify macroscopically in situations where the degree of patina formation on tools was minimal, or when scavenged implements were used but not modified by additional retouch (e.g., Callow, 1986, pp. 207–208; Frame, 1986, p. 355). Moreover, minor differences in patination present at the moment of reworking may become obscured by the total weathering of the artifact in the time span since discard, such as through hydraulic tumbling in fluvial contexts (Shackley, 1974). Finally, scavenged tools often may have been so intensively reworked that no evidence at all for repatinated surfaces remains on them. Although this is likely to have been a rare occurrence, it may be relevant to invasively thinned handaxes or other bifaces.

Amick (2007) addresses the problem of patina formation in relation to the archaeological visibility of artifact use and recycling in the Great Basin Desert (also see Evans & Meggers, 1960; Michels, 1969; Rondeau, 1997; Silliman, 2005). Here, obsidian hydration measurements on a sample of repatinated obsidian projectile points showed a range of 0.1 to 3.9 microns (~100 to 3,900 years) separating the initial discard events from later reuse episodes (Amick, 2007; but cf. Rogers, 2008, for potential problems with obsidian hydration dating). However, at some sites, hydration analyses showed that the tips of points exhibiting no diagnostic evidence of reuse were in some cases chronologically younger than the corresponding stems. This observation suggests that obsidian points may have been scavenged and reworked, often long after their initial discard, but without the effects of weathering producing differentially patinated surfaces (see Amick, 2007, p. 238; for earlier, unpublished studies on this phenomenon, see Moore & Burke, 1992).

In sum, secondary recycling of artifacts within short timescales after discard (i.e., days to months) might in some cases be undetectable through differential patination (Hiscock & Attenbrow, 2005). Conversely, some implements exposed on surface archaeological sites over very long periods of time may have undergone weathering, but of insufficient duration or intensity to produce noticeable patinas. Both possibilities depend on a wide range of environmental conditions and other factors, such as the individual depositional histories of artifacts and the composition and texture of lithic materials (Hurst & Kelly, 1961). The key point is that the absence of differentially weathered scars on a stone tool does not necessarily demonstrate that the knapping actions that produced the scars were temporally contiguous.

OTHER PARAMETERS AFFECTING OUR UNDERSTANDING OF TOOL SCAVENGING AND REUSE

A further issue to consider: In order for scavenging to be an option, hominins obviously must have had access to scavengeable tools (Camilli & Ebert, 1992; Ebert, 1992; McDonald, 1991). To an important extent, therefore, the frequency with which tool scavenging was undertaken in the Lower Paleolithic must have been contingent on the availability and discoverability of abandoned tools on the surface of the landscape. Equally, if evidence for scavenging is to be recognized archaeologically, the discarded artifacts must have been exposed to weathering agents that produced conspicuous surface alterations on the tools prior to their recycling, such that the subsequent reworking events revealed the unaltered interior material. Assuming a constant rate of scavenging, it can be anticipated that the proportion of repatinated implements in assemblages will increase as both ground surface visibility of archaeological sites and the rate of patina formation on discarded artifacts increase.

Accordingly, there must be a threshold of ground surface visibility below which previously discarded artifacts were invisible and scavenging opportunities available to hominins were therefore negligible, such as in highly depositional and/or densely vegetated environments. Conversely, an upper threshold must also exist above which ground surface visibility was more or less total and scavenging opportunities were correspondingly profuse, such as in non-depositional and/or non-vegetated environments. If this upper threshold is combined with very high rates of patina formation, then evidence for differentially patinated tools may be prolific. However, the frequency of tool scavenging behavior, while conditional on the degree of surface visibility of discarded tools, is independent from the rate of patina formation. An opposite scenario can therefore be envisioned in which high rates of tool scavenging occurred in environments with adequate ground surface visibility but, owing to one or more of a host of possible reasons, minimal rates of patina formation. The outcome of this situation would be the presence of a disproportionately low frequency of repatinated artifacts and, hence, the misleading impression that tool scavenging and recycling behavior at this locality was statistically infrequent.

A useful thought experiment is to imagine a scenario in which an ethnoarchaeologist was able to test the intensity of lithic artifact scavenging and reuse among a fictitious group of living stone tool-makers. The researcher, unable to live among these people, could only monitor their behavior indirectly by surveying a series of discrete surface scatters of flaked stone artifacts left by these people in particular parts of the landscape to which they regularly returned. In the case of each scatter, she planted a series of controls by coating the exterior surfaces of some artifacts with a paint that is detectable only under ultraviolet light. Thus, when she returned to each scatter, she could determine which controls had been scavenged and reworked simply by observing which items displayed fresh scars intruding into the covertly painted surfaces—in essentially the same manner as repatinated stone artifacts. Some stone tool scatters were seeded with more controls than others, in order to replicate vagaries in artifact patination rates.

We can imagine the results. To begin with, the ethnoarchaeologist is likely to discover that many environmental and behavioral variables affect the parameters of the experiment. For example, a certain proportion of planted controls would be lost because of high rates of vegetation growth and/or sedimentation in the environment. These controls would be either buried or concealed from view and thus unavailable for the tool-makers to scavenge, or reworked controls would become lost. Behavioral variables would also affect the results: Some controls would be missing from scatters because they were carried away by the tool-makers and discarded elsewhere, including at other monitored lithic scatters. Similarly, even with an array of environmental factors taken into account, controls may not be affected evenly across the monitored sites. Human land-use patterns, such as mobility and duration and intensity of site occupation and reuse, as well as distance to raw material sources, variation in site function, foraging schedules, and so on, may all exert an influence on the rate of tool scavenging.

In short, it may be difficult if not impossible for the experiment to determine the nature and intensity of artifact scavenging based simply on the amount of reworked controls observed. Would high proportions of reworked controls indicate very high rates of artifact scavenging at some sites, or that these scatters were less affected by taphonomic processes? If only a few reworked controls were present at sites where large numbers were initially planted, would this reflect loss of controls through burial processes, or more complex behavioral factors? In scatters where it was possible to plant only a few controls, did intensive scavenging and reuse of items occur without affecting any of the items so marked? An obvious lesson would be that the proportion of controls released into scatters has major bearing on these problems. A key conclusion may therefore be the design of a more exhaustive research program in which every visible artifact in a scatter was numbered, measured, and recorded in meticulous detail, making it possible to monitor far more precisely what was happening to controls over time. In so doing, it would also be possible to document the various ways in which discards were affected by reuse, such as varying rates of reduction in the volume and dimensions and subsequent transformations of artifacts, objects disappearing and reappearing in the same or different scatters, and so on.

Archaeologists obviously are unable to draw on such invaluable observations, and so the interpretation of repatinated artifacts in archaeological assemblages, in one sense, will always be problematic—but then, so will our understanding of prehistoric tool recycling and reuse more generally. Regardless of whether repatinated specimens are present or not, prehistoric tool scavenging and recycling may be tantamount to what we gloss as "normal" lithic reduction.

CONCLUSION

Davidson and McGrew (2005) argue that the scavenging and reuse of discarded stone artifacts played a major role in human evolution because hominin interactions with the residues of past tool-using activities provided a means of transmitting and generating new behavior. In this chapter, we showed that differentially patinated stone tools (unequivocal proxies of tool scavenging and reuse behavior) occur in Lower Paleolithic assemblages, thus confirming the basic tenets of Davidson and McGrew's (2005) model. Ideally, we would be able to trace developments through time in the nature and distribution of repatinated artifacts as proxies of cognitive and behavioral changes in hominin evolution. From the earliest appearance of these items in African Oldowan assemblages, we might expect to see changes in tool scavenging behavior, such as increased scavenging during the Acheulean and Middle Paleolithic periods when highly retouched tool forms (handaxes) amenable to intensive reuse emerge (e.g., Soressi & Hays, 2003). As we have seen, however, when stone tools enter the archaeological record, they acquire characteristics of external weathering under complex and often inconsistent circumstances. It is unlikely that the proportion of repatinated specimens in individual assemblages, whether low or high, provides us with accurate and reliable insight into the degree and intensity of tool scavenging in the hominin past.

The inevitable conclusion to be drawn from this is that while differential patination provides an unequivocal archaeological proxy of tool scavenging, it cannot be considered to accurately reflect its abundance in particular environments.

Abandoned stone artifacts found on exposed archaeological sites are likely to have comprised an important culturally generated resource that was not tied to the environmental densities of stone. We may surmise, therefore, that if proxy evidence for artifact scavenging is present *at all* in early hominin assemblages—that is, in the form of differentially patinated implements—then, regardless of their frequency in given assemblages, tool scavenging and reuse is likely to have been routine behavior. The occasional repatinated artifact may be merely the tip of the iceberg. For example, given the marginal number of cases observed at Boxgrove and the remarkable availability of flint there, some further parameters could be examined to understand the controlling factors behind artifact scavenging in particular landscapes at particular times. Previous research indicated that the relationship between raw material availability and the degree of reuse in the form of resharpening may not be spatially or chronologically simple if the degree of edge resharpening present in individual assemblages is used as a proxy for intensity of reuse (Emery, 2009). Boxgrove handaxes show a relatively unique pattern for the Lower Paleolithic of intensive edge resharpening, which may offer further support for the hidden scope of reuse and recycling. Further avenues for research could involve the examination of differing skill levels of technical actors (Pigeot, 1990), and the difficulty of early-stage biface reduction, or "roughing out" (Winton, 2004), as these variables seem to play an important part in biface *chaînes opératoires* at this site and perhaps at others, too (Leroyer, 2016).

So what are the implications of this discussion for Wynn's decades-old quest to understand how our cognition diverged from that expected of the last common ancestor between chimpanzees and us? It has been suggested that by at least 4.4 million years ago our hominin ancestors (or close relatives) had evolved the cognitive ability to glean social information from "reading" other band members' trackways, including not just the direction of travel but also the physical condition, mental state, and intentions of the individuals who made the tracks, and whether or not they were known to the observer (Shaw-Williams, 2014). This capacity is apparently unique to hominins: Chimpanzees and other apes use scent trails and airborne odors to find unseen targets; they take no notice of trackways. It is thought that simple trackway reading involved hominins being able to imagine themselves in the "body-and-mind" of an absent agent, a form of imaginative self-projection that may have provided a foundation for the evolution of more complex cognition. In a similar vein, Davidson (2010) suggests that memories of emotions experienced by hominins during prior acts of stone tool manufacture and use may have been prompted by the later sight of the discarded products, allowing hominins to reflect on the patterns of their own behavior. Hominin trackways can last for days or weeks, but they eventually disappear (Shaw-Williams, 2014). By contrast, the discarded products of hominin knapping—stone tools and scatters of lithic debris left lying in the landscape, themselves a form of trackway—may be virtually indestructible: So long as the objects are not buried or obscured by undergrowth, these telltale signs of past human behavior can remain visible in the environment for years, perhaps even indefinitely.

Imagine if hominin trackways were like that: if all of the myriad indentations and physical traces left in the landscape by the passage of hominins became fossilized like the famous Laetoli footprints and were thus a permanent part of hominin environments. For hominins attuned by selection pressures to the social information inherent in trackways, this cumulative record of past behavior would represent an epic narrative or "story" extending far beyond the limits of their comprehension. One can speculate that this is essentially what the surface archaeological record of lithic reduction and use would have been like for them. Hominins would have lived their lives surrounded by a complex and ever-growing repository of information about earlier knapping events. Based on various lines of evidence, we have argued that Lower Paleolithic hominins, where possible, would have exploited this resource—perhaps intensively—to satisfy their immediate needs for stone. However, over the course of human evolutionary history, this record evidently became more than merely a handy source of reusable tools. Much like trackway reading, at some point in the deep past our ancestors (or close hominin relatives) must have acquired the ability to read these remnants of prior knapping events and imagine themselves in the body-and-mind of the unseen knappers who had made and left behind still-visible artifacts. They also developed the capacity to conceptualize the story of abandoned stone tools in the landscape as a whole, and to form mental images of how *all* these material objects had come into existence through the agency of beings who had inhabited the world before them, or at least who were no longer visible in their immediate surroundings. At this stage, perhaps, knappers were able to conceive of discarded stone tools not merely as useful components of their natural environment, but as inherently cultural objects with a past and a present. The fundamental issue to consider in the background of Thomas Wynn's research is when in time such an ability first arose in the hominin family, and why, and whether this complex way of perceiving the world has only ever been the province of *Homo sapiens*.

ACKNOWLEDGMENTS

A great many individuals fielded our queries about repatinated artifacts and provided valuable data and insights. For this, we thank K. Akerman, N. Ashton, N. Barton, F. Blaser, A. Bouzouggar, D. Cliquet, I. de la Torre, H. Dibble, H. Djema, H. Fluck, J.-M. Gouedo, R. Hosfeld, E. Hovers, A. Jelinek, C. Juby, S. Kuhn, B. Lefevre, L. Lloyd-Smith, J. McNabb, S. McPherron, J. Pelegrin, M. Petraglia, T. Reynolds, B. Scott, A. Shaw, G. Sharon, S. Soriano, F. Wenban-Smith, R. Wikell, Y. Zaidner, and A. Turq. The late R. Jacobi brought to our attention the writings of Worthington Smith and Sturge on repatinated artifacts, for which we are grateful. We also thank I. Davidson and M. Moore for their thoughtful comments on earlier drafts of this chapter. Brumm's research at the British Museum was funded by an Australian Research Council postdoctoral research fellowship based at the University of Wollongong, and a postdoctoral research fellowship (2007–2009) at the McDonald Institute for Archaeological Research, University of Cambridge. Brumm gratefully acknowledges N. Ashton for his hospitality at the British Museum (Franks House, Department of Prehistory and Europe) and for his permission to study and photograph material in the Sturge collection at Franks House.

REFERENCES

Ackerman, R. E. (1964). Lichens and the patination of chert in Alaska. *American Antiquity*, 29(3), 386–387.

Amick, D. S. (2007). Investigating the behavioral causes and archaeological effects of lithic recycling. In S. P. McPherron (Ed.), *Tools versus cores: Alternative approaches to stone tool analysis* (pp. 223–252). Newcastle, UK: Cambridge Scholars Publishing.

Andersen, H. H., & Whitlow, H. J. (1983). Wear traces and patination on Danish flint artefacts. *Nuclear Instruments and Methods in Physics Research, 218*(1), 468–474.

Andrefsky, W. (1994). Raw-material availability and the organization of technology. *American Antiquity, 59*(1), 21–34.

Andrefsky, W. (2009). The analysis of stone tool procurement, production, and maintenance. *Journal of Archaeological Research, 17*(1), 65–103.

Ashton, N. M. (1998a). The spatial distribution of the flint artefacts and human behaviour. In N. M. Ashton, S. G. Lewis, & S. Parfitt (Eds.), *Excavations at the Lower Palaeolithic site at East Farm, Barnham, Suffolk 1989–94* (pp. 251–258). London: British Museum.

Ashton, N. M. (1998b). The technology of the flint assemblages In. In N. M. Ashton, S. G. Lewis, & S. Parfitt (Eds.), *Excavations at the Lower Palaeolithic site at East Farm, Barnham, Suffolk 1989–94* (pp. 205–235). London: British Museum.

Ashton, N. M., Lewis, S. G., Parfitt, S. A., Candy, I., Keen, D., Kemp, R., . . . White, M. J. (2005). Excavations at the Lower Palaeolithic site at Elveden, Suffolk, UK. *Proceedings of the Prehistoric Society, 71*, 1–61.

Ashton, N. M., Lewis, S. G., Parfitt, S. A., Penkman, K. E. H., & Coope, G. R. (2008). New evidence for complex climate change in MIS 11 from Hoxne, Suffolk, UK. *Quaternary Science Reviews, 27*(7), 652–668.

Aumassip, G. (2004). *Préhistoire du Sahara et de ses abords. T1: Au temps des chasseurs, le paléolithique*. Paris: Maisonneuve & Larose.

Baker, C. M. (1978). The size effect: An explanation of variability in surface artifact assemblage content. *American Antiquity, 43*(2), 288–293.

Baker, T. (2007). Recycling. Paleoindian & other archaeological stuff. Retrieved from http://www.ele.net/recycling/diffsurf.htm

Bamforth, D. B. (1986). Technological efficiency and tool curation. *American Antiquity, 51*(1), 38–50.

Barkai, R., Gopher, A., & Shimelmitz, R. (2006). Middle Pleistocene blade production in the Levant: An Amudian assemblage from Qesem Cave, Israel. *Eurasian Prehistory, 3*(2), 39–74.

Bar-Yosef, O., & Goren-Inbar, N. (1993). *The lithic assemblages of 'Ubeidiya: A Lower Palaeolithic site in the Jordan Valley*. Jerusalem: Hebrew University of Jerusalem.

Beck, C., Taylor, A. K., Jones, G. T., Fadem, C. M., Cook, C. R., & Millward, S. A. (2002). Rocks are heavy: Transport costs and Paleoarchaic quarry behavior in the Great Basin. *Journal of Anthropological Archaeology, 21*(4), 481–507.

Binford, L. R. (1979). Organization and formation processes: Looking at curated technologies. *Journal of Anthropological Research, 35*(3), 255–273.

Boucher de Perthes, J. (1999a). *Antiquités celtiques et antédiluviennes. T II: L'âge de la pierre, les preuves—1857*. Clermont-Ferrand: Éditions Paléo.

Boucher de Perthes, J. (1999b). *Antiquités celtiques et antédiluviennes. T III: La galerie des silex—1867*. Clermont-Ferrand: Éditions Paléo.

Bradley, R., & Clayton, C. (1987). The influence of flint microstructure on the formation of microwear polishes. In G. de G. Sieveking & M. H. Newcomer (Eds.), *Human uses of flint and chert: Proceedings of the Fourth International Flint Symposium held at Brighton Polytechnic 10–15 April 1983* (pp. 81–89). Cambridge, UK: Cambridge University Press.

Breuil, H. (1954). Prolégomènes à une classification préhistorique—discours du président entrant. *Bulletin de la Société préhistorique française, 51*(1), 7–15.

Brown, A. (1893). On the continuity of the Palaeolithic and Neolithic periods. *Journal of the Anthropological Institute of Great Britain and Ireland, 22*, 65–98.

Burkitt, M. C., Paterson, T. T., & Mogridge, C. J. (1939). The Lower Palaeolithic industries near Warsash, Hampshire. *Proceedings of the Prehistoric Society, 51*(1), 39–50.

Burroni, D., Donahue, R. E., Pollard, A. M., & Mussi, M. (2002). The surface alteration features of flint artefacts as a record of environmental processes. *Journal of Archaeological Science, 29*(11), 1277–1287.

Cackler, P. R., Glascock, M. D., Neff, H., & Chiarulli, B. M. (1999). Effects of weathering on the coloration of chert and its implications for provenance studies. *Lithic Technology, 24*(2), 81–90.

Callow, P. (1986). Raw materials and sources. In P. Callow & J. M. Cornford (Eds.), *La Cotte de St Brelade 1961–1978: Excavations by C. B. M. McBurney* (pp. 203–211). Norwich, UK: Geo Books.

Camilli, E. L., & Ebert, J. I. (1992). Artifact reuse and recycling in continuous surface distributions and implications for interpreting land use patterns. In J. Rossignol & L. Wandsnider (Eds.), *Space, time, and archaeological landscapes* (pp. 113–136). New York: Plenum Press.

Clarkson, C. (2007). *Lithics in the land of the Lightning Brothers: The archaeology of Wardaman country, Northern Territory*. Canberra: ANU E-Press.

Clemente-Conte, I. (1997). Thermal alterations of flint implements and the conservation of microwear polish: Preliminary experimental observations. In A. R. Millán & M. A. Bustillo (Eds.), *Siliceous rocks and culture* (pp. 399–410). Granada, Spain: Universidad de Granada.

Close, A. E. (1996). Carry that weight: The use and transportation of stone tools. *Current Anthropology, 37*(3), 545–553.

Collins, M. B. (1993). Comprehensive lithic studies: Context, technology, style, attrition, breakage, use-wear and organic residues. *Lithic Technology, 18*(1/2), 87–94.

Davidson, I. (2010). Stone tools and the evolution of hominin and human cognition. In A. Nowell & I. Davidson (Eds.), *Stone tools and the evolution of human cognition* (pp. 185–205). Boulder, CO: University Press of Colorado.

Davidson, I., & McGrew, W. C. (2005). Stone tools and the uniqueness of human culture. *Journal of the Royal Anthropological Institute of Great Britain and Ireland, 11*(4), 793–817.

de Mortillet, G. (1885). *Le préhistorique: Antiquité de l'homme. Deuxième édition revue et complétée*. Paris: C Reinwald.

de Mortillet, P. (1906). Coup de poing chelléen à deux patines. *Bulletin de la Société préhistorique française, 3*(6), 231–232.

de Pradenne, A. V. (1935). "Fossil tradition" in stone implements. *Antiquity, 9*(33), 74–83.

Ebert, J. I. (1992). *Distributional archaeology*. Albuquerque: University of New Mexico Press.

Elston, R. G. (1992). Modeling the economics and organization of lithic procurement. In R. G. Elston & C. Raven (Eds.), *Archaeological investigations at Towawihi, a Great Basin quarry, Part 1: The periphery* (pp. 31–47). Elko, NV: Bureau of Land Management Elko Resource Area.

Emery, K. (2009). *A re-examination of variability in handaxe form in the British Palaeolithic (doctoral dissertation)*. London: University College London.

Evans, C., & Meggers, B. J. (1960). A new dating method using obsidian. Part II: An archaeological evaluation of the method. *American Antiquity, 25*(4), 523–537.

Evans, J. (1872). *The ancient stone implements, weapons and ornaments of Great Britain*. London: Longmans, Green, Reader, and Dyer.
Frame, H. (1986). Microscopic use-wear traces. In P. Callow & J. M. Cornford (Eds.), *La Cotte de St Brelade 1961–1978: Excavations by C. B. M. McBurney* (pp. 353–362). Norwich, UK: Geo Books.
Frison, G. C. (1968). A functional analysis of certain chipped stone tools. *American Antiquity, 33*(2), 149–155.
Goodwin, A. J. H. (1960). Chemical alteration (patination) of stone. *South African Archaeological Bulletin, 15*(59), 67–76.
Gopher, A., Barkai, R., Shimelmitz, R., Khalaily, M., Lemorini, C., Heshkovitz, I., & Stiner, M. C. (2005). Qesem Cave: An Amudian site in Central Israel. *Journal of the Israel Prehistoric Society, 35*, 69–92.
Gould, R. A. (1980). *Living archaeology*. Cambridge, UK: Cambridge University Press.
Gowlett, J. A. J., Harris, J. W. K., Walton, D., & Wood, B. A. (1981). Early archaeological sites, hominid remains and traces of fire from Chesowanja, Kenya. *Nature, 294*(5837), 125–129.
Haslam, M., Hernandez-Aguilar, R. A., Ling, V., Carvalho, S., De la Torre, I., DeStefano, A., . . . Warren, R. (2009). Primate archaeology. *Nature, 460*(7253), 339–344.
Hayden, B. (1977). Stone tool functions in the Western Desert. In R. V. S. Wright (Ed.), *Stone tools as cultural markers: Change, evolution and complexity* (pp. 178–188). Canberra: Australian Institute of Aboriginal Studies.
Henri-Martin, H. (1906). Commentaire sur la communication de Paul de Mortillet, 1906, "Coup de poing chelléen à deux patines." *Bulletin de la Société préhistorique française, 3*(6), 233.
Henri-Martin, H. (1923). *Recherches sur l'évolution du Moustérien dans le gisement de la Quina (Charente). 2e volume: Industrie lithique*. Angoulème: Mémoires de la Société archéologique et historique de la Charente.
Hiscock, P., & Attenbrow, V. (2005). *Australia's Eastern Regional Sequence revisited: Technology and change at Capertee 3*. Oxford: British Archaeological Reports.
Hogg, A. J. (1905). The patination of flint implements. *Man, 5*, 6–7.
Howard, C. D. (1999). River patina on flint artifacts: Features and genesis. *The Plains Anthropologist, 44*, 293–295.
Howard, C. D. (2002). The gloss patination of flint artifacts. *The Plains Anthropologist, 47*(182), 237–287.
Hurst, V. J., & Kelly, A. R. (1961). Patination of cultural flints. *Science, 134*(3474), 251–256.
Kelley, H. (1965). Outils levalloisiens de grande taille. *Bulletin de la Société préhistorique française, 62*(1), 31–47.
Kelly, R. L. (1988). The three sides of a biface. *American Antiquity, 53*(4), 717–734.
Kuman, K. (2001). An Acheulean factory site with prepared core technology near Taung, South Africa. *South African Archaeological Bulletin, 56*(173/174), 8–22.
Laland, K. N., & Brown, G. R. (2006). Niche construction, human behavior, and the adaptive-lag hypothesis. *Evolutionary Anthropology: Issues, News, and Reviews, 15*(3), 95–104.
Laland, K. N., & O'Brien, M. J. (2010). Niche construction theory and archaeology. *Journal of Archaeological Method and Theory, 17*(4), 303–322.
Laland, K. N., Odling-Smee, F. J., & Feldman, M. W. (2001). Cultural niche construction and human evolution. *Journal of Evolutionary Biology, 14*(1), 22–33.
Leakey, M. D. N. (1971). *Olduvai Gorge. Vol. 3, Excavations in Beds I and II, 1960–63*. Cambridge, UK: Cambridge University Press.

Leroyer, M. (2016). *Palethnologie acheuléenne: De la technologie bifaciale à l'organisation de la subsistance collective. Etude du site de Boxgrove, Eartham Pit (West Sussex, Angleterre) et de deux sites du cours moyen de la Seine.* Doctoral thesis, Université de Paris 1 en anthropologie et archéologie, Paris.

Le Tensorer, J.-M. (Ed.). (1997). *Les premiers hommes du désert syrien Fouilles Syrio-Suisses à Nadaouiyeh Aïn Askar 18 fevr.–1er septembre.* Paris: Muséum national d'Histoire naturelle.

Ludwig, B. V, & Harris, J. W. K. (1998). Towards a technological reassessment of East African Plio-Pleistocene lithic assemblages. In M. D. Petraglia & R. Korisettar (Eds.), *Early human behaviour in global context: The rise and diversity of the Lower Palaeolithic record* (pp. 84–107). London: Routledge.

Marder, O., Milevski, I., & Matskevich, Z. (2006). The handaxes of Revadim Quarry: Typo-technological considerations and aspects of intra-site variability. In N. Goren-Inbar & G. Sharon (Eds.), *Axe age: Acheulian tool-making from quarry to discard* (pp. 223–242). London: Equinox.

McDonald, M. M. A. (1991). Systematic reworking of lithics from earlier cultures in the Early Holocene of Dakhleh Oasis, Egypt. *Journal of Field Archaeology, 18*(2), 269–273.

Medina, M. (2007). *The world's scavengers: Salvaging for sustainable consumption and production.* Lanham, MD: Altamira Press.

Meeks, N. D., Sieveking, G. de G., Tite, M. S., & Cook, J. (1982). Gloss and use-wear traces on flint sickles and similar phenomena. *Journal of Archaeological Science, 9*(4), 317–340.

Metcalfe, D., & Barlow, K. R. (1992). A model for exploring the optimal trade-off between field processing and transport. *American Anthropologist, 94*(2), 340–356.

Michels, J. W. (1969). Testing stratigraphy and artifact reuse through obsidian hydration dating. *American Antiquity, 34*(1), 15–22.

Moore, M. W., & Burke, T. D. (1992). *Cultural resources inventory and evaluation: Truckee River Flood Control Project, Washoe and Storey Counties, Nevada, summary report.* Sacramento, CA: Department of the Army, Corps of Engineers.

Mora, R., de la Torre, I., & Martínez-Moreno, J. (2004). Middle Palaeolithic mobility and land use in the Southwestern Pyrenees: The example of Level 10 in La Roca dels Bous (Noguera, Catalunya, Northeast Spain). In N. J. Conard (Ed.), *Settlement dynamics of the Middle Paleolithic and Middle Stone Age* (Vol. 2, pp. 415–435). Tübingen, Germany: Kerns Verlag.

Odell, G. H. (2001). Stone tool research at the end of the millennium: Classification, function, and behavior. *Journal of Archaeological Research, 9*(1), 45–100.

Odling-Smee, F. J., Laland, K. N., & Feldman, M. W. (1996). Niche construction. *The American Naturalist, 147*(4), 641–648.

Ophoven, C. (1938). Quelques cas de patines intéressantes. *Bulletin de la Société préhistorique française, 35*(10), 31–47.

Pappu, S., & Akhilesh, K. (2006). Preliminary observations on the Acheulian assemblages from Attirampakkam, Tamil Nadu. In N. Goren-Inbar & G. Sharon (Eds.), *Axe age: Acheulian tool-making from quarry to discard* (pp. 155–180). London: Equinox.

Pigeot, N. (1990). Technical and social actors: Flintknapping specialists and apprentices at Magdalenian Etiolles. *Archaeological Review from Cambridge, 9*(1), 126–141.

Polikreti, K. (2007). Detection of ancient marble forgery: Techniques and limitations. *Archaeometry, 49*(4), 603–619.

Pope, G. A., Meierding, T. C., & Paradise, T. R. (2002). Geomorphology's role in the study of weathering of cultural stone. *Geomorphology, 47*(2), 211–225.

Pope, M. I. (2002). *The significance of biface-rich assemblages: An examination of behavioural controls on lithic assemblage formation in the Lower Palaeolithic.* Doctoral dissertation, University of Southhampton, Southampton, UK.

Pope, M. I., & Roberts, M. B. (2005). Observations on the relationship between Palaeolithic individuals and artefact scatters at the Middle Pleistocene site of Boxgrove, UK. In C. Gamble & M. Porr (Eds.), *The hominid individual in context: Archaeological investigations of Lower and Middle Palaeolithic landscapes, locales and artefacts* (pp. 81–97). London: Routledge.

Pope, M. I., Russel, K., & Watson, K. (2006). Biface form and structured behaviour in the Acheulean. *Lithics: The Journal of the Lithic Studies Society, 27*, 44–57.

Potts, R. (1988). *Early hominid activities at Olduvai.* New York: Aldine de Gruyter.

Preston, B. (2000). The functions of things: A philosophical perspective on material culture. In P. Graves-Brown (Ed.), *Matter, materiality and modern culture* (pp. 22–49). London: Routledge.

Prestwich, J. (1860). On the occurrence of flint-implements: Associated with the remains of animals of extinct species in beds of a late geological period, in France at Amiens and Abbeville, and in England at Hoxne. *Philosophical Transactions of the Royal Society of London, 150*, 277–317.

Purdy, B. A., & Clark, D. E. (1987). Weathering of inorganic materials: Dating and other applications. *Advances in Archaeological Method and Theory, 11*, 211–253.

Roberts, M. B., & Parfitt, S. A. (Eds.). (1999). *Boxgrove: A Middle Pleistocene hominid site at Eartham Quarry, Boxgrove, West Sussex.* London: English Heritage.

Rogers, A. K. (2008). Obsidian hydration dating: Accuracy and resolution limitations imposed by intrinsic water variability. *Journal of Archaeological Science, 35*(7), 2009–2016.

Rondeau, M. F. (1997). Technology as context for obsidian hydration studies. *Lithic Technology, 22*(1), 86–98.

Rottländer, R. (1975). The formation of patina on flint. *Archaeometry, 17*(1), 106–110.

Rudner, I., & Rudner, J. (1954). A local Late Stone Age development. *South African Archaeological Bulletin, 9*(35), 103–107.

Rutot, A. L. (1906). *La géologie appliquée à la démonstration de l'authenticité des silex taillés paléolithiques de la vallée de la Haine.* Bruxelles: Hayez.

Schiffer, M. B. (1972). Archaeological context and systemic context. *American Antiquity, 37*(2), 156–165.

Schiffer, M. B. (1987). *Formation processes of the archaeological record.* Albuquerque: University of New Mexico Press.

Schiffer, M. B., Downing, T. E., & McCarthy, M. (1981). Waste not, want not: An ethnoarchaeological study of reuse in Tucson, Arizona. In R. A. Gould & M. B. Schiffer (Eds.), *Modern material culture: The archaeology of us* (pp. 67–86). New York: Academic Press.

Schmalz, R. F. (1960). Flint and the patination of flint artifacts. *Proceedings of the Prehistoric Society, 26*, 44–49.

Shackley, M. L. (1974). Stream abrasion of flint implements. *Nature, 248*(5448), 501–502.

Shackley, M. L. (1989). "Glazed" (silica coated) flint artifacts from the Azraq Basin, eastern Jordan. In L. Copeland & F. F. Hours (Eds.), *The hammer on the rock: Studies in the Early Palaeolithic of Azraq, Jordan* (pp. 399–402). Oxford, UK: British Archaeological Reports.

Shaw-Williams, K. (2014). The social trackways theory of the evolution of human cognition. *Biological Theory, 9*(1), 16–26.

Silliman, S. W. (2005). Obsidian studies and the archaeology of 19th-century California. *Journal of Field Archaeology, 30*(1), 75–94.

Smith, R. A. (1929). Palaeolith found at West Bognor. *Sussex Archaeological Collection, 70*, 196–197.

Smith, W. G. (1894). *Man the primeval savage: His haunts and relics from the hilltops of Bedfordshire to Blackwall*. London: Edward Stanford.

Soressi, M., & Hays, M. A. (2003). Manufacture, transport, and use of Mousterian bifaces: A case study from the Périgord (France). In M. Soressi & H. L. Dibble (Eds.), *Multiple approaches to the study of bifacial technologies* (pp. 125–147). Philadelphia: University of Pennsylvania Museum of Archaeology and Anthropology.

Soriano, S. (2000). *Outillage bifacial et outillage sur éclat au Paléolithique ancien et moyen: Coexistence et interaction (doctoral dissertation)*. Paris: Université de Paris X.

Stout, D., Semaw, S., Rogers, M. J., & Cauche, D. (2010). Technological variation in the earliest Oldowan from Gona, Afar, Ethiopia. *Journal of Human Evolution, 58*(6), 474–491.

Sturge, W. A. (1911). The chronology of the Stone Age. *Proceedings of the Prehistoric Society of East Anglia, 1*(1), 43–105.

Thompson, E. (2009). Acheulean artifact accumulation and early hominin land use, Garden Route Casino Road, Pinnacle Point, South Africa. *Geoarchaeology, 24*(4), 402–428.

Turner, R., & Wymer, J. J. (1987). An assemblage of Palaeolithic hand-axes from the Roman religious complex at Ivy Chimneys, Witham, Essex. *Antiquaries Journal, 67*(1), 44–60.

Viereck, A. (1964). An examination of patina on stone tools with regard to possible dating. *South African Archaeological Bulletin, 19*(74), 36–37.

Villa, P. (1983). *Terra Amata and the Middle Pleistocene archaeological record of southern France*. Berkeley, CA: University of California Press.

Warren, S. H. (1902). I—On the value of mineral condition in determining the relative age of stone implements. *Geological Magazine, 9*(3), 97–105.

Wells, T., Hancock, G., & Fryer, J. (2008). Weathering rates of sandstone in a semi-arid environment (Hunter Valley, Australia). *Environmental Geology, 54*(5), 1047–1057.

Winton, V. (2004). *A study of Palaeolithic artefacts from selected sites on deposits mapped as clay-with-flints of southern England*. Oxford, UK: Archaeopress.

Woodcock, A. (1981). *The Lower and Middle Palaeolithic periods in Sussex*. Oxford, UK: British Archaeological Reports.

Wymer, J. J., & Singer, R. (1993). Flint industries and human activity. In R. Singer, B. G. Gladfelter, & J. J. Wymer (Eds.), *The Lower Paleolithic site at Hoxne, England* (pp. 74–128). Chicago, IL: University of Chicago Press.

Wynn, T., & McGrew, W. C. (1989). An ape's view of the Oldowan. *Man, 24*(3), 383–398.

8

FLAKE-MAKING AND THE "COGNITIVE RUBICON"
INSIGHTS FROM STONE-KNAPPING EXPERIMENTS

Mark W. Moore

A necessary first step in any evolutionary analysis is the identification of what is peculiarly human. (Wynn, 2002, p. 392)

INTRODUCTION

The manufacture of stone tools is one of the distinguishing characteristics of the hominin lineage. Not only are stone tools ubiquitous on archaeological sites spanning some 3.3 million years ago (Mya) (Harmand et al., 2015), but stone tool manufacture is an acquired skill that continues to challenge the mental and motor-action abilities of modern-day humans (Whittaker, 2004). Because of this, stone tools are a key source of information for exploring the mental evolution of our hominin ancestors, as demonstrated by the provocative and influential explanatory models developed by Thomas Wynn and his colleagues. Wynn's research interprets the minimum abilities required to make the stone tools found on early sites through consideration of tool morphology combined with the constraints of the stone-flaking process. Principles of cognitive psychology and neurobiology are invoked to assess the mental abilities necessary to work within those constraints to produce the artifacts seen at various points in the early hominin archaeological record (e.g., Wynn, 1979, 1981, 2002, Wynn & Coolidge, 2004, 2016, Wynn & McGrew, 1989). The epistemological link between the constraints of stone flaking and mental processes are the "middle-range" insights provided by modern flintknappers. These insights are anecdotal (e.g., Edwards, 2001, Pelegrin, 1993, Toth, 1985), the result of controlled experiments (e.g., Dibble & Whittaker, 1981, Pelcin, 1997), or a combination of the two (e.g., Eren, Bradley, & Sampson, 2011, Kelterborn, 2003, Newcomer, 1971, Pelegrin, 2006, Schlanger, 1996, van Peer, 1992).

Cognitive archaeologists have looked to Wynn's models to frame the debate and to set research priorities and directions, and I have drawn on my own anecdotal and experimental knowledge of flintknapping to model the "design space" of stone tool-making to test or refine aspects of Wynn's models (Moore, 2007, 2010, 2011; Moore & Perston, 2016). The design space model parses stone flaking into two hierarchical levels: first, the mental conceptions and motor actions necessary to produce flakes individually (the "flake unit"), and second, the mental conceptions and motor actions necessary to combine flakes units in order to make tools (Moore, 2010). An important implication of the design space model is that, because of the tight physical limits on fracture mechanics combined with hominin physical capabilities, simple combinations of flake units can, in theory, produce complex-looking tools or attributes without a deliberate intention to do so, and without higher-order cognitive skills (Moore, 2010, 2011). Complex-looking tools or attributes are referred to as "spandrels" (Moore, 2010, p. 23), following Gould and Lewontin's (1979) use of the architectural term for inevitable features that erroneously appear to be created for a specific purpose. Testing of the spandrels hypothesis was begun through a series of stone-flaking experiments explicitly designed to disrupt the cognitive processes of a modern flintknapper, and the results demonstrated that certain early stone tools and attributes can, in fact, be parsimoniously explained as spandrels (Moore & Perston, 2016). However, the subject of these experiments was the combination of flake units to make tools. The flake unit itself was viewed as a black box in experimental terms, with the focus of design on the higher level of tool production.

This chapter considers the design space implications of the flake unit in the context of Wynn's views on the innovation of stone flaking among hominins. Hominins appear to have understood the fundamental mechanics of flaking soon after stone tools appeared in the archaeological record (de la Torre, 2010; Toth, Schick, & Semaw, 2006). These earliest tool-makers engaged in controlled flaking to produce well-struck flakes, with low incidences of telltale characteristics of uncontrolled flaking, such as small relative flake sizes, collapsed flake platforms, abundant shatter fragments, battered and crushed core edges, and right-angle platforms (Toth, Schick, Savage-Rumbaugh, Sevcik, & Rumbaugh, 1993, p. 89). We can therefore infer at this early stage an intention on the part of hominins to flake stone for tools (Davidson & McGrew, 2005) and mastery in striking flakes from cores. Wynn is unimpressed by the cognitive significance of this development, arguing that the cognitive complexity of producing the well-struck flakes seen in the earliest assemblages was within the capabilities of modern-day apes and other primates (Wynn, 1981, 2002, pp. 394, 398; Wynn & McGrew, 1989; also see Marchant & McGrew, 2005). In this view, the "cognitive Rubicon" (Wynn & Coolidge, 2016) to more modern-like human mental capabilities was crossed much later, with the emergence of bifacial handaxes in the Early Acheulean (Wynn, 2002), ca. 1.75 Mya (Beyene et al., 2013). The issue is explored here through two thought experiments that go to the heart of stone-flaking design space and clarify the importance of the flake unit for controlled flaking. I argue that the initial invention in stone flaking was, in fact, a cognitive Rubicon exclusive to hominins, and this has important implications for the spatial and working memory aspects of Wynn's models of hominin cognitive development.

MODERN FLINTKNAPPING AND STONE-FLAKING DESIGN SPACE: TWO THOUGHT EXPERIMENTS

The design space model had its genesis in two thought experiments involving modern flintknapping demonstrations. The audience for the demonstration in each thought experiment is naïve, observing flintknapping for the first time, and the flintknapper's intent is to make a Late Acheulean handaxe with three-dimensional symmetry and

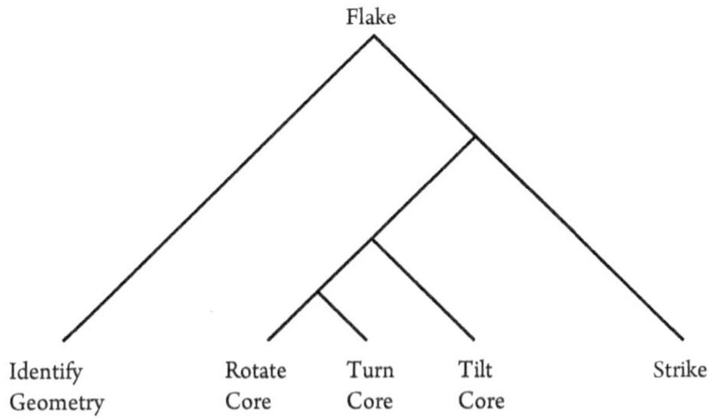

Figure 8.1. Moore's (2010) model of the basic flake unit, showing the hierarchical organization of visual search and motor-action tasks required for controlled stone flaking. The modified tree structure follows Greenfield (1991). Image by the author.

an S-twist (see Wynn, 2000, pp. 395–397). Typically, the audience to a flintknapping demonstration does not interrupt, and the flintknapper demonstrates their expert technical knowledge and know-how to produce a stone tool. But in the first thought experiment, a naïve audience member continually disrupts the flintknapper by dictating precisely where he or she wants each flake to be removed from a stone cobble. And in the second thought experiment, the naïve audience member initially allows the flintknapper to choose where they want to remove the flake, but then says, "Don't remove that flake; remove a different flake," forcing the flintknapper to shift attention to a different spot on the cobble.[1] So, in these two circumstances, would the flintknapper be able to produce a Late Acheulean handaxe?

The thought experiments expose the fundamental division in two sets of technical knowledge required to make complex stone tools. The first set is the mental identifications and motor actions required to strike an individual flake from a core (Figure 8.1; also see Moore, 2010). The success of these identifications and actions is determined by the physics governing stone fracture. For successful stone flaking, the flintknapper must be able to identify a three-dimensional configuration consisting of an area of high mass on one face of the core with an adjacent striking surface oriented at an acute angle to the high mass. To act on this, the flintknapper rotates and tilts the core to deliver an oblique or glancing blow to the striking surface, or *platform*, so that the force is directed through the high mass. The blow is delivered with a hammerstone held in the dominant hand. The strength of the blow must be modulated relative to the nature of the configuration identified on the core, including the size of the intended flake: If the blow is too hard relative to the size of the core, the platform or flake may shatter, and if the blow is not hard enough, the flake may result in a step fracture or may not be created at all. Modulation relative to core size is done by altering the strength of the blow and/or choosing a hammerstone with greater or lesser mass. The mental and

[1] This occurred at a handaxe-making demonstration by flintknapper Nicholas Toth, who was interrupted by the audience member (Iain Davidson) who asked this question. I thank Iain for passing on this anecdote.

motor-action knowledge marshaled to remove a single flake is called the "flake unit" (Moore, 2007, 2010).

In the first thought experiment, the naïve audience member interfered in the flintknapper's attempt to marshal the technical knowledge of platforms and high mass that the flake unit requires. Stone-flaking success was entirely disrupted, and the process was akin to haphazardly bashing two rocks together. Well-struck flakes rarely, if ever, occurred, and a Late Acheulean handaxe was not produced.

In the second thought experiment, the naïve audience member intervened after the flintknapper made their mental identifications but before the motor actions completed the flake unit. In response to the audience member, the flintknapper shifted attention to elsewhere on the core and repeated the mental identifications and removed a flake, but not the flake that was originally intended. Although well-struck flakes were successfully made during the demonstration, the order in which they were struck was not controlled by the flintknapper—the order was effectively randomized—and a Late Acheulean handaxe was not produced. This highlights the second aspect of technical knowledge required to make complex stone tools: Flake removals need to be arranged sequentially in certain ways to achieve complex results. To create a complex stone tool like a Late Acheulean handaxe, flakes units must be arranged hierarchically according to a plan of action, with the flintknapper thinking many flake removals ahead in order to achieve the desired result (Moore, 2010, 2015). In the thought experiment, disruption of the plan of action prevented the flintknapper from thinking ahead to successfully make the handaxe, but it did not prevent the production of well-struck flakes.

In using modern experiments as sources of inferences, flintknappers follow the typical demonstration, where the flintknapper is unhindered in calling on both aspects of his or her technical knowledge to produce the stone tool. If the experimental products and by-products closely match the prehistoric versions, then the less tangible actions of the flintknapper—aspects of flake units and plans of action—are inferred to be similar to that of past flintknappers. This reasoning involves three warranting arguments about equivalencies between the present and past: First, the physics of stone fracture are equivalent; second, hominin body plans are equivalent; and third, cognitive abilities are equivalent. These warranting arguments are uncontroversial for the experimental replication of tools from later prehistory made by behaviorally modern humans. Critiques of these flintknapping experiments tend to focus on empirical issues of pattern matching or epiphenomena rather than questioning these uniformitarian principles (e.g., Clark, 2002, Thomas, 1986).

However, most experiments that replicate stone tools made by non-modern hominins are also structured in this way. Because of the first warranting argument, the plans of action used by modern flintknappers to make objects like Late Acheulean handaxes and Levallois flakes are thought to be accurate reflections of non-modern hominins' plans of actions, but, as with experiments into the complex tools of modern humans, the third warranting argument is required to sustain the interpretive link between past and present. Although the reconstructed plans of action (or associated brain images) from replication experiments like these are deconstructed for insights into evolving hominin cognition, these models can only recapitulate the thought processes of the modern flintknapper because, by using an unfettered modern flintknapper as the link, the third warranting argument is implicit. Returning to the second thought experiment, the naïve audience member constrains the modern flintknapper by randomizing flake removals and preventing their plan of action, in effect removing the third warranting argument implied by the flintknapping demonstration; reduction effectively begins anew with each flake removed (see Moore & Perston, 2016, p. 4). A core was produced, but not a Late Acheulean handaxe.

What did that core look like? To explore this question, we designed a series of experiments—referred to as "spandrels experiments"—to enact a version of the second thought experiment (Moore & Perston, 2016). But before turning to the results of these actual experiments, it is important to consider why the first thought experiment resulted in abject failure. How is controlled flintknapping different from simply bashing rocks together? Addressing this requires further exploration of the inner workings of the flake unit.

VISUAL SEARCH ASPECTS OF THE FLAKE UNIT

In the first thought experiment, disrupting the flake unit resulted in failure of the demonstration. Why is this so? A pervading stereotype in popular culture and among some archaeologists is that simple stone flaking is, in fact, little more than bashing two rocks together. This was the outcome of the first thought experiment, when the naïve audience member—not the flintknapper—chose all of the locations to strike the stone. In contrast, in the second thought experiment, the flintknapper had scope to choose another suitable platform (just not the one identified by the naïve audience member), and flakes were successfully produced. So what part of the flintknapper's knowledge was disrupted in the first thought experiment? It would appear that disruption occurred, first and foremost, in the visual attention aspects of the flake unit that identify suitable geometric configurations on a core (Wynn & Coolidge, 2016).

Visual attention in search tasks by modern humans has been modeled in detail by cognitive psychologists. According to feature integration theory, it is impossible to process all the information in a field of view; thus, attention is deployed to search for a target configuration among the noise of distracting configurations (Quinlan, 2003; Treisman & Gelade, 1980). Attention occurs in two phases: a *pre-attentive* phase, where basic features of the field of view are identified (independent of attention), and the *attentive* phase, where inputs are more extensively processed. The pre-attentive phase provides the information for the attentive phase through the use of a set of guiding representations: a control module composed of empirical attributes that are derived from the visual pathway but which, for reasons involving sequencing of the two phases, must be situated outside of it (Wolfe, 2007; Wolfe & Horowitz, 2004). This control module guides pre-attentive inputs using categorical information about attributes, such as color, orientation, size, shape, topology, and curvature (see Wolfe & Horowitz, 2004, Table 1). The guiding process allows the selection of *targets*, or combinations of attributes, from among the background noise of *distractors*.

Some search tasks are easy, but others are not. Easy, efficient searching is a function of the degree of difference between targets and distractors, referred to as *saliency*: A large difference is a "strong" signal, a small difference is a "weak" signal. The signal is further compromised if there is a great deal of variation among distractors, and if the difference between the target and distractor is not defined by a single unique feature (Hout & Goldinger, 2015). Also, the way in which distractors vary is an important issue: For instance, the search is complicated if distractors are adjacent to the target feature in the visual field. Despite this, even complex field searches can become fast and effortless—"automatic"—by "priority learning," which pulls attention directly to the target item without the need for searching (Christie, Livingstone, & McDonald, 2014; also see Land & Hayhoe, 2001; Land & McLeod, 2000). The target is said to "pop out" from the visual field (Bravo & Nakayama, 1992). Automaticity can be independent from attention (pre-attentive automaticity) or linked to it (post-attentive automaticity) (Logan, 1992; Wolfe, 2016).

Post-attentive automaticity is conditioned by prior experience (and can be learned) and is governed by memory: "Whereas novice performers attend to the various steps of the algorithm they execute to produce a solution, automatic performers attend to the solutions that memory provides" (Logan, 1992, p. 321; also see Dowd & Mitroff, 2013). Although encoding empirical attributes into the control module is obligatory when attention is applied, thereby building memory strength, attention is required to invoke the memory retrieval process; conversely, if information is not readily retrievable from memory, automaticity is not possible (Logan, 1992, p. 336). Attention is directed at a certain level of organization, with all levels beneath it performed automatically (Vallacher & Wegner, 1987). Attention is delivered to the highest level that allows task completion without shifting to lower levels (a proxy of skill), although errors can force attention to lower levels. As proficiency at automaticity increases, attention shifts up the hierarchical levels of task organization. Another aspect of visual search and post-attentive processing is *inhibition*: Familiar targets automatically attract attention and can divert attention from the task at hand, disrupting hierarchical proficiency.

The ability to identify a suitable striking platform to remove a flake from a core is a problem of visual attention. This is the *geometric identification* element of the flake unit (Figure 8.1; also see Moore, 2010), and the search target is composed of a configuration of geometric attributes consisting of (1) a flat surface (2) oriented at an acute angle to (3) a convex surface (4) located adjacent to the flat surface. The flat surface is the striking platform, and the acceptable parameters for it range from slightly concave to moderately convex, whereas the convex surface is the high mass on the core face and must be moderately to markedly convex; thus, the two surfaces are asymmetrical. An acceptable angle between the surfaces must be below 90 degrees and above 35–45 degrees (Whittaker, 1994), but to successfully target the high mass, a range to 75–80 degrees is optimal (Toth et al., 1993, p. 89). These are continuous variables that form a multi-attribute search target wherein the hominin flintknapper must discriminate two intersecting surfaces of specific but different shapes, and a correct orientation of the surfaces to each other in three-dimensional space. The desirable shape configurations on a stone (suitable platforms) often differ subtly from undesirable ones (unsuitable platforms), resulting in a "weak signal."

Further, shape distractors vary considerably and are distributed continuously around acceptable configurations. Rather than a single unique feature, the difference between a suitable platform and an unsuitable one is defined by multiple shape attributes. Because of these factors, the search task for identifying suitable platforms on a stone is a very difficult one. Acting on the configuration involves rotating the core held in the non-dominant hand in such a way that the platform is in a position to strike, which removes the core face from the direct view of the flintknapper. Modern flintknappers visualize the mass so that the blow is delivered behind it and the force directed into it, a process that can be augmented by tactile input from the fingers holding the core. Despite the complexity of the visual search task and mental visualization, skilled modern flintknappers demonstrate post-attentive automaticity in striking flakes by effortlessly accomplishing the search task and removing large, well-struck flakes in rapid succession. The degree of automaticity is a function of experience, and it demonstrates that information about appropriate platform configurations is readily retrievable from memory. It is this post-attentive automaticity in removing individual flakes that frees the modern flintknapper to deliberately shape stones through carefully organized sets of flake removals (cf. Moore, 2015).

Is stone fracture by modern non-human primates analogous to the application of the flake unit? Chimpanzees use percussive technology to crack nuts with stone hammers on wood or stone anvils (Neufuss, Humle, Cremaschi, & Kivell, 2016; Whiten et al., 2001), and Old World macaques use stones as hammers and in play, sometimes involving percussion (Marchant & McGrew, 2005, p. 341; Wynn, Hernandez-Aguilar, Marchant, & McGrew, 2011, p. 186). In a case of technological convergence with apes (Visalberghi & McGrew, 1997), both macaques (Gumert, Kluck, & Malaivijitnond, 2009) and New World capuchin monkeys (Visalberghi et al., 2007) use stones in percussive technology to crack nuts. Capuchins also engage in forceful percussion using hammers on fixed cobbles, apparently to produce dust or powdered lichens, which they ingest (Proffitt et al., 2016). These various percussion activities result in damage to the stones, such as the fracturing of stone anvil edges from mishits (Marchant & McGrew, 2005) or shattering of the percussion stones, and some of these fragments show the empirical attributes used by archaeologists to identify deliberate stone flaking (e.g., Mercader et al., 2007; Proffitt et al., 2016, Figure 2). In all observed cases, however, the production of flakes and cores was an accidental by-product of percussion, not the purposeful production of flakes (Peacock, 1991; Wiśniewski, Badura, Salamon, & Lewandowski, 2014). It is clear, then, that the visual search aspects of the flake unit, and the subsequent motor actions that act on it, are not practiced by wild stone-using chimpanzees or monkeys.

In another study, a skilled modern flintknapper, Nicholas Toth, attempted unsuccessfully over a number of years to teach a captive bonobo, Kanzi, controlled stone flaking through the application of the flake unit. Kanzi was able break stones, but most pieces lacked the attributes of controlled flaking evident in the earliest hominin assemblages (Toth et al., 1993). Part of Kanzi's difficulty may have been related to biomechanical limitations of bonobos in holding stones for forceful blows (however, see Neufuss et al., 2016), but crucially, Kanzi was unable to identify acute platform angles between the platform surface and core face (de Beaune, 2004, p. 141; Toth et al., 1993, p. 89). This shows that even with sustained direct instruction, captive apes, like their wild cousins, are not capable of the visual search necessary to enact the flake unit.

It is instructive to consider the process of teaching controlled flake-making to modern human subjects, the acknowledged masters of copying (Henrich & McElreath, 2007). Students find the process difficult and frustrating—certainly not effortless or intuitive—despite extensive demonstration and direct instruction on the necessary geometric and motor actions required. Most students require thousands of repetitions to master the flake unit, but, importantly, it is rarely beyond their capabilities (MWM, personal observation).

Returning to our first thought experiment, the failure to produce controlled stone flaking was due to the naïve audience member disrupting the flintknapper's visual attention. The result was similar to that of non-human stone use, with stone-on-stone percussion—stone bashing—creating shatter and perhaps the occasional flake but rarely well-struck flakes. The flake unit is not possible without visual attention, and the components of visual attention require cognitive abilities that exceed those of other primates. Next I will show that the mental aspects inherent to the flake unit are similar to the mental aspects invoked by Wynn (2002) to explain the evolution of symmetry in stone tools. But first I will review how rote application of flake units will produce some of those symmetries unintentionally.

COMBINATIONS OF FLAKE UNITS AND STONE-FLAKING SPANDRELS

In the second thought experiment, visual attention was unimpaired—well-struck flakes were produced—but a Late Acheulean handaxe did not result. Why might this be the case? A series of 59 novel experiments were undertaken to test whether this would, in fact, be the outcome and to explore what might be produced instead (Moore & Perston, 2016). Each experiment involved the reduction of one large cobble or flake blank, which required multiple flake removals. For each flake removal, the experienced modern flintknapper[2] employed a visual search and identified and numbered all of the suitable geometric configurations on the stone. One configuration (platform) was chosen using a random number generator, mimicking the naïve audience member in the second thought experiment. The flintknapper removed as large a flake as possible from the platform. Removing the flake inevitably changed the geometry of the stone, so the flintknapper repeated the visual search to identify new platforms (and cross off ones that were no longer viable), the next platform was selected randomly, the flake struck, and so on. The experiment ended when the stone dropped below an arbitrarily defined size threshold. The flintknapper's application of the flake unit was unimpeded—otherwise, as in our first thought experiment, controlled flaking would not have occurred—but, since platforms were chosen randomly, the flintknapper's propensity for goal-directed thinking ahead was impaired in the same way as seen in the second thought experiment. The morphology of the core after each flake removal was recorded and analyzed—an assemblage of 1,115 cores—to explore the kinds of objects produced and compare them to the archaeological record of early stone flaking (Moore & Perston, 2016).

[2] For consistency, one flintknapper conducted all of the experiments. The flake unit was considered an experimental black box, and the expert flintknapper was allowed to use all of his know-how to successfully identify platforms and remove individual flakes (Moore & Perston, 2016, pp. 3–6). Variation in expertise among highly skilled flintknappers tends to be expressed in the conceptualization and execution of flake removal sequences, rather than the removal of individual flakes, and flake removal sequences were explicitly prevented by the experimental protocols. For this reason, the experiments limit the potential unconscious biases of the flintknapper. It is predicted that repetition of the experiments using a different experienced flintknapper, following the same strict protocols, will produce similar results.

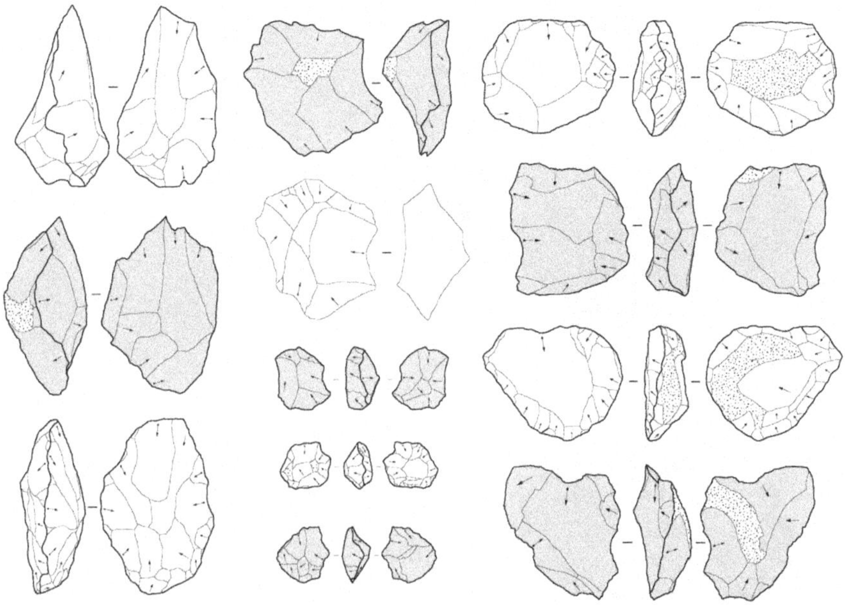

Figure 8.2. Comparison of stone artifacts and spandrels (after Moore & Perston, 2016, Figure 19). The non-shaded stone tools were made by early hominins in Africa. They are traditionally thought to have been the outcome of deliberate design. The shaded objects were produced in experiments with protocols that prevented intentional design. Despite this, the spandrels mimic "designed" tools, including proto-bifaces or handaxes (left), discoidal biface choppers (center), and prepared Levallois cores (right). Image by the author.

Our experiments produced cores and flakes displaying aspects of the ostensibly intentional technologies that mark the major early milestones in the standard story of technological and cognitive evolution (Figure 8.2). This included cores with bifacially flaked edges opposite cortical margins ("choppers"), bifacially flaked discoidal cores, and multiplatform cores similar to polyhedrons. Bifacial flaking was inevitable and occurred within 12 blows in all of the experiments. The shapes of these cores changed in patterned ways through the reduction process, including the consistent progression toward plateaus in biface width-to-thickness and length-to-width ratios (1.8–1.9 and 1.33–1.34, respectively), even though there was no intention to achieve these shapes. The experimental cores were classified into types defined by Isaac (1977), and the proportion of "large tools" in the experimental assemblage places it in the Early Acheulean (after Isaac, 1977, p. 112, Figure 37). Objects that can be classified as proto-bifaces or crude handaxes were produced by the experiments (Figure 8.3), as were cores with "predetermined" flake removals (Figure 8.4)—a key attribute of the Levallois method *sensu lato*. These stone artifact types and attributes mimic those thought by many archaeologists to have been produced by goal-directed, intention-driven stone flaking, but they were instead the outcome of the mechanics that govern stone fracture combined with a simple flake-removal algorithm—the flake unit—applied repetitively to the same cores (Moore & Perston, 2016). Given that visual attention precedes motor action in the operation of the flake unit, the emergence of

Figure 8.3. Handaxe-like proto-biface produced in the spandrels experiments (after Moore & Perston, 2016, Figure 16). Flakes were conjoined to reconstruct this core. Scale 50 mm. Image by the author.

stone-flaking spandrels in hominin evolution appears to be the result of cognitive developments in visual attention.

Nevertheless, although the experimental assemblage mimicked some stone artifact forms and attributes seen in the Oldowan and Early Acheulean, they differed from certain Late Acheulean artifacts in ways relevant to Wynn's model of cognitive evolution. For instance, although all of the experimental cores were reduced bifacially, and relatively crude handaxes with approximate bilateral symmetry were created, handaxes with "congruent" symmetry were not. This was because (1) random platform selection

Figure 8.4. Core and "predetermined" flake produced in the spandrels experiments (after Moore & Perston, 2016, Figure 12). Scale 50 mm. Image by the author.

forced random core reorientations; (2) the flintknapper was prevented from working recursively to conduct prior flaking to improve striking platforms for later flaking (see Moore, 2010, pp. 20–22); and (3) core shaping was prevented by random core reorientations and the mandate to remove as large a flake as possible. The production of congruently symmetrical, teardrop-shaped Late Acheulean handaxes will not occur without eliminating the maximization protocol and randomized platform selection, and perhaps allowing platform preparation by recursive flaking (Moore & Perston, 2016). The degree to which the spandrels experiments unintentionally shaped cores and made them symmetrical relates directly to the empirical evidence used by Wynn (1979, 1981, 2002, Wynn & Coolidge, 2004, 2016) to model cognitive evolution. I will turn to this next, exploring the issue of a "cognitive Rubicon" between an ape adaptive grade and a human one, and provide an alternative model for some of the trends identified by Wynn in the archaeological record.

FLAKE-MAKING AND THE "COGNITIVE RUBICON"

In Wynn's model, the earliest stone cores and retouched flakes reflect relatively primitive spatial concepts of boundary and proximity. A boundary divides a spatial field into two realms, and hominins demonstrated a concept of boundary in creating bifacial edges on stone cores. Proximity, or "nearby-ness," is inferred by the way hominins struck stone cobbles in more or less the same place. The "cognitive Rubicon" (Wynn & Coolidge, 2016) in hominin evolution occurred when spatial concepts of symmetry became a "transformational rule" in stone tool manufacture (Wynn, 2000, p. 138), as seen in the handaxes of the Early Acheulean in Africa, ca. 1.75 Mya (Beyene et al., 2013). The creation of simple and approximate reflectional symmetry on these objects indicates the development of "frame independence, or the ability to see past the constraints imposed by the visual array" (Wynn, 2002, p. 395). In addition, disc-shaped cores in these assemblages suggest trimming to make all of the diameters roughly equal, which implies a simple concept of spatial amount. The breakthrough implied by early handaxes was that spatial abilities such as boundary, proximity, and amount were coordinated with the neural network for shape recognition to impose repetitive shapes onto cores (Wynn, 2002, p. 395). A second breakthrough occurred in the Late Acheulean, ca. 500,000 years ago, with the emergence of teardrop-shaped handaxes with symmetrical "congruency," or quantitative duplication (as opposed to approximation) in the mirrored sides. The congruent symmetries were extended to cross sections as well as plan shape, and, in later examples, to "broken symmetry" (such an S-twist to the handaxe profile). To achieve an S-twist, the hominin flintknapper used sophisticated geometric estimations to visualize how flaking would change the core from a mix of perspectives, including ones not immediately visible. A conception of space as three-dimensional positions was used to organize stone-flaking actions—a problem of central processing, as opposed to shape recognition and spatial assessment (Wynn, 2002).

This complex central processing relied on working memory to underpin the plans of action necessary to coordinate motor actions to impose complex three-dimensional symmetry. Working memory in modern humans (Coolidge & Wynn, 2005; Wynn & Coolidge, 2004) consists of a decision-making central executive to focus attention by inhibiting stimuli that are irrelevant to the goal, maintaining the relevant stimuli, and updating incoming information. The central executive is supported by two systems, a phonological loop for verbal and sound stimuli, and a visuospatial sketchpad for integrating and temporarily storing visual ("what") and spatial ("where") information. Tool-making mostly or exclusively engages the visuospatial sketchpad rather than the phonological loop. Unconscious long-term memory underpins this conscious structure and includes declarative memory for facts and verbal information, and non-declarative procedural memory for nonverbal motor skills. Procedural memories form slowly through motor-action repetition, are resistant to interference, and rarely disappear once acquired. They tend to be very narrowly focused to specific actions or domains. Expert performance at activities like stone flaking is achieved by the creation of associative rules, or retrieval structures, that allow fast and reliable access to procedural memories as chunks. Chunks are precise, algorithm-like patterns of motor actions that can be deployed without conscious direction. By doing so, the expert need not "reason through the relationships anew" (Coolidge & Wynn, 2005, p. 19). Through extensive practice spanning years and tens of thousands of iterations, experts develop powerful retrieval structures and memory chunks of increasing breadth. The tools and materials themselves can be retrieval structures, or external memory "scaffolds." A degree of working memory is implicated in the hierarchical steps necessary to produce congruently symmetrical Late Acheulean handaxes, but it is also reflected in the hierarchical flaking steps necessary to produce the "preferential" flake in the Levallois method *sensu stricto* (Wynn & Coolidge, 2004).

Wynn's model is based on the cognitive aspects of removing multiple flakes from a core, rather than flakes individually, because the skills reflected in the flake unit do not represent a cognitive Rubicon but "merely a variant on the basic ape adaptive pattern, with no obvious leap in intellectual ability required" (Wynn, 2002, p. 394). But, as our first thought experiment has shown, controlled stone flaking is not possible without mastering the flake unit, and, paradoxically, five aspects of Wynn's model of cognitive development can be seen in the operation of the flake unit:

(1) The visual search aspects of the flake unit require seeing past the constraints of the visual array to identify the appropriate configurations for successful flake production. This is a form of "frame independence."
(2) The strike location on the platform surface is a function of the size of the high mass to be removed by the blow. To remove large areas of high mass, the platform must be struck farther in from the edge (a relatively deeper platform), relative to removing small areas of high mass (a shallower platform). This is a measure of spatial amount.
(3) The disparate features of the two platform surfaces and their relationship to each other ("acuteness") involve geometric estimation and visualization of space as a series of three-dimensional positions. The flintknapper, in orienting the core for a strike, must also estimate the position of the core face outside the visual field—a "hidden perspective." These are elements of a Euclidean sense of space.
(4) The flintknapper must re-attend to the morphology of the core each time a flake is removed, maintaining the relevant stimuli provided by the spatial geometry and updating their incoming information. These are aspects of the executive functions of working memory.
(5) Removing well-struck flakes is contingent on constant repetition and leads to expert performance.

The algorithm that links flake units together in series (Moore, 2010) is a key aspect of the long-term procedural memory component of working memory. The quantity of flaking material at early hominins sites—"There seems to have been much more knapping than was necessary to acquire sharp flakes" (Davidson & McGrew, 2005, p. 810)—may be indirect evidence that hominins did not need to consciously re-evaluate the geometric relationships in the flake unit each time a flake was struck. All five of these features are in place with the early emergence of stone flaking in the archaeological record (see de la Torre, 2010).

Wynn's model does not emphasize the significance of removing flakes individually and instead focuses on exploring goal-directed behavior through the removal of multiple flakes from a core, as reflected in artifacts present in the archaeological record, such as handaxes, discoidal cores, and Levallois cores. The spandrels experiments explicitly prevented goal-directed behavior by randomizing platform selection, yet produced cores (e.g., handaxes, discoidal cores, and Levallois cores *sensu lato*) or key defining attributes (bifacial flaking, approximate symmetry in plan and section, removal of a "preferential" flake) of these artifact classes (Moore & Perston, 2016). This is because the motor actions and mental evaluations dictated by fracture mechanics invariably produce convergences in hominin assemblages that differ empirically from random flaking in nature (eoliths) and uncontrolled fracture by other primates (chimpanzees, bonobos, and monkeys). It is reasonable to infer that since these lithic spandrels—core forms and attributes—emerged from experiments that restricted goal direction to the removal of single flakes, it is likely that they were also produced unintentionally by early hominin flintknappers. This is supported empirically by the occasional occurrence of these ostensibly late core forms in very early assemblages, such as proportionately rare handaxes in the Developed Oldowan (de la Torre & Mora, 2014) and the Levallois method *sensu lato* in Oldowan (de la Torre, Mora, Domínguez-Rodrigo, de Luque, & Alcalá, 2003) and other early assemblages (Nowell & White, 2010, p. 73). Yet the spandrels produced in the experiments failed to demonstrate congruent symmetry seen in Late Acheulean handaxes and the hierarchical flaking strategy of the Levallois method *sensu stricto* (Moore & Perston, 2016, p. 27), supporting Wynn's argument that these technologies resulted from hominin goal-direction. Given that many of the necessary cognitive abilities are implied by the architecture of the flake unit and thus were available to hominin flintknappers long before goal direction emerged, what was the process that shifted spandrels from unintentional by-products to the goals of reduction?

One possibility is that these spandrels, such as the unintended combinations of attributes archaeologists recognize as proto-bifaces or crude handaxes, became the cues hominin flintknappers used as retrieval structures for core reduction recipes. Early hominin flintknappers were experts at controlled removal of flakes from cores. Initially, with a focus of attention on individual flakes, the retrieval structure for the flake-removal chunk in procedural memory was the immediate morphology of the core identified in the visual search—an internal aspect of the flake unit itself. The visual search received a relatively stronger signal for angular configurations like those seen on tabular stones, broken cobbles, and larger flakes, in contrast to, for instance, rounded river cobbles, because angular stones are more likely to have readily distinguishable and continuous acute platform edges. When more than one flake was removed from a core, this may have channeled the application of flake units in unifacial "sets" or "chains" (Moore, 2010) on contiguous platforms that often occur on angular stones, such as the perimeter of flake blanks (Moore & Perston, 2016, pp. 8–9), although unrestricted platform selection led to the production of bifacial edges on stones of all shapes.

Sharp edges on flakes were used for cutting, as were unifacial and bifacial edges on cores, and some of these cores were invariably (and unintentionally) symmetrical. In the spandrels experiments, elongated, handaxe-like "proto-bifaces" with approximate reflectional symmetry or "global bilateral symmetry" (after Wynn, 2000, p. 123) composed about 5% of the total experimental core assemblage, although 43% of flake blank and 45% of cobble reductions produced at least one handaxe-like proto-biface over their reduction history (Moore & Perston, 2016, p. 27). These approximately symmetrical, handaxe-like spandrels could serve as cues for procedural memory know-how in a similar way that Coolidge and Wynn (2004, pp. 65–67) suggest Neandertals may have used prismatic blades as priming in attempts to emulate modern human blade-making. In this context, the cognitive shift suggested by the increase in numbers of crude handaxes in the Early Acheulean may have resulted from two developments: the elaboration of cognitive skills inherent in removing flakes individually (the flake unit) to removing flakes in series, combined with an inhibition against striking suitable platforms at the ends of the cores.

The removal of flakes in series involved, at least initially, conscious attention to flake-removal sets, and a related expansion of visual search procedures to target geometric configurations that could be manipulated by flake-removal sets, to recreate the attributes or morphologies of spandrels. With repetition, the requirements and operation of flake-removal sets was itself locked into procedural long-term memory (as seen in modern human flintknappers). Wynn (2002, p. 395; also see Wynn, 1985, p. 40, 1993, p. 315) suggests that application of flake-removal sets might have been prompted by motivation to mirror shapes of core edges through simple copying; these shapes may have served as procedural memory retrieval structures. Inhibition was also necessary: The spandrels experiments show how handaxe morphology can emerge and then disappear during reduction as platforms are randomly chosen at the ends of cores (Moore & Perston, 2016, Fig. 18). This suggests that hominins making elongated bifaces (and cleavers with an unflaked ends) practiced inhibition—a central executive function not clearly present prior to this in the ways flakes were removed serially from cores.

Together, these factors suggest at least an incipient degree of intentionality in early handaxe production—an intention to "emulate" spandrels—although, as Wynn (2002, p. 395) suggests, practicing these nonverbal rules does not necessarily mean that early handaxes and cleavers were objects held "in the mind." It is possible that early handaxes were produced through a focus on certain attributes rather than overall form, and repetitive forms hitchhiked with these attributes as a consequence of fracture mechanics; in that sense, early handaxes could be a sort of higher-order spandrel, resulting from rote application of flake-removal sets and driven by retrieval cues, rather than a deliberately produced type in the way archaeologists usually conceive of them. Both "common" cores and spandrels cores were presumably used by early hominins, and the motivation to emulate them might have been from their improved functional performance (Nowell & Chang, 2009, pp. 82–83). Alternatively, expert performance at advanced working memory tasks requires a learning process focused on procedural memory algorithms (Coolidge & Wynn, 2005), resulting in frequent repetition of spandrel attributes and morphologies—and perhaps their emulation—as a spin-off from the dynamics of social interaction (Wynn, 1993, p. 315; Wynn, 1995, pp. 19–20).

A similar process could have also led to the various Levallois-like core reduction processes seen sporadically in very early stone tools assemblages and more frequently in producing large flake blanks for Acheulean handaxe production (e.g., Clark, 2001; Sharon, 2009; Sharon & Beaumont, 2006). In the spandrels experiments, a large "preferential" flake similar to that seen in Levallois reduction occurred in 24% of the cobble and 13% of the flake blank reductions (Moore & Perston, 2016, p. 20). Preferential flakes were larger than the flakes that preceded them (Figure 8.4), and the attributes of the flakes and/or preferential scars on cores may have served as cues for procedural memory in a manner similar to handaxe spandrels. As discussed previously, the angular configurations on the perimeter of flake blanks offer clear visual targets and the conditions for application of flake-removal sets, an aspect of handaxe production. Thus, a similar cognitive process using spandrels as retrieval cues may underpin the development of Levallois-like methods to emulate blanks that are ideal for applying additional flake-removal sets, and those flake-removal sets in turn used handaxe-like spandrels as retrieval cues that led to the production of handaxes.

Wynn (2002, p. 397) argues that the sophisticated congruent and multidimensional symmetries seen in some Late Acheulean handaxes, and the hierarchical structure of the Levallois Method *sensu stricto*, indicate that the products were "categories" in the minds of the hominin flintknappers and, as categories, they must reside in declarative long-term memory, outside of procedural long-term memory. As such, they serve as semantic memory goals or knowledge (Wynn & Coolidge, 2004, pp. 473–477; also see Wynn, 1993, p. 315). This, in turn, implies the activation of the phonological loop in producing these tools, an aspect of cognition not necessarily required in striking flakes individually or in marshaling flake-removal sets to make Early Acheulean handaxes. Once these products were seen as goals, the memory capacity freed by deploying the preceding algorithms—the flake unit and flake-removal sets—could be marshaled for larger, more complex chunking in procedural memory and greater attention capacity in the central executive. Retrieval structures and material cues could be filed in long-term memory relative to the goal itself—a visualization "held in the mind" (Russell, 1996)—rather than triggered in response to cues offered by spandrels. The ability to incorporate aspects of the phonological loop and declarative memory into a procedural task "was exapted" by language (Coolidge & Wynn, 2005, p. 13). Once this was achieved, "[n]o more complex form of stone knapping *ever* appears" (Coolidge & Wynn, 2005, p. 16, emphasis in original): The subsequent enhancements in working memory resulting in fully modern cognition (Wynn & Coolidge, 2016) were not necessary to produce the range of stone tools seen in later prehistory.

In summary, then, the thought experiments that introduced this essay combined with the results of the spandrels experiments (Moore & Perston, 2016) suggest two things. First, aspects of the early stone tool types we recognize as archaeologists were not "invented" in the usual sense but instead provided fertile ground—cues or scaffolds—for the stimulation of repetition. Eventually, with contributions from evolving language structures, scaffolds for retrieving procedural memory became stone-working goals similar to those envisioned by modern human flintknappers when they replicate Late Acheulean handaxes. And second, many of the basic cognitive elements to support later mental developments, particularly in working memory, are internal to the flake unit and show a complexity beyond the cognitive capabilities of other primates. As such, these cognitive elements are reflected in the expertise in removing flakes that emerged as hominins crossed the cognitive Rubicon that separates us from our primate cousins.

CONCLUSION

This study began with two thought experiments, one that disrupts a modern flintknapper before they can marshal the resources to remove a flake, and another that disrupts the flintknapper's declarative knowledge and procedural know-how. The first ends in disaster—not even flakes are created—and the second ends in relatively advanced core morphologies, but not the modern flintknapper's goal. Thomas Wynn's narrative of cognitive evolution (1979, 1981, 2002, Wynn & Coolidge, 2004, 2016) begins with the implications of the removal of multiple flakes from cores, in effect diving in at the second thought experiment, with less consideration for the implications of the first. Certain key abilities identified by Wynn for later stages of cognitive evolution can be seen at the outset of controlled stone flaking, some 3.3 Mya, in the necessary elements for the removal of individual flakes.

Our experiments show that stone-flaking spandrels, or stone objects and attributes that appear to be deliberately designed, can be produced unintentionally in core reduction (Moore & Perston, 2016). Spandrels are spin-offs from the internal workings of the flake unit itself, yet they anticipate the tools implicated in later breakthroughs in cognitive evolution. Spandrels may have served initially as scaffolds to direct the early expansion in working memory suggested by Wynn's model, and later to serve as retrieval structures for procedural memory. From that point they proliferated and entered declarative memory as stone-flaking "goals," freeing working memory for the development of the exceptionally complex stone tools that followed. This explains important aspects of the early archaeological record—the precocious appearance of advanced stone tools amid much simpler technologies (the early examples are spandrels) and their subsequent increase through time (the later examples are scaffolds), culminating in fully modern levels of complexity (the modern human examples are goals). The "cognitive Rubicon" in hominin evolution was crossed with the appearance of the flake unit, which provided the material basis for cognitive developments through the Pleistocene to the emergence of cognitively modern humans.

ACKNOWLEDGMENTS

The "spandrels" stone-flaking experiments were supported by a grant from the Australian Research Council (DP1096558). I thank Yinika Perston and Mathew Smith for their help with the experimental work, and Iain Davidson and two anonymous reviewers for their comments on an earlier draft of this chapter.

REFERENCES

Beyene, Y., Katoh, S., WoldeGabriel, G., Hart, W. K., Uto, K., Sudo, M., . . . Asfaw, B. (2013). The characteristics and chronology of the earliest Acheulean at Konso, Ethiopia. *Proceedings of the National Academy of Sciences of the United States of America, 110*(5), 1584–1591.

Bravo, M. J., & Nakayama, K. (1992). The role of attention in different visual-search tasks. *Attention, Perception, and Psychophysics, 51*(5), 465–472.

Christie, G. J., Livingstone, A. C., & McDonald, J. J. (2014). Searching for inefficiency in visual search. *Journal of Cognitive Neuroscience, 27*(1), 46–56.

Clark, J. D. (2001). Variability in primary and secondary technologies of the Later Acheulian in Africa. In S. Miliken & J. Cook (Eds.), *A very remote period indeed: Papers on the Palaeolithic presented to Derek Roe* (pp. 1–18). Oakville, CT: Oxbow Books.

Clark, J. E. (2002). Failure as truth: An autopsy of Crabtree's Folsom experiments. In J. E. Clarke & M. B. Collins (Eds.), *Folsom technology and lifeways (lithic technology special publication No. 4)* (pp. 191–208). Tulsa, OK: University of Tulsa.

Coolidge, F. L., & Wynn, T. (2004). A cognitive and neuropsychological perspective on the Châtelperronian. *Journal of Anthropological Research, 60*, 55–73.

Coolidge, F. L., & Wynn, T. (2005). Working memory, its executive functions, and the emergence of modern thinking. *Cambridge Archaeological Journal, 15*(1), 5–26.

Davidson, I., & McGrew, W. C. (2005). Stone tools and the uniqueness of human culture. *Journal of the Royal Anthropological Institute of Great Britain and Ireland, 11*(4), 793–817.

de Beaune, S. A. (2004). The invention of technology: Prehistory and cognition. *Current Anthropology, 45*(2), 139–162.

de la Torre, I. (2010). Insights on the technical competence of the Early Oldowan. In A. Nowell & I. Davidson (Eds.), *Stone tools and the evolution of human cognition* (pp. 45–65). Boulder, CO: University Press of Colorado.

de la Torre, I., & Mora, R. (2014). The transition to the Acheulean in East Africa: An assessment of paradigms and evidence from Olduvai Gorge (Tanzania). *Journal of Archaeological Method and Theory, 21*(4), 781–823.

de la Torre, I., Mora, R., Domínguez-Rodrigo, M., de Luque, L., & Alcalá, L. (2003). The Oldowan industry of Peninj and its bearing on the reconstruction of the technological skills of Lower Pleistocene hominids. *Journal of Human Evolution, 44*(2), 203–224.

Dibble, H. L., & Whittaker, J. C. (1981). New experimental evidence on the relation between percussion flaking and flake variation. *Journal of Archaeological Science, 8*(3), 283–296.

Dowd, E. W., & Mitroff, S. R. (2013). Attentional guidance by working memory overrides salience cues in visual search. *Journal of Experimental Psychology: Human Perception and Performance, 39*(6), 1786–1796.

Edwards, S. W. (2001). A modern knapper's assessment of the technical skills of the Late Acheulean biface workers at Kalambo Falls. In J. D. Clark (Ed.), *Kalambo Falls prehistoric site* (Vol. 3, pp. 605–611). Cambridge, UK: Cambridge University Press.

Eren, M. I., Bradley, B. A., & Sampson, C. G. (2011). Middle Paleolithic skill level and the individual knapper: An experiment. *American Antiquity, 76*(2), 229–251.

Gould, S. J., & Lewontin, R. C. (1979). The spandrels of San Marco and the Panglossian paradigm: A critique of the adaptationist programme. *Proceedings of the Royal Society of London B: Biological Sciences, 205*(1161), 581–598.

Greenfield, P. M. (1991). Language, tools and brain: The ontogeny and phylogeny of hierarchically organized sequential behavior. *Behavioral and Brain Sciences, 14*(4), 531–595.

Gumert, M. D., Kluck, M., & Malaivijitnond, S. (2009). The physical characteristics and usage patterns of stone axe and pounding hammers used by long-tailed macaques in the Andaman Sea region of Thailand. *American Journal of Primatology, 71*(7), 594–608.

Harmand, S., Lewis, J. E., Feibel, C. S., Lepre, C. J., Prat, S., Lenoble, A., . . . Roche, H. (2015). 3.3-million-year-old stone tools from Lomekwi 3, West Turkana, Kenya. *Nature, 521*(7552), 310–315.

Henrich, J., & McElreath, R. (2007). Dual-inheritance theory: The evolution of human cultural capacities and cultural evolution. In R. I. M. Dunbar & L. Barrett (Eds.), *Oxford handbook of evolutionary psychology* (pp. 555–570). Oxford, UK: Oxford University Press.

Hout, M. C., & Goldinger, S. D. (2015). Target templates: The precision of mental representations affects attentional guidance and decision-making in visual search. *Attention, Perception and Psychophysics, 77*(1), 128–149.

Isaac, G. L. (1977). *Olorgesailie: Archaeological studies of a Middle Pleistocene lake basin in Kenya.* Chicago, IL: University of Chicago Press.

Kelterborn, P. (2003). Measurable flintknapping. In K. G. Hirth (Ed.), *Mesoamerican lithic technology: Experimentation and interpretation* (pp. 120–131). Salt Lake City, UT: University of Utah Press.

Land, M. F., & Hayhoe, M. (2001). In what ways do eye movements contribute to everyday activities? *Vision Research, 45*(25), 3559–3565.

Land, M. F., & McLeod, P. (2000). From eye movements to actions: How batsmen hit the ball. *Nature Neuroscience, 3*(12), 1340–1345.

Logan, G. D. (1992). Attention and preattention in theories of automaticity. *American Journal of Psychology, 105*(2), 317–339.

Marchant, L. F., & McGrew, W. C. (2005). Percussive technology: Chimpanzee baobab smashing and the evolutionary modelling of hominin knapping. In V. Roux & B. Bril (Eds.), *Stone knapping: The necessary conditions for a uniquely hominin behaviour* (pp. 341–350). Cambridge, UK: McDonald Institute for Archaeological Research.

Mercader, J., Barton, H., Gillespie, J., Harris, J., Kuhn, S. L., Tyler, R., & Boesch, C. (2007). 4,300–year-old chimpanzee sites and the origins of percussive stone technology. *Proceedings of the National Academy of Sciences of the United States of America, 104*(9), 3043–3048.

Moore, M. W. (2007). Lithic design space modelling and cognition in *Homo floresiensis*. In A. C. Shalley & D. Khlentzos (Eds.), *Mental states: Nature, function and evolution* (pp. 11–33). Amsterdam: John Benjamins.

Moore, M. W. (2010). "Grammars of action" and stone flaking design space. In A. Nowell & I. Davidson (Eds.), *Stone tools and the evolution of human cognition* (pp. 13–43). Boulder, CO: University Press of Colorado.

Moore, M. W. (2011). The design space of stone flaking: Implications for cognitive evolution. *World Archaeology, 43*(4), 702–715.

Moore, M. W. (2015). Bifacial flintknapping in the Northwest Kimberley, Western Australia. *Journal of Archaeological Method and Theory, 22*(3), 913–951.

Moore, M. W., & Perston, Y. (2016). Experimental insights into the cognitive significance of early stone tools. *PLoS One, 11*(7), 1–37.

Neufuss, J., Humle, T., Cremaschi, A., & Kivell, T. L. (2016). Nut-cracking behaviour in wild-born, rehabilitated bonobos (*Pan paniscus*): A comprehensive study of hand-preference, hand grips and efficiency. *American Journal of Primatology, 79*(2), 1–16.

Newcomer, M. H. (1971). Some quantitative experiments in handaxe manufacture. *World Archaeology, 3*(1), 85–94.

Nowell, A., & Chang, M. L. (2009). The case against sexual selection as an explanation of handaxe morphology. *PaleoAnthropology, 2009*, 77–88.

Nowell, A., & White, M. J. (2010). Growing up in the Middle Pleistocene: Life history strategies and their relationship to Acheulian industries. In A. Nowell & I. Davidson (Eds.), *Stone tools and the evolution of human cognition* (pp. 67–82). Boulder, CO: University Press of Colorado.

Peacock, E. (1991). Distinguishing between artifacts and geofacts: A test case from eastern England. *Journal of Field Archaeology, 18*(3), 345–361.

Pelcin, A. W. (1997). The effect of core surface morphology on flake attributes: Evidence from a controlled experiment. *Journal of Archaeological Science, 24*(8), 749–756.

Pelegrin, J. (1993). A framework for analysing prehistoric stone tool manufacture and a tentative application to some early stone industries. In A. Berthelet & J. Chavaillon (Eds.), *The use of tools by human and non-human primates* (pp. 302–314). Oxford, UK: Clarendon Press.

Pelegrin, J. (2006). Long blade technology in the Old World: An experimental approach and some archaeological results. In J. Apel & K. Knutsson (Eds.), *Skilled production and social reproduction: Aspects of traditional stone-tool technologies: Proceedings of a symposium in Uppsala, August 20–24, 2003* (pp. 37–68). Upsala, Sweden: Societas Archaeologica Upsaliensis.

Proffitt, T., Luncz, L. V, Falótico, T., Ottoni, E. B., De la Torre, I., & Haslam, M. (2016). Wild monkeys flake stone tools. *Nature, 539*(7627), 85–88.

Quinlan, P. T. (2003). Visual feature integration theory: Past, present, and future. *Psychological Bulletin, 129*(5), 643–673.

Russell, J. (1996). Development and evolution of the symbolic function: The role of working memory. In P. Mellars & K. R. Gibson (Eds.), *Modelling the early human mind* (pp. 159–177). Cambridge, UK: McDonald Institute for Archaeological Research.

Schlanger, N. (1996). Understanding Levallois: Lithic technology and cognitive archaeology. *Cambridge Archaeological Journal, 6*(2), 231–254.

Sharon, G. (2009). Acheulian giant-core technology: A worldwide perspective. *Current Anthropology, 50*(3), 335–367.

Sharon, G., & Beaumont, P. B. (2006). Victoria West: A highly standardized prepared core technology. In N. Goren-Inbar & G. Sharon (Eds.), *Axe age: Acheulian tool-making from quarry to discard* (pp. 181–199). London: Equinox.

Thomas, D. H. (1986). Points on points: A reply to Flenniken and Raymond. *American Antiquity, 51*(3), 619–627.

Toth, N. P. (1985). The Oldowan reassessed: A close look at early stone artifacts. *Journal of Archaeological Science, 12*(2), 101–120.

Toth, N. P., Schick, K. D., Savage-Rumbaugh, E. S., Sevcik, R. A., & Rumbaugh, D. M. (1993). *Pan* the tool-maker: Investigations into the stone tool-making and tool-using capabilities of a bonobo (*Pan paniscus*). *Journal of Archaeological Science, 20*(1), 81–91.

Toth, N. P., Schick, K. D., & Semaw, S. (2006). A comparative study of the stone tool-making skills of *Pan, Australopithecus,* and *Homo sapiens*. In N. P. Toth & K. D. Schick (Eds.), *The Oldowan: Case studies into the Earliest Stone Age* (pp. 155–222). Gosport, IN: Stone Age Institute Press.

Treisman, A. M., & Gelade, G. (1980). A feature-integration theory of attention. *Cognitive Psychology, 12*(1), 97–136.

Vallacher, R. R., & Wegner, D. M. (1987). What do people think they're doing? Action identification and human behavior. *Psychological Review, 94*(1), 3–15.

van Peer, P. (1992). *The Levallois reduction strategy (monographs in world archaeology no. 13)*. Madison, WI: Prehistory Press.

Visalberghi, E., Fragaszy, D. M., Ottoni, E. B., Izar, P., De Oliveira, M. G., & Andrade, F. R. D. (2007). Characteristics of hammer stones and anvils used by wild bearded capuchin monkeys (*Cebus libidinosus*) to crack open palm nuts. *American Journal of Physical Anthropology, 132*(3), 426–444.

Visalberghi, E., & McGrew, W. C. (1997). *Cebus* meets *Pan*. *International Journal of Primatology, 18*(5), 677–681.

Whiten, A., Goodall, J. van L., McGrew, W. C., Nishida, T., Reynolds, V., Sugiyama, Y., . . . Boesch, C. (2001). Charting cultural variation in chimpanzees. *Behaviour, 138*, 1481–1516.

Whittaker, J. C. (1994). *Flintknapping: Making and understanding stone tools*. Austin, TX: University of Texas Press.

Whittaker, J. C. (2004). *American flintknappers: Stone Age art in the age of computers*. Austin, TX: University of Texas Press.

Wiśniewski, A., Badura, J., Salamon, T., & Lewandowski, J. (2014). The alleged Early Palaeolithic artefacts are in reality geofacts: A revision of the site of Kończyce Wielkie 4 in the Moravian Gate, South Poland. *Journal of Archaeological Science, 52*, 189–203.

Wolfe, J. M. (2007). Guided search 4.0: Current progress with a model of visual search. In W. D. Gray (Ed.), *Integrated models of cognitive systems* (pp. 99–119). New York: Oxford University Press.

Wolfe, J. M. (2016). Visual search revived: The slopes are not that slippery: A reply to Kristjansson. *I-Perception, 3*, 1–6.

Wolfe, J. M., & Horowitz, T. S. (2004). What attributes guide the deployment of visual attention and how do they do it? *Nature Reviews: Neuroscience, 5*(6), 495–501.

Wynn, T. (1979). The intelligence of later Acheulean hominids. *Man, 14*, 371–391.

Wynn, T. (1981). The intelligence of Oldowan hominids. *Journal of Human Evolution, 10*(7), 529–541.

Wynn, T. (1985). Piaget, stone tools and the evolution of human intelligence. *World Archaeology, 17*(1), 32–43.

Wynn, T. (1993). Two developments in the mind of early *Homo*. *Journal of Anthropological Archaeology, 12*(3), 299–322.

Wynn, T. (1995). Handaxe enigmas. *World Archaeology, 27*(1), 10–24.

Wynn, T. (2000). Symmetry and the evolution of the modular linguistic mind. In P. Carruthers & A. Chamberlain (Eds.), *Evolution and the human mind: Modularity, language and meta-cognition* (pp. 113–139). Cambridge, UK: Cambridge University Press.

Wynn, T. (2002). Archaeology and cognitive evolution. *Behavioral and Brain Sciences, 25*(3), 389–402.

Wynn, T., & Coolidge, F. L. (2004). The expert Neandertal mind. *Journal of Human Evolution, 46*(4), 467–487.

Wynn, T., & Coolidge, F. L. (2016). Archaeological insights into hominin cognitive evolution. *Evolutionary Anthropology: Issues, News, and Reviews, 25*(4), 200–213.

Wynn, T., Hernandez-Aguilar, R. A., Marchant, L. F., & McGrew, W. C. (2011). "An ape's view of the Oldowan" revisited. *Evolutionary Anthropology, 20*(5), 181–197.

Wynn, T., & McGrew, W. C. (1989). An ape's view of the Oldowan. *Man, 24*(3), 383–398.

9

STONE TOOLS AND SPATIAL COGNITION

Derek Hodgson

INTRODUCTION

Spatial perception and cognition refers to the updating and encoding of spatial information during movement. In tool-making, such capacities are realized in visuospatial and visuomotor abilities that are fundamental to understanding cognitive evolution. The elaborate processes by which these abilities unfold in the cortex with regard to tool-making and tool use have become clearer, thanks to advances in brain scanning techniques. The intricacies by which this unfolds are complex, so it is necessary to explore the various neural pathways in some detail in order to assess their relevance to the spatial aspects of tool-making.

Initially, two main pathways were identified, the ventral and dorsal pathways, which are outlined in depth in this chapter (Milner & Goodale, 2006; Ungerleider & Mishkin, 1982). These areas are fundamental to different aspects of tool-making/use and therefore have consequences for clarifying the relationship between ancient stone tools and the timeline of cognitive evolution. The main purpose of this analysis is to examine the various pathways involved to show how they can shed light on this issue. Part 1 presents in detail the neural pathways that underpin visuospatial capacities and motor skills in the context of making and using tools. Part 2 investigates implications for how stone tools, as understood through the archaeological record, relate to the underlying neural structures identified.

To achieve this, it will be necessary to draw on evidence from a number of sources, including the cerebral structures of extant simians, brain scans of expert stone knappers, and ideas from neuroscience and cognitive evolution. Hopefully, these diverse sources will allow us to map the probable relationship between visuospatial abilities and the various techno-complexes as they appeared in the archaeological record. Based on the insights gained from analyzing the neural pathways identified, the following hypothesis will be defended: (1) Making and using Oldowan and Early Acheulean tools initially depended on the "blind" dorsal pathway; (2) parts of this pathway became enhanced, leading to the more refined tools of the Late Acheulean; and (3) only after the conscious ventral stream underwent full integration with the dorsal stream did this facilitate the making of hafted and composite tools.

THE DORSAL AND VENTRAL STREAMS

Anatomy and Function of the Dorsal and Ventral Pathways

The central aspect of the argument presented here is that both the ventral and dorsal streams needed to interact and coordinate their specialized functions in order for complex tools to be produced, as proposed by Wynn (2002, 2014; Wynn & Coolidge, 2016) and Hodgson (2005, 2007, 2009a, 2009b, 2012). At higher levels of processing, this integration is realized in the "visuospatial sketchpad" (where spatial and visual information is integrated in working memory), whereby the dorsal pathway interacts with the ventral stream, and associated multiple neural networks interconnect (including, for example, the cerebellum and prefrontal cortex for sensorimotor learning; Taylor & Ivry, 2014). The visuospatial sketchpad is therefore a multifaceted cognitive interface involving an array of interacting systems (Logie & van der Meulen, 2008).[1]

Visuospatial information is processed in the human brain via the "blind" dorsal pathway ("vision for action") for manual dexterity and coordination when interacting with objects in a number of different ways. Because the dorsal stream is blind, visuospatial functions are sublimated so that manual dexterity can be performed automatically with minimal reflection that would hinder smooth ballistic action. Conversely, the conscious ventral system, which transforms visual input for encoding the enduring aspects of objects, as well as their spatial relations (Goodale & Milner, 2013; Milner & Goodale, 2006), is essential for making and using complex tools that require conceptual and semantic input (Buxbaum, Shapiro, & Coslett, 2014). As a result, damage to these pathways leads to specific difficulties when interacting with objects (Buxbaum, 2001; Cronin-Golomb & Hof, 2004; Jakobson, Archibald, Carey, & Goodale, 1991; Jeannerod, 1986; Milner & Goodale, 2006; Perenin & Vighetto, 1988; Ungerleider & Mishkin, 1982). For example, deficits along the ventral stream lead to *apperceptive agnosia*—the inability to distinguish visual shapes and identify or discriminate different visual stimuli—as well as *associative agnosia*, where knowledge of an object's use is preserved, but the object is mistaken for another (Farah, 2004; Warrington & James, 1988). Conversely, damage to the dorsal stream leads to problems with online manual dexterity (Sakreida et al., 2016). Bi and colleagues (2015) similarly found a dissociation between tool use and tool concept networks of the dorsal and ventral pathways. The dorsal and ventral channels are therefore fundamental to understanding the way implicit visuospatial/motor coordinates function and interact with explicit perceptual criteria (Binkofski & Buxbaum, 2013; Milner & Goodale, 2006, 2008).

Reorganization of Early Visual Cortex

Before exploring in detail the dorsal and ventral systems, it is necessary to investigate whether changes to the primary areas of the visual system display any

[1] The term *visuospatial sketchpad* derives from Baddeley's (Baddeley, 2007; Baddeley & Hitch, 1974) model of working memory. However, the term has the impression of a unitary phenomenon associated with a specific brain area. In fact, the underlying neural processes seem to be implicated in different parts of the brain and appear much more complex than Baddeley first envisaged (see e.g., Logie & Van Der Meulen, 2008).

differences with simians, as changes to the higher pathways may have led to a reformatting of primary input areas. Several studies reveal that this is indeed the case, especially in the way the parvo stream (which mediates high-acuity information) and the magno stream (which is specialized for encoding movement) interact in the primary visual cortex, or V1. These areas reveal distinct differences in cortical organization when macaques are compared to higher primates and when the latter are compared to modern humans. Specifically, a mesh-like structure exists in specific cortical layers of humans that is absent in chimpanzees (Hodgson, 2007; Preuss, Qi, & Kaas, 1999), which suggest a novel interaction of pathways in humans even in these initial layers.

In humans, some of the slightly later visual regions also indicate differences when compared to those of non-human primates. For example, the visuospatial characteristics of area V3a appear to have reversed position with V3 (Tootell et al., 1997; Vanduffel et al., 2001). In addition, the medial parietal occipital cortex (mPOC) represents a crucial region within the dorsal stream where several pathways for visuospatial processing arise (Kravitz, Saleem, Baker, & Mishkin, 2011), with the most important being V6a—identified as a reach area for visual motion and action implicated in manipulating objects (Pitzalis, Fattori, & Galletti, 2015). This region, which is divided into separate functional areas, thus contains many spatially tuned, arm movement–related cells (Fattori, Gamberini, Kutz, & Galletti, 2001; Fattori, Kutz, Breveglieri, Marzocchi, & Galletti, 2005) that are extremely sensitive to wrist position (Fattori et al., 2009), especially anteriorly to V6a in V6Ad, which also encodes grip (Pitzalis et al., 2015). Though the functions of V6 and V6a in humans bear obvious similarities to those of macaques, they appear to have undergone displacement in humans (Pitzalis et al., 2015).

These differences become more pronounced when occipital-parietal areas are considered (i.e., where the dorsal stream itself divides into two pathways, the dorsal-dorsal and ventrodorsal streams that project toward the superior parietal area and intraparietal sulcus, respectively). Areas V3a, V6, and V6a serve as a gateway for dorsal-dorsal projections to the superior parietal area for online manual dexterity, whereas the ventrodorsal pathway projects through the middle temporal visual area (MT) to the intraparietal area for visuospatial expertise, which further projects to the ventral premotor region and ultimately area 46 in prefrontal cortex (Figure 9.1). In contrast, the ventral pathway ascends to the inferotemporal area via V4 and lateral occipital cortex (LOC) for conscious visuoperceptual processing. The basic processing streams for the three pathways are outlined in Figure 9.1. The layout in the brain of the dorsal-dorsal and ventrodorsal pathways is illustrated in Figure 9.2 for macaques and Figure 9.3 for humans (details of the ventral stream will be addressed shortly).

THE INTRAPARIETAL SULCUS (IPS)

3D Shape from Motion

We need to explore in detail the minutiae of IPS in order to provide evidence for the differences between humans and higher simians that can inform us of the possible

(A) **Dorsal-dorsal pathway**

V1 ⟶ V3a ⟶ V6 ⟶ V6a ⟶ SPL ⟶ Premotor (dorsal)

(B) **Ventrodorsal pathway**

V1 ⟶ MT/MST ⟶ IPS (LIP, VIP, AIP) ⟶ Premotor (Ventral, F5, and F4) ⟶ Prefrontal area 8a and 46

(C) **Ventral pathway**

V1 ⟶ V4 ⟶ [MT] ⟶ LOC ⟶ Inferotemporal area

Figure 9.1. The three separate visual pathways in the human cortex. (A, top) The dorsal-dorsal pathway. (B, middle) The ventrodorsal pathway. (C, bottom) The ventral pathway. Abbreviations: AIP, anterior intraparietal sulcus; IPS, intraparietal sulcus; LIP, lateral intraparietal sulcus; LOC, lateral occipital cortex; MST, medial superior temporal; MT, middle temporal visual area; SPL, superior parietal lobe; V, visual; VIP, ventral intraparietal sulcus. Image by the author.

neural substrates promoting early tool-making. Though information from primary visual cortex to the superior parietal area (along the dorsal-dorsal pathway) and IPS (along the ventrodorsal pathway) projects similarly in humans and monkeys, differences are observed in the more complex processing and resources devoted to shape in human IPS. This is particularly pertinent to three-dimensional (3D) shape from motion, which demands much more processing than is the case for two-dimensional (2D) shape. Thus, the encoding of 3D shape from motion has been found

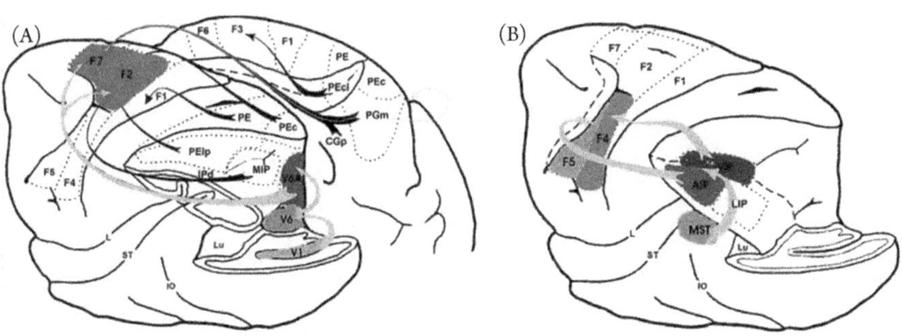

Figure 9.2. The basic "primitive" dorsal system in the macaque brain. (A, left) Dorsal-dorsal stream. (B, right) Ventrodorsal stream. Abbreviations: AIP, anterior intraparietal sulcus; CGp, posterior cingulate gyrus; F, motor areas; IO, inferior occipital sulcus; IPd, intraparietal dorsal; LIP, lateral intraparietal sulcus; Lu, lunate sulcus; MIP, Medial intraparietal; MST, medial superior temporal; PE, PEc, PEcl, PEIp, PGm, posterior parietal cortical regions; ST, Superior Temporal; V, visual. Images previously published as Figure 1 (p. 223) in Binkofski and Buxbaum (2013), Two action systems in the human brain, *Brain and Language*, and republished with the permission of Elsevier.

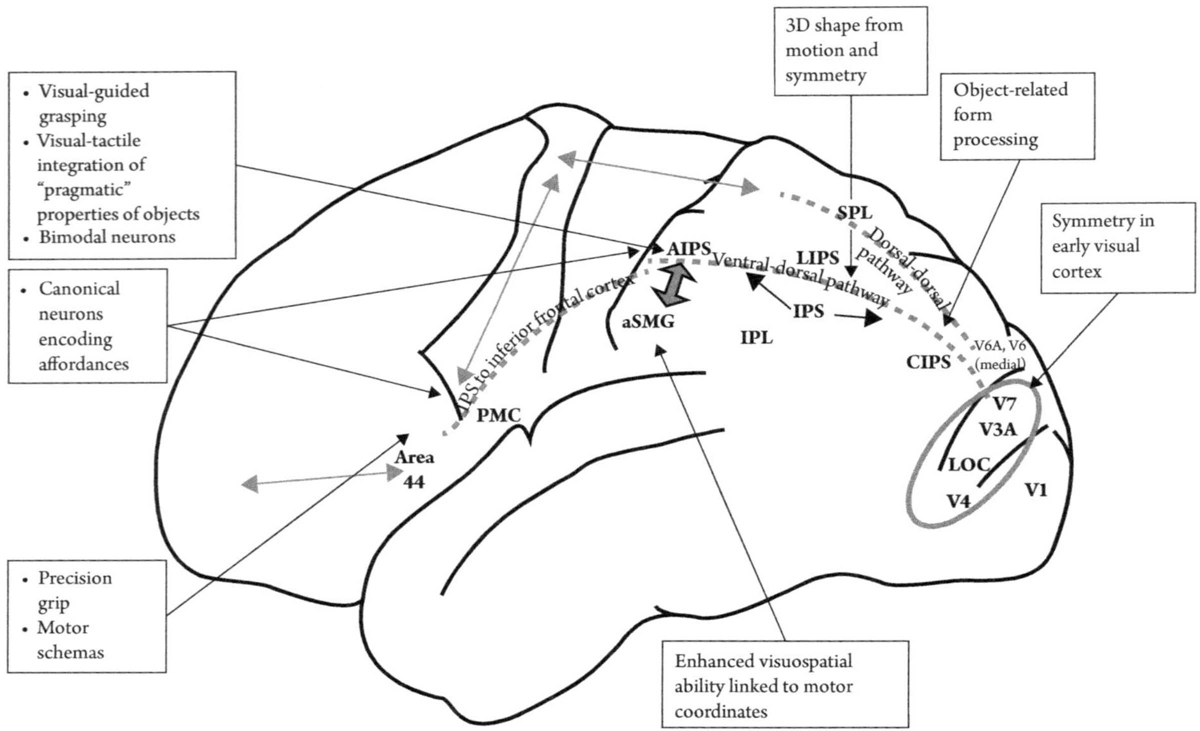

Figure 9.3. Human dorsal-dorsal and ventrodorsal pathways: the enhanced "primitive" dorsal system. Abbreviations: AIPS, anterior intraparietal sulcus; aSMG, anterior supramarginal gyrus; IPL, inferior parietal lobe; IPS, intraparietal sulcus (A, anterior; L, lateral; C, caudal); LOC, lateral occipital cortex; SPL, superior parietal lobule; V, visual; vPMC, ventral premotor cortex. Lines: The wide gray line between the AIPS and aSMG indicates IPS-to-aSMG connectivity as an initial link to the IPL; unlabeled solid gray lines indicate other related pathways. Image by the author.

to be much more pronounced and widespread in human IPS than in non-human primates (see the footnote for a detailed account of these differences).[2]

Visuospatial Aspects of Prehension in IPS

As well as containing sensory neurons for specific shapes and orientations, the IPS (Figure 9.3) is also predisposed toward visually guided grasping and contains motor neurons activated during specific hand movements (Burgess, Jeffery, & O'Keefe, 1999). Indeed, the ventrodorsal pathway through IPS is where visuospatial and motor integration occurs, which allows non-human primates to grip objects using the entire hand; in humans, however, this area facilitates a precision grip with enhanced capacities. This occurs thanks to the interfacing of the forward (anterior) IPS with premotor and motor cortex (as well as somatosensory areas) (Sakreida et al., 2016). This region also contains canonical and bimodal neurons where the affordances of objects and the extension of the body are facilitated (canonical neurons fire at the sight of a graspable object and not when an object is reached for and held, whereas bimodal neurons facilitate the integration of both somatosensory and visual information for incorporating tools into an overall body schema that enables the extension of reaching space). These neurons play an important role in the dorsal stream as part of an arch, referred to as a non-conscious "primitive" system (Stout & Chaminade, 2007) shared by primates (Figures 9.2 and 9.3) where tasks are performed automatically in response to the affordances suggested by objects. Arguably, and this is a key point, together with 3D shape from motion, it was subtle, incremental changes to these neurons in hominin IPS that facilitated the production of Oldowan and Early Acheulean tools.

[2] Orban and colleagues (2006) detailed the differences whereby human IPS contains four motion-sensitive regions—ventral IPS (VIPS), parieto-occipital IPS (POIPS), dorsal IPS medial (DIPSM), and dorsal IPS anterior (DIPSA)—involved in realizing 3D structure from motion. Simian IPS, however, comprises only one motion-sensitive area (VIPS), which is not as sensitive to 3D shape from motion. In addition, human IPS has four regions encoding 2D shape and three representations for central vision (for high-acuity perception), whereas simian IPS includes only two shape-sensitive regions and one central representation. This is confirmed by the finding that, in macaques, IPS shape sensitivity seems more devoted to 2D coordinates than the corresponding area in humans, which is more tuned to 3D criteria (Denys et al., 2004; Orban et al., 2006; Orban, Sunaert, Todd, Van Hecke, & Marchal, 1999; Orban, van Essen, & Vanduffel, 2004). This is reinforced by the fact that shape in this region is more cue invariant in humans (i.e., has a greater ability to deal with variability; Denys et al., 2004). Research with the Japanese monkey (*Macaca fuscata*) corroborates the significance of this pathway in that the lateral intraparietal sulcus (LIPS), which responds visually to 3D objects, receives projections from earlier occipital areas from which axons project to the anterior intraparietal area (AIP, which responds to hand movements for grasping 3D objects using a precision grip) that interact with motor and premotor areas (Davare, Rothwell, & Lemon, 2010; Nakamura et al., 2001). Correspondingly, Vanduffel and colleagues (2002) compared the activation elicited by 2D and 3D arrays in simians and humans and confirmed that 3D shape from motion provoked widespread activation along human IPS (Orban et al., 2006; Todd, 2004). Thus, although latent 3D information may exist in non-human primates (especially in the anterior area of IPS, notably AIP and LIP; Grafton, 2010), this is enhanced in humans. In this regard, Durand and colleagues (2007) proposed that the more anterior areas of IPS (i.e., AIP and LIP) may also assist in discriminating 3D shape from background cues. These findings have interesting consequences for tool use in that in order to produce tools, especially where shape profile is important—as is the case with Acheulean tools—the ability to process 3D shape from motion is crucial.

In sum, there is a reciprocal relationship between the underlying neural structures and tool production. This gave rise to subtle alterations to the structures that were closely aligned to the affordances implicit in objects, which may explain the conservatism of Oldowan and *Early* Acheulean tools. Thus, early tool-making may have been closely associated with implicit cognitive procedures tied to an intimate engagement with the materials undergoing transformation.

Enhanced Intraparietal Links with Supramarginal Gyrus

The dorsal pathway—leading up to and including both the intraparietal area (in the ventrodorsal stream) and superior parietal region (in the dorsal-dorsal stream)—mediates egocentric reference frames (i.e., is viewer-centered), whereas the ventral system is allocentric (or object-centered) (James, Humphrey, Gati, Menon, & Goodale, 2002; Kravitz et al., 2011). Allocentric processing is of particular interest for the social aspects of tool-making, covered in greater detail later.

Human IPS has an increased processing capacity for encoding 3D shape from motion, which is also specialized for reaching and grasping objects in peripersonal space. This suggests that parts of the human intraparietal area have also undergone reorganization. The forward parieto-prefrontal circuit is also pertinent with regard to tool-making, as this area has strong reciprocal connections emanating from IPS along the ventrodorsal pathway to prefrontal cortex (areas 8A and 46) for executive control of visuospatial/motor parameters. These regions have important links to the inferior parietal lobe (IPL) related to the conscious knowledge of tools and which serve as a multimodal circuit that facilitates spatial information as part of a network encoding semantic knowledge for tools (Mahon, Schwarzbach, & Caramazza, 2010; Martin, 2007; Tranel, Kemmerer, Adolphs, Damasio, & Damasio, 2003). This circuit may be derived in humans, especially as Peeters and colleagues (2009; Peeters, Rizzolatti, & Orban, 2013; also see Caruana et al., 2017) suggest the anterior supramarginal gyrus (aSMG) in the human IPL area (Figure 9.3) seems to be absent in non-human primates.

Although reciprocal connections linking the conscious visual ventral stream and the IPS have been found in macaques (Webster, Bachevalier, & Ungerleider, 1994), the difference with humans seems to reside in the level of processing involved. This relates to the more profuse and dense neural interconnections that characterize the human cortex (Rilling et al., 2008), especially with regard to the tracts linking ventral and dorsal streams (Bi et al., 2015; Hecht et al., 2013, 2015) that converge on the aSMG (Peeters et al., 2009, 2013).

Relationship of IPS to Ventral Pathway

The ventral stream projects to the temporal lobe through the middle temporal gyrus for conceptual and semantic aspects of using, identifying, and manipulating tools in relation to motion (Beauchamp & Martin, 2007; Martin, 2007; Tranel et al., 2003), suggesting a considerable enhancement of the ventral stream for carrying out such tasks. As stipulated, lesions to this area cause specific perceptual deficits in humans

Table 9.1. Functions of the Ventral, Ventrodorsal, and Dorsal-Dorsal Streams

Ventral Stream (Perception for Vision)	*Ventrodorsal Stream (Perception for Action)*	*Dorsal-Dorsal Stream (Perception for Action)*
• Explicit/conscious	• Implicit/blind	• Implicit/blind
• Slow	• Moderately slow	• Immediate/fast/automatic
• Offline	• Somewhat offline	• Online
• Longest latency (memory)	• Long latency (minutes)	• Short latency (msec)
• Identification of objects	• Grasp (near)	• Grasp (far)
• Object knowledge	• Object use/functional	• Structural components
• Bilateral	• Left hemisphere	• Bilateral
• Object constancy	• Stable affordances	• Variable affordances
• Sustained	• Lagged time-course	• Continuously updated
• Viewpoint independent	• Viewpoint dependent	• Viewpoint dependent
• Allocentric	• Egocentric	• Egocentric
• Object parts	• Primary axis	• Primary axis
• Symmetry	• Symmetry	
	• 2D/3D shape from motion	

Note: Data compiled by the author.

(Buxbaum et al., 2014), which are absent in non-human primates (Milner & Goodale, 2006). The merging of inputs from the ventral and dorsal streams by way of the aSMG may therefore have facilitated the recruitment of conceptual aspects of tool use (as part of the frontoparietal system) that led to an enhanced visuospatial/motor circuit (Watson & Buxbaum, 2015). The different functions of the two dorsal sub-streams and the ventral stream are set out in Table 9.1.

Tool Use and Differences in Human and Simian Dorsal Stream

The differences observed between human and simian IPS are reflected in the aforementioned bimodal neurons, which respond to somatosensory and visual aspects of hand movements. In Japanese monkeys trained to use tools, bimodal neurons are located in the anterior IPS. Crucially, in one study, these neurons remained unimodal in untrained simians (Hihara et al., 2006). Moreover, in the trained group, 14 days of challenging tool-use drills were required before the bimodal neurons became apparent. In effect, when trained to retrieve objects using tools, simians display increased activation in IPS where somatosensory and visual information is integrated in the form of bimodal neurons that integrate tools into a body schema for the extension of reaching space (Hihara et al., 2006; Iriki, Tanaka, & Iwamura, 1996; Ishibashi, Hihara, & Iriki, 2000; Maravita & Iriki, 2004). Because bimodal neurons are absent in untrained monkeys and intense focused training is needed to recruit such neurons, the associated capacity and behavioral correlates may only be available temporarily, if at all, in wild populations. Thus, the findings of Hihara and colleagues (2006) suggest that in artificial (laboratory) situations, monkeys are able to develop a modicum of visuospatial skills for using tools as revealed in the reformulation of the neural substrates in IPS.

Correspondingly, canonical neurons have been found in area F5 (area 44 in humans) of the ventral premotor cortex as well as anterior IPS of the macaque brain (Grèzes, Armony, Rowe, & Passingham, 2003). These neurons become active when a graspable object is viewed, but not when it is reached for and held. Accordingly, canonical neurons appear to mediate the pragmatic affordances suggested by an object in relation to potential grip (Grèzes et al., 2003), which substantiates that this area functions by way of implicit/embodied processes within the dorsal system. Such neurons may underlie the ability of simians to engage in simple tool use in the wild.

The preceding analysis suggests that the tool skills potentially available to higher non-human primates trained in artificial conditions may represent the initial alterations to the underlying cortical substrates of pre-Oldowan hominins when they first began to make and use tools (Kibunjia, 1994; Toth & Schick, 2009). These skills include the basic throwing and rudimentary percussion skills that allowed crude stone tools to be produced (Schick et al., 1999; Toth, Schick, Savage-Rumbaugh, Sevcik, & Rumbaugh, 1993) and were primarily derived from the movements of the shoulders and elbows, as opposed to the wrists and fingers (Ambrose, 2001). In order to produce Oldowan tools, changes were required to IPS that necessitated extra resources be devoted to 3D shape from motion, so that this could interface with somatosensory and motor criteria by way of bimodal and canonical neurons. Although relatively crude and "ape-like" (Wynn, 2010), Oldowan tools were marginally more complex than those produced by modern chimpanzees and capuchins (Schick & Toth, 2001), and they therefore required greater neural resources for processing 3D shape from motion. As this processing remained within the dorsal pathway, implicit procedures continued to dominate. That is, tool-making continued to be closely aligned and scaffolded by the actual materials (Malafouris, 2010a, 2010b), in the sense of being constrained by the prevailing affordances provided within the dorsal stream.

These differences suggest a derived visuospatial/motor pathway in humans, particularly for processing 3D form, arising from V3A and V6/V6A to IPS and eventually projecting to the ventral premotor area. In this regard, as part of the posterior parietal area, IPS seems to have undergone a progressive expansion and reorganization in humans (Bruner, 2010; Husain & Nachev, 2007; Orban, van Essen, & Vanduffel, 2004; Roland, 1993; Zilles & Palomero-Gallagher, 2001); based on morphological variations of the brain, Bruner (2010) proposed that important alterations occurred in IPS. Arguably, this led to the expansion of the inferior parietal area (Peeters et al., 2009, 2013), thus providing the neural capacity for the ability to produce a greater range of tools during the Middle Paleolithic/Stone Age.

In short, Oldowan tools remained constrained by neurostructures tethered to an enhanced "primitive" visuospatial/visual motor system that, although similar to the system in extant chimpanzees, benefitted from significant "tuning" of the basic network. Given that this system operated implicitly, the visuospatial coordinates required to make Oldowan and Early Acheulean tools were closely aligned to the prevailing affordances of the actual materials undergoing manipulation.

Symmetry

A further difference in humans compared to non-human primates involves sensitivity to symmetry. Sasaki and colleagues (Sasaki, Vanduffel, Knutsen, Tyler, & Tootell, 2005) found that responses of V3A and later extrastriate regions to symmetrical stimuli are much more potent in humans than in simians. They also established that more symmetrical stimuli evoked greater activity in V3A and adjacent areas (LOC or lateral occipital complex, V7 and V4 in the early ventral pathway). As responses to symmetry were absent from areas prior to V3A (though a marginal response to symmetry was found in V3), such responses appear restricted to extrastriate areas (i.e., regions beyond V1 or primary visual cortex) with extensive receptive fields that integrate information from broader areas of the visual array (Beck, Pinsk, & Kastner, 2005). Crucially, symmetry-related responses in macaque V3A, V4, and posterior inferior temporal area were found to be relatively weak in monkeys in comparison to those in humans (Sasaki et al., 2005)—recall that V3A and V3 appear reversed in humans compared to V3A and V3 in simians. Sasaki and colleagues (2005) also found that the extrastriate area responds to symmetrical stimuli varying in size, which suggests a generalized propensity for such shapes. From the visual cortex through to IPS, symmetry is therefore capable of being processed preconsciously (Gurd, Fink, & Marshall, 2002; Hodgson, 2009a, 2009b, 2011; Wagemans, 1997).

Visuospatial coordinates, 3D shape from motion, symmetry, and other basic aspects of form thus appear to be processed by the blind dorsal "where/how" pathway (perception for action) for proactively engaging with objects. This proceeds by way of IPS as part of the implicit ventrodorsal pathway, which has significantly greater processing capacity in a number of respects in humans than in non-human primates. These findings suggest that the human ability to process symmetry and 3D shape from motion is related to the need to encode the detailed visual information involved in making and manipulating tools, an ability that is probably derived from and reciprocally related to the prerequisites of making Oldowan and Early Acheulean tools.

NEUROIMAGING AND EARLY TOOL-MAKING

The significance of the assimilation of visuospatial and motor streams for understanding the cognitive aspects of tool-making is supported by a series of studies carried out by Stout and colleagues (Stout, Apel, Commander, & Roberts, 2014; Stout & Chaminade, 2007; Stout, Toth, Schick, & Chaminade, 2008; Stout, Toth, Schick, Stout, & Hutchins, 2000). These studies show through brain scans that when novice stone tool knappers make Oldowan-like tools, occipital areas are activated together with the superior parietal area and regions within IPS (the dorsal-dorsal and ventrodorsal streams). This is taken to reflect the early, basic visuomotor skills related to immediate solutions for completing tasks that depend on the "primitive" dorsal stream (Figure 9.3). When expert knappers produce Acheulean tools, however, though similar regions are recruited to those activated in Oldowan tools (Stout et al., 2000), activation of these regions has been found to be reduced, whereas forward areas—including the inferior parietal area—become more active (Stout et al., 2008, 2014). However, the prefrontal cortex (more specifically the inferior frontal gyrus) did not become fully active until Late Acheulean tools were produced (Stout et al.,

2014), which suggests that Oldowan and, to a lesser extent, Early Acheulean tools do not involve *extensive* planning. These findings illustrate that as tasks become progressively challenging and knappers improve expertise, more forward multimodal areas of the brain are recruited (Putt, Wijeakumar, Franciscus, & Spencer, 2017). This suggests an integration of various abilities relating to the aforementioned neural pathways along the dorsal pathway as part of an enhanced ventrodorsal stream (including IPS and ventral premotor cortex).

These studies lead to the conclusion that expertise in making stone tools involves greater assimilation of information along the ventrodorsal stream to recruit the enhanced visuospatial and motor abilities that converge in the supramarginal gyrus (SMG) in the inferior parietal cortex (Peeters et al., 2013; Stout et al., 2008). As stipulated, aSMG appears to be a key derived area in humans where information from anterior IPS of the ventrodorsal stream interfaces with the ventral pathway (see Figure 9.4), a finding recently supported by a neuroimaging study of Acheulean tool-making, by Putt and colleagues (Putt et al., 2017).[3] Such assimilation may have been essential to produce the increasingly complex symmetries of Late Acheulean bifaces. This will have occurred through reciprocal links between the ventrodorsal, ventral, and premotor systems that eventually led to stable interconnections developing between AIP and SMG and frontoparietal circuits (Sakreida et al., 2016), including the inferior frontal gyrus (for planning and control). This is corroborated by the fact that, as well as IPS, aSMG becomes active in "simple" tool use in humans—a cortical area that remains inactive in non-human primates trained to use tools (Peeters et al., 2009, 2013). Again, this reflects the constraints of the "primitive" pathway in non-human primates, the existence of which in humans is corroborated by the finding that patients with a damaged ventral stream tend to grasp, for example, a hammer as if it were a rudimentary stick or rod (Goodale & Milner, 2013). In other words, the functional/conceptual aspects of a tool are unappreciated.

The dorsal pathway, therefore, seems to be concerned with the primary axis of objects in the sense that it is unable to integrate different functional parts of a tool appropriately. Neuroimaging further confirms that, when humans plan and use everyday tools, areas of the ventral stream in the left hemisphere are activated, in addition to the dorsal pathway (Johnson-Frey, Newman-Norlund, & Grafton, 2005) (Figure 9.5). Johnson-Frey and colleagues (2005, p. 1) suggest that "this left lateralized network constitutes a neural substrate for the interaction of semantic and motoric representations upon which meaningful skills depend." Thus, it is not surprising that Stout and colleagues found that SMG and the frontoparietal area became more engaged when expert knappers produced tools. Appositely, white matter tracts connecting these pathways are more extensive in humans along the superior longitudinal fasciculus, compared to pathways in macaques and chimpanzees (Bi et al., 2015; Hecht et al., 2013, 2015; Stout & Hecht, 2017). Hodgson (2007, 2012) has suggested that these tracts and associated brain regions have important consequences for understanding the evolution of tool-making.

[3] Interestingly, Putt and colleagues (2017) found the purported relationship between language and the making of Acheulean handaxes to be questionable.

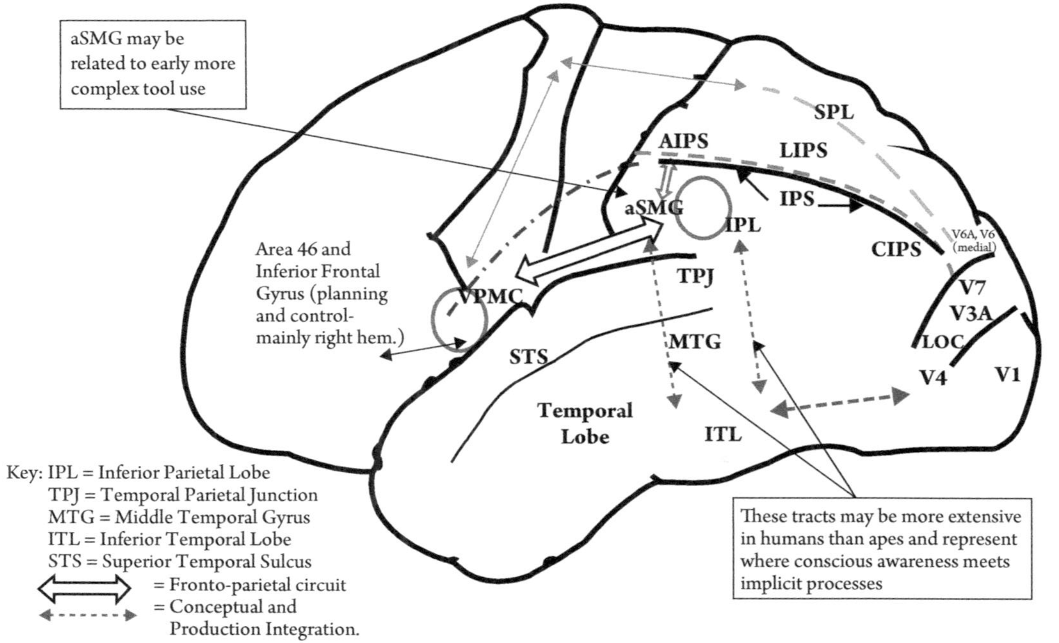

Figure 9.4. The full array of interconnecting pathways in the human cortex implicated in making and using tools. The ventral conscious "what" pathway (thick gray dashed lines with arrows) associated with conceptual/semantic abilities has been added to the enhanced "primitive" circuit illustrated in Figure 9.3, indicating the integration of conceptual and production abilities. The two ellipses (near the vPMC and between the aSMG and IPL) define important areas of interconnectivity. Abbreviations: aSMG, anterior supramarginal gyrus; IPS, intraparietal sulcus (A, anterior; L, lateral; C, caudal); ITL, inferior temporal lobe; LOC, lateral occipital cortex; MTG, middle temporal gyrus; SPL, superior parietal lobule; STS, superior temporal sulcus; TPJ, temporal parietal junction; vPMC, ventral premotor cortex. Lines: The long gray line between the vPMC and aSMG indicates the frontoparietal circuit. Other abbreviations and lines as previously specified in Figure 9.3, except where indicated otherwise. Image by the author.

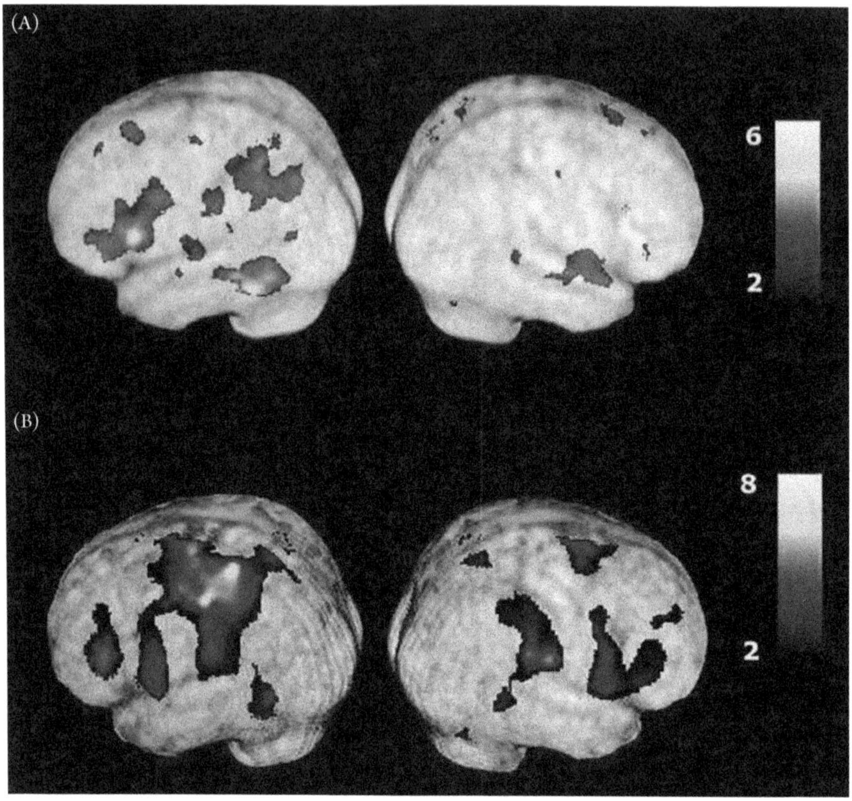

Figure 9.5. Neuroimaging of human cortex illustrating activation of left hemisphere network identified in the text that was found to be more activate when tool use was observed. (A, top) Preparing random hand movements. (B, bottom) Tool actions. Previously published as Figure 2 (p. 684) in Johnson-Frey et al. (2005), A distributed left hemisphere network active during planning of everyday tool use skills, *Cerebral Cortex*, and republished with the kind permission of the author and Oxford University Press.

IMPLICATIONS FOR THE ARCHAEOLOGY OF TOOLS

Evolution and Tools

The fact that a much greater age for the existence of *H. habilis* than had previously been envisaged has been proposed (Spoor et al., 2015) suggests that the lineage dates to before 2.3 million years ago (Mya). This is consistent with the recent finding that Oldowan-like tools, referred to as the "Lomekwian" (LOM3), date to at least 3.3 Mya (Harmand et al., 2015). Pre-Oldowan tools of this order provide support for marginally enhanced hand control related to the initial reorganization of somatosensory, premotor/motor cortex, and visual areas, as well as the cerebellum and spinal tract, before 3.3 Mya (Harmand et al., 2015). This is consistent with Holloway's (2008) observation that the lunate sulcus was already displaced in Australopithecus. It also accords with Wynn's proposal that the basic spatial abilities of Oldowan hominins were *marginally* advanced from those of pre-hominins (Wynn, Tierson, & Palmer, 1996). These attributes indicate the early hominin brain was primed to produce tools, a capability that initially evolved from the basic ability to use tools found in non-human primates (Wynn, 2010; Wynn et al., 1996; also see Chapter 1 in this volume). It should be noted that although bonobos such as Kanzi are able to learn to knap stones to produce tools in artificial situations, their skills still fall short of those required to make Oldowan tools (Toth & Schick, 2009). This tallies with the earlier observation of the enhancement of bimodal neurons in simian IPS when taught to use tools (Hihara et al., 2006). The main issue is to determine how the various tracts identified in the cerebral cortex underwent reorganization and enhancement in tandem with the different levels of expertise, when ancestral species passed from an earlier implicit stage to an explicitly programmed one, as more complex and variable tools began to appear.

Shultz, Nelson, and Dunbar (2012) have argued for extensive periods of relative stasis in brain expansion interspersed with episodes of rapid growth, appearing with *H. habilis* and followed by *H. ergaster.* Then there seems to have been a very gradual increase in brain size from early to late *H. erectus,* followed by a further rapid change in *H. heidelbergensis,* with a final phase occurring in modern humans around 100 thousand years ago (Kya). Some of these rapid changes arose in tandem with new technocomplexes: For example, *H. ergaster* (or early *H. erectus*) appeared around 1.8 Mya, about the same time as Acheulean tools. However, despite the gradual increase in brain size of *H. erectus,* from this date and during the Acheulean, innovation was relatively conservative (Lycett & Gowlett, 2008; Sharon, Alperson-Afil, & Goren-Inbar, 2011). This needs to be tempered by the ocurrence of some development in technique and refinement which appears to have taken place during the latter phases of the Acheulean, from about 700 Kya onward (Clark, 2001; Gowlett, 2006; Stout, 2011; Wynn, 2002; Wynn & Coolidge, 2016; however, some commentators dispute this, as for example, McNabb & Cole, 2015).

The consensus that a trend toward smaller, more refined tools took place during the Late Acheulean (Clark, 2001; Gowlett, 2006; Stout et al., 2014; Wynn, 2002, 2014) seems to accord with the appearance of *H. heidelbergensis,* a species with a larger brain. This leaves the long period from the beginning of the Acheulean up to *H. heidelbergensis* to account for, as there was relatively little change evident in tools during this time. This creates a mismatch between the gradual expansion of the brain and the apparent lack of change in tool shape and types. However, recent evidence from Ethiopia has established that an increase in refinement and symmetry in stone tools did, in fact, occur from about 1.75 Mya up until 850 Kya (Beyene, Asfaw, Sano, & Suwa, 2015; Beyene et al., 2013) and subsequently (Iovita et al., 2017). This is borne out by another site in Ethiopia from a similar time period, where a concomitant increase in symmetry, refinement, and standardization is evident (Gallotti, Raynal, Geraads, & Mussi, 2014). Beyene and colleagues (2015, p. 72) have concluded that the "better made handaxes and cleavers of this time period tend to be standardized in morphology exhibiting a substantially thin, 3-dimensionally symmetric form with fine straight edges." In addition, Le Tensorer (2009) reported finding over 8,000 handaxes from Syria dating to around 500 Kya, with the majority displaying a remarkable consistency in symmetry. Thus, although some researchers refute any progression in refinement, these reasonably secure examples confirm the reality of a progressive tendency, which suggests the cognitive ability of *H. erectus* did increase over time. In this respect,

Acheulean spheroids, thought to have been used as projectile weapons, also underwent a measure of refinement (Wilson, Zhu, Barham, Stanistreet, & Bingham, 2016). In fact, Lorblanchet and Bahn (2017) have suggested that the increasingly perfection of symmetry of spheroids first appears in the Oldowan and proceeds throughout the Acheulean and may have been a result of a growing interest in the aesthetics of shape.

These insights suggest that the neurostructures on which knapping depends became incrementally interdependent and, along with the gradual increase in brain size, a permanent feature of the brain in later *H. erectus*. This, however, does not imply a strict genetically determined scenario but, rather, the gradual accrual of technology-related genes that structured development leading to an adaptive, extended phenotype, probably by way of the Baldwin effect (Baldwin, 1896; Corbey, Jagich, Vaesen, & Collard, 2016; Hodgson, 2012; Sznajder, Sabelis, & Egas, 2012). This will have involved a feedback loop that engendered an increasing efficiency to produce tools through learning that promoted the selective advantage of such a trait, which through successive approximation filtered through the average population by way of genetic assimilation (Waddington, 1953). Indeed, it has recently been confirmed that spatial ability exists as a distinct cognitive domain structured by genetic components (Shakeshaft et al., 2016). These traits will have been supplemented by the advantages that accrue from niche construction, where changes to the immediate environment are beneficial in a changing fitness landscape (Laland & Sterelny, 2006). In effect, Beyene and colleagues (2013, 2015) suggest that visuospatial abilities improved during the specified 900,000-year period with subtle advances in stone tool refinement, which thereby provides a proxy for ascertaining the minimum cognitive competence (Wynn, 2002) during this period.[4]

Refinement in tools during the Acheulean, however, is not restricted to handaxes, as *H. heidelbergensis* increasingly made other tool types with different materials. For example, bones and antlers were utilized to make tools at Boxgrove and Bilzingsleben (Mania & Mania, 2005; Pope & Roberts, 2005; Wenban-Smith, 1999), and wood was used for the Schöningen "spears" (Thieme, 2005), which confirms that a range of materials was being exploited. Moreover, hafted tools began to appear around 500 Kya (Wilkins, Schoville, Brown, & Chazan, 2012). These findings indicate that basic visuospatial abilities were being supplemented by applying such abilities to manipulating new materials to produce more complex tools. Moreover, the interest in shape seems to have become detached from utilitarian concerns, as evidenced by the appearance of exceptionally large handaxes that display an over-concern for symmetry (Hodgson, 2011). The fact that handaxes, rather than picks or cobbles, began to be made with elephant bone suggests an interest in particular materials that goes beyond the practical, possibly reflecting ritual purposes (Hodgson, 2009a, 2009b, 2011; Zutovski & Barkai, 2016). Moreover, across a range of sites, only elephant bones were used to produce bifaces, despite the availability of other large animal bones. Pertinently, some later

[4] *Minimum cognitive competence* refers to the possibility that the maker of a tool may be more cognitively capable than what an artifact reveals (Wynn, 2002). In comparison, "maximum cognitive competence" (McNabb & Cole, 2015) implies that a particular artifact indicates a ceiling in cognitive ability. Wynn's definition appears to be more useful, as we can infer only the basic level of ability from an archaeological horizon that is exceedingly sparse.

handaxes display extraordinary features, including broken symmetry, characteristics that appear in the archaeological record with greater frequency after 500 Kya (Wynn, 2000, 2002, 2014). As Wynn (2014) has suggested, this attests to the appearance of an "inner virtual space" where internally generated images could be manipulated to guide action.

We could therefore assume that during the Early Acheulean, the gradual growth in brain size was paralleled by the subtle reorganization of the cortex in the ways described. Thus, during the Oldowan and Early Acheulean, the slow change in the refinement of tool morphology may have been a function of the gradually evolving neural substructures specified. This suggests that subtle changes occurred to these subsystems before they became manifest on a larger scale. However, it was not until the underlying white matter tracts facilitated the full integration of the ventral "what" ("vision for perception") and dorsal "where/how" ("vision for action") pathways (Milner & Goodale, 2008) that a greater range of tool types and a more complex *chaîne opératoire* occurred (Hodgson, 2007, 2012; Stout & Hecht, 2017).

In sum, Oldowan tools were mediated by the enhanced "primitive" blind dorsal pathway that functioned relatively independently of the consciously defined "what" ventral stream. Precisely because this primitive tract is blind, the materials undergoing modification provided the cues that scaffolded eventual outcomes. That is, interaction with the materials was necessary during the actual tool-making procedure, possibly based on episodic event perception (Donald, 2007), where action is inextricably tethered to the materials concerned.

DISCUSSION

Although some scholars claim there was little or no development in shape or tool refinement from the beginning of the Acheulean onward (McNabb & Cole, 2015), increasing evidence suggests otherwise (Hodgson, 2015; Stout & Hecht, 2017; Wynn & Coolidge, 2016), as indicated by the recent empirical studies cited here. Taken together with the facts that a greater range of materials was employed to make tools and that an interest developed in using elephant bones to make handaxes during the Late Middle Pleistocene, this signposts a growing concern for non-functional aspects of tools. Such insights indicate that during the Early Acheulean, the visuospatial tract underwent modification, especially along the ventrodorsal pathway within IPS, which recruited extra functions compared to the same area in non-human primates. Yet during the Early Acheulean, tool-making continued to be constrained by the material undergoing transformation and, in this sense, was almost completely embodied. It was not until the Late Acheulean that the intraparietal area and ventral pathway began to interface more fully by way of the SMG—thereby facilitating a mental template for creating refined tools—when a proto-aesthetic interest in tool shape arose. In this way, the visuospatial sketchpad was enhanced, not only as a result of the inherent contingencies described, but also by interfacing with other areas of the higher cortex, especially the ventral stream and frontoparietal areas. Interestingly, Logie and van der Meulen (2008) suggest that the modern visuospatial sketchpad should be divided into two areas. One is referred to as "the visual cache," which temporarily stores degradable information regarding form and color; the other, known as "the inner scribe," is a temporary spatial store used to plan movement where rehearsal of information from the visual cache can potentially take place. These two systems have been implicated in the left and right hemispheres, respectively, leading to the possibility of links with the aforementioned "what ventral" and "where/how dorsal" brain tracts. This finding has implications for Wynn's (2002, 2014) and Hodgson's (2005, 2009a, 2009b, 2011, 2012) proposal that such tracts were becoming more integrated from around 700 Kya, when Acheulean tools became more standardized and symmetrical.

The issue of why handaxes became so important with regard to shape profile rather than other Acheulean tools such as cleavers may be related to the fact that handaxes are worked bifacially, whereas cleavers are knapped unifacially. As a result, handaxes undergo rotation more often than cleavers, thereby facilitating greater awareness of symmetry (Hodgson, 2011); the same can be said of spheroids. The outline contour of handaxes is also perceptually less complex than that of cleavers, thereby increasing saliency. This is supported by the fact that when a modern human involved in the active manual control of production views an object, visual recognition and learning are expedited (Harman, Humphrey, & Goodale, 1999; James, Humphrey, & Goodale, 2001). Active exploration of such views thus promotes learning of the 3D structure of objects (e.g., allocentric perception related to the ability for mental rotation and an implicit understanding of Euclidean geometry; Wynn, 1989, 2010).

The perceived lack of a direct link between increases in brain size and tool complexity during the Early to Middle Acheulean is countered by the proposal that such a link may in fact exist, for two reasons. First, there were subtle changes in the refinement of tools, but these have been underestimated, as suggested by the findings of Beyene and colleagues (2013, 2015) and of others (Gallotti et al., 2014; Iovita et al., 2017). The gradual increase in *H. erectus* brain size was partially driven by the incremental enhancement of neural structures in the dorsal stream that culminated in the intermeshing of this stream with the ventral pathway. It was only when this intermeshing became fully operational during the Late Acheulean and subsequently that more complex tools—including hafted and composite tools—could be produced, though composite tools may have also required greater input from sociocultural factors. The allocentric (non-egocentric) frame of reference needed to produce complex tools, available from 500 Kya onward, allowed objects to be visualized from various perspectives. The ability to deal with such perspectives may have had implications for the social aspects of tool-making, in that it could facilitate learning from the behaviors of others (Iriki, 2006), especially when it involved pantomiming (Frey, 2008).

CONCLUSION

Four phases of neurocognition seem to have occurred with regard to tool-making and tool use compared to that of extant non-human primates:

(1) Limited tuning of the "primitive" dorsal system along IPS associated with Oldowan tools
(2) Augmentation of this "primitive" pathway during the Early Acheulean, thanks to an enhanced ventrodorsal system projecting along IPS to ventral premotor cortex

(3) Further enhancement of the ventrodorsal stream within IPS during the Late Acheulean, with improved links to premotor and inferior parietal area, especially through aSMG and inferior frontal gyrus
(4) Addition of a conceptual/semantic domain with the appearance of composite tools during the Middle Paleolithic/Stone Age, driven by increased links to the "what" ventral pathway and inferior frontal gyrus

In contrast to non-human primates, extra visuospatial coordinates may therefore have been drafted into the IPS of Australopithecus/*H. habilis* from a basic primate "primitive" ventrodorsal pathway that facilitated the production of Oldowan tools. This was later enhanced with visual functions, for example, 3D shape from motion, that promoted the making of Early Acheulean tools by *H. erectus*. As these increments increasingly became interconnected with the somatosensory and motor systems, hominins were thus able to manipulate and shape objects with greater acumen. The ventrodorsal and dorsal-dorsal streams provided a means whereby information could be processed implicitly that was closely tied to the object concerned in that these streams remained embodied in a way that led to the conservatism of Oldowan and Acheulean tools.

Later Acheulean tools involved increasing enhancement of IPS and recruitment of IPL by way of aSMG that engendered a greater concern for symmetry and coordinates relating to the processing of 3D shape from motion. Finally, the IPL began to play an important role in the production of more complex tools, as mediated by conceptual imperatives linked to overt awareness of the ventral pathway; this would have driven a reorganization of this part of the cortex that probably reached a "tipping point" with *H. heidelbergensis* (Dubreuil, 2010). When the ventral stream began to engage fully with the dorsal pathway, this may have led to the ability to produce and use a greater range of tools. Thus, thanks to the reorganization of the neural pathways that occurred in response to the reciprocal processes associated with making tools, the human brain came to realize a distinctive architecture that allowed visual information on a number of levels to be combined, including for visuospatial, visuomotor, and overt (conscious) visual capacities.

ACKNOWLEDGMENTS

Many thanks to Karenleigh A. Overmann and Frederick L. Coolidge for their help in facilitating the editorial process. Thanks are also due to anonymous reviewers for comments that assisted in improving the manuscript. Finally, my wholehearted thanks to Tom Wynn for his long-term support and encouragement in guiding my research to the level of this chapter.

REFERENCES

Ambrose, S. H. (2001). Paleolithic technology and human evolution. *Science, 291*(5509), 1748–1753.

Baddeley, A. D. (2007). *Working memory, thought, and action*. New York: Oxford University Press.

Baddeley, A. D., & Hitch, G. J. (1974). Working memory. In G. H. Bower (Ed.), *The psychology of learning and motivation: Advances in research and theory* (Vol. 8, pp. 47–89). New York: Academic Press.

Baldwin, J. M. (1896). A new factor in evolution. *American Naturalist, 30*(354), 441–451.

Beauchamp, M. S., & Martin, A. (2007). Grounding object concepts in perception and action: Evidence from fMRI studies of tools. *Cortex, 43*(3), 461–468.

Beck, D. M., Pinsk, M. A., & Kastner, S. (2005). Symmetry perception in humans and macaques. *Trends in Cognitive Sciences, 9*(9), 405–406.

Beyene, Y., Asfaw, B., Sano, K., & Suwa, G. (Eds.). (2015). *Konso-Gardula research project. Vol. 2, Archaeological collections: Background and the early Acheulean assemblages. Bulletin 48.* Tokyo: University of Tokyo.

Beyene, Y., Katoh, S., WoldeGabriel, G., Hart, W. K., Uto, K., Sudo, M., . . . Asfaw, B. (2013). The characteristics and chronology of the earliest Acheulean at Konso, Ethiopia. *Proceedings of the National Academy of Sciences of the United States of America, 110*(5), 1584–1591.

Bi, Y., Han, Z., Zhong, S., Ma, Y., Gong, G., Huang, R., . . . Caramazza, A. (2015). The white matter structural network underlying human tool use and tool understanding. *Journal of Neuroscience, 35*(17), 6822– 6835.

Binkofski, F., & Buxbaum, L. J. (2013). Two action systems in the human brain. *Brain and Language, 127*(2), 222–229.

Bruner, E. (2010). Morphological differences in the parietal lobes with the human genus: A neurofunctional perspective. *Current Anthropology, 51*(1), S77–S88.

Burgess, N. E., Jeffery, K. J., & O'Keefe, J. E. (1999). The hippocampal and parietal foundations of spatial cognition. In N. E. Burgess, K. J. Jeffery, & J. E. O'Keefe (Eds.), *The hippocampal and parietal foundations of spatial cognition* (pp. 3–20). Oxford, UK: Oxford University Press.

Buxbaum, L. J. (2001). Ideamotor apraxia: A call to action. *Neurocase, 7*(6), 445–458.

Buxbaum, L. J., Shapiro, A. D., & Coslett, H. B. (2014). Critical brain regions for tool-related and imitative actions: A componential analysis. *Brain, 137*(7), 1971–1985.

Caruana, F., Avanzini, P., Mai, R., Pelliccia, V., LoRusso, G., Rizzolatti, G., & Orban, G. A. (2017). Decomposing tool-action observation: A stereo-EEG study. *Cerebral Cortex, 27*(8), 4229–4243.

Clark, J. D. (2001). Variability in primary and secondary technologies of the Later Acheulian in Africa. In S. Miliken & J. Cook (Eds.), *A very remote period indeed: Papers on the Palaeolithic presented to Derek Roe* (pp. 1–18). Oakville, CT: Oxbow Books.

Corbey, R., Jagich, A., Vaesen, K., & Collard, M. (2016). The Acheulean handaxe: More like a bird's song than a Beatles' tune? *Evolutionary Anthropology: Issues, News, and Reviews, 25*(1), 6–19.

Cronin-Golomb, A., & Hof, P. R. (Eds.). (2004). *Vision in Alzheimer's disease.* Basel: Karger Medical and Scientific Publishers.

Davare, M., Rothwell, J. C., & Lemon, R. N. (2010). Causal connectivity between the human anterior intraparietal area and premotor cortex during grasp. *Current Biology, 20*(2), 176–181.

Denys, K., Vanduffel, W., Fize, D., Nelissen, K., Peuskens, H., Van Essen, D., & Orban, G. A. (2004). The processing of visual shape in the cerebral cortex of human and nonhuman primates: A functional magnetic resonance imaging study. *Journal of Neuroscience, 24*(10), 2551–2565.

Donald, M. (2007). The slow process: A hypothetical cognitive adaptation for distributed cognitive networks. *Journal of Physiology—Paris, 101*(4), 214–222.

Dubreuil, B. (2010). Paleolithic public goods games: Why human culture and cooperation did not evolve in one step. *Biology and Philosophy, 25*(1), 53–73.

Durand, J. B., Nelissen, K., Joly, O., Wardak, C., Todd, J. T., Norman, J. F., . . . Orban, G. A. (2007). Anterior regions of monkey parietal cortex process visual 3D shape. *Neuron, 55*(3), 493–505.

Farah, M. J. (2004). *Visual agnosia* (2nd ed.). Cambridge, MA: MIT Press.

Fattori, P., Breveglieri, R., Marzocchi, N., Filippini, D., Bosco, A., & Galletti, C. (2009). Hand orientation during reach-to-grasp movements modulates neuronal activity in the medial posterior parietal area V6A. *Journal of Neuroscience, 29*(6), 1928–1936.

Fattori, P., Gamberini, M., Kutz, D. F., & Galletti, C. (2001). "Armreaching" neurons in the parietal area V6A of the macaque monkey. *European Journal of Neuroscience, 13*(12), 2309–2313.

Fattori, P., Kutz, D. F., Breveglieri, R., Marzocchi, N., & Galletti, C. (2005). Spatial tuning of reaching activity in the medial parieto-occipital cortex (area V6A) of macaque monkey. *European Journal of Neuroscience, 22*(4), 956–972.

Frey, S. H. (2008). Tool use, communicative gesture and cerebral asymmetries in the modern human brain. *Philosophical Transactions of the Royal Society of London. Series B, Biological Sciences, 363*(1499), 1951–1957.

Gallotti, R., Raynal, J.-P., Geraads, D., & Mussi, M. (2014). Garba XIII (Melka Kunture, Upper Awash, Ethiopia): A new Acheulean site of the late Lower Pleistocene. *Quaternary International, 343*, 17–27.

Goodale, M. A., & Milner, A. D. (2013). *Sight unseen: An exploration of conscious and unconscious vision* (2nd ed.). Oxford, UK: Oxford University Press.

Gowlett, J. A. J. (2006). The elements of design form in Acheulian bifaces: Modes, modalities, rules and language. In N. Goren-Inbar & G. Sharon (Eds.), *Axe age: Acheulian tool-making from quarry to discard* (pp. 203–222). London: Equinox.

Grafton, S. T. (2010). The cognitive neuroscience of prehension: Recent developments. *Experimental Brain Research, 4*(4), 475–491.

Grèzes, J., Armony, J. L., Rowe, J., & Passingham, R. E. (2003). Activations related to "mirror" and "canonical" neurons in the human brain: An fMRI study. *NeuroImage, 18*(4), 928–937.

Gurd, J. M., Fink, G. R., & Marshall, J. C. (2002). Tacit symmetry detection and explicit symmetry processing. *Behavioral and Brain Sciences, 25*(3), 409.

Harman, K. L., Humphrey, G. K., & Goodale, M. A. (1999). Active manual control of object views facilitates visual recognition. *Current Biology, 9*(22), 1315–1318.

Harmand, S., Lewis, J. E., Feibel, C. S., Lepre, C. J., Prat, S., Lenoble, A., . . . Roche, H. (2015). 3.3-million-year-old stone tools from Lomekwi 3, West Turkana, Kenya. *Nature, 521*(7552), 310–315.

Hecht, E. E., Gutman, D. A., Khreisheh, N., Taylor, S. V, Kilner, J., Faisal, A. A., . . . Stout, D. (2015). Acquisition of Paleolithic toolmaking abilities involves structural remodeling to inferior frontoparietal regions. *Brain Structure and Function, 220*(4), 2315–2331.

Hecht, E. E., Gutman, D. A., Preuss, T. M., Sanchez, M. M., Parr, L. A., & Rilling, J. K. (2013). Process versus product in social learning: Comparative diffusion tensor imaging of neural systems for action execution–observation matching in macaques, chimpanzees, and humans. *Cerebral Cortex, 23*(5), 1014–1024.

Hihara, S., Notoya, T., Tanaka, M., Ichinose, S., Ojima, H., Obayashi, S., . . . Iriki, A. (2006). Extension of corticocortical afferents into the anterior bank of the intraparietal sulcus by tool-use training in adult monkeys. *Neuropsychologia, 44*(13), 2636–2646.

Hodgson, D. (2005). More on Acheulean tools. *Current Anthropology, 46*(4), 647–650.

Hodgson, D. (2007). The reorganisation of primary visual cortex and extrastriate areas of the human brain in relation to evolution and behavioural indicators. Paper presented at Centre for Archaeology of Human Origins, Southampton University.

Hodgson, D. (2009a). Evolution of the visual cortex and the emergence of symmetry in the Acheulean techno-complex. *Comptes Rendus Palevol, 8*(1), 93–97.

Hodgson, D. (2009b). Symmetry and humans: Reply to Mithen's "sexy handaxe theory." *Antiquity, 83*(319), 195–198.

Hodgson, D. (2011). The first appearance of symmetry in the human lineage: Where perception meets art. *Symmetry, 3*(1), 37–53.

Hodgson, D. (2012). Hominin tool production, neural integration and the social brain. *Human Origins, 1*, 41–64.

Hodgson, D. (2015). The symmetry of Acheulean handaxes and cognitive evolution. *Journal of Archaeological Science: Reports, 2*, 204–208.

Holloway, R. L. (2008). The human brain evolving: A personal retrospective. *Annual Review of Anthropology, 37*, 1–19.

Husain, M., & Nachev, P. (2007). Space and the parietal cortex. *Trends in Cognitive Sciences, 11*(1), 30–36.

Iovita, R., Tuvi-Arad, I., Moncel, M.-H., Despriée, J., Voinchet, P., & Bahain, J.-J. (2017). High handaxe symmetry at the beginning of the European Acheulian: The data from La Noira (France) in context. *PLoS One, 12*(5), e0177063.

Iriki, A. (2006). The neural origins and implications of imitation, mirror neurons and tool use. *Current Opinion in Neurobiology, 16*(6), 660–667.

Iriki, A., Tanaka, M., & Iwamura, Y. (1996). Coding of modified body schema during tool use by macaque postcentral neurons. *NeuroReport, 7*(14), 2325–2330.

Ishibashi, H., Hihara, S., & Iriki, A. (2000). Acquisition and development of monkey tool-use: Behavioural and kinematic analyses. *Canadian Journal of Physiology and Pharmacology, 78*(11), 958–966.

Jakobson, L. S., Archibald, Y. M., Carey, D. P., & Goodale, M. A. (1991). A kinematic analysis of reaching and grasping movements in a patient recovering from optic ataxia. *Neuropsychologia, 29*(8), 803–809.

James, K. H., Humphrey, G. K., & Goodale, M. A. (2001). Manipulating and recognizing virtual objects: Where the action is. *Canadian Journal of Experimental Psychology, 55*(2), 111–120.

James, T. W., Humphrey, G. K., Gati, J. S., Menon, R. S., & Goodale, M. A. (2002). Differential effects of viewpoint on object-driven activation in dorsal and ventral streams. *Neuron, 35*(4), 793–801.

Jeannerod, M. (1986). The formation of finger grip during prehension: A cortically mediated visuomotor pattern. *Behavioural Brain Research, 19*(2), 99–116.

Johnson-Frey, S. H., Newman-Norlund, R., & Grafton, S. T. (2005). A distributed left hemisphere network active during planning of everyday tool use skills. *Cerebral Cortex, 15*(6), 681–695.

Kibunjia, M. (1994). Pliocene archaeological occurrences in the Lake Turkana basin. *Journal of Human Evolution, 27*(1/3), 159–171.

Kravitz, D. J., Saleem, K. S., Baker, C. I., & Mishkin, M. (2011). A new neural framework for visuospatial processing. *Nature Reviews: Neuroscience, 12*(4), 217–230.

Laland, K. N., & Sterelny, K. (2006). Perspective: Seven reasons (not) to neglect niche construction. *Evolution, 60*(9), 1751–1762.

Le Tensorer, J.-M. (2009). The image before the image: Considerations on the colloquium. *Anthropologie, 113*(5), 1005–1017.

Logie, R. H., & van Der Meulen, M. (2008). Fragmenting and integrating visuospatial working memory. In J. F. Brockmole (Ed.), *The visual world in memory* (pp. 1–32). Hove, UK: Psychology Press.

Lorblanchet, M., & Bahn, P. (2017). *The first artists: In search of the world's oldest art.* London: Thames and Hudson.

Lycett, S. J., & Gowlett, J. A. J. (2008). On questions surrounding the Acheulean "tradition." *World Archaeology, 40*(3), 295–315.

Mahon, B. Z., Schwarzbach, J., & Caramazza, A. (2010). The representation of tools in left parietal cortex is independent of visual experience. *Psychological Science, 21*(6), 764–771.

Malafouris, L. (2010a). Metaplasticity and the human becoming: Principles of neuroarchaeology. *Journal of Anthropological Sciences, 88*, 49–72.

Malafouris, L. (2010b). The brain-artefact interface (BAI): A challenge for archaeology and cultural neuroscience. *Social Cognitive and Affective Neuroscience, 5*, 264–273.

Mania, D., & Mania, U. (2005). The natural and socio-cultural environment of *Homo erectus* at Bilzingsleben, Germany. In C. Gamble & M. Porr (Eds.), *The hominid individual in context: Archaeological investigations of Lower and Middle Palaeolithic landscapes, locales and artefacts* (pp. 98–114). London: Routledge.

Maravita, A., & Iriki, A. (2004). Tools for the body (schema). *Trends in Cognitive Sciences, 8*(2), 79–86.

Martin, A. (2007). The representation of object concepts in the brain. *Annual Review of Psychology, 58*, 25–45.

McNabb, J., & Cole, J. (2015). The mirror cracked: Symmetry and refinement in the Acheulean handaxe. *Journal of Archaeological Science: Reports, 3*, 100–111.

Milner, A. D., & Goodale, M. A. (2006). *The visual brain in action* (2nd ed.). Oxford, UK: Oxford University Press.

Milner, A. D., & Goodale, M. A. (2008). Two visual systems re-viewed. *Neuropsychologia, 46*(3), 774–785.

Nakamura, H., Kuroda, T., Wakita, M., Kusunoki, M., Kato, A., Mikami, A., . . . Itoh, K. (2001). From three-dimensional space vision to prehensile hand movements: The lateral intraparietal area links the area V3A and the anterior intraparietal area in macaques. *Journal of Neuroscience, 21*(20), 8174–8187.

Orban, G. A., Claeys, K., Nelissen, K., Smans, R., Sunaert, S., Todd, J. T., . . . Vanduffel, W. (2006). Mapping the parietal cortex of human and non-human primates. *Neuropsychologia, 44*(13), 2647–2667.

Orban, G. A., Sunaert, S., Todd, J. T., Van Hecke, P., & Marchal, G. (1999). Human cortical regions involved in extracting depth from motion. *Neuron, 24*(4), 929–940.

Orban, G. A., van Essen, D., & Vanduffel, W. (2004). Comparative mapping of higher visual areas in monkeys and humans. *Trends in Cognitive Sciences, 8*(7), 315–324.

Peeters, R. R., Rizzolatti, G., & Orban, G. A. (2013). Functional properties of the left parietal tool use region. *NeuroImage, 78*, 83–93.

Peeters, R. R., Simone, L., Nelissen, K., Fabbri-Destro, M., Vanduffel, W., Rizzolatti, G., & Orban, G. A. (2009). The representation of tool use in humans and monkeys: Common and uniquely human features. *Journal of Neuroscience, 29*(37), 11523–11539.

Perenin, M.-T., & Vighetto, A. (1988). Optic ataxia: A specific disruption in visuomotor mechanisms. I. Different aspects of the deficit in reaching for objects. *Brain, 111*(3), 643–674.

Pitzalis, S., Fattori, P., & Galletti, C. (2015). The human cortical area V6 and V6a. *Visual Neuroscience, 32,* 1–15.

Pope, M. I., & Roberts, M. B. (2005). Observations on the relationship between Palaeolithic individuals and artefact scatters at the Middle Pleistocene site of Boxgrove, UK. In C. Gamble & M. Porr (Eds.), *The hominid individual in context: Archaeological investigations of Lower and Middle Palaeolithic landscapes, locales and artefacts* (pp. 81–97). London: Routledge.

Preuss, T. M., Qi, H., & Kaas, J. H. (1999). Distinctive compartmental organization of human primary visual cortex. *Proceedings of the National Academy of Sciences of the United States of America, 96*(20), 11601–11606.

Putt, S. S., Wijeakumar, S., Franciscus, R. G., & Spencer, J. P. (2017). The functional brain networks that underlie Early Stone Age tool manufacture. *Nature Human Behaviour, 1,* 1–8.

Rilling, J. K., Glasser, M. F., Preuss, T. M., Ma, X., Zhao, T., Hu, X., & Behrens, T. E. J. (2008). The evolution of the arcuate fasciculus revealed with comparative DTI. *Nature Neuroscience, 11*(4), 426–428.

Roland, P. E. (1993). *Brain activation.* New York: Wiley.

Sakreida, K., Effnert, I., Thill, S., Menz, M. M., Jirak, D., Eickhoff, C. R., . . . Binkofski, F. (2016). Affordance processing in segregated parieto-frontal dorsal stream sub-pathways. *Neuroscience and Biobehavioral Reviews, 69,* 89–112.

Sasaki, Y., Vanduffel, W., Knutsen, T., Tyler, C., & Tootell, R. (2005). Symmetry activates extrastriate visual cortex in human and non-human primates. *Proceedings of the National Academy of Sciences of the United States of America, 102*(8), 3159–3163.

Schick, K. D., & Toth, N. P. (2001). Paleoanthropology at the millennium. In G. M. Feinman & T. D. Price (Eds.), *Archaeology at the millennium: A sourcebook* (pp. 39–108). New York: Kluwer Academic.

Schick, K. D., Toth, N. P., Garufi, G., Savage-Rumbaugh, E. S., Rumbaugh, D. M., & Sevcik, R. A. (1999). Continuing investigations into the stone tool-making and tool-using capabilities of a bonobo (*Pan paniscus*). *Journal of Archaeological Science, 26*(7), 821–832.

Shakeshaft, N. G., Rimfeld, K., Schofield, K. L., Selzam, S., Malanchini, M., Rodic, M., . . . Plomin, R. (2016). Rotation is visualisation, 3D is 2D: Using a novel measure to investigate the genetics of spatial ability. *Scientific Reports, 6*(30545), 1–10.

Sharon, G., Alperson-Afil, N., & Goren-Inbar, N. (2011). Cultural conservatism and variability in the Acheulian sequence of Gesher Benot Ya'aqov. *Journal of Human Evolution, 60*(4), 387–397.

Shultz, S., Nelson, E., & Dunbar, R. I. M. (2012). Hominin cognitive evolution: Identifying patterns and processes in the fossil and archaeological record. *Philosophical Transactions of the Royal Society of London. Series B, Biological Sciences, 367*(1599), 2130–2140.

Spoor, F., Gunz, P., Neubauer, S., Stelzer, S., Scott, N., Kwekason, A., & Dean, M. C. (2015). Reconstructed *Homo habilis* type OH 7 suggests deep-rooted species diversity in early *Homo. Nature, 519*(7541), 83–86.

Stout, D. (2011). Stone toolmaking and the evolution of human culture and cognition. *Philosophical Transactions of the Royal Society of London. Series B, Biological Sciences, 366*(1567), 1050–1059.

Stout, D., Apel, J., Commander, J., & Roberts, M. B. (2014). Late Acheulean technology and cognition at Boxgrove, UK. *Journal of Archaeological Science, 41,* 576–590.

Stout, D., & Chaminade, T. (2007). The evolutionary neuroscience of tool making. *Neuropsychologia, 45*(5), 1091–1100.

Stout, D., & Hecht, E. E. (2017). Evolutionary neuroscience of cumulative culture. *Proceedings of the National Academy of Sciences of the United States of America, 114*(30), 7861–7868.

Stout, D., Toth, N. P., Schick, K. D., & Chaminade, T. (2008). Neural correlates of Early Stone Age tool-making: Technology, language and cognition in human evolution. *Philosophical Transactions of the Royal Society of London. Series B, Biological Sciences, 363*(1499), 1939–1949.

Stout, D., Toth, N. P., Schick, K. D., Stout, J., & Hutchins, G. (2000). Stone tool-making and brain activation: Positron emission tomography (PET) studies. *Journal of Archaeological Science, 27*(12), 1215–1223.

Sznajder, B., Sabelis, M. W., & Egas, M. (2012). How adaptive learning affects evolution: Reviewing theory on the Baldwin effect. *Evolutionary Biology, 39*(3), 301–310.

Taylor, J. A., & Ivry, R. B. (2014). Cerebellar and prefrontal cortex contributions to adaptation, strategies, and reinforcement learning. *Progress in Brain Research, 210*, 217–253.

Thieme, H. (2005). The Lower Palaeolithic art of hunting: The case of Schoningen 13 II-4, Lower Saxony, Germany. In C. Gamble & M. Porr (Eds.), *The hominid individual in context: Archaeological investigations of Lower and Middle Palaeolithic landscapes, locales and artefacts* (pp. 115–132). New York: Routledge.

Todd, J. T. (2004). The visual perception of 3D shape. *Trends in Cognitive Sciences, 8*(3), 115–121.

Tootell, R. B. H., Mendola, J. D., Hadjikhani, N. K., Ledden, P. J., Liu, A. K., Reppas, J. B., . . . Dale, A. M. (1997). Functional analysis of V3A and related areas in human visual cortex. *Journal of Neuroscience, 17*(18), 7060–7078.

Toth, N. P., & Schick, K. D. (2009). The Oldowan: The tool making of early hominins and chimpanzees compared. *Annual Review of Anthropology, 38*, 289–305.

Toth, N. P., Schick, K. D., Savage-Rumbaugh, E. S., Sevcik, R. A., & Rumbaugh, D. M. (1993). *Pan* the tool-maker: Investigations into the stone tool-making and tool-using capabilities of a bonobo (*Pan paniscus*). *Journal of Archaeological Science, 20*(1), 81–91.

Tranel, D., Kemmerer, D., Adolphs, R., Damasio, H., & Damasio, A. R. (2003). Neural correlates of conceptual knowledge for actions. *Cognitive Neuropsychology, 20*(3/6), 409–432.

Ungerleider, L. G., & Mishkin, M. (1982). Two cortical visual systems. In D. J. Ingle, M. A. Goodale, & R. J. W. Manfield (Eds.), *Analysis of visual behavior* (pp. 549–586). Cambridge, MA: MIT Press.

Vanduffel, W., Fize, D., Mandeville, J. B., Nelissen, K., Van Hecke, P., Rosen, B. R., . . . Orban, G. A. (2001). Visual motion processing investigated using contrast agent-enhanced fMRI in awake behaving monkeys. *Neuron, 32*(4), 565–577.

Vanduffel, W., Fize, D., Peuskens, H., Denys, K., Sunaert, S., Todd, J. T., & Orban, G. A. (2002). Extracting 3D from motion: Differences in human and monkey intraparietal cortex. *Science, 298*(5592), 413–415.

Waddington, C. H. (1953). Genetic assimilation of an acquired character. *Evolution, 7*(2), 118–126.

Wagemans, J. (1997). Characteristics and models of human symmetry detection. *Trends in Cognitive Sciences, 1*(9), 346–352.

Warrington, E. K., & James, M. (1988). Visual apperceptive agnosia: A clinico-anatomical study of three cases. *Cortex, 24*(1), 13–32.

Watson, C. E., & Buxbaum, L. J. (2015). A distributed network critical for selecting among tool-directed actions. *Cortex, 65*, 65–82.

Webster, M. J., Bachevalier, J., & Ungerleider, L. G. (1994). Connections of inferior temporal areas TEO and TE with parietal and frontal cortex in macaque monkeys. *Cerebral Cortex, 4*(5), 470–483.

Wenban-Smith, F. F. (1999). Knapping technology. In M. B. Roberts & S. A. Parfitt (Eds.), *Boxgrove: A Middle Pleistocene hominin site at Eartham Quarry, Boxgrove, West Sussex* (pp. 384–395). London: English Heritage.

Wilkins, J., Schoville, B. J., Brown, K. S., & Chazan, M. (2012). Evidence for early hafted hunting technology. *Science, 338*(6109), 942–946.

Wilson, A. D., Zhu, Q., Barham, L., Stanistreet, I., & Bingham, G. P. (2016). A dynamical analysis of the suitability of prehistoric spheroids from the Cave of Hearths as thrown projectiles. *Science Reports, 6*(30614), 1–9.

Wynn, T. (1989). *The evolution of spatial competence.* Chicago, IL: University of Illinois Press.

Wynn, T. (2000). Symmetry and the evolution of the modular linguistic mind. In P. Carruthers & A. Chamberlain (Eds.), *Evolution and the human mind: Modularity, language and meta-cognition* (pp. 113–139). Cambridge, UK: Cambridge University Press.

Wynn, T. (2002). Archaeology and cognitive evolution. *Behavioral and Brain Sciences, 25*(3), 389–402.

Wynn, T. (2010). The evolution of human spatial cognition. In F. L. Dolins & R. W. Mitchell (Eds.), *Spatial cognition, spatial perception: Mapping the self and space* (pp. 213–236). Cambridge, UK: Cambridge University Press.

Wynn, T. (2014). Commentary on "Extended mind and visuo-spatial integration: Three hands for the Neandertal lineage." *Journal of Anthropological Sciences, 92*, 291–293.

Wynn, T., & Coolidge, F. L. (2016). Archeological insights into hominin cognitive evolution. *Evolutionary Anthropology: Issues, News, and Reviews, 25*(4), 200–213.

Wynn, T., Tierson, F. D., & Palmer, C. T. (1996). Evolution of sex differences in spatial cognition. *American Journal of Physical Anthropology, 101*(S23), 11–42.

Zilles, K., & Palomero-Gallagher, N. (2001). Cyto-, myelo-, and receptor architectonics of the human parietal cortex. *Neuroimage, 14*(1), S8–S20.

Zutovski, K., & Barkai, R. (2016). The use of elephant bones for making Acheulian handaxes: A fresh look at old bones. *Quaternary International, 406*, 227–238.

10

TESTING MODELS OF HANDEDNESS IN STONE TOOLS

Natalie Uomini and Lana Ruck

INTRODUCTION

Unique traits of *Homo sapiens*, such as bipedal locomotion, large and complex brains, culture, and language, are often cited—by professionals and the public—as what make us human. One often-ignored aspect of human uniqueness, however, is handedness, although recent research has begun to highlight its importance in our evolution. Between 85% and 90% of living humans are right-handed, and extreme, population-level hand dominance is considered by many to be a unique part of what makes us *us* (Annett, 1985). In order to determine if this is true, researchers must investigate the benefits, uniqueness, and evolutionary origins of handedness in our species, asking when, how, and why the rightward bias in hand preference evolved.

The right- versus left-hand dichotomy was referenced in the historic record as early as 4,000 years ago, and human societies have described and disputed the nature of handedness in literature, philosophy, and science (Corballis, 1991). From decades of research in multiple disciplines, including psychological and brain sciences, primatology, and archaeology, we have learned much about handedness in extant humans, its manifestations in non-human primates, and its evolutionary history from fossil evidence. Some of the most significant findings of these studies include the following. Handedness is not the only lateralized trait in our species, as it relates to generalized hemispheric specialization and, importantly, the lateralization of many language functions in modern human brains (Cai & van der Haegen, 2015). The inheritance of handedness is a combination of weak genetic components and chance, although the genetic mechanisms are still elusive (Kavaklioglu, Ajmal, Hameed, & Francks, 2016; Paracchini & Scerri, 2017). Developmental pressures are also incredibly important (Fagard, 2013). The strong, population-wide rightward bias in human hand preference is unique among primates, but hand preference itself is not (Uomini, 2009b). Goal-directed, differential bimanual coordination varies widely within the primate lineage, and increasing manipulation complexity is linked with the increasing importance of object manipulation, including tool-making and tool-use (Heldstab et al., 2016). There are many ways to assess individual handedness using the paleoarchaeological record, including methods based on hominin skeletal and dental remains (Uomini, 2011) and lithics (Ruck, Broadfield, & Brown, 2015). According to these studies, predominant right-handedness dates back at least as far as the late stages of the Lower

Paleolithic, and the rightward shift in hand preference likely extends further back (Mosquera, Geribàs, Bargalló, Llorente, & Riba, 2012).

Although we have learned much about human handedness, it is an incredibly complex trait, and many aspects of the rightward bias in its distribution, along with its links to other human lateralities, remain largely a mystery. By further investigating manual behaviors in living humans and other animals, studying their relationship to traits like tool manufacture and use, language, and even visuospatial acuity, and attempting to link these data with the fossil record, we can better understand the timeline and pressures that resulted in our species, the "lopsided apes" (Corballis, 1991).

In this chapter, we focus specifically on how handedness relates to stone tool manufacture throughout hominin evolution. After summarizing major works on assessing handedness in lithic materials, we will describe three dominant hypotheses regarding the evolution of handedness—the social learning hypothesis, the fighting hypothesis, and the task complexity hypothesis—and discuss how they would manifest in the paleoarchaeological record. We conclude with a critical assessment of the cognitive archaeologists' ability to evaluate these hypotheses and other open questions and suggest future directions for research on the evolution of handedness.

COGNITIVE ARCHAEOLOGY AND THE EVOLUTION OF HANDEDNESS

A central theme of this volume asks: What can archaeology tell us about the evolution of the human mind? And while handedness is a relatively narrow topic within that scope, those interested in its evolution are faced with the same problems as any cognitive archaeologist: We have no direct access to the brains and behaviors of our hominin ancestors. Because many of the behaviors we are most interested in, such as cognition and language, are not directly preserved, paleoanthropologists must develop methodologies that allow them to glean as much information as possible from small samples and secondary sources, or proxies (Cashmore, Uomini, & Chapelain, 2008; Toth & Schick, 1986; Wynn, 2002). The paucity of the hominin fossil record is a primary obstacle in studying handedness and cognitive evolution in the hominin lineage, but this has not stopped researchers from trying. There are several methods for studying hominin handedness, both directly and indirectly, and we will focus here on those using lithic technology.

The earliest modern-day speculation about determining handedness from lithic materials was by Sergei A. Semenov (1964), in his experimental studies on various Paleolithic tools. Lithic-based approaches to hominin handedness were otherwise nonexistent until Nick Toth (1985) investigated correlates of handedness in simple cobble-and-flake technologies. Informed by his experimental replications, Toth argued that the orientation of cortex on successively removed lithic flakes can be used to infer handedness from hominin tool assemblages. Using lithic assemblages from Koobi Fora, Kenya (dated to between 1.9 and 1.5 million years ago), Toth argued for initial evidence of high rates of right-handed knappers in early hominin populations, even including *H. habilis* and *H. erectus* (Toth, 1985). While this publication inspired much excitement about the possibility of early hominins being behaviorally more similar to us than previously thought, debates regarding the integrity of Toth's method ensued (Patterson & Sollberger, 1986; Pobiner, 1999). Over the decades, several attempts have been made to reassess Toth's technique, with variable success (Ruck et al., 2015).

Outside of Toth's (1985) study, for decades no attempts were made to identify handedness in lithic materials. However, in a monograph on La Cotte de Saint Brelade, a Paleolithic site in Jersey, Jean M. Cornford (1986) published evidence of handedness in lithic resharpening techniques. There is extensive literature on differential retouch of lithic materials, particularly of hafted tools, and evidence of "prehensility" in Upper Paleolithic tools reinforces investigations of handedness and lithic remains. In recent years, two additional published methods, applicable only to later technologies as they specifically use features on microburins and ground stone artifacts, have shown success in assessing handedness biases in various populations (Dominguez-Ballesteros, 2016; Peresani & Miolo, 2012). However, much like skeletal studies, the assemblages that could be studied regarding prehensility, retouch, and hafting are often dated to later in time, when established evidence of right-handedness is less strongly debated.

In 2001, Gordon Rugg and Maureen Mullane (2001) published a new method based on the skew of the cone of percussion on lithic flakes. Using flake *débitage* from eight knappers with mixed experience levels, Rugg and Mullane correctly identified knapper handedness for 75% of the flakes. However, as they did not analyze any Paleolithic assemblages, no direct information on hominin handedness resulted from their study. Their approach has also been criticized, primarily because relatively few flakes can be assessed using their technique.

Natalie Uomini (2001, 2008) made a concerted effort to assess hominin handedness in Paleolithic materials. In 2001 Uomini attempted to reproduce Toth's (1985) and Rugg and Mullane's (2001) studies and applied their techniques to Paleolithic assemblages from two British Lower Paleolithic sites, Swanscombe and Purfleet, with little success. To infer handedness from a large experimental assemblage and two more British Lower Paleolithic sites, Boxgrove and High Lodge, in addition to the Toth (1985) and Rugg and Mullane (2001) approaches Uomini (2008) used a third approach derived from Cornford's (1986) work on tranchet flakes. Using the experimental assemblage, Uomini had variable success using Toth's cortex model and was unable to evaluate Rugg and Mullane's cone of percussion (CoP) method statistically. She also found that the tranchet flake method was not constrained by knapper handedness (Uomini, 2008). In an attempt to explain the weaknesses of each method, Uomini found that flintknapping was highly stylistic and individualized; rotating and percussing cores was unique to each knapper and often incompatible with the assumptions of the handedness models.

Amèlia Bargalló and Marina Mosquera (2014) published an extended system for determining handedness in lithic *débitage*. Their system includes over 10 features of flake *débitage* that may be used for assessing handedness, all derived from another experimentally created assemblage made by novice knappers. This work, as well as that of Toth and Rugg and Mullane, was reassessed and reviewed in depth by Ruck and colleagues (2015), using an experimental assemblage created by expert knappers. Much like Bargalló and Mosquera's work, this analysis revealed that no single trait on lithic flakes is accurate in predicting handedness, but a combination of traits as a whole could predict knapper handedness relatively well. Ruck and colleagues also noted that work on single knapping events rather than lone flakes may be a more fruitful future direction. An additional blind analysis of an experimental assemblage (Daniel, Putt, &

Franciscus, 2016), also derived from existing methods, found that other than the extraction axis, no features reliably indicated handedness to a blind analyst, and that no relationship between these traits and knapper experience or sex existed.

More promisingly, Eder Dominguez-Ballesteros (2016; Dominguez-Ballesteros & Arrizabalaga, 2015) developed an experimentally validated method to determine handedness on individual flakes and successfully applied it to archaeological assemblages. The asymmetry of the parabolic crack on the point of percussion indicates the direction of strike and hence the hand used. The study revealed left-to-right ratios of 3:7 in the Neandertal lithic assemblages studied (Dominguez-Ballesteros, 2016). However, the most remarkably standardized lateralization in a lithic industry has been found in the Acheulean Victoria West cores, which were consistently struck from the same point on the same side of the preform to make cleaver blanks (Sharon & Beaumont, 2006).

In sum, these studies highlight that while handedness may be difficult to detect in lithic materials, there are many avenues for improvement of existing investigative methods, and there is much work to be done toward finding new, quantitatively based techniques. Also clear from these studies is that handedness is a key trait to track in evolutionary time. Currently, skeletal and other forms of data do not date back as far as the initial right shift in hominin hand preference. Furthermore, as we discuss in the following sections, stone tool-making itself was likely a significant influence on the predominance of right-handedness, as well as on many cognitive functions, throughout hominin evolution. Therefore, stone tool manufacture—and direct evidence of it in lithic materials—is still the most abundant and robust supply of available data, invaluable for cognitive archaeologists interested in the evolutionary histories of these traits. We next discuss three hypotheses that we believe are the most amenable to archaeological testing using the lithic record of tool-making.

THE SOCIAL LEARNING HYPOTHESIS

The social learning hypothesis (SLH) places skill acquisition at its core (Bradshaw & Nettleton, 1982; Steele & Uomini, 2009; Uomini, 2009b). Social learning is an important adaptation in modern humans, and while some evidence of it exists in extant apes (Sanz & Morgan, 2016), the amount of information that humans learn via social inputs is staggering. According to the SLH, increased reliance on social learning for tool-making and tool-use promoted a homogenization of hand preferences. Selection for social learning, cooperation, and complex tool behaviors led to the evolution of population-level handedness when socially learned tool-making became essential for survival.

The SLH is based on predictions that have been tested experimentally. For example, learning a skilled task should be easier when the learner has the same hand preference as that of the demonstrator. This was shown by Michel and Harkins (1985) in a knot-tying experiment where 86 participants learned from either a left- or a right-handed instructor. Matching hand preferences greatly facilitated successful knot-tying. Further experiments on different tasks found the same effect (Rohbanfard & Proteau, 2011), but see Uomini and Lawson's (2017) study for opposite results. To date, there are no other published studies that have tested the SLH specifically in stone tool-making or other archaeologically relevant materials, so more data are needed to resolve this question. Furthermore, additional data from decades of research in developmental and social psychology specifically point to technological analogs for the evolution of social learning systems. Overall, studies suggest that hands—and particularly hands manipulating objects and tools—provide some of the most high-fidelity and salient platforms for information transfer. How this system would have interacted specifically with handedness rates through time is an open question, but we believe that many aspects of the SLH can be studied using lithic technologies (Sterelny, 2012).

THE FIGHTING HYPOTHESIS

The fighting hypothesis (FH) was originally proposed to account for the constant ratio of left- to right-handers over deep time in human prehistory, via the mechanism of negative frequency-dependent selection (Raymond, Pontier, Dufour, & Møller, 1996). Namely, it aims to explain the fact that left-handers have always represented about 10% of the human species as far back as we can trace into deep time (Cashmore et al., 2008; Steele & Uomini, 2005; Uomini, 2011). To date, the FH is the only hypothesis to address the precise evolutionary mechanism needed to maintain this stable handedness polymorphism (i.e., the existence of both right- and left-handers at stable proportions). The FH proposes that left-handers have an advantage in being a minority within predominantly right-handed populations. This negative frequency-dependent advantage is balanced out by a second selection mechanism, the costs of being left-handed (Faurie, Uomini, & Raymond, 2016), which include increased birth defects, cognitive deficits, and health problems in left-handers. Since the fitness costs of left-handedness have not caused a total extinction of left-handers, there must be a benefit that outweighs the costs.

The FH suggests that the advantages of minority left-handedness are most strongly expressed in fighting. During combat, left-handers gain a surprise advantage over opponents, because most opponents are more used to fighting against right-handers thanks to the lower chances of encountering left-handers (because of their minority proportion in the general population). The FH thus predicts that left-handers will be more common and/or more successful in interactive sports (Groothuis, McManus, Schaafsma, & Geuze, 2013). We consider interactive sports to be a valid analogy for the kinds of prehistoric combat situations that hominins would have encountered. This prediction is well supported by data from sports science (Loffing & Hagemann, 2012) and partially supported by data from the Ultimate Fighting Championship (Pollet, Stulp, & Groothuis, 2013). Interestingly, the sinistral advantage is stronger the closer opponents are physically to each other; tennis players show a weaker effect than boxers or ping-pong players (Faurie & Raymond, 2013). More data are needed from real fights—though, quite understandably, such data are hard to get!

Specifically for hominin evolution, the FH is relevant for societies in which better fighters had higher reproductive success. For territorial hominin tribes defending their boundaries, or for hominins migrating into new areas occupied by other groups or species, fighting ability would have been critical to survival (e.g., in chimpanzees, see Boesch et al., 2008; in prehistoric humans, see Gaudzinski-Windheuser & Roebroeks, 2011, and Lahr et al., 2016). The few existing data from traditional human societies who still practice hand-to-hand combat show a strong link between rates of left-handedness and high levels of violence (Faurie & Raymond, 2005), as well as a correlation between hand strength (regardless of direction) and number of children in men

(Schaafsma, Geuze, Lust, Schiefenhövel, & Groothuis, 2013). Future studies yielding data consistent with a higher reproductive advantage for successful left-handed fighters would lend further support to the FH.

THE TASK COMPLEXITY HYPOTHESIS

The task complexity hypothesis (TCH) states that task complexity, such as that inherent in tool manufacture, increases the expression of handedness. The TCH is very well supported by task-specific hand preference data in a wide range of studies on humans, monkeys, and non-human apes. These show that strength of hand preferences varies by the nature of the task, including with more difficult or complex tasks (Cashmore et al., 2008; Mosquera et al., 2012; Rogers, 2009; Uomini, 2009b). Corballis (1998) first proposed that non-humans show weaker population handedness trends than humans because they simply do not engage in tasks as complex as those of humans. Here we define difficult tasks as those requiring more time to learn because they consist of motor actions that are acquired through motor learning (Steele & Uomini, 2009; also see Heldstab et al., 2016, for a more nuanced classification of motor manipulations in primates). Here we define complex tasks as cognitively complex because they require more steps to achieve or more parts to combine (Fairlie & Barham, 2016; Pelegrin & Roche, 2017; Rugg, 2011; Uomini, 2009b; Wynn & Coolidge, 2010). A longer sequence of steps, or sub-routines, to achieve a goal must be remembered in long-term memory (Wynn & Coolidge, 2010); combining a greater number of parts must be managed simultaneously in conscious attention. In both of these cases the complexity is conceptual but all the components of the task must still be executed motorically, which requires long-term working memory (Wynn & Coolidge, 2004). For the purposes of this chapter, we consider the TCH to cover tasks that are difficult, complex, or both.

While the TCH relates task features to the strength of handedness, it does not say anything about direction—left versus right. Indeed, the effect goes both ways. Skilled manual actions are executed proficiently by strong left- and right-handers, regardless of their hand preference. Less complex tasks can be done by either hand (ambidexterity). For example, in non-literate societies, where hand preferences are not affected by learning to write, simple actions such as non-tool-using activities are equally distributed between the right and left hands, with no individual or group preference (Marchant, McGrew, & Eibl-Eibesfeldt, 1995).

The TCH is especially relevant to the recognition of expertise in the archaeological record. While simple actions, like those just outlined, range in both degree and direction of manual recruitment, increasing task complexity drives additional laterality in a biased manner. Given specialization biases for limb movement, visuospatial attention, and sequencing events that predate primates, Papadatou-Pastou (2011) has proposed that this direction slightly favors a right-hand-precise/left-hand-support system for complex bimanual tasks. Thus, the TCH can be used to link this directional bias to changes in either technological complexity or manufacture difficulty through time. Analysis by Wynn and Coolidge (2010) shows that Levallois knapping required holding both visuospatial core configurations (models) and motor actions (procedures) in long-term memory, each of which were activated by cues in working memory during the knapping sequence. In contrast, Late Acheulean biface knapping required fewer sub-routines, fewer elements, and fewer cues, which meant a lighter load on long-term memory (Wynn & Coolidge, 2010). Complex prehistoric technology is analogous to expert performance (Wynn, Haidle, Lombard, & Coolidge, 2017). With such cognitive archaeology analysis, some prehistoric activities can be ordered along a gradient of complexity. This means they can potentially serve as a foundation on which to build testable predictions about the TCH.

DISCUSSION AND FUTURE DIRECTIONS

For an evolutionary account, the SLH relies on two assumptions: (1) Socially learned tool-making was essential for survival, and (2) the benefit gained by a same-handed learner over a different-handed learner was large enough to select for matching hand preferences in the population. Tool-making became essential for hominin survival at some point in our distant past. This probably was true of *H. erectus*, the first species to intensify the dependence on technology to obtain food (Ungar, Grine, & Teaford, 2006). These are the ideal conditions for evolution to select for efficient acquisition of tool-making ability. There is good evidence that social learning greatly enhances the acquisition of basic stone knapping skills (Morgan et al., 2015; Uomini, 2009a). Therefore, the assumptions of the SLH were likely to have been met at least with *H. erectus*. Indeed, the oldest skeletal evidence for a hand preference is found in a *H. ergaster* individual (Uomini, 2009a).

The prediction of the SLH that can be tested using data from the archaeological record is that tools requiring more social learning should show stronger signals of right-handedness. There are several stages of inference required to test this prediction. First, a tool must be reliably associated with social learning. Ideally, a finer scale of analysis would distinguish different degrees of social learning associated with different tool types, ranging from observation to interactive intentional teaching (Uomini & Lawson, 2017). Second, the tool must be shown to be made by a right- or left-hander. Third, the overall pattern of tool assemblages must show greater numbers of right-handed individuals making the more socially learned tools.

In contrast, the FH seeks evidence of left-handers. The association between opponents' physical distance and the strength of the left-handed advantage suggests that handedness polymorphism evolved during hand-to-hand fighting. We propose that this occurred in the earliest stages of human evolution before long-range weapons were invented. Thus, the FH would predict a higher rate of left-handedness in assemblages of artifacts used as handheld weapons, as compared to projected weapons such as spear-throwers or hafted arrowheads.

This prediction is built on several stages of inference that would need to be determined from the archaeological record. First, the artifacts (whether lithic pieces or traces of hafting, for instance) must be convincingly shown to be handheld weapons. Second, usage data must demonstrate that the weapon was used by a left-hander (perhaps via measures of prehensility). Alternatively, if the weapon's *manufacturer* is shown to be left-handed, then a third stage must assume the manufacturer was the same as the user, to link the left-hander with the actual fighting act.

Regarding the TCH, it is clearly the case that task complexity is a factor in the expression of handedness. The most complex manual tasks that humans perform are in music, sign language, and tool-making. Since the first two domains currently have very little archaeological evidence, we will focus here on tool manufacture. The TCH would predict that more complex archaeological finds, such as lithic industries demanding many stages of reduction or highly skilled knapping techniques, should show stronger handedness. The stages of inferences needed to test this prediction are first, the entire sequence of manufacture must have been completed by the same individual, and second, the artifact must be shown to result from a complex sequence, action, or combination. The definition of what constitutes a "complex" artifact is necessarily relative to others, since there is no objective limit of a complex versus non-complex tool. Thus, the most valuable analysis is one that compares different artifacts on their complexity and handedness, showing a clear correlation of hand preference *strengths* (not right or left directions) with complexity levels.

The most complex stone technologies involve long hierarchies of sub-goals, long-term working memory, flexible error correction, and propositional reasoning (Pelegrin & Roche, 2017; Wynn & Coolidge, 2010). To test the TCH, we would need good methods to determine handedness in Oldowan, Acheulean, and Levallois assemblages. As reviewed earlier, there are currently few analysis methods that reliably connect individual knappers to their knapped pieces. Thus, in order to find working methods for assessing handedness in these technologies it would be most fruitful to explore methods for identifying individual knapping events—and individual knappers—in comingled assemblages. Continued experimental work, with explicit recruitment of both right- and left-handed subjects with some level of tool-making familiarity or expertise, would be a necessary direction to move forward. An impetus for work of this kind is that some level of certainty can be achieved with assemblages from well-preserved taphonomic contexts where this information is readily available (Bargalló, Mosquera, & Lorenzo, 2018; Dominguez-Ballesteros, 2016).

If the stages of inference for any hypothesis are to be shown in the archaeological record, they will depend on individual data. Therefore, individual-level hand preferences are determined first. Determination of group-level handedness would then require a series of artifacts from the same age, all showing individual hand preferences. They must not necessarily consist of the same type of artifacts as long as their interpretation is accurate. However, we always face the problem of matching the individual to the artifact. Currently, a group-level analysis can be carried out only in cases where different individuals can be recognized. With methods pending for parsing out comingled assemblages, work toward identifying sites with well-preserved knapping scatters, high proportions of refits, or otherwise identifiable knapping events should be catalogued as priorities for analysis. Until reliable mixed *débitage*–based methods are found, analyses of these assemblages (using methods that work on known knapping events) can help us develop techniques for assessing changes in population- or species-wide right-shift trends, which is a largely unaddressed topic in the field.

Given that the SLH predicts more right-handedness in socially learned artifacts, the FH predicts more left-handedness in handheld weapons, and the TCH predicts strong handedness to both directions in complex technologies, we believe that the prospects are good for recovering handedness in the archaeological record. The key will be to work forward from the predictions. As in any good study of evolutionary cognitive archaeology (Wynn, 2009; Wynn et al., 2017), it will be crucial to clearly articulate each step in the chain of inferences made and to conscientiously integrate the data from multiple disciplines. With the ground-breaking approach pioneered by Thomas Wynn, we can squeeze minds from stones.

ACKNOWLEDGMENTS

Natalie Uomini is supported by the Max Planck Society. We thank an anonymous reviewer for helpful comments on an earlier version of the chapter.

REFERENCES

Annett, M. (1985). *Left, right, hand and brain: The right shift theory*. London: LEA Publishers.

Bargalló, A., & Mosquera, M. (2014). Can hand laterality be identified through lithic technology? *Laterality: Asymmetries of Body, Brain and Cognition, 19*(1), 37–63.

Bargalló, A., Mosquera, M., & Lorenzo, C. (2018). Identifying handedness at knapping: An analysis of the scatter pattern of lithic remains. *Archaeological and Anthropological Sciences, 10*(3), 587–598.

Boesch, C., Crockford, C., Herbinger, I., Wittig, R. M., Moebius, Y., & Normand, E. (2008). Intergroup conflicts among chimpanzees in Taï National Park: Lethal violence and the female perspective. *American Journal of Primatology, 70*(6), 519–532.

Bradshaw, J. L., & Nettleton, N. C. (1982). Language lateralization to the dominant hemisphere: Tool use, gesture and language in hominid evolution. *Current Psychology, 2*, 171–192.

Cai, Q., & Van der Haegen, L. (2015). What can atypical language hemispheric specialization tell us about cognitive functions? *Neuroscience Bulletin, 31*, 220–226.

Cashmore, L., Uomini, N. T., & Chapelain, A. (2008). The evolution of handedness in humans and great apes: A review and current issues. *Journal of Anthropological Sciences, 86*, 7–35.

Corballis, M. C. (1991). *The lopsided ape: Evolution of the generative mind*. Oxford, UK: Oxford University Press.

Corballis, M. C. (1998). Evolution of language and laterality: A gradual descent? *Cahiers de Psychologie Cognitive, 17*(6), 1148–1155.

Cornford, J. M. (1986). Specialized resharpening techniques and evidence of handedness. In P. Callow & J. M. Cornford (Eds.), *La Cotte de St Brelade 1961–1978: Excavations by C. B. M. McBurney* (pp. 337–351, 413–414). Norwich: Geo Books.

Daniel, C., Putt, S. S., & Franciscus, R. G. (2016). Investigating other causes for stone flake features attributed to handedness. In *81st Annual Meeting of the Society for American Archaeology*, April 6–10, 2016, Orlando, FL. tDAR id: 404840. http://doi.org/10.6067/XCV8VX0J9W

Dominguez-Ballesteros, E. (2016). *Evaluación arqueológica del proceso de lateralización Determinación cuantitativa de la lateralidad poblacio- nal, desde los neandertales hasta la actualidad*. PhD Thesis, University of the Basque Country, Vitoria-Gasteiz.

Dominguez-Ballesteros, E., & Arrizabalaga, A. (2015). Flint knapping and determination of human handedness: Methodological proposal with quantifiable results. *Journal of Archaeological Science: Reports, 3*, 313–320.

Fagard, J. (2013). The nature and nurture of human infant hand preference. *Annals of the New York Academy of Sciences, 1288*(1), 114–123.

Fairlie, J. E., & Barham, L. S. (2016). From *chaîne opératoire* to observational analysis: A pilot study of a new methodology for analysing changes in cognitive task-structuring strategies across different hominin tool-making events. *Cambridge Archaeological Journal, 26*(4), 643–664.

Faurie, C., & Raymond, M. (2005). Handedness, homicide and negative frequency-dependent selection. *Proceedings of the Royal Society of London B: Biological Sciences, 272*(1558), 25–28.

Faurie, C., & Raymond, M. (2013). The fighting hypothesis as an evolutionary explanation for the handedness polymorphism in humans: Where are we? *Annals of the New York Academy of Sciences, 1288*(1), 110–113.

Faurie, C., Uomini, N. T., & Raymond, M. (2016). Origins, development and persistence of laterality in humans. In F. Loffing, N. Hagemann, B. Strauss, & C. MacMahon (Eds.), *Laterality in sports: Theories and applications* (pp. 11–30). San Diego, CA: Elsevier Academic Press.

Gaudzinski-Windheuser, S., & Roebroeks, W. (2011). On Neanderthal subsistence in last interglacial forested environments in Northern Europe. In N. J. Conard & J. Richter (Eds.), *Neanderthal lifeways, subsistence and technology: One hundred fifty years of Neanderthal study* (pp. 61–71). Dordrecht, The Netherlands: Springer.

Groothuis, T. G. G., McManus, I. C., Schaafsma, S. M., & Geuze, R. H. (2013). The fighting hypothesis in combat: How well does the fighting hypothesis explain human left-handed minorities? *Annals of the New York Academy of Sciences, 1288*(1), 100–109.

Heldstab, S. A., Kosonen, Z. K., Koski, S. E., Burkart, J. M., Van Schaik, C. P., & Isler, K. (2016). Manipulation complexity in primates coevolved with brain size and terrestriality. *Scientific Reports, 6*(24528), 1–9.

Kavaklioglu, T., Ajmal, M., Hameed, A., & Francks, C. (2016). Whole exome sequencing for handedness in a large and highly consanguineous family. *Neuropsychologia, 93*(Part B), 342–349.

Lahr, M. M., Rivera, F., Power, R. K., Mounier, A., Copsey, B., Crivellaro, F., . . . Foley, R. A. (2016). Inter-group violence among early Holocene hunter-gatherers of West Turkana, Kenya. *Nature, 529*(7586), 394–398.

Loffing, F., & Hagemann, N. (2012). Side bias in human performance: A review on the left-handers' advantage in sports. In T. Dutta, M. K. Mandal, & S. Kumar (Eds.), *Bias in human behaviour* (pp. 163–182). Hauppage, NY: Nova Science.

Marchant, L. F., McGrew, W. C., & Eibl-Eibesfeldt, I. (1995). Is human handedness universal? Ethological analyses from three traditional cultures. *Ethology, 101*(3), 239–258.

Michel, G. F., & Harkins, D. A. (1985). Concordance of handedness between teacher and student facilitates learning manual skills. *Journal of Human Evolution, 14*(6), 597–601.

Morgan, T. J. H., Uomini, N. T., Rendell, L. E., Chouinard-Thuly, L., Street, S. E., Lewis, H. M., . . . Laland, K. N. (2015). Experimental evidence for the co-evolution of hominin tool-making teaching and language. *Nature Communications, 6*(6029), 1–18.

Mosquera, M., Geribàs, N., Bargalló, A., Llorente, M., & Riba, D. (2012). Complex tasks force hand laterality and technological behaviour in naturalistically housed chimpanzees: Inferences in hominin evolution. *The Scientific World Journal, 2012*, 1–12.

Papadatou-Pastou, M. (2011). Handedness and language lateralization: Why are we right-handed and left-brained? *Hellenic Journal of Psychology, 8*, 248–265.

Paracchini, S., & Scerri, T. (2017). Genetics of human handedness and laterality. In L. J. Rogers & G. Vallortigara (Eds.), *Lateralized brain functions: Methods in human and non-human species* (pp. 523–552). New York: Humana Press/Springer.

Patterson, J. B., & Sollberger, L. W. (1986). Comments on Toth's right-handedness study. *Lithic Technology, 15*(3), 109–111.

Pelegrin, J., & Roche, H. (2017). L'humanisation au prisme des pierres taillées. *Comptes rendus Palevol, 16*(2), 175–181.

Peresani, M., & Miolo, R. (2012). Small shifts in handedness bias during the Early Mesolithic? A reconstruction inferred from microburin technology in the eastern Italian Alps. *Journal of Anthropological Archaeology, 31*(1), 93–103.

Pobiner, B. L. (1999). The use of stone tools to determine handedness in hominids. *Current Anthropology, 40*(1), 90–92.

Pollet, T. V, Stulp, G., & Groothuis, T. G. G. (2013). Born to win? Testing the fighting hypothesis in realistic fights: Left-handedness in the Ultimate Fighting Championship. *Animal Behaviour, 86*(4), 839–843.

Raymond, M., Pontier, D., Dufour, A.-B., & Møller, A. P. (1996). Frequency-dependent maintenance of left handedness in humans. *Proceedings of the Royal Society of London B: Biological Sciences, 263*(1377), 1627–1633.

Rogers, L. J. (2009). Hand and paw preferences in relation to the lateralized brain. *Philosophical Transactions of the Royal Society of London. Series B, Biological Sciences, 364*(1519), 943–954.

Rohbanfard, H., & Proteau, L. (2011). Effects of the model's handedness and observer's viewpoint on observational learning. *Experimental Brain Research, 214*(4), 567–576.

Ruck, L., Broadfield, D. C., & Brown, C. T. (2015). Determining hominid handedness in lithic debitage: A review of current methodologies. *Lithic Technology, 40*(3), 171–188.

Rugg, G. (2011). Special issue: Innovation and the evolution of human behavior: Quantifying technological innovation. *PaleoAnthropology, 2011*, 154–165.

Rugg, G., & Mullane, M. (2001). Inferring handedness from lithic evidence. *Laterality: Asymmetries of Body, Brain and Cognition, 6*(3), 247–259.

Sanz, C. M., & Morgan, D. B. (2016). Rethinking chimpanzee tool use: Niche construction and developmental bias in maintaining technological traditions among African apes. *American Journal of Physical Anthropology, 159*, 279–280.

Schaafsma, S. M., Geuze, R. H., Lust, J. M., Schiefenhövel, W., & Groothuis, T. G. G. (2013). The relation between handedness indices and reproductive success in a non-industrial society. *PLoS One, 8*(5), 1–10.

Semenov, S. A. (1964). *Prehistoric technology: An experimental study of the oldest tools and artefacts from traces of manufacture and wear.* (M. W. Thompson, Trans.). London: Cory, Adams and Mackay.

Sharon, G., & Beaumont, P. B. (2006). Victoria West: A highly standardized prepared core technology. In N. Goren-Inbar & G. Sharon (Eds.), *Axe age: Acheulian tool-making from quarry to discard* (pp. 181–199). London: Equinox.

Steele, J., & Uomini, N. T. (2005). Humans, tools and handedness. In V. Roux & B. Brill (Eds.), *Stone knapping: The necessary conditions for a uniquely hominin behaviour* (pp. 217–239). Cambridge, UK: McDonald Institute for Archaeological Research.

Steele, J., & Uomini, N. T. (2009). Can the archaeology of manual specialization tell us anything about language evolution? A survey of the state of play. *Cambridge Archaeological Journal, 19*(1), 97–110.

Sterelny, K. (2012). *The evolved apprentice: How evolution made humans unique.* Cambridge, MA: MIT Press.

Toth, N. P. (1985). Archaeological evidence for preferential right-handedness in the Lower and Middle Pleistocene, and its possible implications. *Journal of Human Evolution, 14*(6), 607–614.

Toth, N. P., & Schick, K. D. (1986). The first million years: The archaeology of protohuman culture. *Advances in Archaeological Method and Theory, 9*, 1–96.

Ungar, P. S., Grine, F. E., & Teaford, M. F. (2006). Diet in early *Homo*: A review of the evidence and a new model of adaptive versatility. *Annual Review of Anthropology, 35*, 209–228.

Uomini, N. T. (2001). *Lithic indicators of handedness: Assessment of methodologies and the evolution of laterality in hominids.* MSc Thesis, University of Durham, England.

Uomini, N. T. (2008). *In the knapper's hands: Testing markers of laterality in hominin lithic production, with reference to the common substrate of language and handedness.* Doctoral thesis, University of Southampton, UK.

Uomini, N. T. (2009a). Prehistoric left-handers and prehistoric language. In S. A. de Beaune, F. L. Coolidge, & T. Wynn (Eds.), *Cognitive archaeology and human evolution* (pp. 37–55). Cambridge, UK: Cambridge University Press.

Uomini, N. T. (2009b). The prehistory of handedness: Archaeological data and comparative ethology. *Journal of Human Evolution, 57*(4), 411–419.

Uomini, N. T. (2011). Handedness in Neanderthals. In N. J. Conard & J. Richter (Eds.), *Neanderthal lifeways, subsistence and technology: One hundred fifty years of Neanderthal study* (pp. 139–154). Heidelberg: Springer.

Uomini, N. T., & Lawson, R. (2017). Effects of handedness and viewpoint on the imitation of origami-making. *Symmetry, 9*(9), 182.

Wynn, T. (2002). Archaeology and cognitive evolution. *Behavioral and Brain Sciences, 25*(3), 389–402.

Wynn, T. (2009). Hafted spears and the archaeology of mind. *Proceedings of the National Academy of Sciences of the United States of America, 106*(24), 9544–9545.

Wynn, T., & Coolidge, F. L. (2004). The expert Neandertal mind. *Journal of Human Evolution, 46*(4), 467–487.

Wynn, T., & Coolidge, F. L. (2010). How Levallois reduction is similar to, and not similar to, playing chess. In A. Nowel & I. Davidson (Eds.), *Stone tools and the evolution of human cognition* (pp. 83–104). Boulder, CO: University of Colorado Press.

Wynn, T., Haidle, M. N., Lombard, M., & Coolidge, F. L. (2017). The expert cognition model in human evolutionary studies. In T. Wynn & F. L. Coolidge (Eds.), *Cognitive models in Palaeolithic archaeology* (pp. 21–43). Oxford, UK: Oxford University Press.

11

EARLY CONVERGENT CULTURAL EVOLUTION
ACHEULEAN GIANT CORE METHODS OF AFRICA

Gonen Sharon

INTRODUCTION

In evolutionary biology, the term *convergent evolution* is the process of independent evolution of similar traits in different biological species, resulting from adaptation to separate ecosystems. Examples are the wings of insects, birds, and bats, or the elongated faces and long sticky tongues of various anteaters inhabiting different continents. Darwin (1859) recognized this mechanism, which in recent years has been applied to the study of human cultural evolution (for discussion and references, see Mesoudi, Whiten, & Laland, 2004). D'Arcy Thompson (see Gould, 1971) and many others have discussed the physical principles and biological mechanisms behind convergent evolution forms. Applying the convergent evolution principle to the study of cognition and behavior is, nevertheless, difficult. Emery and Clayton (2004) studied similar tool use and memory-related strategies in apes and corvids. They claimed that while clearly different, the highly intelligent brains of apes and corvids produced similar use of tools and other intelligence-based behaviors. The evolution of human behavior and culture is even more complicated. The primary examples cited for the principle of cultural convergent evolution are the invention of writing in the Egyptian, Mesopotamian, and Chinese cultures and the innovation of agriculture in diverse parts of the world (e.g., Fuller et al., 2014). Agricultural practice, however, is an aggregate of many actions, strategies, and functions.

Finding a more specific technological example of cultural convergent evolution is harder than one would expect. The primary difficulty, from a global perspective, is to prove that there was no contact between the producers or developers of the relevant tool or technique living in different regions. This difficulty is inherent in Diamond's (1997) discussion of the development of writing, as all Euro-Asian writing systems can be traced to the system invented by the Sumerians during the fourth millennium BC in Mesopotamia. The best example of a specific development is, perhaps, eyeshades invented to shield against the sun's harmful rays. Eyeshades are critical in environments where the snow reflects the sun's rays and can cause blindness. Eyeshades were made of bone in the Inuit culture (e.g., Norn, 1996), bronze in the plains of Central Asia (seventh–ninth century, Tomb No. 227, Astana, Turfan, Xinjiang Uygur Autonomous Region Museum Collection), and wool in Tibet. The need for eye protection led to a nearly identical invention, albeit produced from very different raw materials.

Examples of convergent cultural evolution from the earliest stages of human cultural development, such as the Lower Paleolithic Acheulean, are challenging to establish. The cultural hallmarks of the Acheulean are its large bifacial tools—the handaxe and the cleaver. These tools were produced by hominins for over one million years, from South Africa to the United Kingdom and from Spain to China. Handaxes and cleavers are astonishingly similar in shape and size throughout the world, and the question of their origin, whether they expanded from a single source or were invented separately in different localities, is one of the primary debates in Lower Paleolithic study (see Sharon, 2007, for references and discussion). While this "biface enigma" (Wynn, 1995) remains unresolved, the study of handaxe and cleaver technology can provide us with important clues.

The research findings presented here suggest that different giant core methods applied by Acheulean biface makers for the production of large flakes are evidence of early convergent cultural evolution. I will focus on the two most sophisticated methods, the Victoria West and the Tabelbala-Tachenghit, from the southern and northern edges of the African continent, as they best support this view.

THE LARGE FLAKE ACHEULEAN

The ability to produce large flakes as blanks for the production of bifacial tools was demarcated long ago by Glynn L. Isaac (1969) as the border between the Developed Oldowan and the Acheulean in Africa. The large flake (LF) phase of the Acheulean was subsequently defined by Sharon (2007), according to the following primary criteria:

(1) The primary lithic technology for manufacturing bifacial tools in these assemblages was the production of large flakes from giant cores.
(2) Acheulean hominins applied a wide variety of systematic, well-planned, and predetermined core methods in LF production, all of them well adapted to the type and shape of the raw material at hand.
(3) A general propensity toward LF production from coarse-grained rock types, rather than from fine-grained raw materials, was observed in LF Acheulean industries.
(4) The study of the size range of bifacial tools supports the definition of an LF as one that exceeds 10 cm in maximal length (Kleindienst, 1961).
(5) Most handaxes and cleavers were shaped with minimal retouch of the ventral surface. The main feature of this shaping procedure was the thinning of the flake blank's bulb of percussion, a technological trait that distinguished LF tools from other Acheulean industries. In these industries, large cutting tools (LCTs) were also frequently produced on flake blanks, but the final stages of tool shaping involved a much higher intensity of retouch on both faces of the tool .
(6) LF assemblages contain significant frequencies of "true" cleavers (i.e., ones made on flakes with an unmodified cutting edge, after Tixier, 1956), although it is impossible to establish a rigid frequency threshold.

The LF Acheulean phase has been identified in Africa, Western Asia (from the Levant to the Caucasus), Arabia, and India. The presence of this Acheulean phase was recently identified further to the east in China (e.g., Kuman, Li, & Li, 2014; Li, Li,

& Kuman, 2014; Li, Li, Kuman, et al., 2014; Yamei et al., 2000; Zhang, Huang, & Wang, 2010). To the west, the LF Acheulean is clearly present in the Iberian Peninsula but has not been observed in Europe north of the Pyrenees (Sharon, 2007; Sharon & Barsky, 2016).

The dating of LF Acheulean sites is, in many cases, challenging. Many of the sites have only relative chronology based on geological and typological considerations (Sharon, 2007). Other sites have better chronology based on radiometric dates that date them to the early Middle Pleistocene or even earlier (Gibbon, Granger, Kuman, & Partridge, 2009; Goren-Inbar et al., 2000; Sharon et al., 2010). It seems that in some regions, primarily in Africa, LF industries persisted into the Late Pleistocene, yet it seems safe to argue that most sites should be dated older than 500,000 years (Sharon, 2007).

Acheulean Cleavers

The cleaver is a distinct bifacial tool that has been traditionally overlooked in discussions of the Acheulean techno-complex as a cultural entity (a recent example is provided in Corbey, Jagich, Vaesen, & Collard, 2016). This oversight is most likely due to the fact that cleavers are absent from Europe beyond the Pyrenees, the homeland of pioneering prehistorians who defined the Acheulean according to what they discovered in Europe (e.g., Bordes, 1961). In fact, the cleaver is probably the tool most characteristic of the LF Acheulean (Sharon, 2007; also see Figure 11.1), and the Acheulean "desire" for cleavers is the primary cultural evolutionary force behind the technological innovations presented here.

Cleaver Definition

The term *cleaver* (*hachereau* in French) is the subject of a long and ongoing debate. In his comprehensive study of Acheulean cleavers, Mourre (2003) rightly pointed out that the debate originated in the disagreement between Francophone prehistorians who supported a minimalist definition for cleavers (see later discussion) and English-speaking scholars who favored a more inclusive classification, identifying all bifacially knapped tools with a transverse cutting edge as cleavers (i.e., the "bifacial cleavers" of Bordes, 1961). The cleaver has been documented in Spain, the Levant, and in India, where it was as prevalent as the handaxe. Nevertheless, the region most abundant in Acheulean cleavers is Africa, the source of the tool's identification (for further discussion, see Sharon, 2007). According to Tixier (1956; also see Balout & Tixier, 1957), two elements define a cleaver. First, cleavers are exclusively flake tools. Hence, LCTs with a transverse cutting edge that were made on non-flake blanks (such as a cobble or a flat slab) are not cleavers. Second, a cleaver's cutting edge was never shaped through secondary retouch but was always formed by the joint between the ventral and dorsal faces of the cleaver flake. This edge can be either cortical or the margin of the scar formed on the original large core at the start of the blank extraction process (Figure 11.1). Additional definitions have been offered by many scholars, the most notable being the one by Roe (1994, pp. 151–153), but Tixier's (1956) definition remains the most accurate and will be followed here.

Figure 11.1. Cleavers of the world. (1) STIC, Morocco; (2) Tachenghit, Sahara; (3) Sidi Zin, Algeria; (4) Tegglihalli, India; (5) Chirki, India; (6) Hunsgi V, India; (7) Ternifin, Algeria; (8) Grotte des Ours, Morocco; (9) Gesher Benot Ya'aqov, Israel; (10) NBA, Israel; (11) NBA, Israel; (12) Isimila, Tanzania; (13) Pniel 6, Vaal River, South Africa; (14) Pniel 6, Vaal River, South Africa; (15) Doorenlaacte, Vaal River, South Africa; (16) Riverview, Vaal River, South Africa; (17) Pniel 7, Vaal River, South Africa; (18) Isimila, Tanzania. Image by the author.

Acheulean Giant Core Methods

The primary technological characteristic of the LF Acheulean is the production of large flakes (>10 cm; see Kleindienst, 1962) from giant cores to be used as biface blanks. Many different giant core methods have been distinguished in the various LF Acheulean regions. Sharon (2007, 2009) defined and described as many as seven core methods: Levallois, slab slicing, bifacial, Entame, Kombewa, Victoria West, and Tabelbala-Tachenghit. The giant core methods are indicators of great innovation on the part of the Acheulean biface makers. The biface makers designed excellent technological solutions for the production of large flakes from the raw material available in their different regions. They exploited the shape and size of the rock types in their site vicinities and used them efficiently to produce many large flakes suitable in shape and size for the production of handaxes and cleavers. Most of the flake blanks produced required only minimal further shaping into finished tools (Madsen & Goren-Inbar, 2004; Sharon, 2007, 2008, 2009). Of these different core methods, two stand out as the most sophisticated and technologically advanced: the Victoria West and the Tabelbala-Tachenghit. These core methods are described next in detail.

VICTORIA WEST CORE METHOD

F. J. Jansen (1926) first described cores belonging to this tradition in the region of the town of Victoria West, South Africa. Goodwin and van Riet Lowe (Goodwin, 1929, 1953; Goodwin & van Riet Lowe, 1929; van Riet Lowe, 1935, 1937) subsequently elaborated on Jansen's description of the Victoria West cores and core method. According to them, Victoria West cores are medium-size cores (150–250 mm in maximal dimension) from which a single, large side-struck flake was removed (Figure 11.2) for the purpose of Acheulean LCT blank production in many central South African sites. Later, Sharon and Beaumont (2006) suggested a new understanding of this technology. These research efforts notwithstanding, a comprehensive overview

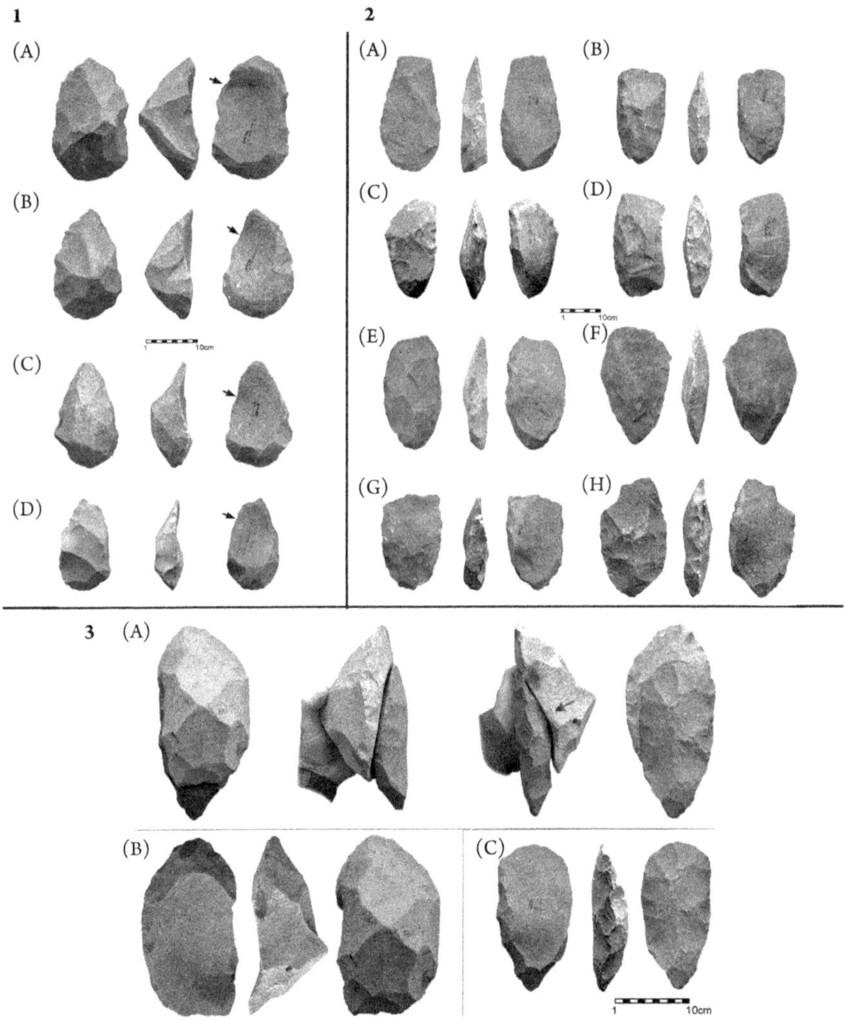

Figure 11.2. Victoria West core method. (1) Victoria West cores from Canteen Kopie; (2) Victoria West cleavers from Vaal River Acheulian sites; (3) Victoria West core and cleaver method. A = core and cleaver combined; B = core; C = cleaver. Image by the author.

of Victoria West technology and its products has never been presented, in part because Acheulean sites that contain Victoria West tools and cores have never been extensively excavated or published. The presence of similar technology has been asserted for the sites of many regions, including India (Corvinus, 1983; for references, also see Pappu & Akhilesh, 2006), Northwest Africa (Biberson, 1961; Clark, 1992), and the Sahara (Alimen, 1978). Nevertheless, since a clear definition for the Victoria West core method has not been available, it has been difficult to verify these claims, and it now seems that most of them should be rejected.

Here, evidence for convergent cultural evolution is based primarily on the assemblage of Victoria West cores collected at the site of Canteen Kopje, whose rich Stratum 2a lithic assemblage has been interpreted as representing the remnants of an Acheulean biface workshop (Beaumont, 1990; McNabb, 2001; McNabb & Beaumont, 2012; Wilkins, Pollarolo, & Kuman, 2010). These Victoria West type I cores (Figure 11.2 (1)) are uniform in size and morphology and made of relatively fine-grained andesite, the raw material most common in the majority of Acheulean sites in the lower Vaal River Basin (Sharon, 2007; Sharon & Beaumont, 2006).

Core/Preform Preparation

The preparation of the Victoria West type I core was highly sophisticated and a great deal of work was invested in the process. The terminology of the Levallois volumetric approach is applicable to both faces of the core in this method. Boëda (1995) defined these surfaces as the *débitage* (or flaking) surface and the preparation of striking platform surface. The two faces are markedly asymmetrical, creating the typical section of the Victoria West core (Figure 11.2). The scar pattern on the *débitage* face exemplifies the knowledge and energy evident in the preparation of Victoria West cores. A carefully planned radial scar pattern was achieved through the removal of well-spaced, well-arranged, shallow, and thin flakes. This typical pattern can be seen on all cores and on the dorsal face of some of the cleavers that were removed from Victoria West cores (Figure 11.2 (2)). The mean total number of scars per core is very high in comparison to other LF Acheulean blank production core types.

Removal of a Cleaver Flake from a Victoria West Core

The uniformity that was exercised in Victoria West core preparation continued into biface flake blank extraction. All of the cores in the assemblage under study were struck from an identical point on the same face of the preform. They were struck at a similar distance from the preform edge and from the same direction (Figure 11.2).

Predetermination of a Victoria West Flake Blank

The morphology of a Victoria West blank was predetermined both by the morphology of its core and by the location of the blow on the core. During flake removal, the preform (core), shaped in the form of a rough biface, was held with the tip pointing toward the knapper and the striking platform preparation surface facing up. When removed, the LF carried with it the tip of the bifacially designed preform. Figure 11.2 (3) illustrates this unique reduction practice; the core in the figure originated from the site of Canteen Kopje, while the cleaver was collected in Pniel 6a. It should be noted that the reduction sequence described here is the final stage in Victoria West core production, after which what remains of the core is abandoned. Many of the Vaal River cleavers were produced from earlier production stages, probably from much larger cores. In short, we do not yet have full understanding of the entire reduction sequence of the Victoria West core method.

Dominance of Cleavers

The Victoria West core reduction sequence was intended for the production of large flakes that were considered suitable as blanks for cleavers. This is also evident from the marked presence of cleavers in most of the Vaal River assemblages and the high frequency of their typical, Victoria West features (striking platforms, shape of butt and direction of blow; Sharon, 2007; Sharon & Beaumont, 2006). Handaxes were also made from Victoria West blanks, though to a lesser degree. Although all of the assemblages under study came from surface collection, the dominance of cleavers is notable in all Vaal River Acheulean sites. The dominance of cleavers reflected in the Vaal River assemblages is not a common phenomenon among Acheulean industries worldwide (Ranov, 2001; however, also see Roche, Brugal, Lefevre, Ploux, & Texier, 1988 for similar type frequencies at Isenya). The dominance of cleavers and the extensive use of a core method designed for cleaver blank production at the Vaal River Acheulean sites cannot as yet be fully explained. There may have been a particular utilitarian demand for these LCTs or, alternatively, the innovation of the controlled Victoria West technology may have prompted knappers to produce more efficient, pre-planned cleaver blanks.

TABELBALA-TACHENGHIT CORE METHOD

The Tabelbala-Tachenghit core method was first defined by the Abbé Breuil (1930) and later described in detail by Tixier (1956), Champault (1951, 1956, 1966), and Alimen (1978). On the basis of Saharan geological formation correlations, the method was ascribed to the Middle Acheulean (Alimen, 1978; Clark, 1992). Of the small sample of Tachenghit cleavers studied here, only a few were identified with certainty as deriving from the Tabelbala-Tachenghit core method (Figure 11.3); other cleavers in the sample display only partial Tabelbala-Tachenghit morphological attributes (Figure 11.3). Of several alternative descriptions for the Tabelbala-Tachenghit core method that have been suggested in the literature, Tixier's (1956) reconstruction is the most frequently quoted. Alimen (1978, pp. 133–135, Figure 39) claimed that Tachenghit cleavers (type 4) could have been produced by two core methods, the Levallois (type 4a) and the Kombewa (type 4b). In her description of the Levallois Tachenghit cleaver (type 4a), she pointed out that the important stage of striking platform preparation can be observed on the tools.

In light of the analysis of the Victoria West core method presented here, an additional reconstruction for the Tabelbala-Tachenghit core method may now be suggested. Figure 11.3B presents three of the larger cores from the Tachenghit sites. Cores 3b-I and 3b-II demonstrate the removal of a single LF from the *débitage* face, which was totally covered in scar removals, very much like the Victoria West type

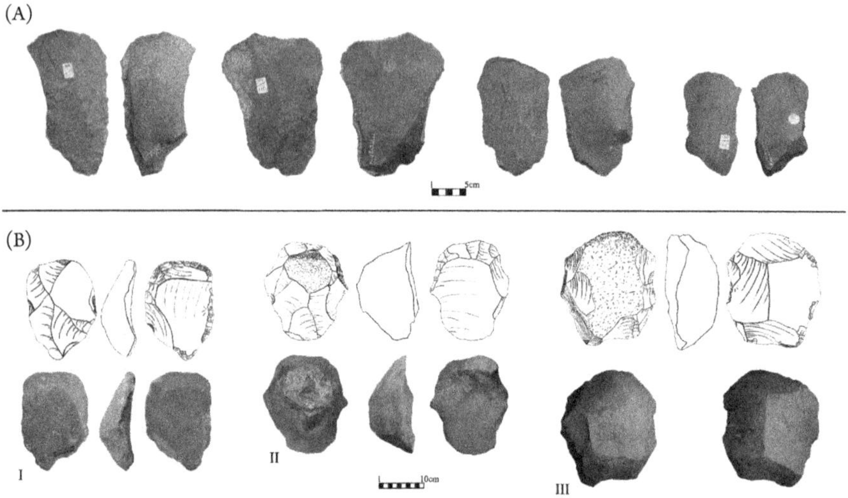

Figure 11.3. Tabelbala-Tachenghit core method. (A) Tabelbala-Tachenghit cleavers; (B) Tabelbala-Tachenghit cores. I, II, III = three of the larger cores. Image by the author.

I technique. The description here of the Tabelbala-Tachenghit core method is very short and fragmentary; very few cores and resulting tools are available for study, with most belonging to surface collections rather than well-documented excavations. Nevertheless, the uniqueness of the cores and cleavers and their specific and clearly evidenced technology enable us to evaluate the Tabelbala-Tachenghit core method as a distinct method. It appears that the Tabelbala-Tachenghit biface makers applied a technology comparable to the Victoria West core method in goal, shape, size, and, in particular, technological economizing and sophistication.

VICTORIA WEST AND THE TABELBALA-TACHENGHIT

The Victoria West and the Tabelbala-Tachenghit core methods share many technological features, but more than this, they share technological concepts and ideas, some of which are summarized here:

- In both methods, the primary goal is the removal of a single, preferred cleaver flake from a relatively small core (small in terms of Acheulean giant cores). After the removal of the desired flake-blank the core is "exhausted" and no additional large flakes can be removed. Hence, the core is discarded at this stage.
- Both core methods follow the *préférentiel* flake approach, as opposed to the *récurrent* concept (Boëda, 1995).
- Two asymmetrical (hierarchical in the terminology used by Boëda, 1995) faces are prepared for the core—the striking platform preparation surface and the flake removal surface.
- The tip of the core is removed during the detachment of the flake blank (Figures 11.2 and 11.3). This results in a flake larger than the scar being left on the core and can be seen in the typical morphology of the resulting flakes.

- Both core methods reveal an advanced degree of pre-planning in stone tool technology. The knappers designed the core stages in advance to dictate the morphology of the resulting flake. They implemented a meticulously planned, consistent chain of removals to produce an LF perfectly suitable in shape and size for the production of a cleaver with minimal (and sometime none) additional shaping needed. Great investment of energy and planning in knapping the core absolved the need for investment in the resulting flake.
- The uniformity in production is evident in the "machine-like" production of cores almost identical in size and morphology. Both core methods also demonstrate uniformity in the direction of the blows intended to remove the blank. The design of the blows is the same and they were struck from the same direction.
- In both core methods, the cleaver is the objective for which the removed flake blanks were pre-designed.

The primary similarity between the Victoria West and Tabelbala-Tachenghit core methods is the sophisticated and advanced technological skill demonstrated by the knappers. The high degree of uniformity of the cores and their end products deserves special emphasis. The cores are greatly similar in shape and size, and the same is true for the cleavers. One can easily identify a Victoria West or Tabelbala-Tachenghit cleaver, and it sometimes seems as if they were "machine made" (Figure 11.2)). They are so uniform in shape and size that a cleaver from Penil 6 can be refitted almost perfectly to a core collected kilometers apart at Canteen Kopje (Figure 11.2 (3)).

Yet, the Victoria West and Tabelbala-Tachenghit core methods are not identical. One fundamental difference between them is that the Tabelbala-Tachenghit flake-striking platforms seem to have been more carefully prepared and isolated from the surface, possibly in an attempt to ensure the accuracy of the blow. They are usually plain and a large incipient cone is visible (Figure 11.3A). On the Victoria West cleavers, the striking platform shows remnants of the bifacially knapped surface of the core. Like the Victoria West core method, it seems that the Tabelbala-Tachenghit core method was restricted geographically, in this case to the small region of the north-western Sahara. We are lacking sufficient data from the Sahara; yet it seems that, unlike the Victoria West core method, the Tabelbala-Tachenghit method was not the dominant or primary core method used. Even within the restricted area in which these cores and cleavers are found, the number of tools that were made by the Tabelbala-Tachenghit core method appears limited.

While both core methods fit comfortably within the definition of the Levallois concept (Boëda, 1995), they are not Levallois cores. Each core method has unique technological features such as precise preparation of the striking platform or removal of the core tip with the flake blank. An additional difference is the size of the core. While some Acheulean giant cores were clearly produced using the Levallois core method, the great majority of Levallois cores are much smaller than the Acheulean giant cores. It can be suggested that the Levallois core method is not fully appropriate for the production of very large flakes and hence was used infrequently. It was not reduced cognitive abilities or technological limitations that prevented the Acheulean knappers from using the Levallois method, as both Victoria West and Tabelbala-Tachenghit are similarly advanced, prepared core technologies. On the contrary, the

Acheulean core methods were better adapted to the production of large flakes from hard-to-knap raw material types (see discussion in Sharon, 2007).

DISCUSSION AND CONCLUSIONS

Application of the principals of evolution to the understanding of human cultural development and behavior can be very fruitful, but requires care. Convergent biological evolution is a well-established principle that, when correctly applied, can contribute to the discussion of the origin of different ideas and technologies in human culture. When studying similar cultural developments that appear in very distinct regions of the world, questions arise regarding their origin. A key question is whether the idea, concept, and technology developed in a single "core" area and then spread to other parts of the world, or whether a parallel need for a tool, subsistence strategy, governmental organization, or communication resulted in the innovation of a similar, yet not identical, solution. The origin of agriculture, writing, and the wheel are examples for discussion of this question (Diamond, 1997).

This question of convergent cultural evolution becomes more complex when moving back in time toward the remote stages of human history and evolution. These are time periods well before written record, and the discussion of ideas and concepts is based on stone tools and a few bones (if we are lucky). Nevertheless, in some cases, the study of lithic technology can result in identification of cultural trends and innovations even in the very early Lower Paleolithic Acheulean.

The study of Acheulean giant core methods provides one of these opportunities. Acheulean hominins were producing bifacial tools, handaxes, and cleavers throughout the vast geographical distribution of this techno-complex. As do many other researchers, I believe that these tools are similar in size, shape, and technology worldwide (Sharon, 2007). Acheulean tool-makers had a cleaver and handaxe "mental template" that determined the lithic technology most suitable for use with available local rock types, shapes, and sizes in the production of these tools. This is particularly evident in the LF Acheulean tradition. The use of large flakes as blanks for the production of bifaces, particularly cleavers, is a highly efficient reduction sequence (Jones, 1994; Madsen & Goren-Inbar, 2004; Sharon, 2007, 2009). In all regions, Acheulean tool-makers seeking to produce cleavers looked at their surrounding environment and applied the best method to produce suitable large flakes for cleaver-blanks. On the one hand, Acheulean knappers were extremely conservative in their tool production. They produced and used similar handaxes and cleavers for hundreds of thousands of years. On the other hand, they demonstrated tremendous innovation and advanced and sophisticated technologies in their efficient use of local raw material to produce their desired tools.

The Victoria West and the Tabelbala-Tachenghit core methods provide the best evidence for such behavior. Both core methods exemplify high dexterity of working in stone. They are sophisticated, demonstrate intimate knowledge of the local rock type properties, indicate a high level of pre-planning, are uniform in shape and size, and even indicate a single, preferred direction of blow, suggesting an asymmetrical stronger hand (not necessarily the right hand).

The Tabelbala-Tachenghit and Victoria West core methods are similar but not quite identical solutions to a mutual need, the need for a cleaver. Both core methods are restricted to a small area in South Africa and the Western Sahara Desert. Not a single Victoria West core or cleaver has ever been found north of the Vaal River, and not a single Tabelbala-Tachenghit core has ever been reported outside of the Western Sahara (Sharon, 2007). They represent two geographically distant, separate solutions that converge into a similar core method.

It is important to emphasize that this technological convergence is observed not at the final tool (cleaver) stage but earlier, at the core/blank production stage. The need for a cleaver is the driving evolutionary force in this case. We are still far from understanding why the cleaver was such a desired tool, produced unchanged for so many generations and in such diverse environments (Figure 11.1). However, cleavers are found in all LF Acheulean sites, and they are similar in shape, size, and production technology. Perhaps just like the need to penetrate ant nests dictated the elongated noses and sticky tongues of anteaters in different continents, the desire for cleavers led to the innovation of different core methods. All giant core methods can actually be understood as examples for convergent cultural evolution, yet the Tabelbala-Tachenghit and Victoria West methods are the clearest examples, as they are advanced, unique, and easy to recognize, designed for the production of cleavers and restricted to a small region.

While the giant core technology provides us with examples of convergent cultural evolution, the end products of the LF Acheulean—handaxes and cleavers—demonstrate the opposite principle. The striking similarity in both size and morphology of LCTs from throughout the geographic and chronological distribution of the LF Acheulean provides clear evidence against the hypothesis suggesting that they are the result of similar, unrelated innovations in the different regions. In light of the data presented here, it is unlikely that tools knapped from such a variety of raw materials by such different core methods would all be shaped into such similar end products. Even if we claim that the makers of the Acheulean LCTs designed their tools for similar functional needs, such similarity in end products is inconceivable. The great adaptive variability in raw material strategies and core methods on the one hand, and the astonishing similarity in the end products on the other, can be explained only in terms of lithic tradition. The makers of the Acheulean LCTs had a clear idea of the shape and size of desired tools that were produced by their lithic cultural tradition. Their sophistication, innovation, and adaptive capabilities are evidenced by the many technological paths used to achieve their objective.

ACKNOWLEDGMENTS

I wish to thank the editors of this volume for inviting me to take part in this important book. Two reviewers read this chapter, made valuable suggestions, and pointed out many pitfalls in the original version. The pitfalls that survived are mine. Amy Klein edited this chapter. This chapter is dedicated to the memory of Peter Beaumont, who generously introduced me to the world of the Vaal River sites and let me look in all of the treasure boxes full of cleavers, handaxes, and cores he excavated during his pioneering research of the South African Acheulean.

REFERENCES

Alimen, M. H. (1978). *L'évolution de l'Acheuléen au Sahara nord-occidental: Saoura, Ougarta, Tabelbala: Atlas*. Paris: Éditions CNRS.

Balout, L., & Tixier, J. (1957). L'Acheuléen de Ternifine. In *Congrès préhistorique de France: Compte rendu de la XV session: Poitiers-Angouleme 15–22 juillet 1956* (pp. 214–218). Paris: Société préhistorique française.

Beaumont, P. B. (1990). Canteen Kopje. In P. B. Beaumont & D. Morris (Eds.), *Guide to the archaeological sites of the Northern Cape* (pp. 14–16). Kimberley: McGregor Museum.

Biberson, P. (1961). *Le Paléolithique inférieur du Maroc atlantique* (Vol. 17). Rabat, Morocco: Service des Antiquités du Maroc.

Boëda, E. (1995). Levallois: A volumetric construction, methods, a technique. In H. L. Dibble & O. Bar-Yosef (Eds.), *The definition and interpretation of Levallois technology* (Vol. 23, pp. 41–65). Madison, WI: Prehistory Press.

Bordes, F. (1961). *Typologie du Paleolithique ancien et moyen*. Paris: Éditions CNRS.

Breuil, H. (1930). Premières impressions de voyage sur la préhistoire sud-africaine. *Anthropologie, 40*, 209–223.

Champault, B. (1951). L'industrie de Tachenghit. In *70ème Congrès de L'Association française pour l'avancement des sciences* (Vol. III, pp. 123–130). Paris: AFAS.

Champault, B. (1956). Nucléus à technique Levalloisienne accompagné d'une partie de ses éclats de taille (industrie de Tachenghit, Sahara algéro-marocain). *Bulletin de la Société préhistorique française, 53*, 267.

Champault, B. (1966). *L'Acheuléen évolué au Sahara occidental. Notes sur l'homme au paléolithique ancien*. Doctoral thesis, Faculté de Lettres, Paris.

Clark, J. D. (1992). The Earlier Stone Age/Lower Paleolithic in Northern Africa and the Sahara. In F. Klees & R. Kuper (Eds.), *New light on the Northeast African past* (pp. 17–38). Köln, Germany: Heinrich-Barth-Institut.

Corbey, R., Jagich, A., Vaesen, K., & Collard, M. (2016). The Acheulean handaxe: More like a bird's song than a Beatles' tune? *Evolutionary Anthropology: Issues, News, and Reviews, 25*(1), 6–19.

Corvinus, G. (1983). *A survey of the Pravara River system in western Maharashtra, India: Volume 2, The excavations of the Acheulian site of Chirki-on-Pravara, India*. Tübingen, Germany: Institut für Urgeschichte.

Darwin, C. (1859). *On the origin of species*. London: Penguin.

Diamond, J. M. (1997). *Guns, germs, and steel: The fates of human societies*. New York: W.W. Norton & Co.

Emery, N. J., & Clayton, N. S. (2004). The mentality of crows: Convergent evolution of intelligence in corvids and apes. *Science, 306*(5703), 1903–1907.

Fuller, D. Q., Denham, T., Arroyo-Kalin, M., Lucas, L., Stevens, C. J., Qin, L., . . . Purugganan, M. D. (2014). Convergent evolution and parallelism in plant domestication revealed by an expanding archaeological record. *Proceedings of the National Academy of Sciences of the United States of America, 111*(17), 6147–6152.

Gibbon, R. J., Granger, D. E., Kuman, K., & Partridge, T. C. (2009). Early Acheulean technology in the Rietputs Formation, South Africa, dated with cosmogenic nuclides. *Journal of Human Evolution, 56*(2), 152–160.

Goodwin, A. J. H. (1929). The Victoria West industry. *The Stone Age Cultures of South Africa, XXVII*, 53–69.

Goodwin, A. J. H. (1953). *Methods in prehistory: An introduction to the discipline of prehistoric archaeology with special reference to South African conditions* (2nd ed.). Cape Town, South Africa: South African Archaeological Society.

Goodwin, A. J. H., & van Riet Lowe, C. (1929). *The Stone Age cultures of South Africa. Annals of the South African Museum* (Vol. 27). Cape Town, South Africa: AMS Press.

Goren-Inbar, N., Feibel, C. S., Verosub, K. L., Melamed, Y., Kislev, M. E., Tchernov, E., & Saragusti, I. (2000). Pleistocene milestones on the out-of-Africa corridor at Gesher Benot Ya'aqov, Israel. *Science, 289*(5481), 944–947.

Gould, S. J. (1971). D'Arcy Thompson and the science of form. *New Literary History, 2*(2), 229–258.

Isaac, G. L. (1969). Studies of early culture in East Africa. *World Archaeology, 1*(1), 1–28.

Jansen, F. J. (1926). A new type of stone implement from Victoria West. *South African Journal of Science, 23*, 818–825.

Jones, P. R. (1994). Results of experimental work in relation to the stone industries of Olduvai Gorge. In M. D. N. Leakey & D. A. Roe (Eds.), *Olduvai Gorge. Volume 5. Excavations in Beds III, IV and the Masek Beds, 1968–1971* (Vol. 5, pp. 254–298). Cambridge, UK: Cambridge University Press.

Kleindienst, M. R. (1961). Variability within the Late Acheulian assemblage in eastern Africa. *The South African Archaeological Bulletin, 16*(62), 35–52.

Kleindienst, M. R. (1962). Components of the East African Acheulian assemblage: An analytic approach. In G. Mortelmans & J. Nenqui (Eds.), *Actes du IVeme Congrès panafricain de préhistoire et de l'étude du quaternaire* (Vol. 40, pp. 81–105). Tervuren, Belgium: Museé Royal.

Kuman, K., Li, C., & Li, H. (2014). Large cutting tools in the Danjiangkou Reservoir region, central China. *Journal of Human Evolution, 76*, 129–153.

Li, H., Li, C., & Kuman, K. (2014). Rethinking the "Acheulean" in East Asia: Evidence from recent investigations in the Danjiangkou Reservoir region, central China. *Quaternary International, 347*(1), 163–175.

Li, H., Li, C., Kuman, K., Cheng, J., Yao, H., & Li, Z. (2014). The Middle Pleistocene handaxe site of Shuangshu in the Danjiangkou Reservoir region, central China. *Journal of Archaeological Science, 52*, 391–409.

Madsen, B., & Goren-Inbar, N. (2004). Acheulian giant core technology and beyond: An archaeological and experimental case study. *Eurasian Prehistory, 2*(1), 3–52.

McNabb, J. (2001). The shape of things to come: A speculative essay on the role of the Victoria West phenomenon at Canteen Kopje, during the South African Stone Age. In S. Milliken & J. Cook (Eds.), *A very remote period indeed: Papers on the Palaeolithic presented to Derek Roe* (pp. 37–46). Oxford, UK: Oxbow Books.

McNabb, J., & Beaumont, P. B. (2012). Excavations in the Acheulean levels at the Earlier Stone Age site of Canteen Koppie, Northern Province, South Africa. *Proceedings of the Prehistoric Society, 78*, 51–71.

Mesoudi, A., Whiten, A., & Laland, K. N. (2004). Perspective: Is human cultural evolution Darwinian? Evidence reviewed from the perspective of *The Origin of Species. Evolution, 58*(1), 1–11.

Mourre, V. (2003). *Implications culturelles de la technologie des hachereaux*. Doctoral thesis, Université de Paris X, Paris.

Norn, M. (1996). *Eskimo snow goggles in Danish and Greenlandic museums, their protective and optical properties* (Vol. 20). Copenhagen: Museum Tusculanum Press.

Pappu, S., & Akhilesh, K. (2006). Preliminary observations on the Acheulian assemblages from Attirampakkam, Tamil Nadu. In N. Goren-Inbar & G. Sharon (Eds.), *Axe age: Acheulian tool-making from quarry to discard* (pp. 155–180). London: Equinox.

Ranov, V. A. (2001). Cleavers: Their distribution, chronology and typology. In S. Milliken & J. Cook (Eds.), *A very remote period indeed: Papers on the Palaeolithic presented to Derek Roe* (pp. 105–113). Oxford, UK: Oxbow Books.

Roche, H., Brugal, J.-P., Lefevre, D., Ploux, S., & Texier, P.-J. (1988). Isenya: État des recherches sur un nouveau site acheuléen d'Afrique orientale. *African Archaeological Review, 6*(1), 27–55.

Roe, D. A. (1994). A metrical analysis of selected sets of handaxes and cleavers from Olduvai Gorge. In M. D. Leaky & D. A. Roe (Eds.), *Olduvai Gorge. Volume 5. Excavations in Beds III, IV and the Masek Beds, 1968–1971* (pp. 146–234). Cambridge, UK: Cambridge University Press.

Sharon, G. (2007). *Acheulian large flake industries: Technology, chronology, and significance.* Oxford, UK: Archaeopress.

Sharon, G. (2008). The impact of raw material on Acheulian large flake production. *Journal of Archaeological Science, 35*(5), 1329–1344.

Sharon, G. (2009). Acheulian giant-core technology: A worldwide perspective. *Current Anthropology, 50*(3), 335–367.

Sharon, G., & Barsky, D. (2016). The emergence of the Acheulian in Europe—A look from the east. *Quaternary International, 411*, 25–33.

Sharon, G., & Beaumont, P. B. (2006). Victoria West: A highly standardized prepared core technology. In N. Goren-Inbar & G. Sharon (Eds.), *Axe age: Acheulian tool-making from quarry to discard* (pp. 181–199). London: Equinox.

Sharon, G., Feibel, C. S., Alperson-Afil, N., Harlavan, Y., Feraud, G., Ashkenazi, S., & Rabinovich, R. (2010). New evidence for the northern Dead Sea Rift Acheulian. *PaleoAnthropology*, 79–99.

Tixier, J. (1956). Le hachereau dans l'Acheuléen nord-africain. Notes typologiques. In *Congrès Préhistorique de France: Compte rendu de la XV session: Poitiers-Angouleme 15–22 juillet 1956* (pp. 914–923). Paris: Société préhistorique française.

van Riet Lowe, C. (1935). Implementiferous gravels of the Vaal River at Riverview Estates. *Nature, 136*(3428), 53–56.

van Riet Lowe, C. (1937). The archaeology of the Vaal River Basin. In P. G. Söhnge, D. J. L. Visser, & C. Van Riet Lowe (Eds.), *The geology and archaeology of the Vaal River Basin: (Union of South Africa Department of Mines, Geological Survey Memoir No 35)* (pp. 61–134). Pretoria, South Africa: The Government Printer.

Wilkins, J., Pollarolo, L., & Kuman, K. (2010). Prepared core reduction at the site of Kudu Koppie in northern South Africa: Temporal patterns across the Earlier and Middle Stone Age boundary. *Journal of Archaeological Science, 37*(6), 1279–1292.

Wynn, T. (1995). Handaxe enigmas. *World Archaeology, 27*(1), 10–24.

Yamei, H., Potts, R., Baoyin, Y., Zhengtang, G., Deino, A., Wei, W., . . . Weiwen, H. (2000). Mid-Pleistocene Acheulean-like stone technology of the Bose Basin, south China. *Science, 287*(5458), 1622–1626.

Zhang, P., Huang, W., & Wang, W. (2010). Acheulean handaxes from Fengshudao, Bose sites of south China. *Quaternary International, 223*, 440–443.

12

CULTURAL TRANSMISSION FROM THE LAST COMMON ANCESTOR TO THE LEVALLOIS REDUCERS

WHAT CAN WE INFER?

Stephen J. Lycett

INTRODUCTION

Evidence of the human capacity to share and learn information is all around us. It is evident in the tools and devices that we make and use, the houses that we build, the belief systems and political institutions we construct, our dietary preferences, and many other behavioral patterns. From birth, we are continuously exposed to a range of activities and behaviors instigated by others, which eventually come to influence many of the beliefs we hold, the clothes we wear, the way we speak, and the knowledge and skills that we might come to have. This sharing and learning of information thus comes to form distinct traditions or customs, shared between individuals belonging to communities and subcommunities of people that are united by common patterns of behavior, or at least some aspects of their behavior. In other words, the product of this sharing and learning process is what many would refer to as "culture," with the differences between (sub)communities of individuals further being described as "cultural differences" (Boyd & Richerson, 1985; Ellen, Lycett, & Johns, 2013; McGrew, 2004; Mesoudi, 2011).

A particularly striking component of human culture is the capacity to accumulate knowledge that builds successively in relative complexity over generations, or what some have referred to as cultural "ratcheting" (Tomasello, 1999). This capacity can lead to a situation where cultural products have a level of sophistication and intricacy such that only someone drawing directly on the cumulative knowledge of several previous generations could replicate, use, perform, and/or manufacture the cultural entity concerned. Hence, a behavior may have begun relatively simply (e.g., throwing a sharpened stick as a spear), then been modified by a later generation to something more technically complex (e.g., using a spear thrower to throw a sharpened stick that is fletched with feathers), then subsequently modified again to a tradition of behavior that draws on even greater levels of culturally transmitted information (e.g., using a bow to launch projectiles that are fletched and also tipped with carved bone points). We have to be cautious in assuming that this ability to draw on previous generations of cultural knowledge is limited solely to humans; there is some evidence from chimpanzees (*Pan troglodytes*), for instance, for the use of brush-tipped termite ("fishing") probes that are more effective than plain probes, as well as use of tool "kits" composed of multiple tools to undertake activities such as honey-gathering and ant-dipping, which might also be the product of "ratcheting" (Boesch, Head, & Robbins, 2009; Sanz, Call, & Morgan, 2009; Sanz, Schöning, & Morgan, 2010). Nevertheless, despite these potential examples, there are evident differences in the extent to which many human cultural products draw on cumulative cultural knowledge and the types of behaviors that chimpanzees are capable of undertaking. Accordingly, whether the difference is one of substantial degree or kind, many commentators agree that one of the most important characteristics of human culture—and one that has allowed us to become a dominant force in our globe's ecological systems—is our capacity to "ratchet" cultural information accurately and cumulatively over successive generations (Boyd, Richerson, & Henrich, 2011; Coolidge & Wynn, 2005; Dean, Kendal, Schapiro, Thierry, & Laland, 2012; Kempe, Lycett, & Mesoudi, 2014; Kurzban & Barrett, 2012).

How and when though did our social learning abilities arise? One possibility is that the human capacity for culture sprang forth suddenly and entirely without evolutionary precedent with the emergence of our species. This scenario, however, is improbable, for several reasons. First, the capacity to learn socially from others of the same species has been identified in a wide range of different animals, including species of birds, fish, and mammals (e.g., Claidière, Smith, Kirby, & Fagot, 2014; Galef & Laland, 2005; van de Waal, Borgeaud, & Whiten, 2013; Whiten, Caldwell, & Mesoudi, 2016; Zentall, 2012). This makes it unlikely that human social learning capacities—while impressive and potentially unique in specific ways—are *entirely* without some form of evolutionary precursor. Most notably, over the last few decades there has been a wealth of new data indicating that species of other great apes, including chimpanzees (*P. troglodytes*), bonobos (*P. paniscus*), gorillas (genus *Gorilla*), and orangutans (genus *Pongo*), have the capacity to learn information socially (e.g., Byrne, Hobaiter, & Klailova, 2011; Dindo, Stoinski, & Whiten, 2010; Hopper, Lambeth, Schapiro, & Whiten, 2015; Jaeggi et al., 2010; Luef & Pika, 2013; McGrew, 2004, 2010b; Whiten, 2010; Whiten, McGuigan, Marshall-Pescini, & Hopper, 2009; Whiten & van Schaik, 2007). This in particular suggests that human cultural capacities are an outcome of an evolutionary process that involved building on capabilities that are rooted deeply within our great ape ancestry.

Second, psychologists have identified and labeled a range of different social learning mechanisms (see e.g., Whiten, Horner, Litchfield, & Marshall-Pescini, 2004), which leads to the possibility that the deployment of, and ability to use, these different mechanisms evolved during the course of human evolution (Lycett, 2010b, p. 261; Tomasello, 2009, pp. 218–219; Whiten, Schick, & Toth, 2009). In one sense, *social learning* can be defined relatively straightforwardly as the learning (i.e., gaining of information) that is obtained from observing the behavior or the outcomes of behavior of others (Heyes, 1994). However, several quite distinct mechanisms by which watching and learning, either directly or indirectly, from others have been identified. *Stimulus enhancement*, for example, is one of the simplest means by which one individual might socially learn information that influences their behavior (Thorpe, 1963). Under this mechanism of learning, the behavioral effect of one individual on another is indirect: A learner does not directly copy the behavior of another individual but, rather, is influenced in her own behavioral patterns through greater exposure to a stimulus that is instigated by another individual's behavior. Having been exposed to the stimulus, the learner then pieces much of the remaining information about the behavior together by herself and implements it, such that her overall behavioral pattern comes to resemble that of the other individual. Byrne and Russon (1998, p. 669) conceptually illustrate this process (which is closely related to a process referred to as "local enhancement") with the example of how one monkey might learn (hypothetically) how to crack nuts from another because of the stimulus provided by a food source, whereupon attention is directed toward key elements in the process (e.g., nuts, hammer, and anvil), which the monkey subsequently puts into adequate operation through individual learning. Such a process repeated over the course of many individual events could, therefore, lead to an authentic "tradition" of nut-cracking (*sensu* Fragaszy & Perry, 2003), even though no direct copying of actions has taken place. Alternatively, an individual might *emulate* the behavior of another, which has been defined as copying the direct outcomes or evident results of a particular behavior, which are key to performing some aspect of the behavior, but does not itself involve direct copying of another individual's actions (Whiten et al., 2004). An example might be when a tool (i.e., probe) is used to fish for termites and another individual then observes that tool and makes one for herself. The term *imitation* is reserved for instances of learning that involve directly copying the precise actions of another individual in order to bring about the same outcome or result as that witnessed (Schillinger, Mesoudi, & Lycett, 2015; Whiten et al., 2004). Finally, *teaching* can be defined as situations in which a knowledgeable individual deliberately alters his or her behavior in such a way that it facilitates learning in a less knowledgeable individual (Thornton & Raihani, 2010). Of course, these mechanisms of learning are not necessarily mutually exclusive, even in humans, and might therefore be used in combination along with lengthy periods of individual learning (e.g., practice) in order to replicate particular behaviors with any degree of accuracy (Lycett, 2015; Martin, 2000).

A rather crude classification of social learning mechanisms thus recognizes at least four different learning pathways: stimulus enhancement, emulation, imitation, and teaching. It is likely that each of these mechanisms can be further split into finer-grained distinctions about the exact details of learning. Indeed, some make distinctions between imitation and "over-imitation," for instance, which places greater emphasis on which exact elements of others' actions are copied and the extent to which they are copied (e.g., Nielsen, 2012; Whiten, McGuigan, et al., 2009). In the case of teaching, it is also necessary to distinguish between linguistically mediated instruction and nonverbal teaching (e.g., Wynn & Coolidge, 2010, p. 97). However, even this crude taxonomy of learning processes can be useful when considering the array of potential means by which one individual can gain information from another, such that it leads to the repetition of patterns of behavior in others.

From at least 3.3 million years ago (Mya) until well after the emergence of our own species, hominins began to deliberately make stone cutting tools through percussive behavior, often referred to as "knapping" (Harmand et al., 2015). This has left us with a rich record of lithic reduction spread through several million years, over large swathes of Africa and Eurasia (Schick, 1998). In principle, therefore, these stone artifacts and their variable forms over time might potentially tell us about social learning patterns in our extinct ancestors. Studying important phenomena such as social learning, language, or cognition in extinct hominins, however, presents considerable challenges to science. These problems arise because of a basic fact: Neither thoughts, conversations, intelligence, nor instances of social transmission fossilize. This means that we do not have *direct* fossil or artifactual evidence for any of these important features of hominin existence, only indirect (proxy) evidence. Over several decades of work, Tom Wynn has tackled the thorny issue of hominin cognition—one of these unseen but key facets of our evolutionary history—by applying two prudent principles (Wynn, 1979, 1981, 1991, 1995, 1996, 2004; Wynn & Coolidge, 2004, 2010, 2012; Wynn, Tierson, & Palmer, 1996). The first of these is the importance of a comparative approach (i.e., using data from other primates and even more distantly related animals alongside that derived from humans) that sets our archaeological data and the questions we ask in their proper phylogenetic and evidential contexts (Wynn, Hernandez-Aguilar, Marchant, & McGrew, 2011; Wynn & McGrew, 1989). The second is that we should try to invoke only the minimum abilities required to explain the archaeological phenomenon concerned (Wynn, 2002). Put another way, this latter principle emphasizes that we are inevitably placed in a position of risking either false-positive statements about the abilities of hominins (i.e., type I errors) or making false-negative statements (type II errors). By invoking only the minimum cognitive abilities required to explain the archaeological phenomenon concerned, we are at heightened risk of making type II errors (i.e., failing to recognize a behavior or capacity in hominins), but we minimize the occurrence of more serious type I errors (i.e., attributing behaviors and abilities to hominins which they did not in fact possess).

Here, Wynn's two principles are applied to infer what can reasonably be determined about cultural transmission capacities in extinct hominins from the last common ancestor (LCA) to the producers of so-called Levallois reduction sequences and their products. Using this approach and drawing on multiple lines of evidence in each case, it is possible to make some inferences about the social learning mechanisms that might minimally have been employed by hominins when forming the empirical archaeological record we can observe. It is also possible to consider the possibilities that provide extensions to the minimal scenario in each case in terms of additional social learning mechanisms that we might be making false-negative inferences about (i.e., where we are in danger of potentially underestimating what hominins were actually doing). These latter possibilities are not discussed for their own sake, but rather, because they provide a basis for future questions and issues that will need to be examined more rigorously, which are discussed further in the final section of this chapter.

SOCIAL LEARNING IN THE LAST COMMON ANCESTOR OF ALL HOMININS

The comparative approach is essential in respect to a wide range of evolutionary questions (Nunn, 2011), and, of course, in the case of humans our closest living relatives—the genus *Pan*—make an especially important point of reference (Carvalho & McGrew, 2012; Lycett, 2010b; McGrew, 2010a; Whiten, Schick, et al., 2009; Wynn & McGrew, 1989). Making inferences about the evolution of hominin cultural transmission must begin with the comparative approach, particularly when trying to reconstruct what the transmission capacities of the LCA may have been (Lycett, 2010b, 2013). In recent decades, a wealth of data has come from intensive studies of chimpanzee behavior, both in the wild and in captivity, which bears directly on the issue of what the social learning capacities of the LCA looked like. While much of this work has been reviewed in detail by others (e.g., Horner & de Waal, 2009; McGrew, 2010a, 2010b, 2015; Whiten, 2011; Whiten, Schick, et al., 2009), to discuss the issue of social learning in prehistoric hominins effectively, it is important to briefly draw out a few of the key points of this work, as well as illustrate some of the latest findings.

Some of the strongest scientific evidence on this issue is derived from controlled experiments undertaken with captive chimpanzees (Horner & de Waal, 2009; Whiten, 2011; Whiten, Schick, et al., 2009). For instance, the social learning of specific tool-using skills, when faced with tasks in which alternative techniques are possible, has been demonstrated experimentally in chimpanzees (*P. troglodytes*), such that members of the group come to share similar patterns of behavior (Whiten, Horner, & de Waal, 2005). These studies suggest that earlier experiments showing the learning of simple stone flake production (and their subsequent use in a cutting task) by Kanzi, a captive male bonobo (*P. paniscus*), and subsequently by his half-sister Panbanisha were indeed influenced by social learning (Schick et al., 1999; Toth, Schick, Savage-Rumbaugh, Sevcik, & Rumbaugh, 1993; Whiten, Schick, et al., 2009). Equally important are "transmission chain" experiments, which have shown that foraging behaviors can be passed successively with fidelity along chains of up to six chimpanzees (Horner, Whiten, Flynn, & de Waal, 2006), as well as experiments demonstrating the transmission of multiple experimental "traditions" among different captive groups (Whiten et al., 2007).

In terms of the mode of social learning utilized by chimpanzees in these types of task, stimulus enhancement might be involved because so many of these experiments involved food as a reward (e.g., Whiten et al., 2005). However, the matching of contrasting behaviors in different groups involved in these experiments (e.g., Horner & de Waal, 2009; Whiten et al., 2005) shows that stimulus enhancement alone cannot explain the group-patterned (as opposed to more random) exhibition of these behaviors across individuals in the experimental groups (Whiten, 2011). Indeed, the ability of chimpanzees to emulate (i.e., be influenced by others' behavior in replicating the behavior) is now well established experimentally (e.g., Hopper, Lambeth, Schapiro, & Whiten, 2008; Horner & Whiten, 2005; Nagell, Olguin, & Tomasello, 1993; Price, Lambeth, Schapiro, & Whiten, 2009). Whether the genus *Pan* has a capacity to engage in imitation (i.e., the copying of behavioral actions), and the extent of any such capacity, is more controversial, although experimental evidence suggests that chimpanzees may indeed copy behavioral actions in specific situations (Bonnie, Horner, Whiten, & de Waal, 2007; Fuhrmann, Ravignani, Marshall-Pescini, & Whiten, 2014; Horner et al., 2006; Tomasello, Savage-Rumbaugh, & Kruger, 1993; Whiten, McGuigan, et al., 2009). Even some former critics of the notion that chimpanzees have a capacity to engage in imitation have more recently modified their opinion on this issue in light of experimental evidence (e.g., Tomasello, 2009, p. 215). Some of the most suggestive evidence for imitation has come through so-called "ghost condition" experiments (Whiten, McGuigan, et al., 2009), where chimpanzees are compared in undertaking tasks that involve seeing only the mechanical operations and results of the experimental task versus seeing another chimpanzee perform those actions. At the time of writing, the latest of these experiments has confirmed the important role that seeing another chimpanzee or human undertake the task has on success, compared to situations involving seeing only a "ghost" (mechanical) display of the task (Hopper et al., 2015). Other experimental studies have provided evidence of motor mimicking during observational learning, which is also suggestive when considering the possibility of imitation in chimpanzees (Fuhrmann et al., 2014). Nevertheless, whether chimpanzees have a capacity for imitation, and in particular the extent of chimpanzee imitation, is still a matter for debate.

It is not, however, only studies from captivity that can help shed light on the probable social learning capabilities of the LCA. Decades-worth of data accumulated by field primatologists attest to the social learning of behaviors by individuals belonging to wild communities of the genus *Pan* (McGrew, 2010b, 2015). A watershed moment in these studies came with Whiten and colleagues' (1999) publication of 39 behavioral variants that differed in their representation across seven of the longest-studied chimpanzee (*P. troglodytes*) communities in East and West Africa. These behaviors are not randomly distributed, as might be expected in the case of individual learning, but form group-general patterns, as might be expected under instances of social learning. More limited evidence from bonobos (*P. paniscus*) suggests that similar patterns may characterize that species of chimpanzee as well (Hohmann & Fruth, 2003). The behaviors listed in Whiten and colleagues' (1999) catalogue of cross-community variability include both tool-use behaviors and social conventions (e.g., specific forms of grooming). Detailed studies of specific behaviors within this data set indicate that wild chimpanzees are indeed drawing on the kinds of capacities demonstrated in captive experiments when learning how to undertake these behaviors. For instance, an experiment at an outdoor laboratory at Bossou (Guinea) indicated that the presence of a knowledgeable individual was key in the spread of a behavior involving cracking a novel species of experimentally introduced nut (Biro et al., 2003). Other work at Bossou has demonstrated the influence of social learning on tool-use behaviors. Most notably, Humle, Snowdon, and Matsuzawa (2009) showed that the amount of time that mothers of juveniles (≤5 years of age) engaged in ant-dipping (i.e., the learning opportunity to which the juveniles were exposed) positively influenced a speedier onset of ant-dipping behavior in the novice and led to higher proficiency.

More recently, the spread of behaviors from one community to another has been closely documented (O'Malley, Wallauer, Murray, & Goodall, 2012) through a process similar to that documented in controlled captive experiments (Whiten et al., 2007). Social network analysis has also been used to document the social transmission of tool-use behavior directly (Hobaiter, Poisot, Zuberbühler, Hoppitt, & Gruber, 2014). Recent studies have also shown how distinctive group-level traditions relating to tool-use behaviors are maintained between neighboring chimpanzee communities, despite the frequent migration of females (Luncz, Mundry, & Boesch, 2012), whereupon migrating females adopt behaviors more similar to that of their new community (Luncz & Boesch, 2014, 2015). These studies, which by using neighboring communities thus control for genetic and ecological differences, are consistent with earlier wider-scale analyses indicating that neither genetic differences (Lycett, Collard, & McGrew, 2010) nor ecological differences between communities (Kamilar & Marshack, 2012) are plausible explanations for the cross-community behavioral variation documented at wider geographic scales (Whiten et al., 1999, 2001).

This is a necessarily brief and selective review of a much larger literature. However, key points to draw from this are that the LCA had a capacity to learn information socially, particularly through mechanisms such as stimulus enhancement and emulation, and—drawing on parsimony and the behavior of wild chimpanzees—the LCA was in all probability using these skills to engage in a range of important social and food-related activities (McGrew, 2010b; Whiten, Schick, et al., 2009). The parsimony element of this inference is strengthened when evidence from the genus *Pan* is combined with that from gorillas and orangutans (e.g., Byrne et al., 2011; Dindo et al., 2010; Jaeggi et al., 2010; Luef & Pika, 2013; Whiten & van Schaik, 2007), making the likelihood of convergence for these minimal social learning capacities all the more improbable. Some form of basic imitative skill was also possibly present, although this is more contentious. If culture is defined as the presence of multiple traditions shared by multiple individuals within a group through means of social learning, then chimpanzees possess this phenomenon (McGrew, 2004, 2015).

SOCIAL LEARNING IN THE LOMEKWIAN–OLDOWAN PHASE OF STONE TOOL PRODUCTION

Thankfully, through the comparative approach, we do not start from a position of ignorance when addressing questions pertaining to the prehistoric stone tool record. For a number of years, the origins of stone tool production were thought to date from around 2.6 Mya, and these early assemblages have often been referred to as representing the "Oldowan" industry (Semaw et al., 1997, 2003), a term derived from their initial discovery at Olduvai (at the time, Oldoway) Gorge in the 1930s (Leakey, 1936). Oldowan artifacts, consisting of simple flakes and cores, have subsequently been found at sites distributed from North Africa to South Africa, and were produced for well over one million years (Schick & Toth, 2006). Some time ago, Wynn and McGrew (1989) laid out the case that there was little, if anything, beyond the cognitive capacities of the genus *Pan* evident in the Oldowan, a position later strengthened (Wynn et al., 2011), most notably by the knapping capabilities of Kanzi the bonobo, albeit following demonstration and continued encouragement by humans for several weeks (Toth et al., 1993).

The recent evidence for the production of stone cutting tools from Lomekwi 3, West Turkana (Kenya) has pushed the origin of stone flaking back to at least 3.3 Mya (Harmand et al., 2015). Currently, little is known about what precisely motivated these bouts of stone flaking at Lomekwi; although unless we assume that hominins were engaging in stone knapping for their own entertainment, it seems likely that food-related cutting activities had something to do with it, and extractive foraging is known to be a major motivator (although not exclusively) underlying tool use in chimpanzees (McGrew, 1992). By the time Oldowan tool industries were being produced (i.e., from 2.6 Mya onward), direct evidence indicates that stone flake production was an activity related to foraging behavior in the form of cut-marked bones (de Heinzelin et al., 1999; Pickering & Domínguez-Rodrigo, 2006; Semaw, 2006). Since food-connected activities provide strong incentives, stimulus/local enhancement would be a plausible learning mechanism to explain the first million years of the archaeological record. Indeed, experimental work with non-human primates has shown that stimulus enhancement can create traditions (Matthews, Paukner, & Suomi, 2010), emphasizing its potential role in the Oldowan. Individual learning could then refine knapping behavior in terms of the requisite motor skills required to produce some of the knapping sequences evinced in the Oldowan record, which show effective and consistent execution of conchoidal fracturing through direct (stone-on-stone) percussion and strike-platform angle selection (e.g., Delagnes & Roche, 2005; Semaw, 2006; Stout & Semaw, 2006).

However, there are reasons to be cautious of the idea that only social learning via stimulus enhancement, much less individual (trial-and-error) learning alone, adequately accounts for the Oldowan record, given the capabilities of the LCA derived from the comparative approach (Lycett, von Cramon-Taubadel, & Eren, 2016). We know through studies of learning in non-human animals that social learning, as opposed to asocial (i.e., individual) learning, is more likely to occur wherever asocial learning is more costly or hazardous (e.g., Chivers & Smith, 1995; Greggor, McIvor, Clayton, & Thornton, 2016; Kelley, Evans, Ramnarine, & Magurran, 2003; Kendal, Coolen, & Laland, 2004). The reason for this is fairly obvious: In behaviors such as predator avoidance, for instance, it is much better to let someone else experiment with alternative strategies, and learn from the outcome of whether they get eaten or not, than to experiment yourself. A key variable here, therefore, is the inherently hazardous, if not outright dangerous, nature of stone knapping. Modern experimenters have noted (via incursion!) the risks involved in flintknapping, including the severing of tendons in a finger, causing impaired movement despite corrective surgery (Whittaker, 1994), and permanent disabling of an arm through incorrect form (Holmes, 1897, pp. 61, 151). Indeed, many contemporary knappers wear protective goggles and other gear to guard against injuries from flying flakes and chip debris. The ethnographic and ethnohistorical literature on knapping also emphasizes potential injury through laceration, as well as the risk of eye injury (e.g., Arthur, 2010, p. 236; Hampton, 1999, p. 267; Kroeber, 1961, p. 184; Pope, 1918, p. 117). In a Pleistocene context, such injuries could have proven potentially fatal, even through mere infection or injury (laceration) that prevented effective foraging.

With risks such as these—and again, through a comparative approach that informs us of the most pertinent triggers to social learning—it seems improbable that hominins would not be deploying their full contingent of social-learning mechanisms, including emulation, in order to reduce these risks, especially while learning (i.e., doing things improperly). Indeed, a particularly important outcome of knapping is that it leaves a physical record (i.e., artifacts and waste material) from which others can potentially learn via emulation, even weeks or months after the original knapping event. Again, the time-transgressive learning opportunities provided by tool manufacture have also been implicated in chimpanzee learning (Fragaszy et al., 2013) and might especially be drawn on in the case of a hazardous activity such as knapping.

In sum, simple mechanisms of social learning such as stimulus enhancement associated with a foraging activity, plus individual learning to master technique, might account adequately for many components of the earliest phases of the archaeological record. However, considering the hazards and costs associated with knapping, hominins would also likely have utilized their capacity for emulation. There is little evidence of group-specific traditions emerging in the (East) African Oldowan (Stout, Semaw, Rogers, & Cauche, 2010), although cross-assemblage comparison in quantitative terms remains limited. Recent work on chimpanzee tools (Koops, Schöning, Isaji, & Hashimoto, 2015) suggests that looking in greater detail for such differences may, however, prove fruitful.

SOCIAL LEARNING IN THE ACHEULEAN

From around 1.7 Mya, hominins began to produce new forms of stone tools to supplement their Oldowan toolkit, including characteristic handaxes (Beyene et al., 2013; Diez-Martín et al., 2015; Lepre et al., 2011). Perhaps unfortunately, the term *handaxe* can cover a wide range of lithic forms—from a flake-based blank or nodule with only a few flakes removed from an edge and an end to produce an elongated, pointed form, to the more extensively flaked examples that became more common during the Middle Pleistocene (Edwards, 2001; Stout, Apel, Commander, & Roberts, 2014). However, the production of these novel artifacts marked a move away from cores and nodules simply being items that were struck in order to produce flakes (Toth, 1985) to a situation where knapping events were strung together (Figure 12.1) to produce shaped, bifacial, and bilaterally organized stone tools with elongated cutting edges (Gowlett, 2006; Lycett, Schillinger, Eren, von Cramon-Taubadel, & Mesoudi, 2016; Roche, 2005; Wynn, 1995). While these artifacts were first produced in East Africa (Beyene et al., 2013; Diez-Martín et al., 2015), they were subsequently produced over large parts of Africa, Europe, and Asia to form an artifactual pattern, geographically and temporally speaking, that has famously been referred to as the "Acheulean" (Gowlett, 2011; Wynn & Tierson, 1990). A range of evidence demonstrates that handaxes were functional items used in cutting and chopping activities such as butchery and woodworking (Bello, Parfitt, & Stringer, 2009; Domínguez-Rodrigo, Serrallonga, Juan-Tresserras, Alcalá, & de Luque, 2001; Gowlett, 2006; Nowell et al., 2016; Roberts & Parfitt, 1999; Solodenko et al., 2015; Yravedra et al., 2010). If ratcheting is building

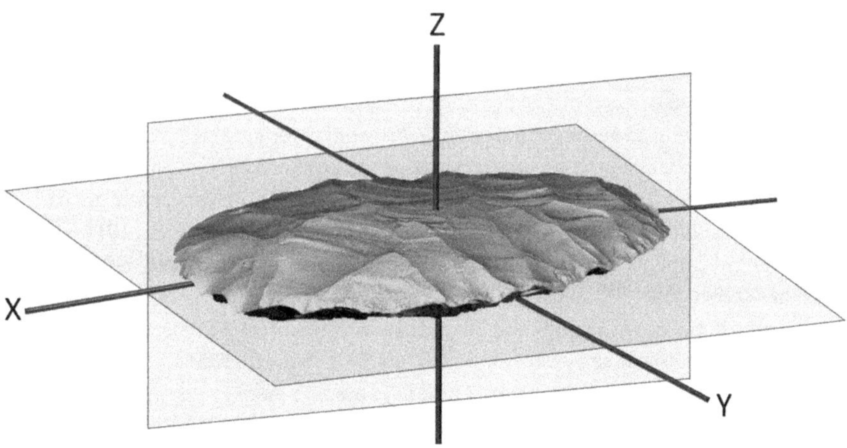

Figure 12.1. Acheulean handaxes are multivariate objects involving manipulation of variables in three planes during their production (Gowlett, 2006; Wynn, 1995, 2002). To produce a bilateral cutting edge, the nodule or flake blank is thinned by invasive flaking (along the z-axis), while maintaining width (y-axis) and overall length (x-axis). This example was knapped by the author and scanned in 3D using a structured light scanner. Image by the author; formulation of the figure inspired by Figure 11 (p. 45) in Inizian and colleagues (1999), *Technology and Terminology of Knapped Stone*, Cercle de Recherches et d'Etudes Préhistoriques.

on prior innovations, then the Acheulean seems to represent an instance of ratcheting on the Oldowan, especially given that handaxes have definite functional zones (i.e., particular performance capabilities) where they can outperform simple flakes of the type used in Oldowan industries (Key & Lycett, 2017b). However, we do not see "runaway ratcheting," such that the Acheulean might stand comparison with later hominin achievements (Shipton & Nielsen, 2015). A significant question that frequently emerges with reference to the Acheulean, therefore, is whether hominins may have been more regularly drawing on social learning skills that extend prominently beyond those possessed by the LCA (Lycett, Schillinger, et al., 2016; Mithen, 1999; Nielsen, 2012; Schillinger et al., 2015; Shipton & Nielsen, 2015).

Some have suggested that the repeated and widespread production of similar artifactual forms implies imitative, rather than solely emulative, learning by the hominins responsible for the Acheulean (Mithen, 1999). As with the Oldowan, however, a combination of stimulus/local enhancement plus results-based (i.e., emulative) copying, whereby hominins were learning from visual aspects of handaxes and/or *débitage* produced by others, may plausibly go some way to explaining the geographically and temporally widespread production of handaxes. Indeed, recent experimental work has shown that even end-product copying can produce definite evolutionary patterns, suggesting that imitation may not be an absolute prerequisite of genuine cultural-evolutionary lineages with trends that are quantifiable (Schillinger, Mesoudi, & Lycett, 2016). Trial-and-error learning is again also likely to have played a part in the learning of handaxe manufacture. However, if injury hazards were a pertinent factor in motivating social learning during the Oldowan, then such costs are likely to have increased, given the more extended reduction sequences involved in handaxe production and the fact that extended periods of practice are generally needed by modern humans to build the skills required to reliably (i.e., repeatedly) manufacture artifacts that bear comparison in three dimensions with many prehistoric examples (Edwards, 2001; Stout, 2005).

However, recent experimental work has shown that copying errors relating to both the size and, in particular, the *shape* characteristics of handaxe form would be subject to rates of copying error, which would make handaxe traditions inherently unstable if they were based solely on emulative copying (Kempe, Lycett, & Mesoudi, 2012; Lycett, Schillinger, Kempe, & Mesoudi, 2015; Schillinger, Mesoudi, & Lycett, 2014). Such work has shown that within just a few experimental "generations" of copying, handaxe form would break down, leading to a relatively rapid erosion of Acheulean-like patterns under such circumstances, even though many of these experiments involved tasks requiring skill levels far below that involved in genuine handaxe manufacture (Kempe et al., 2012; Lycett, Schillinger, et al., 2016; Schillinger et al., 2014, 2016). Imitation, in contrast to emulation, would involve copying not only handaxe forms (i.e., results) but additionally (some of) the actual *behaviors* used by other hominins in the manufacture of their handaxes. Differences in the details and techniques of manufacture, or what Patten (2005) refers to as "process controls," which improve the likelihood of favorable outcomes during knapping (see, e.g., Nonaka, Bril, & Rein, 2010) would make desirable targets for this, particularly through looking at visible payoffs while watching others. Process controls can range from simple methods of platform preparation, such as grinding or faceting that operate at the level of a single flake removal (Stout et al., 2014) or even just holding and gripping a blank in a particular way (Patten, 2005), to extended sequences of actions strategically strung together to perform an operation effectively and safely.

Important here, therefore, is evidence from controlled experiments (Schillinger et al., 2015), which have shown that imitative learning significantly reduces copying error during the production of handaxe-shaped objects compared with emulative learning. Importantly, these differences in learning conditions emerged at the artifactual level, despite the fact that manufacture of the handaxe-shaped (foam) artifacts in these experiments did not require the level of skill involved in real handaxe manufacture, and so learning factors could have reasonably been expected to be less pertinent than in the case of genuine handaxe production (Schillinger et al., 2015). Intriguingly, Morgan and colleagues (2015) have argued that language may play an important role in the learning of stone tool traditions based on experiments involving the learning of simple flake production, where linguistically mediated learning led to quantitative differences in flake products over alternative forms of social learning. However, Putt, Woods, and Franciscus (2014) found no strong effect for verbal versus nonverbal communication in an experiment involving attempts at biface production, suggesting that verbal instruction is not a necessary component in the learning of handaxe manufacture, although imitative learning and emulative learning were not examined in this study alongside teaching. Schillinger and colleagues' (2015) experimental results are important, therefore, in highlighting the statistically significant impact that learning by imitation alone can have on the fidelity of transmission in a tradition requiring shaping (i.e., where artifactual traits are the intentional result of multiple, strategically combined actions), even in the absence of any additional gestural or verbal communication. Given the inherent instability of handaxe traditions that can be expected in an emulative-only form of copying tradition (Kempe et al., 2012; Schillinger et al., 2014, 2016), imitation-based learning may have been an essential element in maintaining the kinds of pattern empirically attested in the Acheulean record (Lycett, Schillinger, et al., 2016; Lycett et al., 2015; Schillinger et al., 2015).

Imitative learning of specific manufacturing details or process controls (*sensu* Patten, 2005) would also explain why morphometric studies of handaxe form have revealed statistically patterned variation among assemblages of handaxes from different regions or sites (e.g., Lycett & Gowlett, 2008; Lycett & von Cramon-Taubadel, 2015; Wynn & Tierson, 1990). Indeed, such patterns remain difficult to explain solely in terms of reduction and/or raw material factors (Eren, Roos, Story, von Cramon-Taubadel, & Lycett, 2014; Lycett & von Cramon-Taubadel, 2015; Sharon, 2008; Shipton & Clarkson, 2015) or even functional differences (Key & Lycett, 2017a), although function and culture are by no means mutually exclusive. It is important to emphasize, however, that such patterned variation does not necessarily indicate adherence to a strict "mental template" with respect to artifactual form, but could indirectly emerge through the learning of artifactual "recipes" that happen to differ between groups, or through the unconscious copying of behaviors that only inadvertently affect artifactual form in subtle—but statistically detectable—ways (Lycett, 2010a, p. 212; Schillinger, Mesoudi, & Lycett, 2017). Again, controlled experiments using what have been referred to as "model artifacts" (somewhat similar in overall approach and philosophy to the use of "model organisms" to study small-scale processes of transmission in biology) have demonstrated the potency of such "recipes" in producing statistically identifiable cultural patterns at the assemblage level (Schillinger et al., 2017).

In sum, study of the Acheulean presents all of the same challenges to inferring social learning as the Oldowan. Nevertheless, it is minimally reasonable to infer that the full capacities of the LCA (i.e., stimulus enhancement learning and emulation) would have been called on in producing the empirical patterns attested in the Acheulean (Kempe et al., 2012). Moreover, there is growing evidence that imitation may have been a more important mechanism of social learning during the Acheulean than is indicated by the Oldowan that preceded it (Schillinger et al., 2015; Shipton & Nielsen, 2015). There is currently insufficient evidence, however, to propose that any form of social learning beyond imitation was an essential feature of Acheulean learning systems.

SOCIAL LEARNING IN THE MIDDLE PALEOLITHIC

The Middle Paleolithic (or in Africa, Middle Stone Age) is an archaeological phase, the onset of which is most frequently defined by the appearance of what have become widely known as "Levallois" reduction sequences (Avraham, 1982; Morgan & Renne, 2008; Tryon & Faith, 2013; Wurz, 2013). Based on current evidence, these new forms of core reduction appeared both in East Africa and Europe around 300 thousand years ago (Kya) (Adler et al., 2014; Moncel, Moigne, Sam, & Combier, 2011; Morgan & Renne, 2008; Tryon & Faith, 2013). Subsequently, Levallois artifacts were produced over large swathes of Africa and Eurasia in a geographic distribution that broadly mirrors that of the Acheulean industries they replaced (Schick, 1998). Levallois industries were manufactured by several species, including Neandertals (*Homo neanderthalensis*) and modern humans (*H. sapiens*), which has meant that discussions of Levallois have frequently been embroiled in debates regarding aspects of later human evolution and cognition (Hublin, 2009; Wynn & Coolidge, 2004, 2010).

Although identification of Levallois reduction sequences and their products has been a frequent subject of debate (e.g., Copeland, 1983; Perpère, 1986; van Peer, 1992), many scholars have aggregated around the idea that Levallois reduction occurs via a specific "volumetric" arrangement of core construction and reduction (Boëda, 1995; Brantingham & Kuhn, 2001; Chazan, 1997; reviewed in Lycett & Eren, 2013b). First outlined by Boëda (1994, 1995), this volumetric concept of Levallois reduction states that the core is first organized bifacially into two hierarchically related surfaces, which intersect at the core's margin, ultimately forming a plane of intersection (Figure 12.2A). "Levallois" flakes are then removed via direct percussion from the ventral surface of the core, which possesses distal and lateral convexities, broadly parallel to the plane of intersection (Figure 12.2A). The production of Levallois reduction sequences has long been considered to potentially reveal signs of planning and forethought on the part of hominins (Bordes, 1950; Chazan, 1997; Schlanger, 1996; Spurrell, 1884; van Peer, 1992; Wynn & Coolidge, 2004, 2010).

Experimental and morphometric analyses of Levallois flakes removed during such reduction sequences have indicated that they possess specific properties that would make them functionally desirable to hominins, including greater overall robusticity, balance, and capacity for retouch (Eren & Lycett, 2012). Moreover, additional analyses of experimental cores have demonstrated that Levallois-style reduction produces average flake angles (in the Levallois flakes) beneficial in providing a useful cutting edge, yet not so acute that they would be friable upon application (Eren & Lycett, 2016). In addition to producing flakes that possess a series of properties that would make them

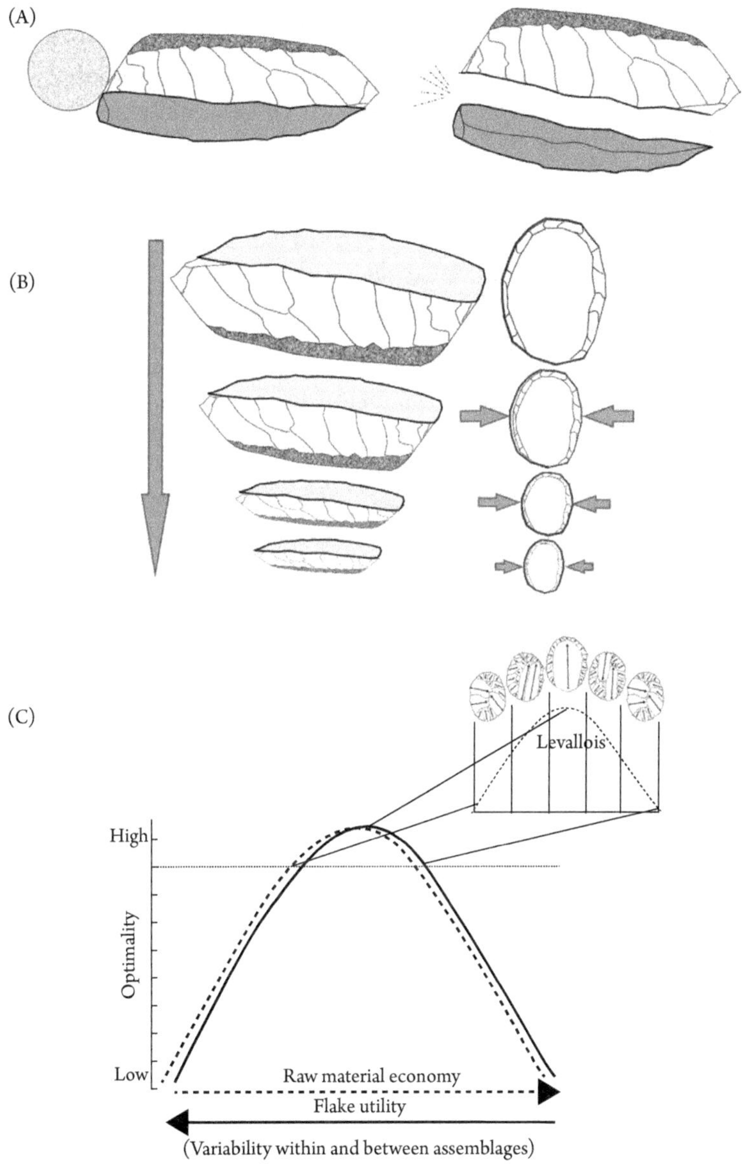

Figure 12.2. Levallois reduction. (A) Core form is established around two hierarchically organized surfaces that meet at a plane of intersection, with removal of "Levallois" flakes from steep-angled striking platforms, broadly parallel to this plane. (B) Distal and lateral convexities and core margin must be managed in three dimensions, iteratively, as core reduction proceeds and the margin migrates downward. (C) Levallois reduction provides benefits in terms of several flake properties and raw material economy (see text for discussion and related references). (A) Image by the author. (B) and (C) were previously published as Figures 1 and 6 (pp. 5 and 10) in Lycett and colleagues (2016), Levallois: Potential implications for learning and cultural transmission capacities, *Lithic Technology*, and republished with the permission of Taylor and Francis Ltd. (www.tandfonline.com).

functionally beneficial, there may also be additional economic benefits to organizing core reduction according to the Levallois concept. Brantingham and Kuhn (2001) used mathematical modeling based on geometric principles to show that core forms coinciding with Boëda's (1995) volumetric concept of Levallois minimized raw material waste while simultaneously maximizing the cumulative cutting edge procured in the Levallois flakes. Lycett and Eren (2013a) subsequently statistically analyzed experimental assemblages to show that key assumptions of these theoretical models could be upheld under conditions using real stones, and they also supported Brantingham and Kuhn's (2001) conclusion that economic considerations might have logically motivated Levallois reduction.

All of this modeling and experimental work is consistent with, and indeed supports, the long-held view derived from studies of archaeological assemblages that Levallois core reduction strategies, and the flakes removed from them, were intentional products (Chazan, 1997; Schlanger, 1996; Spurrell, 1884; van Peer, 1992) that required some degree of forethought and planning on the part of hominins (Wynn & Coolidge, 2004, 2010). While such "classic" cores by no means represent the only or even the dominant type of core form within archaeological Levallois assemblages (see, e.g., Delagnes, 1995; Meignen, Delagnes, & Bourguignon, 2009), such optimality models suggest that reduction schemes that *more closely* resemble classic conceptions of Levallois (even as an average) would provide predictable benefits over other (non-Levallois) reduction strategies (Figure 12.2). Again, we may therefore be seeing some ratcheting on earlier concepts and methods of knapping.

Interestingly, from the viewpoint of social learning, Levallois reduction is frequently considered to be a particularly challenging feat of knapping (Callahan, 1982; Eren, Bradley, & Sampson, 2011; Hayden, 1993; Pelegrin, 2005; Wynn & Coolidge, 2004). Several features of the core form must be successfully manipulated to consistently produce such artifacts (Boëda, 1995; Pelegrin, 2005; Schlanger, 1996; van Peer, 1992; Wynn & Coolidge, 2004, 2010) and thus to procure the optimal benefits of Levallois-style reduction (*sensu* Brantingham & Kuhn, 2001; Lycett & Eren, 2013b). Moreover, three-dimensional geometric morphometric analyses of archaeological Levallois cores have shown that for classic (optimal) Levallois cores to be produced, the knapper must impose a specific margin geometry on the core, which has been shown to vary less than other aspects of core form, even across different regions (Lycett & von Cramon-Taubadel, 2013). This constrained three-dimensional margin geometry would need to be imposed consistently and repeatedly if the sort of knapping scheme outlined by Boëda (1995) and others were to be executed successfully (Figure 12.2B). This suggests a key role for margin geometry—in addition to other features such as distal and lateral convexities—in the successful execution of Levallois knapping schemes.

As with Acheulean handaxes, the extended knapping sequences involved in Levallois, as well as the requirement to successfully learn specific aspects of the Levallois core form and manage them iteratively, imply that—minimally—both emulative and imitative modes of learning would be involved in replicating the sorts of sequences attested archaeologically. All the issues relating to the erosive effects of copying errors in traditions involving shaping (e.g., Lycett, Schillinger, et al., 2016; Schillinger et al., 2015) will have been as important in Levallois industries, if not more so, as they were in Acheulean industries. Indeed, there is experimental evidence that task difficulty has a positive influence on the propensity to use social information, at least in human subjects (Baron, Vandello, & Brunsman, 1996).

The question remains as to whether more complex methods of social learning were necessary for Levallois industries to emerge. Wynn and Coolidge (2010) contended that given the difficulty of Levallois knapping and its relative complexity, active instruction (i.e., teaching), albeit potentially mediated non-linguistically, may have been necessary for successful learning of Levallois reduction. Lycett and colleagues (Lycett, von Cramon-Taubadel, et al., 2016) have suggested that additional support for this contention may come from several further considerations. First, it has been suggested that the relative opaqueness of a task may influence the intensity of the social learning required (e.g., Gergely & Csibra, 2006); in more opaque tasks, extracting the requisite information through observation is more difficult than it is in more transparent tasks (e.g., Horner & Whiten, 2005). In the case of handaxe production, the tool and its functional (i.e., practically useful) properties, such as elongate form and extended cutting edge, emerge as a direct result of the bifacial knapping process targeted directly toward shaping the mass as its end goal. In Levallois reduction, however, while it similarly involves shaping, some of the most useful elements of the exercise (i.e., the Levallois flakes) are ultimately dislocated from the very features responsible for their emergence (i.e., the core mass), making it relatively more difficult to make connections between the multiple features involved (Figure 12.2). Moreover, potential benefits of features such as flake utility and—even worse—raw material economy, may be particularly difficult benefits for an observer to extract, making the subject of the exercise even more opaque, especially if these are statistical *tendencies* of Levallois reduction rather than "absolute" features (Figure 12.2C). Suggestively, recent mathematical models (Fogarty, Strimling, & Laland, 2011) have indicated that active instruction would most likely occur in conditions where any costs imposed on the "teacher" (e.g., time) could be outweighed by the benefits of passing key information to kin *and* where that specific information could not easily be gained by the novice for themselves (i.e., through unguided observational learning). In sum, Levallois reduction may well satisfy these conditions and, hence, be a plausible candidate for active instruction. However, Wynn and Coolidge (2010) and Lycett and colleagues (Lycett, von Cramon-Taubadel, et al., 2016) have stressed that at the current time, the use of teaching (of any form) in the learning of Levallois must be considered a working hypothesis and, as such, must be the subject of future testing.

ONGOING CHALLENGES: MOVING FORWARD

Changing patterns in the lithic record over the last 3.3 million years plausibly attest to changes in our evolving capacities for social learning, as do some of the limits of technological variability seen over the Oldowan, Acheulean, and Levallois stages of stone tool production on a broadly chronological basis. To say, however, that we ideally need to know far more would be a conclusion so obvious as to be banal. Given that several questions have been raised, especially with respect to possible mechanisms of social learning that may be being used at various points by hominins, a far more exciting question to ask is what we might do to learn more in respect to some of these issues.

A notable feature that runs throughout this review is the important role that experiments can play in informing our understanding of the Paleolithic record in terms of social learning. One of the things we ideally need to know more about is the effect of learning in knapping and knapping-like activities by non-human primates. More than 25 years after the initiation of the pioneering experiments with Kanzi (Toth et al., 1993), we still only have limited reference points in this respect, but the importance of such endeavors has been proven (Toth, Schick, & Semaw, 2006). Ethical and safety issues are obviously going to be a factor with respect to undertaking similar experiments with other species of great apes in the future. However, one possibility might be to extend the "model artifact" approach, which has currently been used in experiments involving human participants to study the material outcomes of social learning (e.g., Schillinger et al., 2015, 2016). In the model artifact approach, several aspects of the experiments are similar to those used in social psychology to study learning processes, but the method additionally involves directly studying the material (i.e., artifactual) outcomes in terms of quantitative attributes as much as the actual social learning processes themselves (Schillinger et al., 2016). Extension of this approach to non-human primates might provide us with important insights into the material outcomes of social learning, which could then be applied to the archaeological record. Tasks that mimic certain aspects of knapping and/or the spatial facets of Pleistocene artifacts (Wynn et al., 1996) might be especially useful in this regard.

In addition, but in direct combination with experimental results, studies of cultural evolutionary patterns (*sensu* Lycett, Schillinger, et al., 2016) must be examined. Cultural evolutionary theory provides us with a wide range of predictive hypotheses against which empirical data can be tested for goodness of fit (Lycett, 2015; Mesoudi, 2011). As Wynn and Tierson's (1990) pioneering study demonstrated some time ago, looking at time-space patterns in the Lower Paleolithic with an eye to these kinds of theoretical framework may yet have something profound to tell us. One issue that may merit further consideration is the extent of cultural variability in the Oldowan, especially in light of recent work on chimpanzee (*P. troglodytes*) material culture variability (Koops et al., 2015). One methodological issue here is that we may need better means of quantifying Oldowan variability, although again, recent developments suggest that such studies are not impossible (Reti, 2016).

Both the Lomekwian and Oldowan phases of stone tool production and the Acheulean represent large expanses of time, and a major question will be whether social learning changed *within* these broad techno-complexes as opposed to simply between them, as has been hinted at here. There are indications that Lomekwi 3 represents knapping behaviors different from those exhibited by later hominins (Harmand et al., 2015), and it will be necessary to determine whether this represents any shift in social learning patterns. This again alludes to the importance of a comparative and experimental framework for determining what can and cannot be learned in the absence of social learning by the great apes. Importantly, with respect to issues of potential change within broad technological patterns, is the evidence that handaxe-producing hominins of the later Acheulean had a greater range of knapping skills, including methods of manufacture and platform preparation techniques (Edwards, 2001; Stout et al., 2014). In other words, handaxes at one point in time may have been produced through exploitation of social learning skills that were lacking in earlier Acheulean hominins. Even in the case of Levallois, while this form of reduction may be indicative of specific cognitive and social capacities, we should not necessarily assume that this alone is a basis for assuming equivalency on these matters in all hominins that made Levallois-style reduction patterns (Wynn & Coolidge, 2004); other evidence necessarily comes into play (Wynn & Coolidge, 2010). The tantalizing possibility of teaching in the successful replication of Levallois industries may, however, be one of our best opportunities to identify the earliest evidence of this behavior within the hominin lineage. Addressing this question may therefore be especially important, given better evidence for cumulative cultural evolution following the appearance of these industries and the eventual emergence of both Neandertal and *H. sapiens* populations.

This chapter has deliberately focused on the evidence for social learning capacities over the course of human evolution. However, we must also, of course, tie inferences about social learning to the evidence for evolving cognitive capacities underlying our planning and technological capacities, evolving life history patterns, and the fossil record. Our efforts to understand social learning and its evolutionary basis and context "must therefore be multidisciplinary" (Wynn, 2002, p. 402). Recent endeavors have illustrated the growing potential to do this (e.g., Lycett, Schillinger, et al., 2016; Nowell, 2010; Shipton & Nielsen, 2015; Stout, Hecht, Khreisheh, Bradley, & Chaminade, 2015; Wynn et al., 2011), and if we can learn more about the evolving social learning capacities of our extinct relatives, our ability to tie together these various components of our evolutionary history will only improve.

ACKNOWLEDGMENTS

I am deeply indebted to Metin Eren, Alastair Key, Alex Mesoudi, Kerstin Schillinger, and Noreen von Cramon-Taubadel for numerous valuable conversations and collaborations in recent years that have directly influenced my thinking on the issues discussed in this chapter, although none of them should be held responsible for its content. I thank two anonymous reviewers for their helpful and insightful comments. I am also immensely grateful for the kind invitation to participate in this project. Tom Wynn's work has been an inspiration to many, including myself; I dedicate this chapter to him.

REFERENCES

Adler, D. S., Wilkinson, K. N., Blockley, S., Mark, D. F., Pinhasi, R., Schmidt-Magee, B. A., . . . Gasparian, B. (2014). Early Levallois technology and the Lower to Middle Paleolithic transition in the Southern Caucasus. *Science, 345*(6204), 1609–1613.

Arthur, K. W. (2010). Feminine knowledge and skill reconsidered: Women and flaked stone tools. *American Anthropologist, 112*(2), 228–243.

Avraham, R. (Ed.). (1982). *The transition from the Lower to Middle Palaeolithic and the origins of modern man: International symposium to commemorate the 50th anniversary of excavations in the Mount Carmel Caves by Garrod, D. A. E., University of Haifa, 6–14 October 1980*. Oxford, UK: British Archaeological Reports.

Baron, R. S., Vandello, J. A., & Brunsman, B. (1996). The forgotten variable in conformity research: Impact of task importance on social influence. *Journal of Personality and Social Psychology, 71*(5), 915–927.

Bello, S. M., Parfitt, S. A., & Stringer, C. B. (2009). Quantitative micromorphological analyses of cut marks produced by ancient and modern handaxes. *Journal of Archaeological Science, 36*(9), 1869–1880.

Beyene, Y., Katoh, S., WoldeGabriel, G., Hart, W. K., Uto, K., Sudo, M., . . . Asfaw, B. (2013). The characteristics and chronology of the earliest Acheulean at Konso, Ethiopia. *Proceedings of the National Academy of Sciences of the United States of America, 110*(5), 1584–1591.

Biro, D., Inoue-Nakamura, N., Tonooka, R., Yamakoshi, G., Sousa, C., & Matsuzawa, T. (2003). Cultural innovation and transmission of tool use in wild chimpanzees: Evidence from field experiments. *Animal Cognition, 6*(4), 213–223.

Boëda, E. (1994). *Le concept Levallois: Variabilité des méthodes. Monographie du CRA, 9.* Paris: Éditions CNRS.

Boëda, E. (1995). Levallois: A volumetric construction, methods, a technique. In H. L. Dibble & O. Bar-Yosef (Eds.), *The definition and interpretation of Levallois technology* (Vol. 23, pp. 41–65). Madison, WI: Prehistory Press.

Boesch, C., Head, J., & Robbins, M. M. (2009). Complex tool sets for honey extraction among chimpanzees in Loango National Park, Gabon. *Journal of Human Evolution, 56*(6), 560–569.

Bonnie, K. E., Horner, V., Whiten, A., & de Waal, F. B. M. (2007). Spread of arbitrary conventions among chimpanzees: A controlled experiment. *Proceedings of the Royal Society of London B: Biological Sciences, 274*(1608), 367–372.

Bordes, F. (1950). Principes d'une méthode d'étude des techniques de debitage et de la typologie du Paléolithique ancien et moyen. *Anthropologie, 54*, 19–34.

Boyd, R., & Richerson, P. (1985). *Culture and the evolutionary process.* Chicago: University of Chicago Press.

Boyd, R., Richerson, P. J., & Henrich, J. (2011). The cultural niche: Why social learning is essential for human adaptation. *Proceedings of the National Academy of Sciences of the United States of America, 108*(Suppl. 2), 10918–10925.

Brantingham, P. J., & Kuhn, S. L. (2001). Constraints on Levallois core technology: A mathematical model. *Journal of Archaeological Science, 28*(7), 747–761.

Byrne, R. W., Hobaiter, C., & Klailova, M. (2011). Local traditions in gorilla manual skill: Evidence for observational learning of behavioral organization. *Animal Cognition, 14*(5), 683–693.

Byrne, R. W., & Russon, A. E. (1998). Learning by imitation: A hierarchical approach. *Behavioral and Brain Sciences, 21*(5), 667–684.

Callahan, E. (1982). An interview with flintknapper Jacques Pelegrin. *Contract Abstracts, 3*(1), 62–70.

Carvalho, S., & McGrew, W. C. (2012). The origins of the Oldowan: Why chimpanzees (*Pan troglodytes*) still are good models for technological evolution in Africa. In M. Domínguez-Rodrigo (Ed.), *Stone tools and fossil bones: Debates in the archaeology of human origins* (pp. 201–221). Cambridge, UK: Cambridge University Press.

Chazan, M. (1997). Redefining Levallois. *Journal of Human Evolution, 33*(6), 719–735.

Chivers, D. P., & Smith, R. J. F. (1995). Chemical recognition of risky habitats is culturally transmitted among fathead minnows, *Pimephales promelas* (Osteichthyes, Cyprinidae). *Ethology, 99*(4), 286–296.

Claidière, N., Smith, K., Kirby, S., & Fagot, J. (2014). Cultural evolution of systematically structured behaviour in a non-human primate. *Proceedings of the Royal Society of London B: Biological Sciences, 281*(1797), 20141541.

Coolidge, F. L., & Wynn, T. (2005). Working memory, its executive functions, and the emergence of modern thinking. *Cambridge Archaeological Journal, 15*(1), 5–26.

Copeland, L. (1983). Levallois/non-Levallois determinations in the Early Levant Mousterian: Problems and questions for 1983. *Paléorient, 9*, 15–27.

Dean, L. G., Kendal, R. L., Schapiro, S. J., Thierry, B., & Laland, K. N. (2012). Identification of the social and cognitive processes underlying human cumulative culture. *Science, 335*(6072), 1114–1118.

de Heinzelin, J., Clark, J. D., White, T. D., Hart, W., Renne, P. R., WoldeGabriel, G., . . . Vrba, E. S. (1999). Environment and behavior of 2.5-million-year-old Bouri hominids. *Science, 284*(5414), 625–629.

Delagnes, A. (1995). Variability within uniformity: Three levels of variability within the Levallois system. In H. L. Dibble & O. Bar-Yosef (Eds.), *The definition and interpretation of Levallois technology* (pp. 201–211). Madison, WI: Prehistory Press.

Delagnes, A., & Roche, H. (2005). Late Pliocene hominid knapping skills: The case of Lokalalei 2C, West Turkana, Kenya. *Journal of Human Evolution, 48*(5), 435–472.

Diez-Martín, F., Yustos, P. S., Uribelarrea, D., Baquedano, E., Mark, D. F., Mabulla, A. Z. P., . . . Domínguez-Rodrigo, M. (2015). The origin of the Acheulean: The 1.7 million-year-old site of FLK West, Olduvai Gorge (Tanzania). *Scientific Reports, 5*(17839), 1–9.

Dindo, M., Stoinski, T. S., & Whiten, A. (2010). Observational learning in orangutan cultural transmission chains. *Biology Letters, 7*(2), 181–183.

Domínguez-Rodrigo, M., Serrallonga, J., Juan-Tresserras, J., Alcalá, L., & de Luque, L. (2001). Woodworking activities by early humans: A plant residue analysis on Acheulian stone tools from Peninj (Tanzania). *Journal of Human Evolution, 40*(4), 289–299.

Edwards, S. W. (2001). A modern knapper's assessment of the technical skills of the Late Acheulean biface workers at Kalambo Falls. In J. D. Clark (Ed.), *Kalambo Falls prehistoric site* (Vol. 3, pp. 605–611). Cambridge, UK: Cambridge University Press.

Ellen, R. F., Lycett, S. J., & Johns, S. E. (Eds.). (2013). *Understanding cultural transmission in anthropology: A critical synthesis.* New York: Berghahn.

Eren, M. I., Bradley, B. A., & Sampson, C. G. (2011). Middle Paleolithic skill level and the individual knapper: An experiment. *American Antiquity, 76*(2), 229–251.

Eren, M. I., & Lycett, S. J. (2012). Why Levallois? A morphometric comparison of experimental "preferential" Levallois flakes versus debitage flakes. *PLoS One, 7*(1), 1–10.

Eren, M. I., & Lycett, S. J. (2016). A statistical examination of flake edge angles produced during experimental lineal Levallois reductions and consideration of their functional implications. *Journal of Archaeological Method and Theory, 23*(1), 379–398.

Eren, M. I., Roos, C. I., Story, B. A., von Cramon-Taubadel, N., & Lycett, S. J. (2014). The role of raw material differences in stone tool shape variation: An experimental assessment. *Journal of Archaeological Science, 49*, 472–487.

Fogarty, L., Strimling, P., & Laland, K. N. (2011). The evolution of teaching. *Evolution, 65*(10), 2760–2770.

Fragaszy, D. M., Biro, D., Eshchar, Y., Humle, T., Izar, P., De Resende, B. D., & Visalberghi, E. (2013). The fourth dimension of tool use: Temporally enduring artefacts aid primates learning to use tools. *Philosophical Transactions of the Royal Society of London. Series B, Biological Sciences, 368*(1630), 1–10.

Fragaszy, D. M., & Perry, S. (Eds.). (2003). *The biology of traditions: Models and evidence.* Cambridge, UK: Cambridge University Press.

Fuhrmann, D., Ravignani, A., Marshall-Pescini, S., & Whiten, A. (2014). Synchrony and motor mimicking in chimpanzee observational learning. *Scientific Reports, 4*(5283), 1–7.

Galef, B. G., & Laland, K. N. (2005). Social learning in animals: Empirical studies and theoretical models. *Bioscience, 55*(6), 489–499.

Gergely, G., & Csibra, G. (2006). Sylvia's recipe: The role of imitation and pedagogy in the transmission of cultural knowledge. In N. J. Enfield & S. C. Levenson (Eds.), *Roots of human sociality: Culture, cognition, and human interaction* (pp. 229–255). New York: Berg.

Gowlett, J. A. J. (2006). The elements of design form in Acheulian bifaces: Modes, modalities, rules and language. In N. Goren-Inbar & G. Sharon (Eds.), *Axe age: Acheulian tool-making from quarry to discard* (pp. 203–222). London: Equinox.

Gowlett, J. A. J. (2011). The Empire of the Acheulean strikes back. In J. Sept & D. R. Pilbeam (Eds.), *Casting the net wide: Papers in honor of Glynn Isaac and his approach to human origins research* (pp. 93–114). Cambridge, UK: American School of Prehistoric Research.

Greggor, A. L., McIvor, G. E., Clayton, N. S., & Thornton, A. (2016). Contagious risk taking: Social information and context influence wild jackdaws' responses to novelty and risk. *Scientific Reports, 6*(27764), 1–7.

Hampton, O. W. (1999). *Culture of stone: Sacred and profane uses of stone among the Dani.* College Station, TX: Texas A&M Press.

Harmand, S., Lewis, J. E., Feibel, C. S., Lepre, C. J., Prat, S., Lenoble, A., . . . Roche, H. (2015). 3.3-million-year-old stone tools from Lomekwi 3, West Turkana, Kenya. *Nature, 521*(7552), 310–315.

Hayden, B. (1993). The cultural capacities of Neandertals: A review and re-evaluation. *Journal of Human Evolution, 24*(2), 113–146.

Heyes, C. M. (1994). Social learning in animals: Categories and mechanisms. *Biological Reviews, 69*(2), 207–231.

Hobaiter, C., Poisot, T., Zuberbühler, K., Hoppitt, W. J. E., & Gruber, T. (2014). Social network analysis shows direct evidence for social transmission of tool use in wild chimpanzees. *PLoS Biology, 12*(9), 1–12.

Hohmann, G., & Fruth, B. (2003). Culture in bonobos? Between-species and within-species variation in behavior. *Current Anthropology, 44*(4), 563–571.

Holmes, W. H. (1897). *Stone implements of the Potomac-Chesapeake tidewater province: Fifteenth annual report of the Bureau of Ethnology*. Washington, DC: Government Printing Office.

Hopper, L. M., Lambeth, S. P., Schapiro, S. J., & Whiten, A. (2008). Observational learning in chimpanzees and children studied through "ghost" conditions. *Proceedings of the Royal Society of London B: Biological Sciences, 275*(1636), 835–840.

Hopper, L. M., Lambeth, S. P., Schapiro, S. J., & Whiten, A. (2015). The importance of witnessed agency in chimpanzee social learning of tool use. *Behavioural Processes, 112*, 120–129.

Horner, V., & de Waal, F. B. M. (2009). Controlled studies of chimpanzee cultural transmission. *Progress in Brain Research, 178*, 3–15.

Horner, V., & Whiten, A. (2005). Causal knowledge and imitation/emulation switching in chimpanzees (*Pan troglodytes*) and children (*Homo sapiens*). *Animal Cognition, 8*(3), 164–181.

Horner, V., Whiten, A., Flynn, E. G., & de Waal, F. B. M. (2006). Faithful replication of foraging techniques along cultural transmission chains by chimpanzees and children. *Proceedings of the National Academy of Sciences of the United States of America, 103*(37), 13878–13883.

Hublin, J.-J. (2009). The origin of Neandertals. *Proceedings of the National Academy of Sciences of the United States of America, 106*(38), 16022–16027.

Humle, T., Snowdon, C. T., & Matsuzawa, T. (2009). Social influences on ant-dipping acquisition in the wild chimpanzees (*Pan troglodytes verus*) of Bossou, Guinea, West Africa. *Animal Cognition, 12*(1), 37–48.

Inizian, M.-L., Reduron-Ballinger, M., & Roche, H. (1999). *Technology and terminology of knapped stone: Followed by a multilingual vocabulary Arabic, English, French, German, Greek, Italian, Portuguese, Spanish* (Vol. 5). Paris: Cercle de Recherches et d'Etudes Préhistoriques.

Jaeggi, A. V, Dunkel, L. P., van Noordwijk, M. A., Wich, S. A., Sura, A. A. L., & van Schaik, C. P. (2010). Social learning of diet and foraging skills by wild immature Bornean orangutans: Implications for culture. *American Journal of Primatology, 72*(1), 62–71.

Kamilar, J. M., & Marshack, J. L. (2012). Does geography or ecology best explain "cultural" variation among chimpanzee communities? *Journal of Human Evolution, 62*(2), 256–260.

Kelley, J. L., Evans, J. P., Ramnarine, I. W., & Magurran, A. E. (2003). Back to school: Can antipredator behaviour in guppies be enhanced through social learning? *Animal Behaviour, 65*(4), 655–662.

Kempe, M., Lycett, S. J., & Mesoudi, A. (2012). An experimental test of the accumulated copying error model of cultural mutation for Acheulean handaxe size. *PLoS One, 7*(11), e48333.

Kempe, M., Lycett, S. J., & Mesoudi, A. (2014). From cultural traditions to cumulative culture: Parameterizing the differences between human and nonhuman culture. *Journal of Theoretical Biology, 359*, 29–36.

Kendal, R. L., Coolen, I., & Laland, K. N. (2004). The role of conformity in foraging when personal and social information conflict. *Behavioral Ecology, 15*(2), 269–277.

Key, A. J. M., & Lycett, S. J. (2017a). Influence of handaxe size and shape on cutting efficiency: A large-scale experiment and morphometric analysis. *Journal of Archaeological Method and Theory, 24*(2), 514–541.

Key, A. J. M., & Lycett, S. J. (2017b). Reassessing the production of handaxes versus flakes from a functional perspective. *Archaeological and Anthropological Sciences, 9*(5), 737–753.

Koops, K., Schöning, C., Isaji, M., & Hashimoto, C. (2015). Cultural differences in ant-dipping tool length between neighbouring chimpanzee communities in Kalinzu, Uganda. *Scientific Reports, 5*(12456), 1–8.

Kroeber, T. (1961). *Ishi in two worlds: A biography of the last wild Indian in North America.* Berkeley, CA: University of California Press.

Kurzban, R., & Barrett, H. C. (2012). Origins of cumulative culture. *Science, 335*(6072), 1056–1057.

Leakey, L. S. B. (1936). *Stone Age Africa: An outline of prehistory in Africa.* Oxford, UK: Oxford University Press.

Lepre, C. J., Roche, H., Kent, D. V, Harmand, S., Quinn, R. L., Brugal, J.-P., . . . Feibel, C. S. (2011). An earlier origin for the Acheulian. *Nature, 477*(7362), 82–85.

Luef, E. M., & Pika, S. (2013). Gorilla mothers also matter! New insights on social transmission in gorillas (*Gorilla gorilla gorilla*) in captivity. *PLoS One, 8*(11), e79600.

Luncz, L. V, & Boesch, C. (2014). Tradition over trend: Neighboring chimpanzee communities maintain differences in cultural behavior despite frequent immigration of adult females. *American Journal of Primatology, 76*(7), 649–657.

Luncz, L. V, & Boesch, C. (2015). The extent of cultural variation between adjacent chimpanzee (*Pan troglodytes verus*) communities: A microecological approach. *American Journal of Physical Anthropology, 156*(1), 67–75.

Luncz, L. V, Mundry, R., & Boesch, C. (2012). Evidence for cultural differences between neighbouring chimpanzee communities. *Current Biology, 22*(10), 922–926.

Lycett, S. J. (2010a). Cultural transmission, genetic models, and lithic variability: Integrative analytical approaches. In S. J. Lycett & P. R. Chauhan (Eds.), *New perspectives on old stones: Analytical approaches to Palaeolithic technologies* (pp. 207–234). New York: Springer.

Lycett, S. J. (2010b). The importance of history in definitions of culture: Implications from phylogenetic approaches to the study of social learning in chimpanzees. *Learning & Behavior, 38*(3), 252–264.

Lycett, S. J. (2013). Cultural transmission theory and fossil hominin behaviour: A discussion of epistemological and methodological strengths. In R. F. Ellen, S. J. Lycett, & S. E. Johns (Eds.), *Understanding cultural transmission in anthropology: A critical synthesis* (pp. 102–130). New York: Berghahn.

Lycett, S. J. (2015). Cultural evolutionary approaches to artifact variation over time and space: Basis, progress, and prospects. *Journal of Archaeological Science, 56*, 21–31.

Lycett, S. J., Collard, M., & McGrew, W. C. (2010). Are behavioral differences among wild chimpanzee communities genetic or cultural? An assessment using tool-use data and phylogenetic methods. *American Journal of Physical Anthropology, 142*(3), 461–467.

Lycett, S. J., & Eren, M. I. (2013a). Levallois economics: An examination of "waste" production in experimentally produced Levallois reduction sequences. *Journal of Archaeological Science, 40*(5), 2384–2392.

Lycett, S. J., & Eren, M. I. (2013b). Levallois lessons: The challenge of integrating mathematical models, experiments and the archaeological record. *World Archaeology, 45*(4), 519–538.

Lycett, S. J., & Gowlett, J. A. J. (2008). On questions surrounding the Acheulean "tradition." *World Archaeology, 40*(3), 295–315.

Lycett, S. J., Schillinger, K., Eren, M. I., von Cramon-Taubadel, N., & Mesoudi, A. (2016). Factors affecting Acheulean handaxe variation: Experimental insights, microevolutionary processes, and macroevolutionary outcomes. *Quaternary International, 411*, 386–401.

Lycett, S. J., Schillinger, K., Kempe, M., & Mesoudi, A. (2015). Learning in the Acheulean: Experimental insights using handaxe form as a "model organism." In A. Mesoudi & K. Aoki (Eds.), *Learning strategies and cultural evolution during the Palaeolithic* (pp. 155–166). Tokyo: Springer.

Lycett, S. J., & von Cramon-Taubadel, N. (2013). A 3D morphometric analysis of surface geometry in Levallois cores: Patterns of stability and variability across regions and their implications. *Journal of Archaeological Science, 40*(3), 1508–1517.

Lycett, S. J., & von Cramon-Taubadel, N. (2015). Toward a "quantitative genetic" approach to lithic variation. *Journal of Archaeological Method and Theory, 22*(2), 646–675.

Lycett, S. J., von Cramon-Taubadel, N., & Eren, M. I. (2016). Levallois: Potential implications for learning and cultural transmission capacities. *Lithic Technology, 41*(1), 19–38.

Martin, G. (2000). Stasis in complex artifacts. In J. Ziman (Ed.), *Technological innovation as an evolutionary process* (pp. 90–100). Cambridge, UK: Cambridge University Press.

Matthews, L. J., Paukner, A., & Suomi, S. J. (2010). Can traditions emerge from the interaction of stimulus enhancement and reinforcement learning? An experimental model. *American Anthropologist, 112*(2), 257–269.

McGrew, W. C. (1992). *Chimpanzee material culture: Implications for human evolution*. Cambridge, UK: Cambridge University Press.

McGrew, W. C. (2004). *The cultured chimpanzee: Reflections on cultural primatology*. Cambridge, UK: Cambridge University Press.

McGrew, W. C. (2010a). In search of the last common ancestor: New findings on wild chimpanzees. *Philosophical Transactions of the Royal Society of London. Series B, Biological Sciences, 365*(1556), 3267–3276.

McGrew, W. C. (2010b). New theatres of conflict in the animal culture wars: Recent findings from chimpanzees. In E. V Lonsdorf, S. R. Ross, & T. Matsuzawa (Eds.), *The mind of the chimpanzee: Ecological and experimental perspectives* (pp. 168–177). Chicago, IL: Chicago University Press.

McGrew, W. C. (2015). The cultured chimpanzee: Nonsense or breakthrough? *Human Ethology Bulletin, 30*(1), 41–52.

Meignen, L., Delagnes, A., & Bourguignon, L. (2009). Patterns of lithic material procurements and transformation during the Middle Palaeolithic in western Europe. In B. Adams & B. S. Blades (Eds.), *Lithic materials and Palaeolithic societies* (pp. 15–24). Chichester, UK: Blackwell.

Mesoudi, A. (2011). *Cultural evolution: How Darwinian theory can explain human culture and synthesize the social sciences.* Chicago, IL: Chicago University Press.

Mithen, S. J. (1999). Imitation and cultural change: A view from the Stone Age, with specific reference to the manufacture of handaxes. In H. O. Box & K. R. Gibson (Eds.), *Mammalian social learning: Comparative and ecological perspectives* (pp. 389–399). Cambridge, UK: Cambridge University Press.

Moncel, M.-H., Moigne, A.-M., Sam, Y., & Combier, J. (2011). The emergence of Neanderthal technical behavior: New evidence from Orgnac 3 (Level 1, MIS 8), southeastern France. *Current Anthropology, 52*(1), 37–75.

Morgan, L. E., & Renne, P. R. (2008). Diachronous dawn of Africa's Middle Stone Age: New 40AR/39AR ages from the Ethiopian Rift. *Geology, 36*(12), 967–970.

Morgan, T. J. H., Uomini, N. T., Rendell, L. E., Chouinard-Thuly, L., Street, S. E., Lewis, H. M., . . . Laland, K. N. (2015). Experimental evidence for the co-evolution of hominin toolmaking teaching and language. *Nature Communications, 6*(6029), 1–18.

Nagell, K., Olguin, R. S., & Tomasello, M. (1993). Processes of social learning in the tool use of chimpanzees (*Pan troglodytes*) and human children (*Homo sapiens*). *Journal of Comparative Psychology, 107*(2), 174–186.

Nielsen, M. (2012). Imitation, pretend play, and childhood: Essential elements in the evolution of human culture? *Journal of Comparative Psychology, 126*(2), 170–181.

Nonaka, T., Bril, B., & Rein, R. (2010). How do stone knappers predict and control the outcome of flaking? Implications for understanding early stone tool technology. *Journal of Human Evolution, 59*(2), 155–167.

Nowell, A. (2010). Working memory and the speed of life. *Current Anthropology, 51*(S1), S121–S133.

Nowell, A., Walker, C., Cordova, C. E., Ames, C. J. H., Pokines, J. T., Stueber, D., . . . Al-Souliman, A. S. A. (2016). Middle Pleistocene subsistence in the Azraq Oasis, Jordan: Protein residue and other proxies. *Journal of Archaeological Science, 73*, 36–44.

Nunn, C. L. (2011). *The comparative approach in evolutionary anthropology and biology.* Chicago, IL: Chicago University Press.

O'Malley, R. C., Wallauer, W., Murray, C. M., & Goodall, J. van L. (2012). The appearance and spread of ant fishing among the Kasekela chimpanzees of Gombe: A possible case of intercommunity cultural transmission. *Current Anthropology, 53*(5), 650–663.

Patten, B. (2005). *Peoples of the flute: A study in anthropolithic forensics.* Denver, CO: Stone Dagger Publications.

Pelegrin, J. (2005). Remarks about archaeological techniques and methods of knapping. In V. Roux & B. Bril (Eds.), *Stone knapping: The necessary conditions for a uniquely hominin behaviour* (pp. 23–33). Cambridge, UK: McDonald Institute for Archaeological Research.

Perpère, M. (1986). Apport de la typométrie à la définition des éclats Levallois: L'exemple d'Ault. *Bulletin de La Société Préhistorique Française, 83*, 115–118.

Pickering, T. R., & Domínguez-Rodrigo, M. (2006). The acquisition and use of large mammal carcasses by Oldowan hominins in eastern and southern Africa: A selected review and assessment. In N. P. Toth & K. D. Schick (Eds.), *The Oldowan: Case studies into the Earliest Stone Age* (pp. 113–128). Gosport, IN: Stone Age Institute Press.

Pope, S. T. (1918). *Yahi archery*. Berkeley, CA: University of California Press.

Price, E. E., Lambeth, S. P., Schapiro, S. J., & Whiten, A. (2009). A potent effect of observational learning on chimpanzee tool construction. *Proceedings of the Royal Society of London B: Biological Sciences, 276*(1671), 3377–3383.

Putt, S. S., Woods, A. D., & Franciscus, R. G. (2014). The role of verbal interaction during experimental bifacial stone tool manufacture. *Lithic Technology, 39*(2), 96–112.

Reti, J. S. (2016). Quantifying Oldowan stone tool production at Olduvai Gorge, Tanzania. *PLoS One, 11*(1), e0147352.

Roberts, M. B., & Parfitt, S. A. (Eds.). (1999). *Boxgrove: A Middle Pleistocene hominid site at Eartham Quarry, Boxgrove, West Sussex*. London: English Heritage.

Roche, H. (2005). From simple flaking to shaping: Stone knapping evolution among early hominins. In V. Roux & B. Bril (Eds.), *Stone knapping: The necessary conditions for a uniquely hominin behaviour* (pp. 35–48). Cambridge, UK: McDonald Institute for Archaeological Research.

Sanz, C. M., Call, J., & Morgan, D. B. (2009). Design complexity in termite-fishing tools of chimpanzees (*Pan troglodytes*). *Biology Letters: Animal Behaviour, 5*, 293–296.

Sanz, C. M., Schöning, C., & Morgan, D. B. (2010). Chimpanzees prey on army ants with specialized tool set. *American Journal of Primatology, 72*(1), 17–24.

Schick, K. D. (1998). A comparative perspective on Paleolithic cultural patterns. In T. Akazawa, K. Aoki, & O. Bar-Yosef (Eds.), *Neanderthals and modern humans in western Asia* (pp. 449–460). New York: Plenum Press.

Schick, K. D., & Toth, N. P. (2006). An overview of the Oldowan industrial complex: The sites and the nature of their evidence. In N. P. Toth & K. D. Schick (Eds.), *The Oldowan: Case studies into the earliest Stone Age* (pp. 3–42). Bloomington, IN: Stone Age Institute.

Schick, K. D., Toth, N. P., Garufi, G., Savage-Rumbaugh, E. S., Rumbaugh, D. M., & Sevcik, R. A. (1999). Continuing investigations into the stone tool-making and tool-using capabilities of a bonobo (*Pan paniscus*). *Journal of Archaeological Science, 26*(7), 821–832.

Schillinger, K., Mesoudi, A., & Lycett, S. J. (2014). Copying error and the cultural evolution of "additive" vs. "reductive" material traditions: An experimental assessment. *American Antiquity, 79*(1), 128–143.

Schillinger, K., Mesoudi, A., & Lycett, S. J. (2015). The impact of imitative versus emulative learning mechanisms on artifactual variation: Implications for the evolution of material culture. *Evolution and Human Behavior, 36*(6), 446–455.

Schillinger, K., Mesoudi, A., & Lycett, S. J. (2016). Copying error, evolution, and phylogenetic signal in artifactual traditions: An experimental approach using "model artifacts." *Journal of Archaeological Science, 70*, 23–34.

Schillinger, K., Mesoudi, A., & Lycett, S. J. (2017). Differences in manufacturing traditions and assemblage-level patterns: the origins of cultural differences in archaeological data. *Journal of Archaeological Method and Theory, 24*(2), 640–658.

Schlanger, N. (1996). Understanding Levallois: Lithic technology and cognitive archaeology. *Cambridge Archaeological Journal, 6*(2), 231–254.

Semaw, S. (2006). The oldest stone artifacts from Gona (2.6–2.5 Ma), Afar, Ethiopia: Implications for understanding the earliest stages of stone knapping. In N. P. Toth & K. D. Schick (Eds.), *The Oldowan: Case studies into the Earliest Stone Age* (pp. 43–75). Gosport, IN: Stone Age Institute.

Semaw, S., Renne, P. R., Harris, J. W. K., Feibel, C. S., Bernor, R. L., Fesseha, N., & Mowbray, K. (1997). 2.5-million-year-old stone tools from Gona, Ethiopia. *Nature, 385*(6614), 333–336.

Semaw, S., Rogers, M. J., Quade, J., Renne, P. R., Butler, R. F., Domínguez-Rodrigo, M., . . . Simpson, S. W. (2003). 2.6-million-year-old stone tools and associated bones from OGS-6 and OGS-7, Gona, Afar, Ethiopia. *Journal of Human Evolution, 45*(2), 169–177.

Sharon, G. (2008). The impact of raw material on Acheulian large flake production. *Journal of Archaeological Science, 35*(5), 1329–1344.

Shipton, C., & Clarkson, C. (2015). Handaxe reduction and its influence on shape: An experimental test and archaeological case study. *Journal of Archaeological Science: Reports, 3*, 408–419.

Shipton, C., & Nielsen, M. (2015). Before cumulative culture. *Human Nature, 26*(3), 331–345.

Solodenko, N., Zupancich, A., Cesaro, S. N., Marder, O., Lemorini, C., & Barkai, R. (2015). Fat residue and use-wear found on Acheulian biface and scraper associated with butchered elephant remains at the site of Revadim, Israel. *PLoS One, 10*(3), e0118572.

Spurrell, F. C. J. (1884). On some Palaeolithic knapping tools and modes of using them. *Journal of the Anthropological Institute of Great Britain and Ireland, 13*, 109–118.

Stout, D. (2005). The social and cultural context of stone knapping skill acquisition. In V. Roux & B. Bril (Eds.), *Stone knapping: The necessary conditions for a uniquely hominin behaviour* (pp. 331–340). Cambridge: McDonald Institute for Archaeological Research.

Stout, D., Apel, J., Commander, J., & Roberts, M. B. (2014). Late Acheulean technology and cognition at Boxgrove, UK. *Journal of Archaeological Science, 41*, 576–590.

Stout, D., Hecht, E. E., Khreisheh, N., Bradley, B. A., & Chaminade, T. (2015). Cognitive demands of Lower Paleolithic toolmaking. *PLoS One, 10*(4), 1–18.

Stout, D., & Semaw, S. (2006). Knapping skill of the earliest stone toolmakers: Insights from the study of modern human novices. In N. P. Toth & K. D. Schick (Eds.), *The Oldowan: Case studies into the earliest Stone Age* (pp. 307–320). Gosport, IN: Stone Age Institute Press.

Stout, D., Semaw, S., Rogers, M. J., & Cauche, D. (2010). Technological variation in the earliest Oldowan from Gona, Afar, Ethiopia. *Journal of Human Evolution, 58*(6), 474–491.

Thornton, A., & Raihani, N. J. (2010). Identifying teaching in wild animals. *Learning & Behavior, 38*(3), 297–309.

Thorpe, W. H. (1963). *Learning and instinct in animals.* London: Methuen.

Tomasello, M. (1999). *The cultural origins of human cognition.* Cambridge, MA: Harvard University Press.

Tomasello, M. (2009). Postscript: Chimpanzee culture, 2009. In K. N. Laland & B. G. Galef (Eds.), *The question of animal culture* (pp. 213–221). Cambridge, MA: Harvard University Press.

Tomasello, M., Savage-Rumbaugh, E. S., & Kruger, A. C. (1993). Imitative learning of actions on objects by children, chimpanzees, and enculturated chimpanzees. *Child Development, 64*(6), 1688–1705.

Toth, N. P. (1985). The Oldowan reassessed: A close look at early stone artifacts. *Journal of Archaeological Science, 12*(2), 101–120.

Toth, N. P., Schick, K. D., Savage-Rumbaugh, E. S., Sevcik, R. A., & Rumbaugh, D. M. (1993). *Pan* the tool-maker: Investigations into the stone tool-making and tool-using capabilities of a bonobo (*Pan paniscus*). *Journal of Archaeological Science, 20*(1), 81–91.

Toth, N. P., Schick, K. D., & Semaw, S. (2006). A comparative study of the stone tool-making skills of *Pan, Australopithecus,* and *Homo sapiens.* In N. P. Toth & K. D. Schick (Eds.), *The*

Oldowan: Case studies into the Earliest Stone Age (pp. 155–222). Gosport, IN: Stone Age Institute Press.

Tryon, C. A., & Faith, J. T. (2013). Variability in the Middle Stone Age of Eastern Africa. *Current Anthropology, 54*(S8), S234–S254.

van de Waal, E., Borgeaud, C., & Whiten, A. (2013). Potent social learning and conformity shape a wild primate's foraging decisions. *Science, 340*(6131), 483–485.

van Peer, P. (1992). *The Levallois reduction strategy (monographs in world archaeology no. 13)*. Madison, WI: Prehistory Press.

Whiten, A. (2010). A coming of age for cultural Panthropology. In E. V. Lonsdorf, S. R. Ross, & T. Matsuzawa (Eds.), *The mind of the chimpanzee: Ecological and experimental perspectives* (pp. 87–100). Chicago, IL: Chicago University Press.

Whiten, A. (2011). The scope of culture in chimpanzees, humans and ancestral apes. *Philosophical Transactions of the Royal Society of London. Series B, Biological Sciences, 366*(1567), 997–1007.

Whiten, A., Caldwell, C. A., & Mesoudi, A. (2016). Cultural diffusion in humans and other animals. *Current Opinion in Psychology, 8*, 15–21.

Whiten, A., Goodall, J. van L., McGrew, W. C., Nishida, T., Reynolds, V., Sugiyama, Y., . . . Boesch, C. (1999). Cultures in chimpanzees. *Nature, 399*(6737), 682–685.

Whiten, A., Goodall, J. van L., McGrew, W. C., Nishida, T., Reynolds, V., Sugiyama, Y., . . . Boesch, C. (2001). Charting cultural variation in chimpanzees. *Behaviour, 138*, 1481–1516.

Whiten, A., Horner, V., & De Waal, F. B. M. (2005). Conformity to cultural norms of tool use in chimpanzees. *Nature, 437*(7059), 737–740.

Whiten, A., Horner, V., Litchfield, C. A., & Marshall-Pescini, S. (2004). How do apes ape? *Animal Learning & Behavior, 32*(1), 36–52.

Whiten, A., McGuigan, N., Marshall-Pescini, S., & Hopper, L. M. (2009). Emulation, imitation, over-imitation and the scope of culture for child and chimpanzee. *Philosophical Transactions of the Royal Society of London. Series B, Biological Sciences, 364*(1528), 2417–2428.

Whiten, A., Schick, K. D., & Toth, N. P. (2009). The evolution and cultural transmission of percussive technology: Integrating evidence from palaeoanthropology and primatology. *Journal of Human Evolution, 57*(4), 420–435.

Whiten, A., Spiteri, A., Horner, V., Bonnie, K. E., Lambeth, S. P., Schapiro, S. J., & De Waal, F. B. M. (2007). Transmission of multiple traditions within and between chimpanzee groups. *Current Biology, 17*(12), 1038–1043.

Whiten, A., & van Schaik, C. P. (2007). The evolution of animal "cultures" and social intelligence. *Philosophical Transactions of the Royal Society of London. Series B, Biological Sciences, 362*(1480), 603–620.

Whittaker, J. C. (1994). *Flintknapping: Making and understanding stone tools*. Austin, TX: University of Texas Press.

Wurz, S. (2013). Technological trends in the Middle Stone Age of South Africa between MIS 7 and MIS 3. *Current Anthropology, 54*(S8), S305–S319.

Wynn, T. (1979). The intelligence of later Acheulean hominids. *Man, 14*, 371–391.

Wynn, T. (1981). The intelligence of Oldowan hominids. *Journal of Human Evolution, 10*(7), 529–541.

Wynn, T. (1991). Tools, grammar and the archaeology of cognition. *Cambridge Archaeological Journal, 1*(2), 191–206.

Wynn, T. (1995). Handaxe enigmas. *World Archaeology, 27*(1), 10–24.

Wynn, T. (1996). The evolution of tools and symbolic behaviour. In A. Lock & C. R. Peters (Eds.), *Handbook of human symbolic evolution* (pp. 263–287). Oxford, UK: Clarendon Press.

Wynn, T. (2002). Archaeology and cognitive evolution. *Behavioral and Brain Sciences*, *25*(3), 389–402.

Wynn, T. (2004). Evolutionary developments in the cognition of symmetry. In D. K. Washburn (Ed.), *Embedded symmetries, natural and cultural* (pp. 27–46). Albuquerque, NM: University of New Mexico Press.

Wynn, T., & Coolidge, F. L. (2004). The expert Neandertal mind. *Journal of Human Evolution*, *46*(4), 467–487.

Wynn, T., & Coolidge, F. L. (2010). How Levallois reduction is similar to, and not similar to, playing chess. In A. Nowel & I. Davidson (Eds.), *Stone tools and the evolution of human cognition* (pp. 83–104). Boulder, CO: University of Colorado Press.

Wynn, T., & Coolidge, F. L. (2012). *How to think like a Neandertal.* Oxford, UK: Oxford University Press.

Wynn, T., Hernandez-Aguilar, R. A., Marchant, L. F., & McGrew, W. C. (2011). "An ape's view of the Oldowan" revisited. *Evolutionary Anthropology*, *20*(5), 181–197.

Wynn, T., & McGrew, W. C. (1989). An ape's view of the Oldowan. *Man*, *24*(3), 383–398.

Wynn, T., & Tierson, F. D. (1990). Regional comparison of the shapes of later Acheulean handaxes. *American Anthropologist*, *92*(1), 73–84.

Wynn, T., Tierson, F. D., & Palmer, C. T. (1996). Evolution of sex differences in spatial cognition. *American Journal of Physical Anthropology*, *101*(S23), 11–42.

Yravedra, J., Domínguez-Rodrigo, M., Santonja, M., Pérez-González, A., Panera, J., Rubio-Jara, S., & Baquedano, E. (2010). Cut marks on the Middle Pleistocene elephant carcass of Áridos 2 (Madrid, Spain). *Journal of Archaeological Science*, *37*(10), 2469–2476.

Zentall, T. R. (2012). Perspectives on observational learning in animals. *Journal of Comparative Psychology*, *126*(2), 114–128.

13

THE HANDAXE AESTHETIC

Thomas Wynn and Tony Berlant

INTRODUCTION

It is uncontroversial to attribute modern aesthetic sensibilities to the artists of the African Later Stone Age (LSA) and European Upper Paleolithic (UP). Rock art and mobiliary art from both have always struck archaeologists and art historians alike as familiar, even if there is considerable disagreement about specific interpretations. From an evolutionary perspective, it is unlikely that the modern human aesthetic sensibility emerged fully formed without any prior development. Yet neither LSA nor UP art have clear antecedents in the form of less technically adept and earlier examples. One can argue around this evidential lacuna by positing antecedents that left little or no archaeological record, such as body painting (Barham, 2002; Henshilwood & Dubreuil, 2011). However, even if persuasive, these conclusions do not include descriptions of the art itself, and thus are of little use in a discussion of the developments in aesthetic sense per se. The alternative is to turn to evidence that is available in abundance—stone tools—and try to extract evidence for aesthetic development. Most stone tools were straightforward mechanical devices that hominins made and used to perform work. But some appear to have been "overdetermined"; hominins invested more time and effort in their manufacture than was necessary for their mechanical function. It is this "added value" that has the potential to inform us about aesthetic sensibilities.

This chapter addresses the aesthetic sensibility evident in Acheulean handaxes. The interpretive perspective is that of neuroaesthetics, a comparatively recent specialty in cognitive science that investigates the neurocognitive basis of aesthetic and artistic experience. From this stance an aesthetic experience is "one that allows the beholder to 'perceive-feel-see' an artwork" (Cinzia & Vittorio, 2009). However, aesthetics and art are not co-extensive. Visual art, especially, relies heavily on cultural, social, and historical context. Visual aesthetics, by contrast, encompasses the emotional and evaluative response to objects and scenes (Chatterjee, 2014a; Chatterjee & Vartanian, 2014; Vartanian & Skov, 2014) and thus includes experiences that many would not consider art (e.g., viewing a sunset or the Golden Gate Bridge).

Many archaeologists have suggested that Acheulean handaxes were aesthetic objects (Gowlett, 2006; Hodgson, 2011; Le Tensorer, 2006, 2009; Machin, 2009) and that the hominins who made them overdetermined their form in order to exploit this aesthetic effect. Yet, to our knowledge, only Hodgson (2011, 2015) has explored how this handaxe aesthetic may have differed from the aesthetic experience of modern humans. This is an important evolutionary question, but to address it one needs explicit analytical concepts beyond just a general invocation of "aesthetic." Neuroaesthetics fills this role admirably by parsing aesthetic experience into constituent components, which can in turn be applied to the archaeological record.

THE AESTHETIC MIND

"The psychology of aesthetics . . . aims to identify and describe the psychological mechanisms that allow humans to experience and appreciate a broad variety of objects and phenomena, including utensils, commodities, designs, other people, or nature, in aesthetic terms (beautiful, attractive, ugly, sublime, picturesque, and so on)" (Leder & Nadal, 2014). From the perspective of neuroaesthetics, human aesthetic experience is a complex behavioral phenomenon that draws on multiple neural networks and resources. There is no dedicated neural circuitry for aesthetic experience in the sense of neuron groups that evolved only for this purpose. Instead, multiple neural networks operate together, networks that in each case evolved initially to solve a particular evolutionary problem and which continue to perform these adaptive functions. Aesthetic experience is thus an emergent property of these neural resources working together (Chatterjee & Vartanian, 2014). Neuroaestheticians differ somewhat in how they parse aesthetic cognition into major components, but these differences reflect personal academic history and research emphasis, not significant differences in how the basic phenomenon is understood. Here we will follow Chatterjee (e.g., 2014b), who favors a model with three major components: (1) *sensorimotor*, which includes primary perceptual processing and more "downstream" integration; (2) *emotion-valuation*, which includes pleasure/reward networks and also appraisal (like–dislike); and (3) *meaning-knowledge*, which includes all of the individual history and cultural knowledge that an individual brings to aesthetic experience. Even though the model treats these components as distinct, each is still a distributed neural network in its own right, with engagement of multiple brain regions, and thus there is a degree of overlap that includes shared resources. For example, primary sensory regions appear to play an active role in valuation (Satpute et al., 2015). Indeed, it appears that shared resources are the rule and not the exception in neuronal functioning, and that the relationship between discrete groups of neurons is more important than individual neuron functioning (Anderson, 2010).

Sensorimotor

Primary sensorimotor processing plays an important "bottom-up" role in aesthetic experience (Redies, 2015). The term *primary* refers to the cortical cell groups that receive input from peripheral sense organs; "bottom up" means that these neural processes are pre-attentive, and inaccessible to conscious control or even awareness (which is why one cannot see past the visual effects of an optical illusion). Here we emphasize primary visual processing. It is the best understood and described sensory processing system, and also the basis for most developed theories of aesthetic experience.

Visual information is processed in the primary visual cortex of the occipital lobes where cell groups are sensitive to particular line orientations, colors, and other basic visual qualities. More cell groups are sensitive to potentially salient features of a scene than to potentially less salient features, thus biasing initial visual processing. For example, there are more resources devoted to detecting horizontal and vertical lines that to oblique lines. Not only does the brain devote more resources to these salient orientations, it also attaches positive affect. These orientations are more pleasing to the eye. The visual cortex then assembles these lower level inputs into higher-level shapes and scenes and again biases certain percepts over others. People tend to prefer complex forms over simple forms, symmetrical forms over asymmetrical forms, and forms whose proportions approximate the golden ratio (1.6:1). Arguably these perceptual biases evolved long ago because they provided important information about the content of visual scenes (e.g., symmetry probably indicates a life form), but even in the modern world they continue to play a role in perception and aesthetics (Chatterjee & Vartanian, 2014; Leder & Nadal, 2014; Palmer, Schloss, & Sammartino, 2013; Ramachandran & Hirstein, 1999).

The brain then passes visual information "forward" for further processing. There are two major pathways, a dorsal pathway and a ventral pathway. The dorsal pathway consists mostly of neural resources in the parietal lobes that process spatial information such as left–right, up–down, inside–outside, sequential connectivity, and so on. The ventral pathway includes regions in the temporal lobes, especially the fusiform gyrus, and processes shape information such as faces and, in recent humans, the graphical elements of writing and numerals. And here again the brain biases certain visual patterns. Indeed, certain visual patterns excite the cells in the ventral pathway, yielding what Leder and colleagues (Leder, Belke, Oeberst, & Augustin, 2004) refer to as "implicit memory effects" (p. 495). These effects are salient, incorporate learned information, elicit pleasure, but are pre-attentive.

Four of these implicit effects are especially important in modern aesthetic experience: familiarity, prototypicality, peak shift, and framing. Repeated exposure to a form or scene elicits familiarity, and familiarity is affectively positive. The brain enjoys the familiar more than the novel, at least at a perceptual level. Similarly, prototypical examples elicit a more positive affective response than eccentric examples. The closer an example comes to the ideal form of a class of objects, the higher the positive affect. There is disagreement in cognitive science about just how the brain constructs such categories, but there is no question that it does and that such categories are components of perceptual processing (Carey, 2009). Peak shift is exaggeration of affectively strong effects (Ramachandran & Hirstein, 1999). Political cartoonists exploit this effect heavily when they exaggerate selected features of the face of a public figure. Horror and slasher films also rely on peak shift to elicit shock and fear, so peak shift is not only for pleasure and amusement. A fourth implicit effect was not emphasized by Ramachandran (Ramachandran & Hirstein, 1999) or Leder et al. (2004) but will be important to our analysis. This is *framing* (Palmer et al., 2013). The central axis of many patterns is especially salient and can serve to focus visual attention. Thus, rectangular frames, which focus attention to the central part of the visual field, are so common in Western art that they themselves rarely even engage attention. But they play an important role in aesthetics nonetheless.

Emotion-Valuation

Aesthetic experience includes evaluation. We like something, or we dislike it, or it moves us in neither way. It is this evaluative piece that distinguishes aesthetic experience from other sensory experience, and it has a rather complex cognitive profile that incorporates both bottom-up and top-down resources in the brain. "Aesthetic processing, at its core, can thus be equated with *object-appraisal* processes, resulting in emotions that sit along the spectrum from transcendence to repulsion" (Brown, Gao, Tisdelle, Eickhoff, & Liotti, 2011). Although neuroimaging indicates that resources in the orbitofrontal cortex and the anterior cingulate cortex are intimately involved in appraisal (and also, provocatively, in social cognition), the brain structure that is key to appraisal is the anterior insula, a structure located deep within the lateral sulcus. When Brown and colleagues (2011) compared neural activation for four different kinds of aesthetic judgments—vision, hearing, gustation, and olfaction—the anterior insula was one of the few regions strongly activated in each of them. This brain region has long been recognized as a locus for gustatory processing and thus appears to have been exapted for other kinds of evaluation. One's metaphoric good taste relies on one's actual sense of taste! Appraisal judgments elicit emotions. Indeed, they activate the same pleasure/reward network that is activated in addictive behaviors, including the ventral striatum and the nucleus accumbens. Negative judgments activate negative reward circuits. In general, these resources power desire or avoidance; we want what we like. But Chatterjee (2014b) makes the interesting argument that aesthetic appreciation in the modern world requires liking without wanting, and disliking without avoiding. These divorcings almost certainly result from personal history and learning. Thus top-down resources of attentive cognition play a significant role in what is otherwise a bottom-up process.

The orbitofrontal cortex is the structure most often identified with learned categories of pleasure–avoidance. Through experience one associates categories of percepts with pleasure or avoidance, and thus, as with familiarity, personal history plays an important role in aesthetic evaluation. One likes what one has previously liked.

Meaning-Knowledge

The final component of aesthetic cognition is the meaning-knowledge component. This component includes all of the top-down knowledge resources an observer (or artisan) brings to aesthetic experience. Much of this knowledge consists of memories, of which there are several varieties, each of which has a different neural basis. First there are long-term declarative memories, which are also termed *semantic memories* on the assumption that they exist in a language-amenable format and can be expressed in words. This is an inconvenient definition for evolutionary scholars. Presumably chimpanzees have long-term memories of explicit information (e.g., "panda nuts are edible") that do not exist in a language-amenable format. We prefer the term *explicit memories*. Long-term explicit memories constitute an immense store of information, including information relevant to aesthetic experience: authorship, style, historical context, and explicit symbolic meanings, to name just a few. The second kind of memory is the autobiographical memory, which is the recall of specific episodes in one's own past, such as the first encounter with "The Raft of the Medusa" in the Louvre. Finally, there are procedural memories, including muscle memories. Procedural knowledge is nonverbal and is processed by the brain differently from declarative memories. Nevertheless, procedural knowledge can be a component of aesthetic experience, especially if one has personal history of aesthetic production and can access the "feel" of the process (e.g., painting or stone knapping).

The mind uses these memories to do several things during aesthetic experience. One is to make explicit classifications: This is a painting; it is in the style of early nineteenth-century French Romanticism; it is by Gericault. Such classification itself engages pleasure/reward resources. Memory of historical context also is relevant: The Medusa was a French ship that sank off the coast of West Africa, and the fate of the survivors created a sensation in France at the time. Thus, the more one knows about the painting, the higher the level of affect. The social context of viewing is an additional extrinsic top-down effect. If one views "The Raft of the Medusa" in the company of an art student who hates French Romanticism, one may well experience the painting differently oneself. Attention to context of viewing can have a profound influence on how one evaluates an aesthetic work.

In the modern world an essential cognitive component of aesthetic experience is the "search for meaning" (Leder et al., 2004). Indeed, archaeologists hear the following question often from students and the public when discussing prehistoric art: "What does it mean?" The kind of meaning they intend is symbolic meaning: This image or artifact stands for another thing, or an abstract idea. Modern life around the world is immersed in such symbols, and the category of things we consider art often carries explicit or implied symbolism. Cognitively, attaching a semantic association is a fairly straightforward kind of learning, and much of our explicit knowledge is held in memory in the form of such semantic associations. But there are other kinds of semiotic representations in addition to symbols. Indexes stand for things by association, but do so without the mediation of an arbitrary sign linkage. Instead, the link is direct, as when the color red comes to stand for danger, or STOP. Finally, icons stand for things through physical resemblance. Aesthetic experience exploits all of these kinds of semiotic representation, often at the same time, and observing art requires one to unpack all of it (e.g., a crucifix, which is an icon [dying man], an index [the passion], and a symbol [God's love, etc.]).

The meaning-knowledge component is top-down processing and more or less effortful on the part of the observer or actor. But there is a "training effect," such that experience in top-down appraisal eventually changes bottom-up processes, and aesthetic phenomena that initially induced displeasure or disgust come to induce pleasure. Experts do see in ways that novices do not (Chatterjee, 2014b; Leder et al., 2004; Ramachandran & Hirstein, 1999).

Identifying the several different neural components of aesthetic experience greatly facilitates analysis of the archaeological record. Rather than search for aesthetics in general, one can take the easier route and look for evidence of the individual components. But before we can tackle this kind of analysis it is first necessary to address a more fundamental problem, the emergence of a true tool concept.

THE EMERGENCE OF A TOOL CONCEPT

Before tools could become aesthetisized objects they needed first to acquire an ontological status distinct from other objects. They needed to be a category of thing separate from food, or places, or non-tool objects. We modern humans think about tools as such a category, and this status seems to be something we learn very early on in life.

However, we do not seem to be born with it. Human infants do come with some hard-wired conceptual biases. Even infants who are only a few months old can distinguish between animate objects and inanimate objects. For example, they show surprise when shown a scene in which an inanimate object seems to move of its own accord (Mandler, 2004). Such a conceptual distinction has clear survival value, and it is almost certainly very old in an evolutionary sense. But our "tool concept" has not apparently acquired dedicated, hard-wired neural circuits. We learn it very early on through engagement with tools as infants. Once tools acquired a distinct ontological status they could become a focus of attention and thus available for aesthetic consideration.

Non-human primates do not appear to have a tool concept. Many anthropoid primates use objects as tools and even modify the objects to suit need. Moreover, object manipulation, especially of food, is an important component of anthropoid adaptations that has selected for an extensive neural network in support (Hecht et al., 2013; Orban & Caruana, 2014; Orban et al., 2006). But in non-human primates this network does not include shape recognition or "semantic" regions of the brain. It does for modern humans using tools (Orban & Caruana, 2014). Non-human primates do not seem to give a great deal of attention to the tools themselves. Their focus is on using the objects to help them acquire something they otherwise could not acquire. When they complete a task, they abandon the object on the spot. They may pick it up for a different episode hours or days later, but in the meantime it ceases to exist. Non-human primates do not curate tools.

The earliest stone knappers continued this task-oriented approach to tool use. The very earliest knapped stone tools were simple flakes struck from cores and used to assist in the butchery of scavenged meat (Harmand et al., 2015; Semaw et al., 2003; Toth & Schick, 2006). They also used the resulting cores as bashing tools (Mora & de la Torre, 2005). The knappers paid little attention to characteristics of the tools themselves and abandoned them immediately after use. Eventually, after doing this for several hundred thousand years, the hominins demonstrated activities that may have been precursors to a true tool concept. First, they began to carry raw material and cores considerable distance—up to 13 km in one case (Braun, Harris, & Maina, 2009). Even if this transfer occurred in stages it had a provocative cognitive impact. The objects acquired temporal extension. They existed prior to and, in the case of cores, after an episode of use. Temporal extension leads to object continuance, which is an essential component of tool concept. The knappers also began to attend more closely to attributes of cores. We know this from the refitted cores from Lokalalei 2C, which date to about 2.3 million years ago (Mya) (Delagnes & Roche, 2005). From the refits we know that the knappers examined the cores closely to identify the most productive platform locations. Attention to features of an object was also a prerequisite for distinguishing tools from other objects.

This focus on attributes of cores may have led to the production of the first handaxes. In a revealing bit of experimental archaeology, Moore and colleague (Moore & Perston, 2016) instructed modern knappers to flake cores using a rigid set of instructions to search only for the most optimal platform location for each successive knapping strike without considering future consequences for the core. The most common result, entirely unplanned on the part of the knapper, was the emergence of an elongated core with a bifacial edge (i.e., cores with two of the basic attributes of handaxes). If the Lokalalei knappers paid close attention to cores, and they seem to have done so, and also carried elongated cores with them and then used those cores, the result was the first de facto true tool. Once that occurred, the stage was set for the appearance of handaxes.

John Gowlett (2006) has provided the best available description of a basic handaxe by identifying six "design imperatives" that guided the knappers. These imperatives are primarily ergonomic considerations, features that produced an effective sturdy, hand-held cutting tool. The six are as follows:

(1) *Glob-butt*: Handaxes are thickest toward the base because handiness and effectiveness require that the center of gravity of a tool should lie within the grip.
(2) *Forward extension*: By extending the cutting edge away from the grip, the tool acquires added leverage in addition to a longer cutting edge.
(3) *Edge support*: Trimming the cutting edge bifacially produces a sturdy, sharp cutting tool.
(4) *Lateral extension*: The tool must also have extension perpendicular to the long axis to reduce the tendency to twist during use.
(5) *Thickness adjustment*: The knapper's primary means of adjusting the weight of the tool was through maintaining or reducing thickness.
(6) *Skewness*: Knappers often offset the long axis slightly to accommodate handedness.

By 1.79 Mya, artifacts that meet these design imperatives appeared in the archaeological record of East Africa at the Kenyan site of Kokiselei (Lepre et al., 2011) and the Ethiopian site of Konso (Beyene et al., 2013). Figure 13.1 is a photo of one of the Kokiselei handaxes. The photo very clearly presents glob-butt, forward extension, and lateral extension (the bifacial edge is present but not obvious in the photo).

About the same time as the appearance of these first handaxes hominins also began using a knapping technique that produced large flakes (>10 cm) from cores the size of small boulders (Jones, 1981; Sharon, 2010). These large flakes greatly eased the task of imposing the basic handaxe design imperatives because the bulb of percussion provided a convenient glob-butt, and knappers could produce forward extension and lateral extension with minimum effort. There is an interesting chicken-and-egg linkage between large flakes and handaxes. Did the handaxe imperatives motivate the development of large flake technique, or did large flake technique enable the development of the handaxe? The early site at Konso, at least 1.76 Ma (Beyene et al., 2013), has yielded clear examples of large flake manufacture (for picks and cleavers as well as handaxes), and at this point it is impossible to assign priority.

Handaxes that instantiated the six design imperatives were in place by 1.79 Mya, providing a target tool concept that hominins could exploit for aesthetic effects, and very soon they did.

THE EXPLOITATION OF VISUAL EFFECTS

From almost the very first appearance of handaxes, knappers appear to have exploited visual processing effects to produce pleasing results. In rough chronological order of their earliest clear appearance these visual effects were imposition of basic *Gestalt forms*, especially symmetry; *peak shift*, initially via size exaggeration; *prototypicality* via

Figure 13.1. A 1.79 Mya handaxe from Kokiselei, demonstrating glob-butt, forward extension, and lateral extension (see Lepre et al., 2011). Photo by Thomas Wynn.

regularization of form; *familiarity,* as represented by community styles; and *framing,* use of the handaxe form to focus visual attention on inclusions.

Derek Hodgson (2009, 2011, 2015) has drawn attention to the potential of basic Gestalt forms, and also peak shift, to produce pleasurable effects via stimulation of mu-opioid sensitive cell groups by pattern activations in the primary visual cortex. Good forms are pleasant to look at. Hodgson uses this basic fact of visual processing to ground his "visual resonance theory," which traces this effect through the course of human evolution. One of the most salient Gestalt forms is symmetry, including bilateral symmetry and radial symmetry. Figure 13.2 presents a remarkable handaxe from FLK West at Olduvai Gorge that dates to 1.69 Mya, only a few thousand years more recent than the Kokiselei handaxe in Figure 13.1.

While the Kokiselei handaxe meets all of the design imperatives, it is only vaguely symmetrical. The FLK West handaxe is quite different. Here the symmetry is clear and intended; the knapper trimmed this handaxe so that one side mirrored the other. We must note here that this handaxe is an exception. All of the other FLK West handaxes are similar to Kokiselei and Konso handaxes—basic handheld cutting tools that instantiate the six design imperatives but have only vague bilateral symmetry. This "exceptionalism" is one of the enigmas of handaxe technology, but it was a common phenomenon.

Figure 13.2. Giant (>30 cm) 1.69 Mya handaxe from Olduvai Gorge FLK West that demonstrates clear bilateral symmetry. Previously published as Figure 5 (p. 6) in Diez-Martín et al. (2015), The origin of the Acheulean: The 1.7 million-year-old site of FLK West, Olduvai Gorge (Tanzania), *Scientific Reports*, and distributed under a Creative Commons License.

Bilateral symmetry was not the only symmetry to interest these stone knappers. They also liked spheres (radial symmetry in 3D). Figure 13.3 is a photo of an artifact from Upper Bed II at Olduvai Gorge, a bit more recent in time than FLK West.

Nick Toth (Toth & Schick, 1986) has argued that such "spheroids" emerged unintentionally when hominins used roundish cores repeatedly for bashing, but this explanation rings hollow for artifacts such as the one in Figure 13.3. The sphericality just seems too regular. We maintain that the hominins intended them to be round because the shape was pleasing, whatever their function might have been. The Israeli site of 'Ubeidiya, also from about 1.5 Mya, has yielded huge spheroids, all clearly modified but too large to have been handheld bashing tools. This corroborates our conclusion that 1.5 Mya hominins liked stone balls, for whatever reason.

Not only is the FLK West handaxe bilaterally symmetrical, it is also huge (>30 cm in length), almost out of the range of a handheld tool. Such size exaggeration exploits the peak shift effect. If symmetry is pleasing, giant symmetry will have an even greater visual impact. Peak shift is one of the most salient visual effects exploited by artists in the modern world (Chatterjee, 2014b; Leder et al., 2004; Ramachandran & Hirstein, 1999), and its early appearance makes a strong *prima facie* case that at least some of

Figure 13.3. Spheroid from Upper Bed II, Olduvai Gorge, ca. 8 cm. Photo by Thomas Wynn.

the handaxes were aesthetisized artifacts. Peak shift is not just size exaggeration; it extends to any visual effect, including color and texture. Knappers at Olduvai Gorge 1.4 Mya selected sparkling quartz clasts and modified them into handaxes (the effect is sadly hard to capture in photos), and selection for striking colors and raw materials continued for the next million years (Figures 13.4 and 13.5).

Prototypicality is perhaps the most provocative visual effect evident for handaxes. Prototypical forms are more visually pleasing than eccentric examples. To the trained, Western student of art a melted clock by Dali may be visually appealing, but this response depends on a specific cultural context, and is not a pre-attentive effect. In pre-attentive visual terms it is the prototypical clock that elicits the most positive affect. But what is the prototypical form of a handaxe? It is not just any bilaterally symmetrical tool. Most specialists actually do have a shape in mind when they think of a handaxe, a shape based interestingly not on average handaxes but on the exceptional examples that illustrate academic papers and books. This shape is the teardrop, which in geometric terms is known as a *hemilemniscate* (half of a lemniscate of Bernoulli; see Figure 13.6) (Mason, 1967).

Why would a teardrop become a prototypical shape? After all, it is not one of the basic Gestalt forms. The answer appears to lie in the regularization of the six design imperatives. If one transforms glob-butt, forward extension, and so on into a prototypical shape with regular curves and proportions and bilateral symmetry, the result will be a hemilemniscate or some close approximation. The hemilemniscate thus did not emerge out of nowhere like Athena from the head of Zeus, but resulted from a regularization of the ergonomic imperatives of the handaxe form. However, the shape did not appear immediately; indeed, it was not until perhaps 1 Mya that true hemilemniscate handaxes appeared. Greater regularity of form, with gradual curves and greater symmetry, is a form of peak shift. Prototypicality via peak shift eventually generated

Figure 13.4. (Left) Quartz handaxe from Olduvai Gorge. Photo by Thomas Wynn and Tony Berlant.

hypertrophic forms such as the 780,000-year-old handaxe from Gesher Benot Ya'aqov, Israel shown in Figure 13.7.

Two final visual effects became apparent late in the temporal range of handaxes—familiarity and framing. Familiar forms are more pleasing than exotic or strange forms; thus knappers should tend to have made handaxes that were similar to those made by other knappers of their community. Unfortunately, this is very hard to document archaeologically. The "Handaxe Age" was so long ago that only very rare sites preserve coherent patterns produced by single groups. Handaxe assemblages are more often sets that accumulated over long periods, often thousands of years, and thus not amenable to community identification. There are exceptions, the most spectacular of which is probably Boxgrove, a 500,000-year-old site on the southern coast of Britain. Boxgrove is actually a prehistoric landscape that was very briefly exposed before being covered by a rise in sea level. It preserves localities—and handaxes—produced over a very brief period of time, perhaps only a single generation of 25 years (Roberts & Parfitt, 1999). A single community of hominins (*Homo heidelbergensis*) produced all of the handaxes. And the handaxes are all very, very similar (Figure 13.8), a testament to the attractiveness of familiar patterns.

Figure 13.5. (Right) Conglomerate handaxe from Algeria (25 cm). Photos by Thomas Wynn and Tony Berlant.

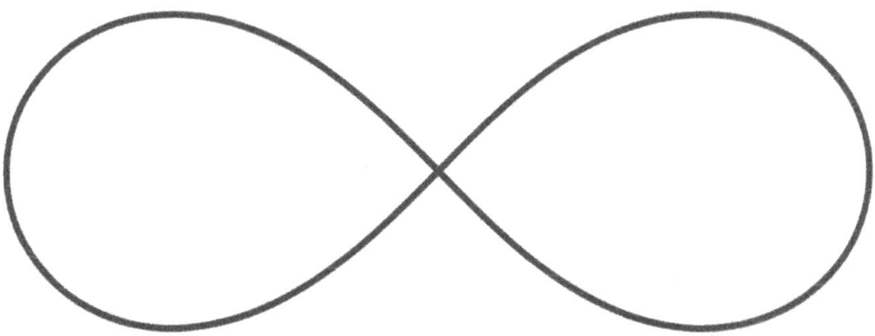

Figure 13.6. Lemniscate of Bernoulli. Image in the public domain.

Figure 13.7. A 780,000-year-old handaxe from Gesher Benot Ya'aqov North Bridge (Goren-Inbar, Werker, & Feibel, 2002). Photo by Thomas Wynn.

Framing exploits the visual focusing power of a regular form. Figure 13.9 is the most famous example of framing. The knappers selected a flint nodule that contained a well-preserved fossil shell. By knapping around the shell, they were able to accentuate it and draw attention to it. The example is not unique. Knappers often used the handaxe to frame crystals, fossils, holes, and, in some cases, faces (Figure 13.10).

THE EXPLOITATION OF EMOTION-VALUATION

Evaluation is inherent to all aesthetic experience in the modern world. We like something or we dislike it; we find something beautiful or we find it repulsive. These are emotionally salient responses and engage the neural resources of the pleasure/reward network. But there are also neural resources devoted to appraisal itself, and unlike sensorimotor effects, appraisal engages both bottom-up and top-down processing. The "automatic" bottom-up resources include the anterior insula and anterior cingulate cortex, resources that initially evolved long ago in support of gustation; "liking" what was healthy and disliking what was not had great survival value. To use another

Figure 13.8. Three 500,000-year-old handaxes from Boxgrove (Roberts & Parfitt, 1999). Photo by Thomas Wynn.

Figure 13.9. Handaxe with inclusion from West Tofts. Reproduced with the permission of the University of Cambridge Museum of Archaeology and Anthropology. Accession no. 1916.82/ Record 2.

Figure 13.10. Cleaver from Algeria. Photo by Tony Berlant.

gustatory metaphor, bottom-up appraisal is one's "gut" response. But appraisal is more than gut response, at least for most of us in the modern world. It also engages memories and knowledge about the world. Judgment of "beauty" is an example. It is an informed response based as much on learned community norms as it is on gut reaction. The neural structure that seems to mediate these responses is the orbitofrontal cortex. Any search for the appraisal piece of aesthetic experience in the evolutionary past must try to differentiate between the bottom-up and top-down processes, if at all possible.

There is no reason to conclude that non-human primates make aesthetic judgments about their tools. Indeed, non-human primates appear entirely indifferent to the appearance of their tools. They do assess whether or not an object can perform a task, but this is not a gustatory, taste-based appraisal (though in such a task-oriented activity the task itself might well be). Early stone knappers appear to have been ape-like in this regard, as in many others (Wynn, Hernandez-Aguilar, Marchant, & McGrew, 2011). The visual appearance of a flake or a core or a hammer was irrelevant in a gustatory sense as long as the tool performed the necessary task. As we discussed earlier, these early knappers did begin to attend more closely to features of cores as an aid to effective knapping, and such focused attention was an important prerequisite, we believe, to the development of a tool concept. With a tool concept in place, tools became available as objects of consideration in their own right, and this consideration soon included a component of aesthetic appraisal.

Overdetermination is the earliest clue to aesthetic appraisal. As early as the handaxe from FLK West (Figure 13.2), a hominin knapper invested more time and energy to achieve a pleasing form than was necessary for its functionality. Why? At a minimum the knapper himself or herself made an appraisal about this artifact. Even a judgment as basic as "this pleases me" is an aesthetic appraisal. And it clearly was not an appraisal of how well the tool might work (the knapper could certainly have made such a judgment as well), it was an appraisal of the form of the handaxe. We think that it is very telling that overdetermination accompanied handaxes from the very beginning. It suggests that aesthetic appraisal was an established component of the way of life of *H. erectus* from the outset. But it is important not to over-interpret overdetermination. What was required for the FLK West example was an individual knapper judging his or her work to be pleasing. There need not have been, in fact probably were not, any community based, top-down components to this assessment. This does not mean, however, that the appraisal must have been entirely idiosyncratic; the knapper may have had another individual in mind.

Exceptionalism is a term we have adopted for a recurrent phenomenon of many handaxe assemblages. In many, many cases the majority of handaxes in an assemblage are bland, mediocre examples that are vaguely symmetrical but otherwise unimpressive in terms of regularity of form or skill in manufacture. But one or two are qualitatively different—overdetermined, with regular proportions and evidence of skilled knapping. This is hard to quantify for a number of reasons. First, museums often put the best examples on display, leaving the hundreds of mediocre examples in dungeon drawers or, worse, disposed of or lost. Second, there is no set standard for what is an exceptional example. Intuitively, the variation in a handaxe assemblage appears to adhere to a normal distribution of size, skill, and form, with the exceptional pieces being true outliers. Figure 13.11 presents a sample of handaxes from Gesher Benot Ya'aqov that contrast dramatically with the prototypical example in Figure 13.7.

The Kathu Pan assemblage includes one of the most beautiful handaxes ever found (Figure 13.12), along with thousands of mundane, mediocre examples (based on personal examination, TW estimates the frequency of exceptional examples at Kathu Pan at well under 1%).

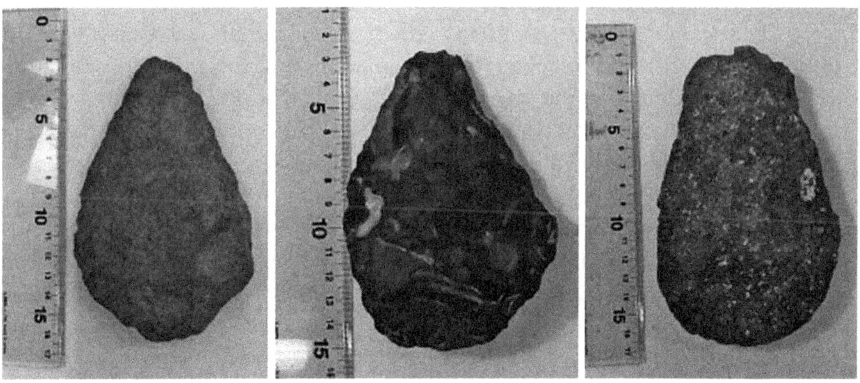

Figure 13.11. Three average handaxes from Gesher Benot Ya'aqov. Compare to Figure 13.7. Photos by Thomas Wynn.

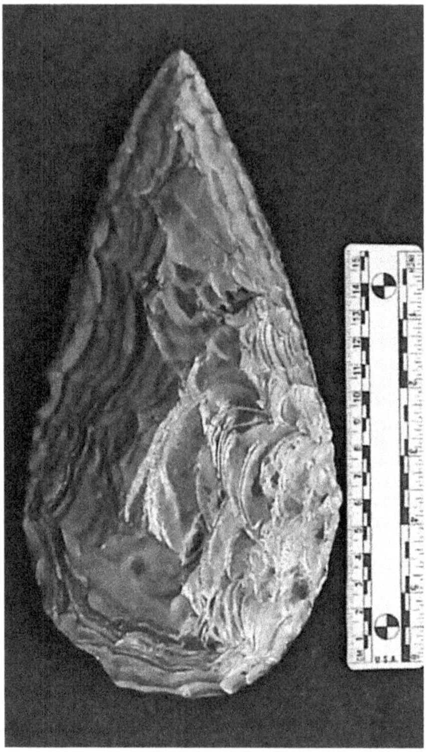

Figure 13.12. Handaxe from Kathu Pan, South Africa. Photo by Thomas Wynn.

Even the FLK West handaxe (Figure 13.2) was dramatically different from all of the other handaxes from the site (Diez-Martín et al., 2015) and also compared to other known handaxes from Lower Bed II at Olduvai Gorge. Thus it appears that exceptionalism was a common feature of handaxe assemblages for the entire duration of the Acheulean. But why? What does it imply about *H. erectus* (and later *H. heidelbergensis*) in general, and aesthetics in particular?

We believe that exceptionalism may reflect something unusual about the social realities of *H. erectus*' life. It suggests that the hominins used material displays in atypical situations, which in turn suggests that the knappers worked for the appraisal of some other individual or individuals, but again, only in unusual circumstances. The exceptional examples are in a sense *over*-overdetermined, and although it is possible that a knapper did it solely for his or her own pleasure, it just seems unlikely. This has implications for theory of mind (ToM). The knapper of one of these exceptional handaxes considered not just his or her own point of view but also what at least one other individual could see. It is impossible at our current state of knowledge to know what these atypical circumstances were. Social dominance and status maintenance are probably the most likely, given the role of visual display in non-human and human social interaction. But because it seems to have been atypical, it is unlikely to have been day-to-day status maintenance. What is important for us is that aesthetic appraisal had become part of social judgment at a very early point in the evolution of the genus *Homo*. Such incorporation of social knowledge and context is a top-down component.

It was not until late in the temporal range of Acheulean handaxes that there is evidence for community standards of aesthetic appraisal. As we discussed earlier in the context of familiarity, it is difficult even to see social communities in the deep past. We can see individual action, and even dyads in the case of exceptionalism, but not communities. By 500,000 years ago there do appear to have been local trends in handaxe shape, but archaeologists have never been able to define anything that would qualify as a style in the sense of artifact styles from recent prehistory (Ashton & White, 2003; Lycett & Gowlett, 2008; Machin, 2009; White, 1998). The Boxgrove handaxes were arguably made by a single community over about a generation, and they are remarkably uniform. But without another example it is impossible to determine whether or not this shape was limited to this community. If we assume for the moment that it was, then the social situation had changed, with community standards replacing (or supplementing) the individual display role. It is certainly telling that all of the Boxgrove handaxes are overdetermined. There was either a community standard or only a few individuals knapped handaxes, which strikes us as unlikely.

In sum, the appraisal component of aesthetic experience was present from the very beginning of handaxe technology. It was initially individual appreciation of form, but the FLK West handaxe suggests that dyadic or even polyadic display occurred in atypical circumstances—aesthetics had taken a social role. The nature of this social role appears to have evolved over the long millennia of handaxe technology, with possibility of community standards based on appraisal emerging by 500,000 years ago. This takes us naturally to the final component of aesthetic experience—cultural and symbolic context.

Meaning-Knowledge

The meaning-knowledge component of aesthetic cognition consists of the explicit and implicit knowledge that an observer or artisan brings to the experience. In the modern world, such knowledge is largely semantic in nature and can be expressed in language, even if it does not always manifest itself linguistically. But explicit awareness of factual knowledge need not be stored as arbitrary symbols. All mammals rely on learned information, but only humans store it in the form of words. Knowing when and how the transition to symbol-based knowledge systems occurred is one of the fundamental questions of paleoanthropology. And the status of *H. erectus* and *H. heidelbergensis*, who were the handaxe artisans, has always been a topic of contention. We will touch on this issue later, but we find another question to be more accessible: To what degree did explicit knowledge play a role in the handaxe aesthetic experience?

Is it possible to detect or infer the use of explicit knowledge in the manufacture of stone tools, handaxes in particular? Like all manual technology, stone knapping is primarily an expert system that relies very heavily on nonverbal procedural knowledge acquired through physical practice (Keller & Keller, 1996; Wynn & Coolidge, 2014). But like all expert systems, explicit knowledge does play a role, and did so for even the earliest stone knappers (e.g., "this raw material flakes well"). Our first question, then, is this: What specific examples of explicit knowledge played a role in early aesthetic experience? From features of the artifacts themselves, we infer at least two for early handaxes and three for late handaxes.

Curation and overdetermination indicate the use of a tool concept, and overdetermination, peak shift, and exceptionalism indicate that knappers considered what other individuals could see and understand. If there were no tool concept, knappers would not have carried finished tools around, and certainly would not have invested extra time and effort into making one. Knappers arguably learned the tool concept as infants raised in the context of a tool-oriented technology; there is no need to posit a semantic category. As a permanent component of the knapper's life-world, tools became available for aesthetic expression, perhaps initially only for personal pleasure via basic Gestalt forms. But exceptionalism and peak shift, evident for the FLK West giant handaxe, suggest that knowledge of others was also a consideration. Peak shift alone would not require appreciation of alternative viewpoints, but exceptionalism would. The knapper used aesthetic effects to display the handaxe, perhaps in order to impress or inform someone else. Thus what the "other" knew became a bit of explicit knowledge used to guide aesthetic expression. It is impossible to know "what" this knowledge was, but we know "that" it was there. This corroborates the arguments of Shipton and Cole (Cole, 2014, 2015; Shipton, 2010) that ToM was essential to handaxe production.

For the first million years of the handaxe age, the social considerations appear to have been direct. Knappers occasionally made exceptional handaxes to influence the understanding or behavior of another individual or individuals. There do not appear to have been community norms in the form of a repeated style. But by 500,000 years ago there were. If Boxgrove is typical, the *H. heidelbergensis* used community norms to govern style. A community norm is more abstract than a personal preference. As a piece of explicit knowledge shared by everyone in the community, it is tempting to conclude that it must have been an item of semantic knowledge. But it is just as possible to learn such a standard without labeling it, though it would require consideration of not just another's knowledge but everyone's knowledge. At a minimum, then, the Boxgrove handaxes required a higher level of ToM, as Cole has argued (Cole, 2014).

Consideration of style brings us to one of the more contentious issues of paleoanthropology—the evolution of symbolic culture. Certainly modern aesthetic experience is embedded in a very complex symbolic milieu, with explicit and implicit meanings pervading most aesthetic productions. Was this true for the handaxe makers? To answer this question it is first necessary to define symbolism more carefully. Here we will opt for the semiotic definition of Peirce (Houser & Kloesel, 1992) that distinguishes between true symbols (arbitrary link between sign and referent), indexes (direct association), and icons (physical resemblance).

There are no grounds for concluding that the handaxes acted as true symbols, in the sense of standing for something else in an arbitrary way (e.g., if handaxes were symbols of the sun). Of course, this would be very difficult to detect archaeologically. An arbitrary link is by its nature somewhat abstract, or at least incorporeal, and thus unlikely to be preserved archaeologically. Indexes are more likely to leave clues. In the modern world, indexical reference via material culture is so ubiquitous that it is arguably as important as language in delivering social information. The choices we make can come to stand for us and the group we belong to. The clothes we wear, the tools we use, and houses we live in all have indexical meaning that others interpret readily. Much of our "immersion in symbolic culture" consists of our immersion in indexical signs.

Handaxes may have played this role. Tools, especially tools underpinned by a tool concept, have a natural indexicality. They stand for their "toolness" and they stand for their use. Simply drawing attention to a specific handaxe would activate the linked associations of its use; gesturing with a handaxe would have had considerable communicative effect. More profoundly, handaxes could stand for the maker/user, and it is here that aesthetics enters the equation. From the very beginning, *H. erectus* overdetermined the shape of handaxes. True, there was a component of personal pleasure in the production of Gestalt forms, peak shift, and prototypicality, but there was also an indexical payoff. Producing a more aesthetically pleasing handaxe enhanced the indexical message. Exceptionalism indicates that this enhancement was exploited intentionally, at least occasionally. Thus handaxes were embedded in a very specific semiotic milieu, even if it was not narrowly symbolic.

Icons are the third variety of sign, and perhaps the easiest for us to find archaeologically. The earliest proffered example of an iconic image in the Paleolithic is the Berekhat Ram figurine, a modified piece of pumice from an Acheulean site in Israel (Goren-Inbar & Peltz, 1995). It dates to about 230,000 years ago. Archaeologists disagree about whether or not the likeness to a woman is close enough to qualify as the earliest icon, but it is certainly close enough to elicit debate! Figure 13.13 presents two handaxes from Bentadjine in Algeria.

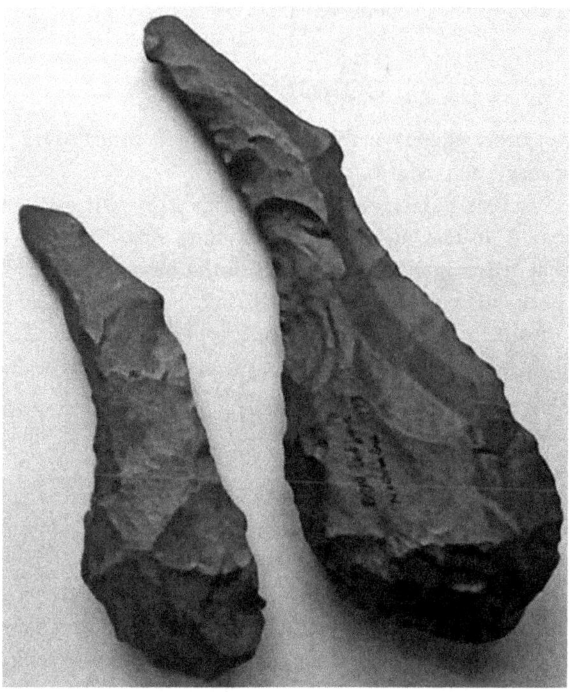

Figure 13.13. Handaxes from Bentadjine, Algeria (larger ca. 25 cm). Photo by Thomas Wynn.

Unfortunately, they are surface finds. However, associated artifacts include handaxes and cleavers whose manufacturing techniques are typical of the final phase of the Acheulean in the region (Alimen, 1978). Both appear to be zoomorphic, the artifact on the right especially so. The contextual problems prohibit enthusiasm, but these handaxes may well indicate that the makers knew and employed iconic reference.

What is perhaps most remarkable about the handaxe age is the almost complete absence of anything other than handaxes that could reasonably be interpreted as a symbolic artifact. At the very end of the time range, *H. heidelbergensis* began to use pigments (Barham, 2002), and even practiced corpse disposal (Carbonell & Mosquera, 2006), but both developments occurred when handaxes had already begun to be supplanted as the primary focus of lithic technology.

The meaning-knowledge component of the handaxe aesthetic appears to have been very different from the meaning-knowledge component of modern aesthetics. There were no abstract meanings to consider, indeed, no symbols in the narrow sense. Knappers did bring to bear knowledge acquired over the course of their lives, and this knowledge may well have included indexical associations. Indeed, the presence of a tool concept, indexical reference, and ToM tells us something very important about the handaxe aesthetic: Material culture had come to mediate social relations, at least to some degree, and aesthetics played an important role. Martín-Loeches (2017) has similarly argued for the importance of pre-symbolic "art," emphasizing that embodied and emotional components of artifact production long preceded symbolic reference. Thus, compared to artifacts of the modern world, the referential role of *H. erectus* and *H. heidelbergensis* artifacts was quite impoverished, and always concrete. It was this component of aesthetic cognition—explicit use of multilayered symbolic reference—that evolved most dramatically between 500,000 years ago and the present.

CONCLUSION

The preceding neuroaesthetics analysis warrants three conclusions concerning the aesthetic implications of handaxes.

First, handaxes were, in fact, aesthetisized artifacts. Here we agree with Hodgson (2009, 2011, 2015). In making handaxes, hominins exploited many of the implicit visual effects that artists continue to exploit in the modern world. These included Gestalt forms, peak shift, prototypicality, familiarity, and framing. In addition, hominin knappers made aesthetic appraisals of the visual appearance of their handaxes, at least occasionally. These appraisals took place in a social context, with aesthetics playing a yet-to-be-understood social role.

Second, the handaxe aesthetic differed from aesthetics as experienced in the modern world. There is no evidence that handaxe aesthetic experience included the rich, multilayered symbolic milieu that is typical of all modern artistic endeavors. Despite its possible role in indexical reference, the handaxe aesthetic was arguably pre-symbolic.

Third, the handaxe aesthetic evolved over the 1.5 million years that handaxes played a central role in hominin lithic technology. Many of the technical developments that occurred in stone knapping gave hominins increasing control over their final products, and the knappers used this control to accentuate aesthetic effects. By half a million years ago, some hominin knappers regularly produced giant handaxes (over 30 cm),

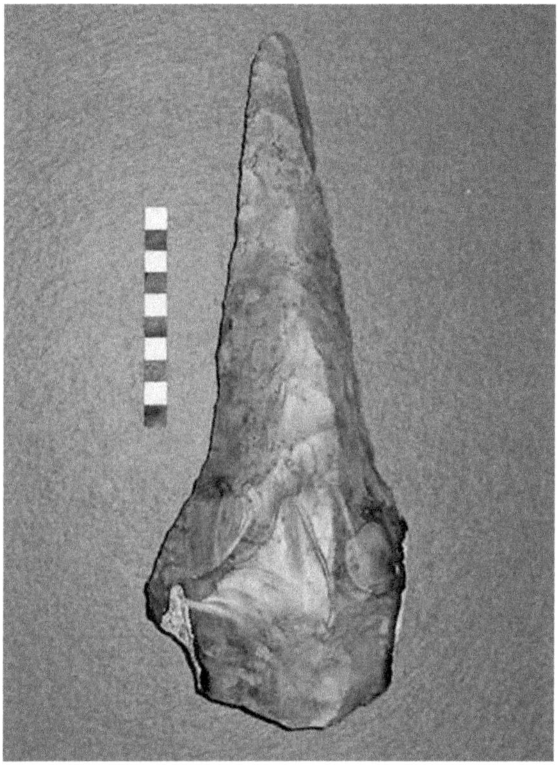

Figure 13.14. Giant hypertrophic ficron handaxe from Cuxton, England. Photo by Thomas Wynn.

others made twisted ovates, and others made hypertropic ficrons such as the Cuxton Giant (Figure 13.14). There is even evidence toward the very end of this period that some knappers invested their handaxes with iconic reference, implying a beginning to the multilayered symbolic reference that eventually blossomed into the modern aesthetic experience as we know it today.

ACKNOWLEDGMENTS

Ms. Anne Kelley contributed to the initial development of this chapter. The research was funded by the Nasher Sculpture Center of Dallas, Texas, in support of an international exhibition, *First Sculpture: Handaxes to Figure Stones*, which ran from January through April 2018. Director Jeremy Strick and the staff of the Nasher have provided enthusiastic support. We have visited over 20 museums in our search for appropriate examples. Artifacts that appear in this publication are located in the following institutions: National Museums of Kenya (Figure 13.1) (we thank Dr. Purity Kurita and Dr. Immanuel Ndjema); National Museums of Tanzania (Figures 13.2 and 13.3) (we thank Dr. Audax Mabulla); Cambridge Museum of Ethnology and Archaeology (Figures 13.4 and 13.9) (we thank Dr. Imogene Gunn); Prehistoric Museum of the Huleh Valley (Figure 13.7) (we thank Dr. Gonen Sharon); The British Museum (Figure 13.8) (we thank Dr. Nick Ashton); Hebrew University, Jerusalem

(Figure 13.11) (we thank Dr. Naama Goren-Inbar); MacGregor Museum, Kimberley, South Africa (Figure 13.12) (we thank Dr. David Morris); Museum of Prehistory, Sauveboeuf, Aubas, France (Figure 13.13) (we thank Mr. Claude Douce); and University of Southampton (Figure 13.14) (we thank Dr. Francis Wenban-Smith).

We also thank two anonymous reviewers for their valuable suggestions.

REFERENCES

Alimen, M. H. (1978). *L'évolution de l'Acheuléen au Sahara nord-occidental: Saoura, Ougarta, Tabelbala*. Paris: Éditions CNRS.

Anderson, M. L. (2010). Neural reuse: A fundamental organizational principle of the brain. *Behavioral and Brain Sciences, 33*(4), 245–313.

Ashton, N. M., & White, M. J. (2003). Bifaces and raw material: Flexible flaking in the British Early Paleolithic. In M. Soressi & H. L. Dibble (Eds.), *Multiple approaches to the study of bifacial technologies* (pp. 109–124). Philadelphia: University of Pennsylvania Museum of Archaeology and Anthropology.

Barham, L. S. (2002). Systematic pigment use in the Middle Pleistocene of South-Central Africa. *Current Anthropology, 43*(1), 181–190.

Beyene, Y., Katoh, S., WoldeGabriel, G., Hart, W. K., Uto, K., Sudo, M., . . . Asfaw, B. (2013). The characteristics and chronology of the earliest Acheulean at Konso, Ethiopia. *Proceedings of the National Academy of Sciences of the United States of America, 110*(5), 1584–1591.

Braun, D. R., Harris, J. W. K., & Maina, D. N. (2009). Oldowan raw material procurement and use: Evidence from the Koobi Fora formation. *Archaeometry, 51*(1), 26–42.

Brown, S., Gao, X., Tisdelle, L., Eickhoff, S. B., & Liotti, M. (2011). Naturalizing aesthetics: Brain areas for aesthetic appraisal across sensory modalities. *NeuroImage, 58*(1), 250–258.

Carbonell, E., & Mosquera, M. (2006). The emergence of a symbolic behaviour: The sepulchral pit of Sima de los Huesos, Sierra de Atapuerca, Burgos, Spain. *Comptes Rendus Palevol, 5*(1), 155–160.

Carey, S. (2009). *The origin of concepts*. Oxford, UK: Oxford University Press.

Chatterjee, A. (2014a). Scientific aesthetics: Three steps forward. *British Journal of Psychology, 105*(4), 465–467.

Chatterjee, A. (2014b). *The aesthetic brain: How we evolved to desire beauty and enjoy art*. Oxford, UK: Oxford University Press.

Chatterjee, A., & Vartanian, O. (2014). Neuroaesthetics. *Trends in Cognitive Sciences, 18*(7), 370–375.

Cinzia, D. D., & Vittorio, G. (2009). Neuroaesthetics: A review. *Current Opinion in Neurobiology, 19*(6), 682–687.

Cole, J. (2014). The identity model: A theory to access visual display and hominin cognition within the Palaeolithic. In R. I. M. Dunbar, C. Gamble, & J. A. J. Gowlett (Eds.), *Lucy to language: The benchmark papers* (pp. 90–107). Oxford, UK: Oxford University Press.

Cole, J. (2015). Handaxe symmetry in the Lower and Middle Palaeolithic: Implications for the Acheulean gaze. In F. Coward, R. Hosfield, M. I. Pope, & F. Wenban-Smith (Eds.), *Settlement, society and cognition in human evolution: Landscapes in mind* (pp. 234–257). Cambridge, UK: Cambridge University Press.

Delagnes, A., & Roche, H. (2005). Late Pliocene hominid knapping skills: The case of Lokalalei 2C, West Turkana, Kenya. *Journal of Human Evolution, 48*(5), 435–472.

Diez-Martín, F., Yustos, P. S., Uribelarrea, D., Baquedano, E., Mark, D. F., Mabulla, A. Z. P., . . . Domínguez-Rodrigo, M. (2015). The origin of the Acheulean: The 1.7 million-year-old site of FLK West, Olduvai Gorge (Tanzania). *Scientific Reports*, *5*(17839), 1–9.

Goren-Inbar, N., & Peltz, S. (1995). Additional remarks on the Berekhat Ram figurine. *Rock Art Research*, *12*, 131–132.

Goren-Inbar, N., Werker, E., & Feibel, C. S. (2002). The Acheulian Site of Gesher Benot Ya'aqov, Israel: The Wood Assemblage. Oxford: Oxbow Books.

Gowlett, J. A. J. (2006). The elements of design form in Acheulian bifaces: Modes, modalities, rules and language. In N. Goren-Inbar & G. Sharon (Eds.), *Axe age: Acheulian tool-making from quarry to discard* (pp. 203–222). London: Equinox.

Harmand, S., Lewis, J. E., Feibel, C. S., Lepre, C. J., Prat, S., Lenoble, A., . . . Roche, H. (2015). 3.3-million-year-old stone tools from Lomekwi 3, West Turkana, Kenya. *Nature*, *521*(7552), 310–315.

Hecht, E. E., Murphy, L. E., Gutman, D. A., Votaw, J. R., Schuster, D. M., Preuss, T. M., . . . Parr, L. A. (2013). Differences in neural activation for object-directed grasping in chimpanzees and humans. *Journal of Neuroscience*, *33*(35), 14117–14134.

Henshilwood, C. S., & Dubreuil, B. (2011). The Still Bay and Howiesons Poort, 77–59 ka: Symbolic culture and the evolution of the mind during the African Middle Stone Age. *Current Anthropology*, *52*(3), 361–400.

Hodgson, D. (2009). Evolution of the visual cortex and the emergence of symmetry in the Acheulean techno-complex. *Comptes Rendus Palevol*, *8*(1), 93–97.

Hodgson, D. (2011). The first appearance of symmetry in the human lineage: Where perception meets art. *Symmetry*, *3*(1), 37–53.

Hodgson, D. (2015). The symmetry of Acheulean handaxes and cognitive evolution. *Journal of Archaeological Science: Reports*, *2*, 204–208.

Houser, N., & Kloesel, C. J. W. (Eds.). (1992). *The essential Peirce*. Bloomington, IN: Indiana University Press.

Jones, P. R. (1981). Experimental implement manufacture and use: A case study from Olduvai Gorge, Tanzania. In J. Z. Young, E. M. Jope, & K. P. Oakley (Eds.), *The emergence of man* (pp. 189–195). London: The Royal Society and the British Academy.

Keller, C. M., & Keller, J. (1996). *Cognition and tool use: The blacksmith at work*. Cambridge, UK: Cambridge University Press.

Leder, H., Belke, B., Oeberst, A., & Augustin, D. (2004). A model of aesthetic appreciation and aesthetic judgments. *British Journal of Psychology*, *95*(4), 489–508.

Leder, H., & Nadal, M. (2014). Ten years of a model of aesthetic appreciation and aesthetic judgments: The aesthetic episode: Developments and challenges in empirical aesthetics. *British Journal of Psychology*, *105*(4), 443–464.

Lepre, C. J., Roche, H., Kent, D. V, Harmand, S., Quinn, R. L., Brugal, J.-P., . . . Feibel, C. S. (2011). An earlier origin for the Acheulian. *Nature*, *477*(7362), 82–85.

Le Tensorer, J.-M. (2006). Les cultures acheuléennes et la question de l'émergence de la pensée symbolique chez *Homo erectus* a partir des données relatives à la forme symétrique et harmonique des bifaces. *Comptes Rendus Palevol*, *5*(1), 127–135.

Le Tensorer, J.-M. (2009). L'image avant l'image: Réflexions sur le colloque. *Anthropologie*, *113*(5), 1005–1017.

Lycett, S. J., & Gowlett, J. A. J. (2008). On questions surrounding the Acheulean "tradition." *World Archaeology*, *40*(3), 295–315.

Machin, A. J. (2009). The role of the individual agent in Acheulean biface variability. *Journal of Social Archaeology, 9*(1), 35–58.

Mandler, J. M. (2004). A synopsis of *The foundations of mind: Origins of conceptual thought. Developmental Science, 7*(5), 499–505.

Martín-Loeches, M. (2017). Art without symbolic mind: Embodied cognition and the origins of visual artistic behavior. In T. Wynn & F. L. Coolidge (Eds.), *Cognitive models in Palaeolithic archaeology* (pp. 113–132). Oxford, UK: Oxford University Press.

Mason, R. J. (1967). Analytical procedures in the Earlier and Middle Stone Age cultures in Southern Africa. In W. W. Bishop & J. D. Clark (Eds.), *Background to evolution in Africa* (pp. 737–764). Chicago, IL: University of Chicago Press.

Moore, M. W., & Perston, Y. (2016). Experimental insights into the cognitive significance of early stone tools. *PLoS One, 11*(7), 1–37.

Mora, R., & de la Torre, I. (2005). Percussion tools in Olduvai Beds I and II (Tanzania): Implications for early human activities. *Journal of Anthropological Archaeology, 24*(2), 179–192.

Orban, G. A., & Caruana, F. (2014). The neural basis of human tool use. *Frontiers in Psychology, 5*, 1–12.

Orban, G. A., Claeys, K., Nelissen, K., Smans, R., Sunaert, S., Todd, J. T., . . . Vanduffel, W. (2006). Mapping the parietal cortex of human and non-human primates. *Neuropsychologia, 44*(13), 2647–2667.

Palmer, S. E., Schloss, K. B., & Sammartino, J. (2013). Visual aesthetics and human preference. *Annual Review of Psychology, 64*, 77–107.

Ramachandran, V. S., & Hirstein, W. (1999). The science of art: A neurological theory of aesthetic experience. *Journal of Consciousness Studies, 6*(6–7), 15–51.

Redies, C. (2015). Combining universal beauty and cultural context in a unifying model of visual aesthetic experience. *Frontiers in Human Neuroscience, 9*(218), 1–20.

Roberts, M. B., & Parfitt, S. A. (Eds.). (1999). *Boxgrove: A Middle Pleistocene hominid site at Eartham Quarry, Boxgrove, West Sussex*. London: English Heritage.

Satpute, A. B., Kang, J., Bickart, K. C., Yardley, H., Wager, T. D., & Feldman Barrett, L. (2015). Involvement of sensory regions in affective experience: A meta-analysis. *Frontiers in Human Neuroscience, 6*, 1–14.

Semaw, S., Rogers, M. J., Quade, J., Renne, P. R., Butler, R. F., Domínguez-Rodrigo, M., . . . Simpson, S. W. (2003). 2.6-million-year-old stone tools and associated bones from OGS-6 and OGS-7, Gona, Afar, Ethiopia. *Journal of Human Evolution, 45*(2), 169–177.

Sharon, G. (2010). Large flake Acheulian. *Quaternary International, 223*, 226–233.

Shipton, C. (2010). Imitation and shared intentionality in the Acheulean. *Cambridge Archaeological Journal, 20*(2), 197–210.

Toth, N. P., & Schick, K. D. (1986). The first million years: The archaeology of protohuman culture. *Advances in Archaeological Method and Theory, 9*, 1–96.

Toth, N. P., & Schick, K. D. (Eds.). (2006). *The Oldowan: Case studies into the earliest Stone Age*. Gosport, IN: Stone Age Institute Press.

Vartanian, O., & Skov, M. (2014). Neural correlates of viewing paintings: Evidence from a quantitative meta-analysis of functional magnetic resonance imaging data. *Brain and Cognition, 87*, 52–56.

White, M. J. (1998). On the significance of Acheulean biface variability in southern Britain. In *Proceedings of the Prehistoric Society* (Vol. 64, pp. 15–44). Cambridge, UK: Cambridge University Press.

Wynn, T., & Coolidge, F. L. (2014). Technical cognition, working memory and creativity. *Pragmatics & Cognition*, 22(1), 45–63.

Wynn, T., Hernandez-Aguilar, R. A., Marchant, L. F., & McGrew, W. C. (2011). "An ape's view of the Oldowan" revisited. *Evolutionary Anthropology*, 20(5), 181–197.

14

THE STORIES STONES TELL OF LANGUAGE AND ITS EVOLUTION

Shelby S. Putt

INTRODUCTION

Language is a complex human behavior guided by a cognitive system of the brain. Because humans are the only animal species to possess the full suite of characteristics that are required for language, we must assume that this cognitive system evolved in the time since chimpanzees and humans last shared a common ancestor over 5 million years ago (Mya). Ancient stone tools in the archaeological record are products of extinct hominin cognition and date possibly as far back as 3.3 Mya. Therefore, it is easy to see why, for decades, researchers have attempted to trace the evolution of language by drawing connections between aspects of language and stone tools.

Regarding language and tool use, Thomas Wynn (1991, p. 191) once wrote, "With few exceptions, all [attempts to incorporate prehistoric artifacts into discussions of language origins] have failed to be persuasive. Their weakness, in my opinion, lies in their lack of theoretical justification for drawing a connection between tool behavior and language." Wynn highlighted a trend among archaeologists to make bold but errant claims about the evolution of language based on speculation about stone tool features that supposedly relied on "cognitive abilities" similar to language (Kitahara-Frisch, 1978). Without an objective method to directly link the cognitive processes of tool-making and language, however, these types of claims would always be unpersuasive. Some might interpret Wynn's statement as dissuasive toward an archaeological investigation into the evolution of language in general. However, anyone who is familiar with Wynn's large body of work on evolutionary cognitive archaeology knows that he is a strong proponent of theory-based archaeological research that draws on findings from the cognitive sciences. So, on the contrary, I believe he meant this statement to be a call to action to other archaeologists to put aside their subjective speculations and instead seek out novel, evidence-based methods to link stone tool-making behaviors to the brain, using established theory from the cognitive sciences.

The field of cognitive neuroscience has indeed developed a theoretical foundation for a link between language and motor processing in the brain. By extension, there is some evidence to indicate that language and stone tool production share a common neural circuit in the inferior frontal cortex (Stout, Toth, Schick, & Chaminade, 2008), leading some researchers to propose that language had a technological origin (Stout & Chaminade, 2012). This would mean that language co-opted the already established hierarchical functions of Broca's area that were necessary for complex tool-making.

But does this hypothesis stand up to rigorous testing? In this chapter, I will synthesize the results from the most recent brain imaging research into experimental stone tool manufacture to establish whether a direct link can be made between language and stone tools, and I will test the plausibility of a technological origin for language in Broca's area and its right hemisphere analog. Thus, I accept Wynn's call to action as I explore what stories stone tools may tell about how the cognitive system that supports language production arose and changed over time.

MOTOR AND TECHNOLOGICAL HYPOTHESES FOR LANGUAGE ORIGINS

Broca's area, which occupies the inferior frontal gyrus (IFG) of the left hemisphere in most right-handed individuals, has been defined traditionally as the center for speech/language production. Recent studies, however, have demonstrated that Broca's area is just one of many parts of the brain that participate in language-processing functions (Vigneau et al., 2011). For example, the right hemisphere analog to Broca's area is involved in the perception of emotional and prosodic information conveyed in speech (Wildgruber et al., 2005). Broca's area in the left hemisphere also participates in non-linguistic behaviors, such as distal manual motor functions (Heiser, Iacoboni, Maeda, Marcus, & Mazziotta, 2003; Higuchi, Chaminade, Imamizu, & Kawato, 2009), working memory (Rottschy et al., 2012), long-term memory (Ranganath, Johnson, & d'Esposito, 2003), smelling familiar odors (Ciumas, Lindström, Aoun, & Savic, 2008), enjoyment of music (Koelsch, Fritz, Müller, & Friederici, 2006), and tactile imagery (Yoo, Freeman, McCarthy, & Jolesz, 2003), to name a few. The fact that Broca's area participates in multiple functions besides language production alone has led some researchers to propose that this area should no longer be considered as a single-purpose functional unit (Fedorenko, Duncan, & Kanwisher, 2012) but instead as a hub for multimodal information processing and integrating hierarchical information.

Humans display a size asymmetry in Broca's area, such that it is larger on the left side than the right side, and this asymmetry has been correlated with language dominance (Amunts et al., 1999). It has also been documented in fossil human endocasts, leading to speculation that extinct human species may have possessed some language capabilities (Broadfield et al., 2001; Tobias, 1983). Great apes, however, possess a homolog to Broca's area in the left hemisphere that displays a similar pattern of morphological asymmetry to that in humans (Cantalupo & Hopkins, 2001), despite not having language. There is some evidence that Broca's area in non-human primates participates in the production of communicative signals (Taglialatela, Russell, Schaeffer, & Hopkins, 2008), but this area is involved also in distal manual motor and domain-general functions (Hepp-Reymond, Hüsler, Maier, & Qi, 1994; Kurata & Tanji, 1986; Rizzolatti, Scandolara, Gentilucci, & Camarda, 1981). In fact, the homolog to Broca's area in macaques contains mirror neurons, brain cells that discharge when both performing a goal-directed action and observing or hearing someone else perform a goal-directed action (Gallese, Fadiga, Fogassi, & Rizzolatti, 1996; Kohler et al., 2002; Rizzolatti et al., 1988). Mirror neurons even respond to actions performed with a tool (Ferrari, Rozzi, & Fogassi, 2005). The likely function of a mirror neuron is thus to represent an action internally for motor learning and to understand the intentions of the action as it relates to oneself.

Mirror neurons are also present in the human brain (Grèzes, Armony, Rowe, & Passingham, 2003; Iacoboni et al., 1999; Koski, Iacoboni, Dubeau, Woods, & Mazziotta, 2003; Mukamel, Ekstrom, Kaplan, Iacoboni, & Fried, 2010; Nishitani & Hari, 2002) but seem to have adopted some novel functions since we shared a common ancestor with macaques. Many of these novel functions were necessary pre-adaptations to language, such as understanding the intentions of oral communicative actions (Buccino et al., 2001), imitation (Iacoboni et al., 1999), processing hierarchically complex sequential information (Molnar-Szakacs, Kaplan, Greenfield, & Iacoboni, 2006), and understanding the intentions and emotions of others (Carr, Iacoboni, Dubeau, Mazziotta, & Lenzi, 2003; Iacoboni et al., 2005). For these reasons, some researchers argue that Broca's area evolved atop the mirror neuron circuit located in the IFG. In other words, humans' capacity for language evolved from the multimodal mirror system that was already present in anthropoid primates, which subserved manual, vocal, and facial behaviors. Rizzolatti and Arbib (1998) were the first to present this idea, naming it the mirror system hypothesis. They proposed that primitive mirror neurons that identify actions, as seen in the macaque, evolved in several stages to support imitation, pantomime, proto-signing, and, eventually, vocal language in the hominin lineage. In other words, this hypothesis asserts that the language-processing operations of Broca's area have a general motor origin, an idea that has been around for a long time (Allot, 2012; Kimura, 1979; Lieberman, 1984; Studdert-Kennedy, 1983).

Some researchers have proposed that tool use in particular provided the motor foundation for language in the brain (Holloway, 1969; Ruck, 2014; Stout & Chaminade, 2012). If technology indeed provided the pre-adaptations for language processing, then we are in luck, because there are many hundreds of thousands of stone tools that have been preserved in the archaeological record that should bear some clues as to the process of the evolution of language. Holloway (1969), for example, directly compared language and stone tool manufacture by hypothesizing that both rely on syntactic rules to govern the order of words or actions to form a meaningful utterance or tool, respectively. The grammar of tool-making, as well as other parallel design features, led him to argue that language and stone tool-making involve similar, if not identical, cognitive processes. Some scholars have expressed doubt about the viability of stone tools as indicators of language evolution, arguing, for example, that the syntax of stone tool-making and that of language are not comparable because the order of tool-making actions is dictated by external conditions rather than internally derived rules (Graves, 1994; Wynn, 1995). There are now efforts being made by neuroarchaeologists to test this technological hypothesis for language origins by replicating the process of Early Stone Age/Lower Paleolithic (ESA/LP) tool manufacture while monitoring the corresponding functional brain activity patterns that occur during this task. If the language functions of Broca's area evolved by co-opting the same cognitive processes already in place for stone tool manufacture, then we should see functional overlap between these two behaviors in this neural substrate (Stout & Chaminade, 2012).

Stout and colleagues (Stout & Chaminade, 2007; Stout, Passingham, Frith, Apel, & Chaminade, 2011; Stout et al., 2008) provided a tantalizing window into the potential cognitive and language capabilities of early *Homo* with their series of brain imaging experiments that examined the neural correlates of ESA/LP tool manufacture. The two stone tool industries that they investigated included the Oldowan and Late

Acheulean. The Oldowan industry first appears in the archaeological record around 2.6 Mya (Semaw et al., 1997) and is characterized as an expedient technique to remove simple flakes from a core. The Acheulean industry, which consists of shaped core tools, first appears around 1.76 Mya and becomes more refined and skillfully made by 0.7 Mya (Beyene et al., 2013; Roche, 2005). These studies demonstrated that Oldowan tool-making involves bilateral ventral premotor areas of the brain, situated just posterior to Broca's area, while Late Acheulean tool-making engages these same areas, as well as *pars triangularis* in the right hemisphere. *Pars triangularis* forms the anterior part of the IFG and is a supramodal processor for hierarchically structured sequential information (Fadiga, Craighero, & D'Ausilio, 2009). It is probably best known for its role in language processing because, as a part of Broca's area in the left hemisphere, it is thought to be involved in the integration of semantic and syntactic information (Vigneau et al., 2006). Stout and Chaminade (2012) relate these results to language having a technological origin occurring some time during the ESA/LP. So, in this time, language could have co-opted the hierarchical processing functions of the IFG that were already in place for carrying out complex actions like Acheulean tool manufacture. It should be noted that while *pars triangularis* in the right hemisphere does participate in some language functions, it is the left hemisphere that handles most language functions. The technological hypothesis would be more convincing if there were evidence for Broca's area activation in the left hemisphere during stone knapping.

We cannot be confident at this point that language in any form had evolved by the time early *Homo* was making stone tools. Stout and colleagues did not consider language instruction, as opposed to nonverbal mimicry, as a variable in their experiments. All the participants learned via verbally delivered instructions at some point in their training, which may not faithfully replicate the conditions of nonverbal skill transmission in the past. My colleagues and I discovered that the process of making Oldowan and Acheulean tools can be learned without the aid of linguistic instruction, based on the results of an experiment that explored the learning differences between participants who were taught with spoken language and those taught without it (Putt, Woods, & Franciscus, 2014; also see Morgan et al., 2015; Ohnuma, Aoki, & Akazawa, 1997). Moreover, the flakes produced under the verbal and nonverbal learning conditions were morphologically different from each other, which suggested to us that different cognitive processes were involved in their production. Subsequently, we demonstrated with a neuroimaging experiment that *pars triangularis* in the right hemisphere was activated during Acheulean tool-making only among the participants who learned the skill verbally (Putt, Wijeakumar, Franciscus, & Spencer, 2017). Both the Oldowan and Acheulean tool-making tasks required increased cognitive control when they were learned *without* language instruction.

TESTING THE TECHNOLOGICAL HYPOTHESIS FOR LANGUAGE ORIGINS

Now that we know that language instruction while learning to knap replicative ESA/LP stone tools affects the pattern of neural activation, it is very important that we reconsider the neuroarchaeological evidence for the evolution of language. It may be beneficial to take a more conservative approach by replicating ESA/LP stone tool-making behaviors that were learned without language instruction. If the language centers of Broca's area or its right hemisphere analog overlap with stone knapping activation among participants who learned to knap via silent imitation (i.e., without verbal input), then this would provide stronger support for the technological hypothesis for language origins, but only if general motor activity does not also overlap with these language centers.

To test this hypothesis, I draw on the results of an experiment that my colleagues and I recently conducted that utilized a neuroimaging technique known as functional near-infrared spectroscopy (fNIRS) to measure the hemodynamic changes occurring in the brains of 33 healthy, right-handed, adult, human subjects (17 females, 16 males; age [mean ± SD] 23.8 ± 7.9 years) as they made ESA/LP tools (see Figure 14.1, Putt et al., 2017; also see Putt, 2016). fNIRS was a useful technique for this purpose because it is less influenced by motion artifact than PET and fMRI and therefore allowed us to measure real-time brain activity as people naturalistically made Oldowan and Acheulean stone tools. Participants were divided into two groups, which determined the context of ESA/LP tool-making skill transmission during their seven 1-hour-long individual training sessions. One group watched instruction videos in which the instructor delivered the lessons with verbal instructions (n = 17). The other group watched the same videos but with the sound turned off so that skills were learned via imitation rather than verbal instructions (n = 16). The instructor's face was not visible in any of the videos, to eliminate linguistic cues. At different points in their learning (after the first, fourth, and seventh training sessions) participants attended three neuroimaging sessions.

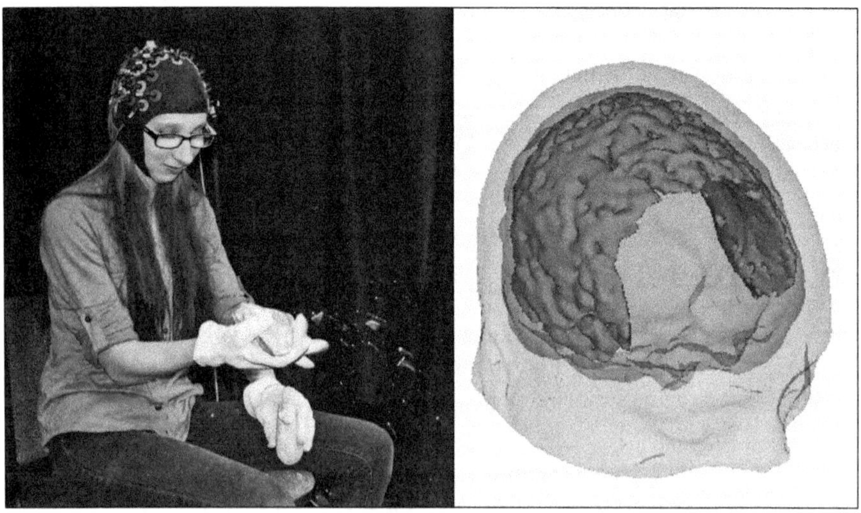

Figure 14.1. Functional brain imaging using functional near-infrared spectroscopy (fNIRS) during naturalistic stone knapping (left) and cortical areas covered by the optode geometry on the custom-made cap (right). Note: All participants in the study were right-handed. The model used for this figure is left-handed and did not participate in the study. Adapted from Figure 1b (p. 2) in Putt et al. (2017), The functional brain networks that underlie Early Stone Age tool manufacture, *Nature Human Behaviour*, and republished with permission of *Nature*.

Neuroimaging sessions included a motor baseline task, an Oldowan task, and an Acheulean task. Each task had a block design that consisted of alternating task and rest blocks. The motor baseline task involved clicking two rocks together to the beat of a metronome without the added element of trying to create flakes. With the metronome absent, participants removed simple flakes from a core during the Oldowan task and shaped a handaxe during the Acheulean task. fNIRS data were acquired at 25 Hz with a 24-channel TechEn CW6 system with wavelengths of 690 nm and 830 nm. Light was delivered to a customized cap via fiber-optic cables (Figure 14.1). fNIRS data were motion corrected and reconstructed in image space within the brain volume prior to statistical analysis (see Putt et al., 2017, for more information on this process).

There are two different ways to determine if knapping and language functionally overlap in the IFG. The first (by-eye) method simply involves visually inspecting a brain map to see where active knapping task clusters lie relative to anatomical regions of the brain that are associated with language processing. This is the method Stout and colleagues (2008) used to determine that Acheulean tool-making engages a language structure in the right IFG, but as we now know, this result was, at least potentially, the consequence of verbal instruction in the learning context (Putt et al., 2017). The second (by-coordinate) method is more rigorous in that it measures the level of knapping-specific activation at the precise location of known language centers. I carried out both by-eye and by-coordinate analyses to check for functional overlap between ESA/LP knapping and language, with a focus on bilateral IFG.

For the first analysis, I identified any active clusters, as determined by a multifactor analysis of variance (ANOVA), that overlapped with 8-mm spheres, which were constructed around the coordinates from a large language-processing meta-analysis (Vigneau et al., 2006, 2011). This meta-analysis included phonological, lexicosemantic, and sentence-processing neuroimaging studies. I then only considered clusters where the nonverbal group displayed a signal change above the threshold of zero, which signifies activation (Putt, 2016). I found significant clusters in *pars triangularis* in both the right and left hemispheres that met these criteria, $F3opd_L$ and $F3t_R$ (codes refer to the labeling system used by Vigneau et al., 2006, 2011; Figure 14.2).

One caveat to keep in mind is that the motor baseline task should not elicit significantly greater activation in these IFG areas than the knapping tasks if we are to argue that their language functions are exapted specifically from functions involved in technological behaviors. Otherwise, a hypothesis for a more general motor origin for language would better fit the data. We can see from the bar plot in Figure 14.2A, however, that rhythmically clicking rocks together during the motor baseline task activates the right IFG ($F3t_R$) to an even greater extent than does knapping. Because one of the main linguistic functions of the right IFG is to comprehend affective prosody (Wildgruber et al., 2005), which is the rhythmic pattern of stress and intonation in language, the large motor baseline signal in the right *pars triangularis* ($F3t_R$) may reflect the timing element of the motor baseline task. It appears that this area plays an important role in translating rhythm into body movement while knapping as well, at least among nonverbally instructed participants; therefore, its increased activation during stone knapping may reflect the timing element of flake removal, for example, initiating and inhibiting movement of the arm based on action goals. It is unclear why this area would be deactivated among verbally instructed participants while knapping; this issue should be explored further.

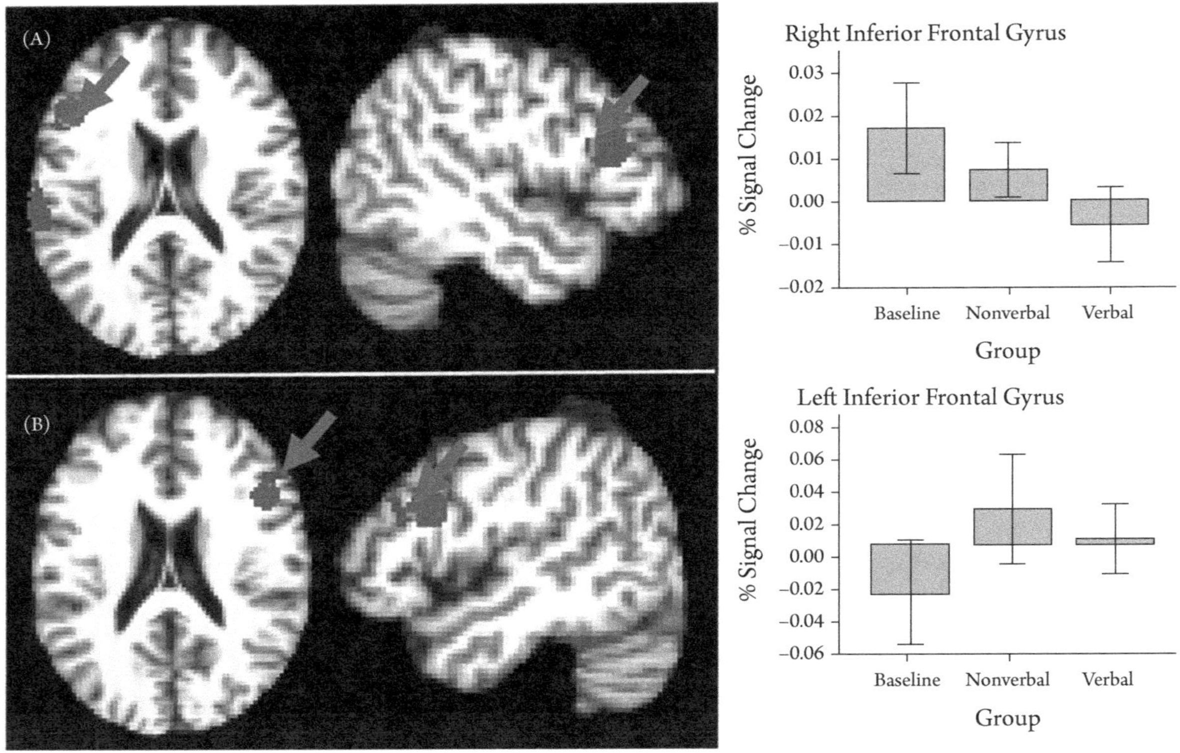

Figure 14.2. Language-processing areas (circles) in the inferior frontal gyrus (IFG) that overlap with lithic reduction activation clusters. Voxels that overlap are represented by lighter gray areas. Displayed are the right pars triangularis from the Group main effect, which overlaps with $F3t_R$ (Vigneau et al., 2011; A) and the left pars triangularis from the Group × Session interaction effect, which overlaps with $F3opd_L$ (B). % Signal change is in μM units. Error bars represent 95% confidence intervals. Previously published as Figure 26 in Putt (2016), *Human brain activity during stone tool production: Tracing the evolution of cognition and language*, doctoral thesis, University of Iowa.

Conversely, the left *pars triangularis* (F3opd$_L$) demonstrates a specialized role for stone knapping (Figure 14.2B; Putt, 2016). The left hemisphere knapping cluster overlaps with an important language center that is known to participate in semantic retrieval and selection (Noesselt, Shah, & Jäncke, 2003), lexical decision tasks (Binder et al., 2003; Perani et al., 1998; Poldrack et al., 1999), comprehension of complex sentences (Ben-Shachar, Palti, & Grodzinsky, 2004; Caplan, 2001; Caplan, Alpert, & Waters, 1999; Constable et al., 2004; Stromswold, Caplan, Alpert, & Rauch, 1996), and detection of grammatical errors (Embick, Marantz, Miyashita, O'Neil, & Sakai, 2000). The activation of this cluster among participants in the nonverbal group during the knapping tasks and its corresponding deactivation during the motor baseline task indicate its involvement in non-motor-related (i.e., cognitive) processes. These results appear to support Holloway's (1969) claim that the grammar of stone toolmaking and the grammar of language may indeed rely on the same cognitive processes. Despite this promising result, we see in Figure 14.2B that the overlap between this knapping cluster and the arbitrary sphere I constructed around the language center coordinates is fairly peripheral, and we cannot be completely confident from this analysis that these behaviors truly overlap.

To address this issue, I now turn to the second analysis that I carried out. I used the known coordinates for language centers in the IFG from the same language meta-analysis (Vigneau et al., 2006, 2011) to extract beta values, which represent the level of change in the knapping signal at the same coordinates. These values were then statistically compared to the values obtained from the rest intervals using a Wilcoxon signed-rank test to determine if knapping significantly activated this specific area of the brain. I found four language centers in the bilateral IFG where the nonverbal group had a significantly higher knapping signal than resting signal (Putt et al., 2017). These areas included two areas in *pars triangularis* (F3t$_R$ and F3td$_R$) in the right hemisphere and ventral *pars triangularis* (F3tv$_L$) and *pars opercularis* (F3opd$_L$) in the left hemisphere (Figure 14.3). F3opd$_L$ is the same left hemisphere region from the first analysis where we encountered overlap that could not be explained by general motor functions.

Once again, it is important to consider the possibility that these areas may only be active during knapping tasks simply because of the motor element of knapping. I found a higher level of motor baseline activation than knapping activation in the right hemisphere regions of interest, as well as in F3tv$_L$. Of these four regions, F3opd$_L$ is the only language center where the knapping signal is significantly higher than the motor baseline signal (Figure 14.3). Moreover, the F3opd$_L$ knapping signal is significantly higher than the resting signal during the Acheulean task, but not during the Oldowan task (Figure 14.4). What this most likely means is that the Acheulean task differs from the Oldowan task in that it places more emphasis on the order of action sequences in relation to meeting the sub-goals of the task, such as platform setup, removing square edges, and thinning and shaping the piece, and the ultimate goal(s) of the task to produce a functional and/or aesthetically pleasing core tool and usable flakes. In this way, Acheulean tool production is analogous and possibly even homologous to language production in that both may utilize the semantic, syntactic, and sentence-level processing functions of F3opd$_L$, discussed earlier.

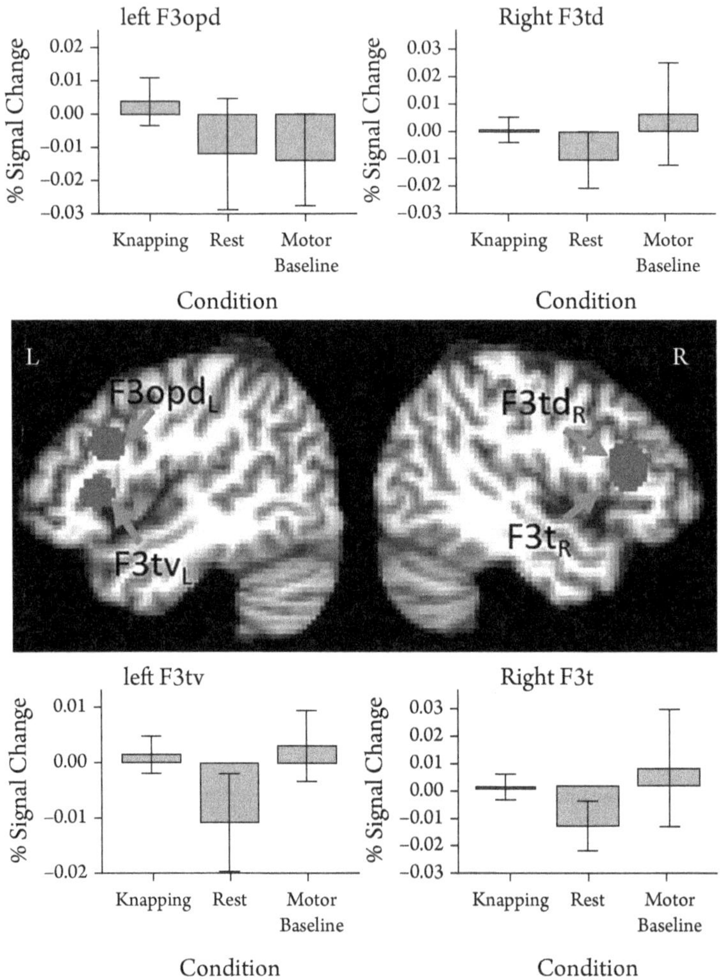

Figure 14.3. Language centers that are significantly activated during knapping tasks relative to rest, after 7 hours of nonverbal training. Note that only in F3opd$_{L}$ is the knapping signal significantly higher than the motor baseline signal, indicating that this area plays a non-motor role in both language and stone knapping. Image by the author.

DISCUSSION

Do these results support a technological hypothesis for language origins? As Stout and Chaminade (2012, p. 76) noted, "any motor activity can be described as a hierarchically structured sequence of behavioral units. The hypothesis of a *special* evolutionary relationship between tool-making and language predicts more particular overlap in information processing demands and/or neuroanatomical substrates between these two behaviors." I interpret this statement to mean that a co-evolutionary relationship between language and technology can only be claimed if we are confident that it is the cognitive behaviors and not the motor behaviors of tool-making that are driving the activation of a neural substrate within Broca's area. Both the by-eye and by-coordinates

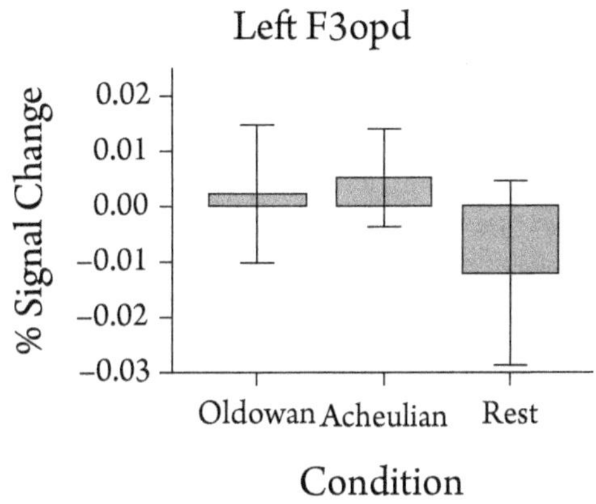

Figure 14.4. Relative activation of F3opd$_L$ during Early Stone Age/Lower Paleolithic (ESA/LP) knapping tasks and rest, after 7 hours of nonverbal training. Activation is significantly higher than rest during the Acheulean knapping task, but not during the Oldowan knapping task. Image by the author.

methods described herein identified F3opd$_L$ as the only language center in bilateral IFG that participates in the non-motor aspects of ESA/LP stone tool manufacture.

The functional and anatomical overlap in this region, therefore, indicates that a co-evolutionary relationship may exist between language and technology, as others have posited (Uomini & Meyer, 2013). This could mean that the cognitive skills needed to make knapped stone tools were later exapted for communication purposes in this segment of Broca's area, contributing eventually to the evolution of complex language, though we cannot completely rule out the possibility that a specialized form of thinking caused by a lifetime of language use resulted in Broca's area activation among our modern human participants. One possibility for how this evolutionary change occurred is that as tool manufacturing processes became more complex over time, the social context of learning technical skills would have become more important to ensure that these skills were transmitted faithfully across generations. Increasing emphasis on the intentional social transmission of tool-making skills (Stout & Chaminade, 2012) and perhaps a highly social tool-making context could have provided the necessary scaffolding in this neuroanatomical region for intentional vocal communication, thereby bridging the tool-making and communication functions that were already present in this area in pre-linguistic hominins.

Language is not a monolithic whole, nor is it confined to one particular point in the brain. Although the functional overlap between language and technology in a segment of Broca's area is an important discovery, it only speaks to the specific functions of that particular region. What about the myriad other language centers in bilateral IFG, or in the rest of the brain for that matter? As we have seen, other language sites in the IFG appear to be recruited to an even greater extent while grasping rocks and performing general arm movements to an assigned pace, especially in the right hemisphere. Why should this be? As I already mentioned, the right IFG has prosodic language functions, which may rely on a more general timing system that mediates auditory sensory memory (Rao et al., 1997). It follows, then, that the right IFG would be recruited during stone knapping behaviors because of the need to accurately time and coordinate arm movements. This region would be even more heavily recruited when the timing of movements is an important goal of the task, which would explain why the hemodynamic signal in this region was so much stronger during the motor baseline task than the stone knapping tasks. It is less clear why F3tv$_L$, which participates in semantic and syntactic language functions (Vigneau et al., 2011), has greater activation during the motor baseline task than the knapping tasks, though this could also relate to the metronome's rhythmic structure. This suggests, however, that the left *pars triangularis* portion of Broca's area also has a motor function, a function usually only attributed to *pars opercularis* (Binkofski & Buccino, 2004).

These results suggest that the language functions of different neuroanatomical regions of bilateral IFG likely have separate evolutionary histories, such that language probably did not evolve all at once as a single package (Corballis, 2010; Jackendoff, 2002). Whereas there is evidence that ESA/LP technologies could have contributed to the evolution of language functions in the left dorsal *pars opercularis* of Broca's area, I theorize that the language functions in other parts of the IFG are probably derived from general motor functions. We can therefore assume that language-processing areas in other parts of the brain probably also have separate evolutionary histories, though this hypothesis remains to be tested.

This brings us back to the question of whether or not stone tools can tell us anything about the language capacity of the individuals who made them. Wynn claimed that any attempts to link stone tools and language are inherently unpersuasive unless theoretically justified. In response, I have relied on the methods and theory of cognitive neuroscience to justify a functional overlap between language and stone tool production in one portion of Broca's area of the brain, which points to these two behaviors relying on a similar cognitive process. As this region participates in the specific syntactic and semantic processing functions of language described earlier, it likely also plays a similar role during stone tool manufacture, allowing the tool-maker to identify action units and place them in the correct order so as to derive meaning from the overall structure of action units. The stronger activation of this area during the Acheulean task implies that these abilities are more important for making Acheulean handaxes than Oldowan tools, which is probably because handaxe production requires a more complex sequence of actions. And as the archaeological record reflects a gradation in operational complexity of tool-making tasks over time, we can infer that these specific functions that take place in this part of Broca's area used for tool production, and possibly language, evolved around 1.8 Mya, about the time that technological complexity shifted. It appears that stone tools do have stories to tell about language and its evolution, but only with a controlled neuroscientific approach can we establish a direct link between stone tools and language in the brain and thus decode their message.

ACKNOWLEDGMENTS

I would like to thank the editors for inviting me to contribute to this volume. I thank Mark Putt and two anonymous reviewers for their helpful comments on an earlier draft. This work would not have been possible without the help of my research team, including John Spencer, Sobanawartiny Wijeakumar, and Robert Franciscus, my research assistants, Danielle Jones and Chloe Daniel, and funding from the Wenner-Gren

Foundation (Grant #8968), the Leakey Foundation, the John Templeton Foundation, Sigma Xi, the Scientific Research Society, the University of Iowa, and AAUW.

REFERENCES

Allot, R. (2012). *The motor theory of language origin: 1989*. Bloomington, IN: Xlibris Corporation.

Amunts, K., Schleicher, A., Bürgel, U., Mohlberg, H., Uylings, H., & Zilles, K. (1999). Broca's region revisited: Cytoarchitecture and intersubject variability. *Journal of Comparative Neurology, 412*(2), 319–341.

Ben-Shachar, M., Palti, D., & Grodzinsky, Y. (2004). Neural correlates of syntactic movement: Converging evidence from two fMRI experiments. *NeuroImage, 21*(4), 1320–1336.

Beyene, Y., Katoh, S., WoldeGabriel, G., Hart, W. K., Uto, K., Sudo, M., . . . Asfaw, B. (2013). The characteristics and chronology of the earliest Acheulean at Konso, Ethiopia. *Proceedings of the National Academy of Sciences of the United States of America, 110*(5), 1584–1591.

Binder, J. R., McKiernan, K. A., Parsons, M. E., Westbury, C. F., Possing, E. T., Kaufman, J. N., & Buchanan, L. (2003). Neural correlates of lexical access during visual word recognition. *Journal of Cognitive Neuroscience, 15*(3), 372–393.

Binkofski, F., & Buccino, G. (2004). Motor functions of the Broca's region. *Brain and Language, 89*(2), 362–369.

Broadfield, D. C., Holloway, R. L., Mowbray, K., Silvers, A., Yuan, M. S., & Màrquez, S. (2001). Endocast of Sambungmacan 3 (Sm 3): A new *Homo erectus* from Indonesia. *The Anatomical Record, 262*(4), 369–379.

Buccino, G., Binkofski, F., Fink, G. R., Fadiga, L., Fogassi, L., Gallese, V., . . . Freund, H.-J. (2001). Action observation activates premotor and parietal areas in a somatotopic manner: An fMRI study. *European Journal of Neuroscience, 13*(2), 400–404.

Cantalupo, C., & Hopkins, W. D. (2001). Asymmetric Broca's area in great apes: A region of the ape brain is uncannily similar to one linked with speech in humans. *Nature, 414*(6863), 505.

Caplan, D. (2001). Functional neuroimaging studies of syntactic processing. *Journal of Psycholinguistic Research, 30*(3), 297–320.

Caplan, D., Alpert, N., & Waters, G. S. (1999). PET studies of syntactic processing with auditory sentence presentation. *NeuroImage, 9*(3), 343–351.

Carr, L., Iacoboni, M., Dubeau, M.-C., Mazziotta, J. C., & Lenzi, G. L. (2003). Neural mechanisms of empathy in humans: A relay from neural systems for imitation to limbic areas. *Proceedings of the National Academy of Sciences of the United States of America, 100*(9), 5497–5502.

Ciumas, C., Lindström, P., Aoun, B., & Savic, I. (2008). Imaging of odor perception delineates functional disintegration of the limbic circuits in mesial temporal lobe epilepsy. *NeuroImage, 39*(2), 578–592.

Constable, R. T., Pugh, K. R., Berroya, E., Mencl, W. E., Westerveld, M., Ni, W., & Shankweiler, D. (2004). Sentence complexity and input modality effects in sentence comprehension: An fMRI study. *NeuroImage, 22*(1), 11–21.

Corballis, M. C. (2010). Mirror neurons and the evolution of language. *Brain and Language, 112*(1), 25–35.

Embick, D., Marantz, A., Miyashita, Y., O'Neil, W., & Sakai, K. L. (2000). A syntactic specialization for Broca's area. *Proceedings of the National Academy of Sciences of the United States of America, 97*(11), 6150–6154.

Fadiga, L., Craighero, L., & D'Ausilio, A. (2009). Broca's area in language, action, and music. *Annals of the New York Academy of Sciences, 1169*(1), 448–458.

Fedorenko, E., Duncan, J., & Kanwisher, N. (2012). Language-selective and domain-general regions lie side by side within Broca's area. *Current Biology, 22*(21), 2059–2062.

Ferrari, P. F., Rozzi, S., & Fogassi, L. (2005). Mirror neurons responding to observation of actions made with tools in monkey ventral premotor cortex. *Journal of Cognitive Neuroscience, 17*(2), 212–226.

Gallese, V., Fadiga, L., Fogassi, L., & Rizzolatti, G. (1996). Action recognition in the premotor cortex. *Brain, 119*(2), 593–609.

Graves, P. (1994). Flakes and ladders: What the archaeological record cannot tell us about the origins of language. *World Archaeology, 26*(2), 158–171.

Grèzes, J., Armony, J. L., Rowe, J., & Passingham, R. E. (2003). Activations related to "mirror" and "canonical" neurons in the human brain: An fMRI study. *NeuroImage, 18*(4), 928–937.

Heiser, M., Iacoboni, M., Maeda, F., Marcus, J., & Mazziotta, J. C. (2003). The essential role of Broca's area in imitation. *European Journal of Neuroscience, 17*(5), 1123–1128.

Hepp-Reymond, M.-C., Hüsler, E. J., Maier, M. A., & Qi, H.-X. (1994). Force related neuronal activity in two regions of the primate ventral premotor cortex. *Canadian Journal of Physiology and Pharmacology, 72*(5), 571–579.

Higuchi, S., Chaminade, T., Imamizu, H., & Kawato, M. (2009). Shared neural correlates for language and tool use in Broca's area. *Neuroreport, 20*(15), 1376–1381.

Holloway, R. L. (1969). Culture: A human domain. *Current Anthropology, 10*(4), 395–412.

Iacoboni, M., Molnar-Szakacs, I., Gallese, V., Buccino, G., Mazziotta, J. C., & Rizzolatti, G. (2005). Grasping the intentions of others with one's own mirror neuron system. *PloS Biology, 3*(3), 1–7.

Iacoboni, M., Woods, R. P., Brass, M., Bekkering, H., Mazziotta, J. C., & Rizzolatti, G. (1999). Cortical mechanisms of human imitation. *Science, 286*(5449), 2526–2529.

Jackendoff, R. (2002). *Foundations of language: Brain, meaning, grammar, evolution.* Oxford, UK: Oxford University Press.

Kimura, D. (1979). Neuromotor mechanisms in the evolution of human communication. In H. D. Steklis & M. J. Raleigh (Eds.), *Neurobiology of social communication in primates: An evolutionary perspective* (pp. 197–220). New York: Academic Press.

Kitahara-Frisch, J. (1978). Stone tools as indicators of linguistic ability in early man. *Kagaku Kisoron Gakkai Annals, 5*(3), 101–109.

Koelsch, S., Fritz, T., Müller, K., & Friederici, A. D. (2006). Investigating emotion with music: An fMRI study. *Human Brain Mapping, 27*(3), 239–250.

Kohler, E., Keysers, C., Umilta, M. A., Fogassi, L., Gallese, V., & Rizzolatti, G. (2002). Hearing sounds, understanding actions: Action representation in mirror neurons. *Science, 297*(5582), 846–848.

Koski, L., Iacoboni, M., Dubeau, M.-C., Woods, R. P., & Mazziotta, J. C. (2003). Modulation of cortical activity during different imitative behaviors. *Journal of Neurophysiology, 89*(1), 460–471.

Kurata, K., & Tanji, J. (1986). Premotor cortex neurons in macaques: Activity before distal and proximal forelimb movements. *Journal of Neuroscience, 6*(2), 403–411.

Lieberman, P. (1984). *The biology and evolution of language.* Cambridge, MA: Harvard University Press.

Molnar-Szakacs, I., Kaplan, J., Greenfield, P. M., & Iacoboni, M. (2006). Observing complex action sequences: The role of the fronto-parietal mirror neuron system. *NeuroImage, 33*(3), 923–935.

Morgan, T. J. H., Uomini, N. T., Rendell, L. E., Chouinard-Thuly, L., Street, S. E., Lewis, H. M., . . . Laland, K. N. (2015). Experimental evidence for the co-evolution of hominin tool-making teaching and language. *Nature Communications, 6*(6029), 1–18.

Mukamel, R., Ekstrom, A. D., Kaplan, J., Iacoboni, M., & Fried, I. (2010). Single-neuron responses in humans during execution and observation of actions. *Current Biology, 20*(8), 750–756.

Nishitani, N., & Hari, R. (2002). Viewing lip forms: Cortical dynamics. *Neuron, 36*(6), 1211–1220.

Noesselt, T., Shah, N. J., & Jäncke, L. (2003). Top-down and bottom-up modulation of language related areas—An fMRI study. *BMC Neuroscience, 4*(1), 13.

Ohnuma, K., Aoki, K., & Akazawa, T. (1997). Transmission of tool-making through verbal and nonverbal communication: Preliminary experiments in Levallois flake production. *Anthropological Science, 105*(3), 159–168.

Perani, D., Paulesu, E., Galles, N. S., Dupoux, E., Dehaene, S., Bettinardi, V., . . . Mehler, J. (1998). The bilingual brain: Proficiency and age of acquisition of the second language. *Brain, 121*(10), 1841–1852.

Poldrack, R. A., Wagner, A. D., Prull, M. W., Desmond, J. E., Glover, G. H., & Gabrieli, J. DE. (1999). Functional specialization for semantic and phonological processing in the left inferior prefrontal cortex. *NeuroImage, 10*(1), 15–35.

Putt, S. S. (2016). *Human brain activity during stone tool production: Tracing the evolution of cognition and language.* Doctoral thesis, University of Iowa, Iowa City.

Putt, S. S., Wijeakumar, S., Franciscus, R. G., & Spencer, J. P. (2017). The functional brain networks that underlie Early Stone Age tool manufacture. *Nature Human Behaviour, 1*, 1–8.

Putt, S. S., Woods, A. D., & Franciscus, R. G. (2014). The role of verbal interaction during experimental bifacial stone tool manufacture. *Lithic Technology, 39*(2), 96–112.

Ranganath, C., Johnson, M. K., & d'Esposito, M. (2003). Prefrontal activity associated with working memory and episodic long-term memory. *Neuropsychologia, 41*(3), 378–389.

Rao, S. M., Harrington, D. L., Haaland, K. Y., Bobholz, J. A., Cox, R. W., & Binder, J. R. (1997). Distributed neural systems underlying the timing of movements. *Journal of Neuroscience, 17*(14), 5528–5535.

Rizzolatti, G., & Arbib, M. A. (1998). Language within our grasp. *Trends in Neurosciences, 21*(5), 188–194.

Rizzolatti, G., Camarda, R., Fogassi, L., Gentilucci, M., Luppino, G., & Matelli, M. (1988). Functional organization of inferior area 6 in the macaque monkey: II. Area F5 and the control of distal movements. *Experimental Brain Research, 71*(3), 491–507.

Rizzolatti, G., Scandolara, C., Gentilucci, M., & Camarda, R. (1981). Response properties and behavioral modulation of "mouth" neurons of the postarcuate cortex (area 6) in macaque monkeys. *Brain Research, 225*(2), 421–424.

Roche, H. (2005). From simple flaking to shaping: Stone knapping evolution among early hominins. In V. Roux & B. Bril (Eds.), *Stone knapping: The necessary conditions for a uniquely hominin behaviour* (pp. 35–48). Cambridge, UK: McDonald Institute for Archaeological Research.

Rottschy, C., Langner, R., Dogan, I., Reetz, K., Laird, A. R., Schulz, J. B., . . . Eickhoff, S. B. (2012). Modelling neural correlates of working memory: A coordinate-based meta-analysis. *NeuroImage*, *60*(1), 830–846.

Ruck, L. (2014). Manual praxis in stone tool manufacture: Implications for language evolution. *Brain and Language*, *139*, 68–83.

Semaw, S., Renne, P. R., Harris, J. W. K., Feibel, C. S., Bernor, R. L., Fesseha, N., & Mowbray, K. (1997). 2.5-million-year-old stone tools from Gona, Ethiopia. *Nature*, *385*(6614), 333–336.

Stout, D., & Chaminade, T. (2007). The evolutionary neuroscience of tool making. *Neuropsychologia*, *45*(5), 1091–1100.

Stout, D., & Chaminade, T. (2012). Stone tools, language and the brain in human evolution. *Philosophical Transactions of the Royal Society of London. Series B, Biological Sciences*, *367*(1585), 75–87.

Stout, D., Passingham, R. E., Frith, C. D., Apel, J., & Chaminade, T. (2011). Technology, expertise and social cognition in human evolution. *European Journal of Neuroscience*, *33*(7), 1328–1338.

Stout, D., Toth, N. P., Schick, K. D., & Chaminade, T. (2008). Neural correlates of Early Stone Age tool-making: Technology, language and cognition in human evolution. *Philosophical Transactions of the Royal Society of London. Series B, Biological Sciences*, *363*(1499), 1939–1949.

Stromswold, K., Caplan, D., Alpert, N., & Rauch, S. (1996). Localization of syntactic comprehension by positron emission tomography. *Brain and Language*, *52*(3), 452–473.

Studdert-Kennedy, M. (1983). *Psychobiology of language*. Cambridge, MA: MIT Press.

Taglialatela, J. P., Russell, J. L., Schaeffer, J. A., & Hopkins, W. D. (2008). Communicative signaling activates "Broca"s' homolog in chimpanzees. *Current Biology*, *18*(5), 343–348.

Tobias, P. V. (1983). Recent advances in the evolution of the hominids with especial reference to brain and speech. In C. Chagas (Ed.), *Working group on recent advances in the evolution of primates, May 24–27, 1982* (pp. 85–140). Vatican City: Pontificial Academy of Sciences.

Uomini, N. T., & Meyer, G. F. (2013). Shared brain lateralization patterns in language and Acheulean stone tool production: A functional transcranial Doppler ultrasound study. *PLoS One*, *8*(8), 1–9.

Vigneau, M., Beaucousin, V., Hervé, P.-Y., Duffau, H., Crivello, F., Houde, O., . . . Tzourio-Mazoyer, N. (2006). Meta-analyzing left hemisphere language areas: Phonology, semantics, and sentence processing. *NeuroImage*, *30*(4), 1414–1432.

Vigneau, M., Beaucousin, V., Hervé, P.-Y., Jobard, G., Petit, L., Crivello, F., . . . Tzourio-Mazoyer, N. (2011). What is right-hemisphere contribution to phonological, lexico-semantic, and sentence processing? Insights from a meta-analysis. *NeuroImage*, *54*(1), 577–593.

Wildgruber, D., Riecker, A., Hertrich, I., Erb, M., Grodd, W., Ethofer, T., & Ackermann, H. (2005). Identification of emotional intonation evaluated by fMRI. *NeuroImage*, *24*(4), 1233–1241.

Wynn, T. (1991). Tools, grammar and the archaeology of cognition. *Cambridge Archaeological Journal*, *1*(2), 191–206.

Wynn, T. (1995). Handaxe enigmas. *World Archaeology*, *27*(1), 10–24.

Yoo, S.-S., Freeman, D. K. I., McCarthy, J. J., & Jolesz, F. A. (2003). Neural substrates of tactile imagery: A functional MRI study. *Neuroreport*, *14*(4), 581–585.

15

IN THREE MINDS
EXTENDING COGNITIVE ARCHAEOLOGY WITH THE SOCIAL BRAIN

Cory Stade and Clive Gamble

INTRODUCTION

Our present ability to engage with the cognition of our early ancestors is in large measure due to the work of Thomas Wynn. In a paper published almost 40 years ago (Wynn, 1979), he proposed a methodology that examined the evolution of cognition in fossil hominins. Wynn's call to arms was revolutionary for a discipline that eschews studying soft human capacities such as intentions and emotions, which do not clearly survive as material evidence. Steven Mithen (1996, pp. 36–37), for example, recalls reading the paper as a student and finding the conclusions and the application of Piaget's ideas inspiring. The discipline of cognitive archaeology has therefore indeed burgeoned from the seeds sown by Wynn and his fusion of archaeology with psychology.

Two things distinguished Wynn's approach from other cognitive excursions into the heads of fossil ancestors: First, there was his familiarity with cognitive science; second, his strict application of these models to archaeological evidence. Much later, he wrote with Fredrick Coolidge (Wynn & Coolidge, 2009) that a sound argument in cognitive archaeology must do three things:

(1) Understand the cognitive ability being scrutinized.
(2) Identify specific actions enabled by this cognitive ability.
(3) Define criteria to identify these cognitive actions using archaeological evidence.

This framework contrasts with the use of archaeological evidence by some evolutionary psychologists, whose "cultivated ignorance of the paleoanthropological record" he has criticized (Wynn, 2009). The three criteria also highlight ill-advised attempts by archaeologists to fit their taxonomy of stone tools (devised to serve other agendas in the nineteenth century) into a sequence of cognitive development (Wynn, 2009).

It is against this background that we explore ways to extend Wynn's vision for an evolutionary cognitive archaeology (ECA). He has identified (Wynn, 2009) three current approaches in deep history: linguistic (Noble & Davidson, 1996), action-oriented (Leroi-Gourhan, 1993), and representational (Mithen, 1996), and we do not disagree with this division. In this chapter, however, we consider three models of the Paleolithic mind—cognitivist, experiential, and anthropological—that underpin such approaches. We would not go so far as to declare the death of the mind as an archaeological pursuit (Gosden, 2010), but we do suggest that the ECA Wynn favors needs to be broadened to include the social and the emotional, alongside the materiality, of hominin existence. What follows is an exploration of possibilities. Our starting point is Wynn's conclusion that cognitive evolution was a temporal mosaic (2009, p. 149). To examine this mosaic further, we adopt the social brain hypothesis and the analytical device of theory of mind (ToM). We avoid matching the evidence to problematic taxonomies such as Gabriel de Mortillet's (1872) classic Paleolithic typology and Grahame Clark's (1969) five technological modes. Instead, we align our argument to Wynn's as follows:

(1) Understanding the cognitive ability being scrutinized in terms of ToM and associated empathetic abilities
(2) Identifying specific actions enabled by this cognitive ability, namely, the emotional scaffolding of social life dependent on cognitive reasoning permitted by ToM
(3) Defining criteria to identify these cognitive actions using archaeological evidence (the materiality of "stuff," and things that *extend* cognition, not *reflect* it)

In this way, we will expand Wynn's ECA using his own rigorous framework. We will do this by encompassing the emotional and the social, which are inseparable and integral to interpreting the surviving materials of hominin minds.

IN THREE MINDS

The first and dominant model of the Paleolithic mind, which we term the *rational mind*, is Cartesian in origin, in that it separates mind from body. It deals in binary distinctions, most notably between subject and object, animate and inanimate, and human and animal. Cartesian rationalism has difficulty in ascribing agency to "things," reserving such a capacity for human actors alone. As a model of the mind and how cognition operates, its greatest success is the triumph of the scientific method and the measurable benefits this has brought to parameters such as medicine and engineering. On the downside, much of this advance has been achieved by circumscribing the domain of scientific knowledge. In particular, the importance of emotion has been downplayed at best and often ignored. This can be seen in recent attempts by some evolutionary scientists to expose religion and faith as unscientific and grounded in emotion. In the same way, music, the most emotive of the arts, has been dismissed as "auditory cheesecake" (Pinker, 1997), for no other reason than it eludes easy rational analysis (Morley, 2014).

We describe the second mind as experiential, or phenomenological. In direct contrast to the dominant rational approach, phenomenology addresses human experience. Experiential approaches put mind and body back together, but, it seems, with the emphasis favoring a bodily perspective. As a result, the power of affect replaces a quantitative understanding of time, place, things, and people with a qualitative feeling. This has led to the rise of an "interpretive archaeology," although currently there are few examples from deep history. Instead, experiential approaches have flourished in the richly populated landscapes of farmers and city states. Put simply, the experiential model of the mind thrives on lots of "stuff" (Miller, 2012), and lots of stuff is something that happens with the advent of agriculture or a sedentary society. One example of how this might work in deep history is provided by Gamble's (2007) study that built on Tilley's (1999) investigation of solid metaphors. A simple distinction was made between objects that contained, or wrapped, and those which acted as instruments. These two broad categories emphasize the bodily experiences of being "in" or acting "on." They work as solid, material metaphors, comparable to more familiar linguistic tropes (Lakoff & Johnson, 1980).

The third mind is anthropological, by which we mean it is relational (emphasizing the relationship between agents, not just the discrete agents themselves). This model further dissolves the dichotomy between mind and body or subject and object by reversing the idea that individuals are isolates. Instead of individuals stopping at the boundary of their skin, their agency bleeds into other people, things, and the surrounding environment. This mind is porous, and it leaks like a sieve. Because it is underpinned by reciprocal agency, the flows through the holes in the sieve are two-way. Instead of acting on a largely passive assemblage of things and places, the world possesses its own agency. Our mind is therefore distributed and our cognition extended (Clark & Chalmers, 1998). The tree we sit under or the stone tool a hominin made is, and was, as much part of the relational mind as the cells of the neocortex. But we must be wary of repeating Cartesian dualisms. Using a relational approach, contrasts have been drawn between Western and Melanesian individuals, for example, which seem to reinforce the binary approach to cognition and mind (LiPuma, 1998). Instead, we need to acknowledge that a single model of the mind—rational, experiential, or relational—does not do justice to the complexities of human cognition.

Our inclination for the present purpose draws us toward the extended mind and a relational model of cognition (Dunbar, Gamble, & Gowlett, 2010; Fowler, 2016), but we recognize that the perspectives from the other two models will be appropriate in other circumstances, and we do not see them as mutually exclusive. What is important, however, are the questions we want to address as part of an ECA. In the rest of this chapter, we will focus on a social and emotional perspective, and ask what cognitive abilities correlated with the increase in brain growth, sociality, and technological complexity over almost four million years of deep human history.

SOCIAL BRAINS, SOCIAL MINDS

The suggestion that human intelligence evolved to support social lives rather than solve ecological problems is not new (e.g., Byrne & Whiten, 1988). But since Dunbar's (1992) observation that primate brain size scales predictably with social group size, the resultant social brain hypothesis has taken center stage within archaeological discussions (Dunbar, Gamble, & Gowlett, 2014). The rigors of surviving and thriving within large, complex groups are demanding, requiring individuals to engage dynamically with a panoply of highly individual and unpredictable social partners.

These group dynamics relate to ToM, the ability to attribute mental states to others, such as emotions, intentions, and knowledge or belief states. This cognitive skill of mentalizing allows us to take the perspective of our social partners, imagining what they may be thinking, and understanding that their thoughts may be different from our own. A ToM is of vital importance for many forms of social interaction that we often take for granted. Complex language, imitation, teaching, and social emotions such as pride, shame, and guilt all require the ability to hypothesize about the thoughts of our social partners. The social brain hypothesis suggests that increased brain size throughout human evolution may index the acquisition of increasingly advanced levels of ToM (Dunbar, 2004, 2007; also see Chapter 17 in this volume).

The insights provided by the social brain hypothesis are a valuable movement toward a social and emotional perspective on Wynn's ECA. However, it belongs firmly within a "cognitive" Paleolithic mind. ToM, as currently defined, refers almost exclusively to processes occurring *within* the brains of individuals. Little explicit focus is placed on the sociocultural context in which this thinking occurs. While the context in which it has been used is often this cognitivist perspective, nothing inherent about ToM prevents it from working within an anthropological conceptual framework. The challenge for an ECA, as we see it, is to define ToM in such a way that it can be relocated within the anthropological mind, focusing instead on the "space between minds" (a relational space), and the processes that tie them together. Whereas two cognitive minds may occupy their own mental worlds and exist independently of one another, anthropological minds can only be fully understood when tethered together, engaged in intersubjective interaction with others and their worlds. This is done through the sharing of mental states and material meaning, extending the boundaries of the mind beyond the confines of the brain as well as the individual.

Emotions act as social glue (Whitehouse & Lanman, 2014), creating the "bonds that bind" people together (Gamble, Gowlett, & Dunbar, 2012). This is done not by individual meaning-making but through a dynamic process of social discourse, with emotions deriving their shape and meaning from the interactions that give rise to them (Boiger & Mesquita, 2015). When engaged in dynamic social interactions, it is not possible for minds to remain contained; they spill out into the world through our emotions, providing constant feedback to social partners about our mental processes. As such, emotions play a significant role in connecting our inner psyches with the outside world. This can be described as "hot" cognition, where we *feel* another person's thoughts in a deeply affective sense, as opposed to the "cold" cognition of mentalizing, with individuals detached from the situation they are evaluating. With relationships felt deeply and affectively, people are bound together more tightly, creating the cohesive social groups that characterize the modern human niche. Anthropological minds are compassionate and altruistic not because they *think* they should be, but because they *feel* the need to be so.

These dynamic social interactions are embedded within, and shaped by, a shared material world. From an early age, children explore the world they share with their caregivers, pointing out objects and deriving pleasure when they are validated with a response (Lewis & Granic, 2010). By jointly attending to an object in this way, two minds are brought into the material world, with social partners anchored physically together. The mind becomes extended, as objects are no long simply one's own but are understood to exist in the worlds of others, replete with their meanings. In this sense, people and things can be seen as thinking jointly together, with objects having their own agency to guide interactions. It is through this participation in the dynamic process of social interaction that objects go beyond their *affordances* (Gibson, 1986) to become material metaphors for action and experience (Gamble, 2007). By thinking *through* things, the material world acquires meaning and significance, with memories and experiences folded into them, coming ultimately to inform the construction of folk psychological categories (i.e., a ToM) and the way we think about the world around us. If emotions are the bonds that bind, objects are the enticement that draws minds out into the world and brings them together in a shared material landscape.

THE MATERIAL DIMENSION

An anthropological model of cognitive archaeology intimately situates materials and minds with gradient boundaries. It acknowledges the agency of tools and objects and their active role in the cognition of individuals. Here it is important to remember that individuals are also objects and therefore function within an extended mind as "things to think through" too. Remembering that individuals are part of an extended mind further strengthens the social and emotional dimension of an ECA.

What this means for the interpretation of material culture is that people and objects think together, and people jointly think with objects in triangulation (between person to person and object). For the archaeologist, material artifacts were once an active part of a human (and its relationships) cognition. The things people used in the past are therefore *cognitive fossils*; viewing things this way legitimizes how evolutionary cognitive archaeologists can surmise about minds in deep time, using objects as its foundational evidence, to reconstruct past abilities and cognitive processes in exactly the way a bone contributes to the reconstruction of physical structure.

This changes the way we view the material record—not as a passive recipient of actions being done to it, but as active parties in the doing and thinking. The next step will be to reapproach the material record while keeping in mind the "doing and thinking" that objects took a role in, and extrapolate information about those activities from them. The bulk of Paleolithic cognitive fossils recovered by archaeologists are stone tools, and these have formed a key element in Wynn's research. He remarked, "Stone tools have always played a role in archaeologists' attempts to document the evolution of human behaviour" (Wynn, 1985, p. 32), but also pointed out that a weakness in archaeological interpretation is the lack of psychological information used to discuss the mechanisms at play. That is where an understanding of emotions and ToM can be brought in to provide robust explanations of cultural material. To do this they must be present throughout the argument and embedded within the methodology. If stone tools are cultural objects that have been learned, and seen to be created and used within a community, then they are wrapped in a social and emotional life.

The earliest known knapped stone tools, at Lomekwi in Kenya, dating from 3.3 million years ago (Mya) (Harmand et al., 2015), are large, robust, igneous tools that experiments conducted by the research team suggest were made by either a passive hammer or bipolar technique (Lewis & Harmand, 2016). In the passive hammer technique, a core is held in both hands and brought down onto an anvil to fracture it. In the bipolar technique, a stone is held on an anvil while another stone is brought down onto it, similar to chimpanzee nut-cracking technologies (Matsuzawa, Humle, & Sugiyama, 2014). The relevance of the knapping technique is the manner in which these skills would have been culturally transmitted between individuals in the community; as we can see from chimpanzee communities, nut-cracking can be learned through social learning processes that do not require the theorizing of mental states that comes with a ToM (Tennie, Call, & Tomasello, 2010). Chimpanzees learn nut-cracking when they are young and have an intense bond with their mother, who passively exposes them over a long period of time to the behaviors and skills involved in acquiring the ability. However, these skills are mostly acquired through the young chimpanzee being in the vicinity of the tools and experiencing their affordances through their own experimentation, learning, for example, that there are nuts within a shell and that they are desirable. Young chimpanzees do not appear to learn by imitating the actions of their mothers (Matsuzawa et al., 2014, but also see Boesch, 1991). Each chimpanzee, then, is effectively "reinventing the wheel" (Tennie et al., 2010), and this prevents the cultural behavior from being transmitted in a faithful enough way that small modifications also can be added and transmitted, thereby never introducing the "ratchet effect" of cumulative culture. Lomekwian tools also have this hallmark and thus do not signal a community with the cultural repertoire of abundant material culture, or a ratchet effect made available by complex social learning.

Returning to the cognitive fossil record, the time when these cognitive fossils begin to evoke a different social interaction taking place would be when more complex tool forms emerge and which appear to have been faithfully replicated according to the ratchet effect resulting from more sophisticated information transfer. While it is perhaps less clear whether core and flake technologies of the Oldowan techno-complex would have been the result of faithful replication (perhaps more of replicated technique than replicated form), handaxes provide a more salient example of a replicated form. These artifacts are the popular example of "stasis" in the human technological record and were used over a million years across three continents. The general plan of a handaxe does not appear to be one that could be learned again and again from nothing (barring a biological impetus, e.g., Corbey, Jagich, Vaesen, & Collard, 2016), even if the learner were exposed to the appropriate materials and functional motivations. Its idiosyncratic form, and the fact that the form itself appears remarkably similar geographically and temporally, hints at a cultural transmission of a particular method of creating a tool to perform a function. This alludes to the importance of the social context in which learning to make a tool such as a handaxe would have taken place. An imitative ability would then necessarily exist, which leads to implications of other abilities such as joint attention and the ability to triangulate attention between people and objects, all abilities that scaffold the understanding of the existence of others' minds. Once these abilities of a quasi-ToM are possessed, they open up a shared mental landscape where meaning is not solely personal; materials and methods now acquire conventions, and the relationships between materials and minds are strengthened and interactive. The material dimension has then increased in quality, and exponentially so.

This relationship can be intensified with the awareness of itself. More complex ToM is defined by the knowledge of the shared mental landscape that comes with knowing that others not only think as you do but know that about you as well. In particular, this ability (and motivation) to intentionally impart knowledge to another can result in teaching. Cognitive fossils that allude to teaching having taken place are those with aspects not easily acquired by mere copying; the "teacher" must make the modification explicit for it to be salient, both in technique and in function. Lithics made by prepared core technology have been suggested more than once by researchers to imply teaching for its successful transmission (Lycett, von Cramon-Taubadel, & Eren, 2016; Morgan et al., 2015). With this type of transmission, a social network is necessarily implied. Teaching also implies affective investment in imparting the knowledge, for it occurs with socially laden meaning, which evokes established empathetic networks.

When learning to knap, the tool becomes the third partner in a process of "inter-subjective triangulation," as it is the subject of joint attention from both the teacher and the pupil. As the object of attention, the thing draws minds together, becoming invested with the meaning and significance of the interpersonal interaction. To the teacher, a handaxe may be an example for demonstration ("here's one I made earlier") for the pupil; it may also be an example of something to achieve and aspire to. For both, the handaxe is emblematic of the relationship, a mnemonic for past interactions in a dynamic and changing relationship. As things become more widely used, they can appear in a multitude of interactions and can come to stand for a variety of different relationships. In this way, things can be seen as the tangible remains left over from the relationships within which they once acted.

THE CAVE IN THE MIND: ECA AND LANDSCAPE

Wynn has focused his analytical gaze on stone tools. With colleague Fredrick Coolidge, he has developed a sophisticated model incorporating psychological research for the evolution of memory that can be referenced to changes in lithic morphology. Many of the chapters in this volume explore these issues. In our extension of the ECA, we want to look at a case study. Our aim in this is to expand the ECA by considering the perspectives of the three minds—rational, experiential, and anthropological. We offer here a preliminary analysis of the recent discovery of early Neandertal constructions inside Bruniquel Cave, France (Jaubert et al., 2016). We rely on the excellent descriptions provided by other researchers, including the video fly-round contained in the report.

The Response of the Rational Mind

Bruniquel was unexpected for three reasons: its age, the species that must have made these constructions, and where they were found. The six, largely circular constructions of intentionally broken speleofacts, overlain in places with traces of fire, are closely dated to 176,000 years ago. This makes them contemporary with some of the oldest anatomically modern human fossils in East Africa (Clark et al., 2003; McDougall, Brown, & Fleagle, 2005). In Europe, their age places these constructions in a severe cold stage (MIS6), when Northern Europe was glaciated and southwest France was a permafrost zone. Currently, the only Eurasian hominin who can be credited with their construction are Neandertals.

The constructions at Bruniquel challenge the model of human cognitive evolution that has been in place since McBrearty and Brooks' (2000) paper sank the previously prevalent idea of a human revolution. Discoveries across Africa and in the time frame from 300,000 to 50,000 years ago (Wadley, 2013) pointed to the slow accumulation of a wide range of modern traits within a single continent (Henshilwood et al., 2002). Bruniquel shows that other geospatial areas and distinct hominin groups were part of this development, a position supported by Neandertals intentionally selecting feathers by color for display (Finlayson et al., 2012) and being interested in eagle claws as ornaments (Radovčić, Sršen, Radovčić, & Frayer, 2015).

The social brain model suggests that we should not be surprised at this cognitive sophistication. Neandertals and Pleistocene humans have very similar personal network sizes, as predicted from their brain sizes. For some (Pearce & Dunbar, 2012; Pearce, Stringer, & Dunbar, 2013), this similarity is instead a difference to be explained by the larger contribution of the visual cortex to overall brain size among Neandertals, itself a response to lower light conditions in northern latitudes. However, the Bruniquel evidence suggests that the slightly smaller network sizes favored by Pearce still involved complex cultural mechanisms to increase the strength of social ties. We see an additional pressure in MIS6 in southern France, where the severity of the landscapes supports resources at levels that necessitated closer cooperation.

The Cartesian mind, therefore, argues that finds such as Bruniquel are selected for by the ecology of survival. At an earlier career stage, Gamble (1982) used landscape to explain the similarity in form of Gravettian female figurines (Gaudzinski-Windheuser & Jöris, 2015; Mussi, 2015). Cooperation becomes essential in such highly stressed environments, and material culture is drawn into the adaptive mix to work the margins on survival. But why make circles from speleothems, and why do this in the deep, dark underground are questions that comprise, to paraphrase Lewis-Williams (2004), the cave in the ECA mind. This approach struggles to come up with a better answer because it does not consider intentions and emotions as part of its remit. Initial reaction, reported in *Nature* under the title "Neandertals Built Cave Structures—And No One Knows Why" (Callaway, 2016), points to problems the rational mind has always had when confronted by something extra-paradigmatic. Ritual, religion, and cave bears have all been put forward. These put a name to the constructions, but not an explanation.

The Response of the Experiential Mind

The discoveries made at Bruniquel suggest a reimagining of the way Middle Paleolithic hominins, Neandertals, engaged with and experienced their world. First, the cave would have been experienced as an aspect of those who created it. Second, time spent in the cave would have been a shared experience of coming together.

Unusual for the time, Bruniquel is a constructed environment, with stalagmites arranged into circular enclosures imposing a structure and order on the natural layout of the cave. As such, the process of constructing Bruniquel was one of hominin agency acting on the world, and an imposition of individual will on natural order. That expression of agency came at the expense of the natural environment; breaking off the stalagmites was an act of destruction that allowed the cave to be reconstituted as something new, something previously only imagined in the minds of those constructing the cave. In this way, those who experienced the alteration of the cave were experimenting, testing the influence of their agency on the world around them.

The hominins that occupied Bruniquel took a dangerous place, dark and home to hibernating carnivores, and imposed on it their own sense of order. The act of delving so deeply within a cave was one of pushing the boundaries of safety, with the constructed space experienced as a liminal zone where ever-present dangers were kept back by hominin agency. This can be seen in the use of fire, a wild and destructive force harnessed by hominins to provide vital and life-sustaining warmth and light (Harris & Sørensen, 2010).

By constructing the enclosures, the inhabitants of Bruniquel created their own niche within an unpredictable world. Wiessner (2014) has shown that in hunter-gatherer communities, the fireside provides a time and place for mundane gossiping, storytelling, and recounting myths. As a result, hearths provide the context to build social identification as well as a group narrative. Bruniquel was a space where the daily grind could be left behind. The process of coming together to alter the cave would also have created an experience of shared endeavor as individuals worked together to create something new. As individuals grew closer together through this process, it may have become apparent that some things existed outside of the group affiliations constructed with the cave itself: a group contained (see Gamble, 2007, for metaphors of containment) within their own constructed world. Bruniquel then sees a group identity emerge from shared action in the world.

The Danish have a word for an emotional state known as *hygge*. There is no direct translation in English, but the closest may be a feeling of comfort and homeliness. It is tempting to imagine the Neandertals of Bruniquel enjoying their own version of *hygge*: no ritual, no function, no cold cognition, but a simple experience of being together within the world. For those who spent time at Bruniquel, the cave would have been a profoundly altering experience, distinct from what lay outside it.

The Response of the Anthropological Mind

Our ability to interpret the activities at Bruniquel so that it satisfies an anthropological view of the mind must emphasize the agency of individuals both on and through materials and each other. To achieve this goal, we need to reconstruct the network of interactions along which social and emotional exchanges were made. The relations between the individuals and the materials act as information conduits along which social bonds tether individuals, and emotions play out—the emphasis is on the spaces between, instead of the discrete players themselves.

The stalagmite construction in the chamber at Bruniquel was probably not built by one individual or even two. Because of its size, we can assume that a number of people worked together, and coordination and social planning have been evoked in papers by Jaubert and colleagues (2016) and Soressi (2016). The size of the structure is also indicative of use by more than just one individual. To be taken as an example of individuals working together, the construction is inferred to have resulted from a common goal, rather than a coincidence of intentions; those intentions were likely not immediately visible or tangible (as an immediate function would likely be obvious to us as naïve observers without recourse to explicit intention of the structure builders as well). The goal was then a structure to execute a goal behavior (or behaviors), and this would implicate shared knowledge. In this case, for the social coordination to be successful, it could be that the coordination required not only shared knowledge but also the understanding of which knowledge is shared, and that behaviors could therefore be directed by an individual(s). If they did not have the ability to assume shared knowledge and reflect on the hypothesized information others held, cooperation could probably have occurred but likely not have been managed. Reflection of the order of "I think that you think that I think" is reflective of third-level ToM and all the cognitive development both ontogenetically and phylogenetically that it entails (de Villiers, 2007; Liddle & Nettle, 2006).

The precise use of the Bruniquel constructions need not be deciphered to evoke these interpretations. Rather, what is important is the evidence for a number of individuals working together; the construction's perceived function would likely have been opaque to the players without explanation and prior shared knowledge, and therefore would have required a demonstration of function or need. This is not a content-based interpretation but rather a relational one. Therefore, notions of the stalagmites being "symbolic," prepared for a ritual purpose or a means to induce social cohesion to confer an adaptive benefit, need not apply. It is the avenues of information and how they tether materials and individuals together that structures a strong cognitive argument for minds and their activity.

CONCLUSION

In this chapter, we attempted a broader approach to the evolutionary cognitive archaeology developed by Thomas Wynn, which marries psychology and archaeology, to address the evolution of the human mind and then extend it to include social and emotional concerns. We examined the mind through three different lenses: a rational, Cartesian influence; an experiential prism guided by a phenomenological approach; and an anthropological framework based on the extended and distributed relationships between things.

Too often, ECA sits within the Cartesian mind, perhaps because the study of cognition is inherently dualistic, separating thought processes from the social worlds in which they occur. However, we believe that this need not be the case. In our anthropological mind, we have shown how the cognitive fossils hominins left behind can also be seen as social artifacts, emblematic of the interactions from which they arose. It is this view that allows us to see cognition most strongly within material artifacts, because they are acknowledged as intimate players in doing and thinking. In this way, we can consider material culture as fossils of cognition because of their previous role in cognitive processes. We have also shown the potential of a grounded, empirically driven experiential approach, which promises new avenues of archaeological enquiry, and addressed the possibility of constructing a truly affective world from the scant remains left behind.

While we do not propose that these are the only ways to think about questions of cognitive archaeology and the evolution of hominin minds, we think that our three minds conceptualize and stretch Wynn's original ECA to include that which is missing, the social and emotional aspects that are so integral to understanding human behavior. When used judiciously, the three minds of an extended ECA can provide archaeologists with an interpretive power beyond what can be achieved by traditional approaches alone, creating a cognitive archaeology that acknowledges how cognition operates by extending and distributing the process of social and emotional relations.

By developing ECA, Wynn laid the groundwork for a truly heuristic study of the past, showing archaeologists that there is much more to a handaxe than meets the eye, and that there is more interpretive power in what researchers can deduce about the behavior and thoughts of their makers. The challenge for the next generation is to continue stretching the boundaries of the archaeological exercise he began. Wynn squeezed interpretations of cognitive activities from ancient stones, and they are not yet wrung dry.

ACKNOWLEDGMENTS

We thank Dan Hunt for his valuable contributions to the discussions and ideas that resulted in this chapter, and to the anonymous reviewers for their comments.

REFERENCES

Boesch, C. (1991). Teaching among wild chimpanzees. *Animal Behaviour, 41*(3), 530–532.

Boiger, M., & Mesquita, B. (2015). A sociodynamic perspective on the construction of emotion. In L. Feldman Barrett & J. A. Russel (Eds.), *The psychological construction of emotion* (pp. 377–398). New York: Guilford Press.

Byrne, R. W., & Whiten, A. (Eds.). (1988). *Machiavellian intelligence: Social expertise and the evolution of intellect in monkeys, apes, and humans*. Oxford, UK: Clarendon Press.

Callaway, E. (2016). Neanderthals built cave structures—and no one knows why. *Nature News*. doi:10.1038/nature.2016.19975

Clark, A., & Chalmers, D. J. (1998). The extended mind. *Analysis, 58*(1), 7–19.

Clark, G. (1969). *World prehistory: A new outline* (2nd ed.). Cambridge, UK: Cambridge University Press.

Clark, J. D., Beyene, Y., WoldeGabriel, G., Hart, W. K., Renne, P. R., Gilbert, H., . . . White, T. D. (2003). Stratigraphic, chronological and behavioural contexts of Pleistocene *Homo sapiens* from Middle Awash, Ethiopia. *Nature, 423*(6941), 747–752.

Corbey, R., Jagich, A., Vaesen, K., & Collard, M. (2016). The Acheulean handaxe: More like a bird's song than a Beatles' tune? *Evolutionary Anthropology: Issues, News, and Reviews, 25*(1), 6–19.

de Mortillet, G. (1872). Classification de l'âge de la pierre. In *Matériaux pour l'histoire primitive et naturelle de l'Homme* (pp. 464–465). Paris: C. Reinwald.

de Villiers, J. (2007). The interface of language and theory of mind. *Lingua, 117*(11), 1858–1878.

Dunbar, R. I. M. (1992). Neocortex size as a constraint on group size in primates. *Journal of Human Evolution, 22*(6), 469–493.

Dunbar, R. I. M. (2004). *The human story: A new history of mankind's evolution*. London: Faber and Faber.

Dunbar, R. I. M. (2007). The social brain and the cultural explosion of the human revolution. In P. Mellars, K. Boyle, O. Bar-Yosef, & C. Stringer (Eds.), *Rethinking the human revolution: New behavioural and biological perspectives on the origin and dispersal of modern humans* (pp. 91–98). Cambridge, UK: McDonald Institute for Archaeological Research.

Dunbar, R. I. M., Gamble, C., & Gowlett, J. A. J. (Eds.). (2010). *Social brain, distributed mind*. London: British Academy.

Dunbar, R. I. M., Gamble, C., & Gowlett, J. A. J. (Eds.). (2014). *Lucy to language: The benchmark papers*. Oxford, UK: Oxford University Press.

Finlayson, C., Brown, K., Blasco, R., Rosell, J., Negro, J. J., Bortolotti, G. R., . . . Llanes, J. M. R. (2012). Birds of a feather: Neanderthal exploitation of raptors and corvids. *PLoS One, 7*(9), 1–9.

Fowler, C. (2016). Relational personhood revisited. *Cambridge Archaeological Journal, 26*(3), 397–412.

Gamble, C. (1982). Interaction and alliance in Palaeolithic society. *Man, 17*, 92–107.

Gamble, C. (2007). *Origins and revolutions: Human identity in earliest prehistory*. Cambridge, UK: Cambridge University Press.

Gamble, C., Gowlett, J. A. J., & Dunbar, R. I. M. (2012). The social brain and the shape of the Palaeolithic. *Cambridge Archaeological Journal, 21*(1), 115–135.

Gaudzinski-Windheuser, S., & Jöris, O. (2015). Contextualising the female image—Symbols for common ideas and communal identity in Upper Palaeolithic societies. In F. Coward, R. Hosfield, M. I. Pope, & F. Wenban-Smith (Eds.), *Settlement, society and cognition in human evolution: Landscapes in mind* (pp. 288–314). Cambridge, UK: Cambridge University Press.

Gibson, J. J. (1986). *The ecological approach to visual perception*. Hillsdale, NJ: Lawrence Erlbaum Associates.

Gosden, C. (2010). The death of the mind. In L. Malafouris & C. Renfrew (Eds.), *The cognitive life of things: Recasting the boundaries of the mind* (pp. 39–46). Cambridge, UK: McDonald Institute for Archaeological Research.

Harmand, S., Lewis, J. E., Feibel, C. S., Lepre, C. J., Prat, S., Lenoble, A., . . . Roche, H. (2015). 3.3-million-year-old stone tools from Lomekwi 3, West Turkana, Kenya. *Nature, 521*(7552), 310–315.

Harris, O. J. T., & Sørensen, T. F. (2010). Rethinking emotion and material culture. *Archaeological Dialogues, 17*(2), 145–163.

Henshilwood, C. S., D'Errico, F., Yates, R., Jacobs, Z., Tribolo, C., Duller, G. A. T., . . . Wintle, A. G. (2002). Emergence of modern human behavior: Middle Stone Age engravings from South Africa. *Science, 295*(5558), 1278–1280.

Jaubert, J., Verheyden, S., Genty, D., Soulier, M., Cheng, H., Blamart, D., . . . Santos, F. (2016). Early Neanderthal constructions deep in Bruniquel Cave in southwestern France. *Nature, 534*(7605), 111–114.

Lakoff, G., & Johnson, M. (1980). *Metaphors we live by*. Chicago, IL: University of Chicago Press.

Leroi-Gourhan, A. (1993). *Gesture and speech*. Cambridge, MA: MIT Press.

Lewis, J. E., & Harmand, S. (2016). An earlier origin for stone tool making: Implications for cognitive evolution and the transition to *Homo*. *Philosophical Transactions of the Royal Society of London. Series B, Biological Sciences, 371*(1698), 1–8.

Lewis, M. D., & Granic, I. (2010). Phases of social–emotional development from birth to school age. In M. Ferrari & L. Vuletic (Eds.), *The developmental relations among mind, brain and education* (pp. 179–212). London: Springer.

Lewis-Williams, D. (2004). *The mind in the cave: Consciousness and the origins of art*. London: Thames & Hudson.

Liddle, B., & Nettle, D. (2006). Higher-order theory of mind and social competence in school-age children. *Journal of Cultural and Evolutionary Psychology, 4*(3/4), 231–244.

LiPuma, E. (1998). Modernity and forms of personhood in Melanesia. In M. Lambek & A. Strathern (Eds.), *Bodies and persons: Comparative perspectives from Africa and Melanesia* (pp. 53–79). Cambridge, UK: Cambridge University Press.

Lycett, S. J., von Cramon-Taubadel, N., & Eren, M. I. (2016). Levallois: Potential implications for learning and cultural transmission capacities. *Lithic Technology, 41*(1), 19–38.

Matsuzawa, T., Humle, T., & Sugiyama, Y. (Eds.). (2014). *The chimpanzees of Bossou and Nimba*. Tokyo: Springer Science & Business Media.

McBrearty, S., & Brooks, A. S. (2000). The revolution that wasn't: A new interpretation of the origin of modern human behavior. *Journal of Human Evolution, 39*(5), 453–563.

McDougall, I., Brown, F. H., & Fleagle, J. G. (2005). Stratigraphic placement and age of modern humans from Kibish, Ethiopia. *Nature, 433*(7027), 733–736.

Miller, D. (2012). *Stuff*. Cambridge, UK: Polity Press.

Mithen, S. J. (1996). *The prehistory of mind: The cognitive origins of art, religion and science.* London: Thames & Hudson.

Morgan, T. J. H., Uomini, N. T., Rendell, L. E., Chouinard-Thuly, L., Street, S. E., Lewis, H. M., . . . Laland, K. N. (2015). Experimental evidence for the co-evolution of hominin tool-making teaching and language. *Nature Communications, 6*(6029), 1–18.

Morley, I. R. M. (2014). A multi-disciplinary approach to the origins of music: Perspectives from anthropology, archaeology, cognition and behaviour. *Journal of Anthropological Sciences, 92*, 147–177.

Mussi, M. (2015). Encoding and decoding the message: The case of the Mid Upper Palaeolithic female imagery. In F. Coward, R. Hosfield, M. I. Pope, & F. Wenban-Smith (Eds.), *Settlement, society and cognition in human evolution: Landscapes in mind* (pp. 275–287). Cambridge, UK: Cambridge University Press.

Noble, W., & Davidson, I. (1996). *Human evolution, language, and mind: A psychological and archaeological inquiry*. Cambridge, UK: Cambridge University Press.

Pearce, E., & Dunbar, R. I. M. (2012). Latitudinal variation in light levels drives human visual system size. *Biology Letters, 8*, 90–93.

Pearce, E., Stringer, C. B., & Dunbar, R. I. M. (2013). New insights into differences in brain organization between Neanderthals and anatomically modern humans. *Philosophical Transactions of the Royal Society of London. Series B, Biological Sciences, 280*(1758), 1–7.

Pinker, S. (1997). *How the mind works*. New York: W. W. Norton & Company.

Radovčić, D., Sršen, A. O., Radovčić, J., & Frayer, D. W. (2015). Evidence for Neandertal jewelry: Modified white-tailed eagle claws at Krapina. *PLoS One, 10*(3), 1–14.

Soressi, M. (2016). Archaeology: Neanderthals built underground. *Nature, 534*(7605), 43–44.

Tennie, C., Call, J., & Tomasello, M. (2010). Evidence for emulation in chimpanzees in social settings using the floating peanut task. *PLoS One, 5*, e10544.

Tilley, C. (1999). *Metaphor and material culture*. Oxford, UK: Blackwell.

Wadley, L. (2013). Recognizing complex cognition through innovative technology in Stone Age and Palaeolithic sites. *Cambridge Archaeological Journal, 23*(2), 163–183.

Whitehouse, H., & Lanman, J. A. (2014). The ties that bind us: Ritual, fusion and identification. *Current Anthropology, 55*(6), 674–695.

Wiessner, P. W. (2014). Embers of society: Firelight talk among the Ju/'hoansi Bushmen. *Proceedings of the National Academy of Sciences of the United States of America, 111*(39), 14027–14035.

Wynn, T. (1979). The intelligence of later Acheulean hominids. *Man, 14*, 371–391.

Wynn, T. (1985). Piaget, stone tools and the evolution of human intelligence. *World Archaeology, 17*(1), 32–43.

Wynn, T. (2009). Whither evolutionary cognitive archaeology? In S. A. de Beaune, F. L. Coolidge, & T. Wynn (Eds.), *Cognitive archaeology and human evolution* (pp. 145–149). Cambridge, UK: Cambridge University Press.

Wynn, T., & Coolidge, F. L. (2009). Implications of a strict standard for recognizing modern cognition in prehistory. In S. A. de Beaune, F. L. Coolidge, & T. Wynn (Eds.), *Cognitive archaeology and human evolution* (pp. 117–127). Cambridge, UK: Cambridge University Press.

16

THE EVOLUTION OF SOCIAL TRANSMISSION IN THE ACHEULEAN

Ceri Shipton

INTRODUCTION

Humans have the most high-fidelity social transmission of any species. Successive generations of humans are able to maintain the knowledge and skills of their forbearers; they are able to share these between each other, and in many cases build on them to achieve new heights (Tennie, Call, & Tomasello, 2009; Tomasello, 1999). This cultural ratchet effect, or cumulative cultural evolution, has given rise to the remarkable technological and artistic prowess of human societies (Dean, Vale, Laland, Flynn, & Kendal, 2014). Cumulative cultural evolution is critically dependent on high transmission fidelity (Lewis & Laland, 2012), and in humans it is our proclivity for imitation and our capacity for verbal teaching that enable this fidelity (Dean, Kendal, Schapiro, Thierry, & Laland, 2012). This chapter presents the case that these capacities arose during the Acheulean, the longest-lived prehistoric culture, which first emerged around 1.75 million years ago (Mya) (Beyene et al., 2013).

The most striking features of the Acheulean are its longevity and ubiquity: The culture lasted for over 1.5 million years. It is found from the Cape of Good Hope to Britain and from the Atlantic seaboard of the Maghreb to the Himalayan foothills. It was produced by multiple hominin species, including *Homo erectus* and *H. heidelbergensis*. One hypothesis for explaining this ubiquity is that Acheulean knapping techniques were genetically determined (Corbey, Jagich, Vaesen, & Collard, 2016). However, this is belied by the diversity of knapping techniques found in the archaeological record, which were tailored to suit specific rock forms (e.g., Sharon, 2007; Shipton, Petraglia, & Paddayya, 2009). An alternative hypothesis is that Acheulean technology was so easy to invent that its characteristic bifacial tool forms—handaxes and cleavers—were repeatedly discovered (Tennie, Braun, Premo, & McPherron, 2016; Tennie, Premo, Braun, & McPherron, 2017). However, this does not explain why the populations of *H. erectus* that left Africa prior to the advent of the Acheulean did not invent it also (Baena et al., 2010; Mgeladze et al., 2011), nor why the earliest examples of Acheulean technology are all in East Africa (Beyene et al., 2013; Diez-Martín et al., 2015; Lepre et al., 2011). Similar forms to handaxes do occur at other times and places in prehistory (e.g., Moore, 2003), but cleavers are a specific tool type (Tixier & Inizian, 1983), perhaps unique to the Acheulean, that do not seem to have been reinvented. In contrast to these hypotheses, here it is argued that the Acheulean was underpinned by capacities for high-fidelity social transmission comparable to our own, while at the same time cultural conservatism and low levels of innovation ensured that it was not superseded by new technologies for hundreds of thousands of years.

HOMININ SOCIAL TRANSMISSION BEFORE THE ACHEULEAN

Robust social transmission has deep antecedents in the primate lineage. Our closest living relatives, the chimpanzees, acquire their most complex technological skills in the wild through social transmission (Koops, Schöning, Isaji, & Hashimoto, 2015; Luncz, Mundry, & Boesch, 2012). Notably, these skills include percussive technologies. Experiments have demonstrated that chimpanzees are capable of imitating, that is, copying both end-states and the processes needed to achieve them (Horner, Whiten, Flynn, & De Waal, 2006; Whiten, McGuigan, Marshall-Pescini, & Hopper, 2009). However, chimpanzees do not display the same propensity for imitation that we do: They do not copy intentional but causally opaque or irrelevant actions (Lyons, Damrosch, Lin, Macris, & Keil, 2011), and they do not readily imitate novel actions that depart from their habitual behaviors (Marshall-Pescini & Whiten, 2008a; Tennie, Call, & Tomasello, 2012).

The earliest stone knapping is thus far only known from the single site of Lomekwi (Harmand et al., 2015; Lewis & Harmand, 2016), and it is not yet clear the extent to which social transmission was involved in its propagation. At least two varieties of knapping were used at the site: the passive hammer and bipolar techniques. Passive hammer is one of the simplest ways to break open a stone—by hitting it against a hard surface—and it is likely in the zone of latent solutions (Tennie et al., 2017) of non-human primate knappers (Savage-Rumbaugh & Fields, 2006; Westergaard & Suomi, 1994). Bipolar flaking, in which the target (a core, in knapping) is stabilized on an anvil and struck by a handheld hammer, is analogous to chimpanzee nut-cracking (Sakura & Matsuzawa, 1991), a technique transmitted through emulation and low-fidelity imitation (Marshall-Pescini & Whiten, 2008b; Whiten et al., 2009).

Oldowan knapping, which begins around 2.6 Mya, may represent social transmission of greater fidelity than that demonstrated by chimpanzees. Passive hammer is notably rare or absent in the Oldowan, where the dominant technique is freehand percussion, a technique in which both core and hammer are held in the hands. At the earliest Oldowan sites, hominins appear to have mastered the basics of this technique (Goldman-Neuman & Hovers, 2009; Stout, Quade, Semaw, Rogers, & Levin, 2005), such as where to strike according to the shape of individual clasts. Oldowan cores were often extensively reduced, involving the correction of knapping errors and the maintenance of edge angles throughout the reduction sequence, behaviors that imply knowledge of the overarching knapping schema in addition to merely identifying the next flake removal (Delagnes & Roche, 2005; de la Torre, 2004; Hovers, 2009). Early Oldowan artifacts do not exhibit the extensive bruising marks characteristic of modern novice knappers, where stones have been hit in places inappropriate for flaking with the result that no flake is detached (Delagnes & Roche, 2005). This suggests that Oldowan knowledge of knapping did not come from trial-and-error learning but was acquired in a more fully formed state.

Kinematic analysis of experimental Oldowan knapping indicates that there is wide inter-individual variability in the specific movements of both experts and apprentices and little correlation between paired experts and apprentices (Parry, Dietrich, & Bril, 2014; Rein, Nonaka, & Bril, 2014). It is not particular movements that are transmitted between knappers but, more likely, the holistic functional dynamics of the body-object-environment: combining appropriate angles on the core, the angle of the blow, and the point of percussion on the platform, as well as higher-order knapping strategies.

At the early Oldowan site of Gona in Ethiopia, all localities contain examples of both bifacial and unifacial flaking; however, two localities near each other (EG10 and EG12) are dominated by unifacial flaking, whereas another further afield (OGS7) is dominated by bifacial flaking (Stout, Semaw, Rogers, & Cauche, 2010). This bias in flaking strategies at different localities suggests that strategies were disseminated through social transmission. Three human novice knappers who were familiar with the appearance of the Oldowan material from Gona were provided with some of the local rock used for knapping, asked to replicate the artifacts, and given no instruction on how to do so (Stout & Semaw, 2006). Their products were compared with those from one of the Gona localities (EG10); the results showed experimental novices were significantly more variable in their reduction strategies than was true of the EG10 archaeological sample, again suggesting social transmission of knapping strategies among the Gona hominins.

A transmission chain experiment with modern knappers examined how different levels of social transmission, ranging from only seeing the flakes to full verbal teaching, influenced the performance of freehand hard-hammer flake production (Morgan et al., 2015). The number of viable flint flakes that knappers were able to produce (i.e., those longer than 2 cm) increased with successive levels of teaching. Basic teaching was enough to produce significant improvements in the proportion of viable flakes in the emulation and imitation conditions. This condition involved manually shaping pupils' grasp of their hammerstone or core, and orienting to allow pupils a clear view of the demonstrator, behavior similar to that observed among wild chimpanzees (Boesch, 1991). Only in the emulation and imitation conditions were there improvements in the proportion of viable flakes between successive iterations of transmission. The experiment by Morgan and colleagues (2015) suggests a social transmission portfolio (cf. Whiten, Horner, Litchfield, & Marshall-Pescini, 2004) that includes basic teaching is enough for novices to begin Oldowan knapping at a level where they can produce viable flakes; when emulation and imitation are added to basic teaching, novices are able to improve their skills with practice.

Despite the conditions required for social transmission of simple Oldowan technology being present in extant great apes, two bonobos who were taught to knap were not able to pick up the task quickly and would likely have given up without repeated encouragement (Savage-Rumbaugh & Fields, 2006). These bonobos are now competent knappers, but they do not achieve the same skill levels as the Gona hominins (Toth, Schick, & Semaw, 2006). While there are known anatomical constraints in the bonobo hand-and-wrist morphology, relative to that of *H. sapiens* and perhaps also Oldowan hominins (Key & Dunmore, 2015; Kivell, 2015; Marzke, 2013), the fidelity of bonobo imitation could also limit their ability to acquire knapping skills, as freehand percussion requires precise orientation of both the core and the hammer strike. The more advanced knapping techniques documented in the Oldowan—for example, the discoidal schema (de la Torre, 2004), platform maintenance (Delagnes & Roche,

2005), and the removal of step terminations (Hovers, 2009)—may well require even greater fidelity of social transmission than that displayed by extant non-human great apes.

SOCIAL TRANSMISSION IN THE ACHEULEAN

Some aspects of the Acheulean suggest that it represented more robust social transmission than the Oldowan. These include the ubiquity of the symmetrical forms imposed on handaxes and cleavers over much of the Old World (Wynn, 2000). The maintenance of finished tool forms, and the various manufacturing methods underpinning their creation in different rock types, requires high-fidelity imitation, in which both the means and the goal are conceived as separate entities and the correspondence between them is adapted and maintained.

At the 800,000-year-old site of Gesher Benot Ya'aqov in the Levant, stratified Acheulean stone tools accumulated over a period of around 50,000 years (Feibel, 2004). These assemblages exhibit constancy in biface size and shape throughout the period, and were consistently produced on large flakes of basalt with the thickest part of the flake, the platform and bulb of percussion, retouched away by a minimal number of flake scars (Sharon, Alperson-Afil, & Goren-Inbar, 2011). This conservatism in biface production technology suggests continuous social transmission of both the method and the end products throughout the roughly 50,000 years that hominins were occupying the site.

Technological organization at the high-integrity Acheulean site of Isampur Quarry in India, estimated to be 1.2 million years old (Blackwell et al., 2001), indicates high-fidelity imitation in the transmission of knapping. Contrasting canalized manufacturing methods were used for the production of handaxes and cleavers, even though other methods were possible, showing both the methods and the corresponding goal were transmitted together (Petraglia, LaPorta, & Paddayya, 1999; Shipton, 2010, 2013; Shipton et al., 2009). Handaxes were made by reducing slabs of medium thickness to leave the principal plane of the finished handaxe parallel to the bedding plane of the original slab (see Figure 16.1). Cleavers were made by striking large flake blanks off thicker slabs, leaving the cleaver oblique to the bedding plane. A small minority of cleavers (5 out of 36 whose manufacturing method was identifiable) were, however, made in the same way as handaxes (Shipton, 2013). This shows that while it was possible to make cleavers by shaping thinner slabs, the normal method of manufacture was to create a thick core from which to strike large flake blanks. The flake-cleaver reduction sequence required several preparatory steps. Without familiarity with the sequence, the purpose of some of these steps is opaque, such as striking smaller flakes down the sides of the slabs to establish the main striking platform (Figure 16.2). Transmitting such a sequence would necessitate high-fidelity imitation of the method, and perhaps even over-imitation (Shipton & Nielsen, 2015).

ACHEULEAN OVER-IMITATION

In this section, new evidence is presented to make the case for over-imitation in the Acheulean. *Over-imitation* (specifically, automatic causal encoding; see Rossano, 2017) describes the human tendency to replicate all of the intentional actions of a

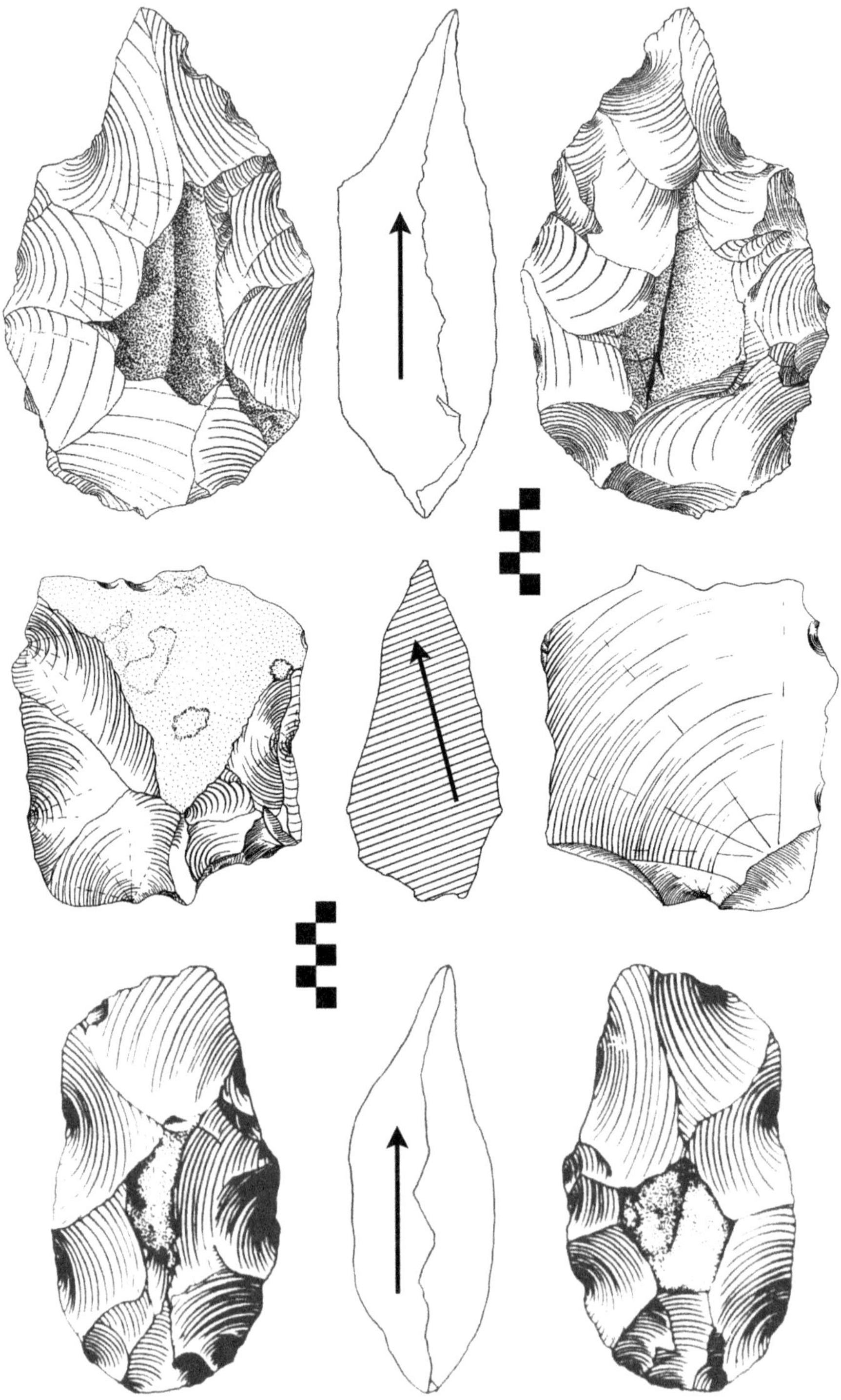

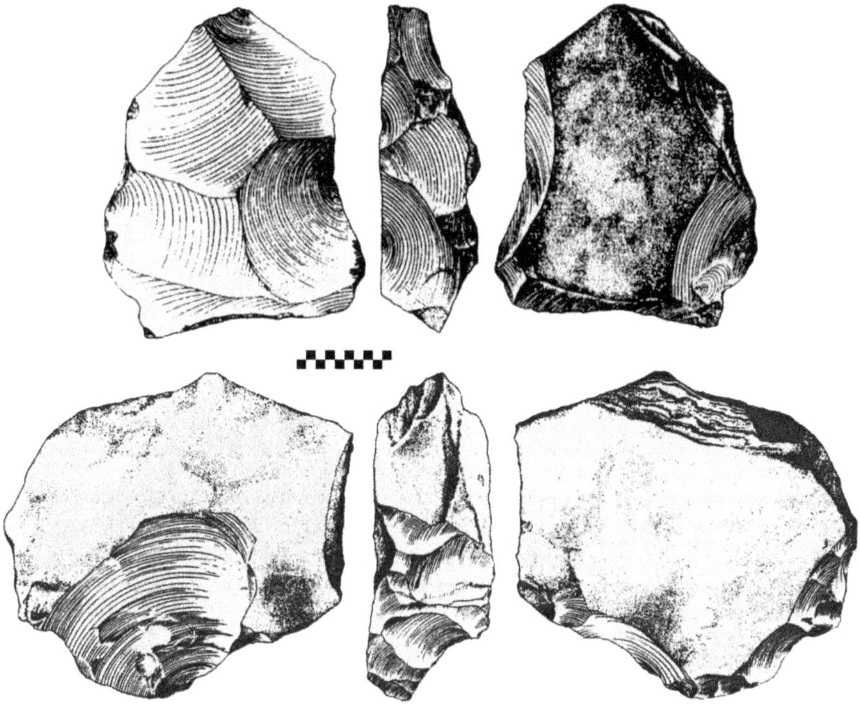

Figure 16.2. Two thick slab cores from Isampur Quarry. The smaller scars struck down the side of the slab, visible in profile and on the underside of the cores on the right, were made to establish a striking platform. This platform was then used to produce the larger flakes needed for cleaver blanks, whose large scars are visible on the upper surface of the cores on the left. Adapted from Paddayya and colleagues (2006), The Acheulian quarry at Isampur, Lower Deccan, India, *Axe Age: Acheulian Tool-Making from Quarry to Discard*, Equinox; and Paddayya (2007), Evolution within the Acheulian in India: A case study from the Hunsgi and Baichbal Valleys, Karnataka, *Bulletin of the Deccan College Research Institute*.

demonstrator, even when their purpose is unclear or steps are redundant (Horner & Whiten, 2005; Lyons et al., 2011; Nielsen, Mushin, Tomaselli, & Whiten, 2014). Humans in fact imitate with greater precision when they do not understand the causal mechanisms of a task (Williamson & Markman, 2006). Four case studies are presented that may represent instances of over-imitation in the Acheulean; these are

---

Figure 16.1. Acheulean bifaces from Isampur Quarry, India. Arrows denote orientation of the bedding plane of the original limestone slab from which the tools were made. Top: Handaxe made on a thinner slab reduced parallel to the bedding plane. Note the flat cortex on both surfaces. Middle: Cleaver made on a large flake struck obliquely to the bedding plane of a thick slab. Note that the cortex is only on one surface and at the distal end. Bottom: Cleaver made on a thinner slab reduced parallel to the bedding plane. Note the flat cortex on both surfaces. Scale bars are centimeters. Adapted from Paddayya (2007), Evolution within the Acheulian in India: A case study from the Hunsgi and Baichbal Valleys, Karnataka, *Bulletin of the Deccan College Research Institute*; Paddayya, Jhaldiyal, and Petraglia (2006), The Acheulian quarry at Isampur, Lower Deccan, India, *Axe Age: Acheulian Tool-Making from Quarry to Discard*, Equinox; and Petraglia and colleagues (1999), The first Acheulian quarry in India: Stone tool manufacture, biface morphology, and behaviors, *Journal of Anthropological Research*.

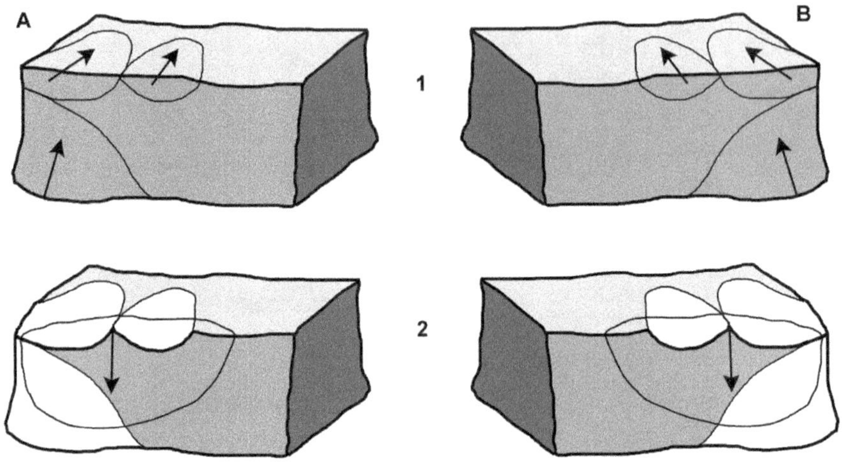

Figure 16.3. Schema for producing cleavers at Chirki. (1) Three preparatory scars struck to create the cleaver bit scar and dihedral platform for the cleaver blank. (2) Striking the cleaver blank from the dihedral platform, (A) the way cleavers were typically produced at Chirki and (B) an equally viable mirrored version on the same core shape that rarely seems to have been used. Image by the author.

the sites of Chirki and Morgaon on the Deccan Plateau in India, the Victoria West facies of the southern African Karoo, and the Tabelbala-Tachengit facies of northwestern Africa.

The Acheulean site of Chirki has uranium series and palaeomagnetism age estimates of over 350,000 years (Kale, 1990) and over 780,000 years (Sangode, Mishra, Naik, & Deo, 2007) respectively. Here at a riverside location, hominins made handaxes and cleavers on local basalt, sometimes shaping smaller weathered cobbles and slabs, and sometimes striking large flake blanks from larger boulders (Corvinus, 1983). A particular manufacturing sequence was used to produce the large flake blanks from which cleavers were made. A large flake was first removed from one edge of a main flaking surface (Figure 16.3). On the opposite edge, a few centimeters offset, a dihedral platform was set up by orthogonal removals, and then a large flake was struck across the main surface from this platform. Where the second large flake met the first scar, the bit of the cleaver was formed. The cleaver bit scar was thus struck in the opposite direction to the cleaver blank flake (Figure 16.4).

In the Chirki cleaver production sequence, both sides of the core were evidently capable of producing large flakes, so the process could have been randomly mirrored on the basalt blocks, resulting in some cleavers with the platform on the right side and some with the platform on the left (Figure 16.3). However, Corvinus (1983, p. 40) noted that "the majority of cleavers at Chirki are detached in this way [from the right] and have the talon [dihedral platform] on their right side, so that it seems quite obvious that this was the intended way of producing cleavers" (see Figure 16.4). In a sample of 57 Chirki flake blank cleavers illustrated by Corvinus (1983), where it was possible to determine the direction of the blank strike (when viewed from the ventral surface, not the dorsal as Corvinus described strike direction), 49 were from the left, only 10 from the right, and 2 were from the proximal end. In a separate analysis of 45

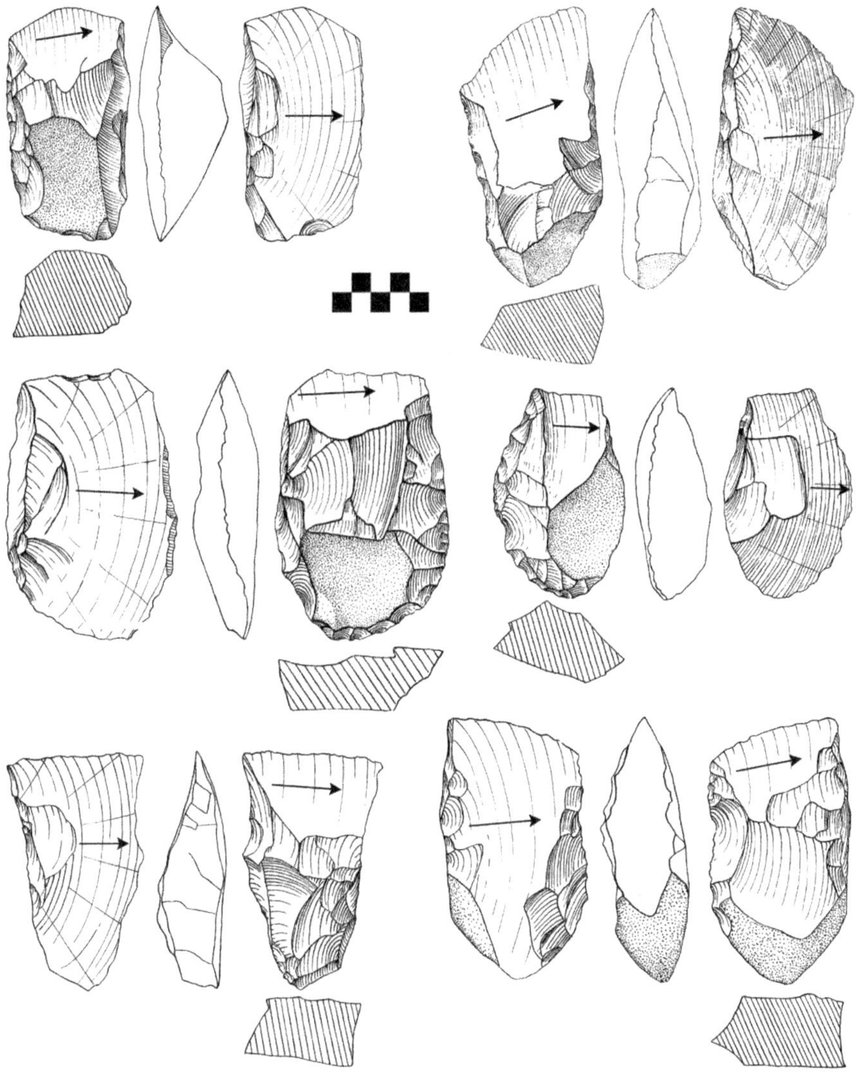

Figure 16.4. Six cleavers from Chirki. Note that the flakes on which they were made were all struck from the left (arrows) when looking at the ventral surface, corresponding to the right side of the dorsal surface, as noted by Corvinus (1983). Note also that the preparatory flake scars used to create the cleaver bits were all struck from the opposite direction—the left (arrows) when looking at the dorsal surface. Image by the author.

Chirki flake blank bifaces, where it was possible to determine the direction of the blank strike (when viewed from the ventral surface, not the dorsal, as Corvinus described strike direction), 15 were from the left and only 3 were from the right (Sharon, 2007). Likewise from Corvinus' illustrations, the bit scar was struck from the left in 32 cases where scar direction is clear, and from the right in just 9 cases (Figure 16.4). There was thus at Chirki a tendency to flake the opposing edges of the main core face in a particular but arbitrary order.

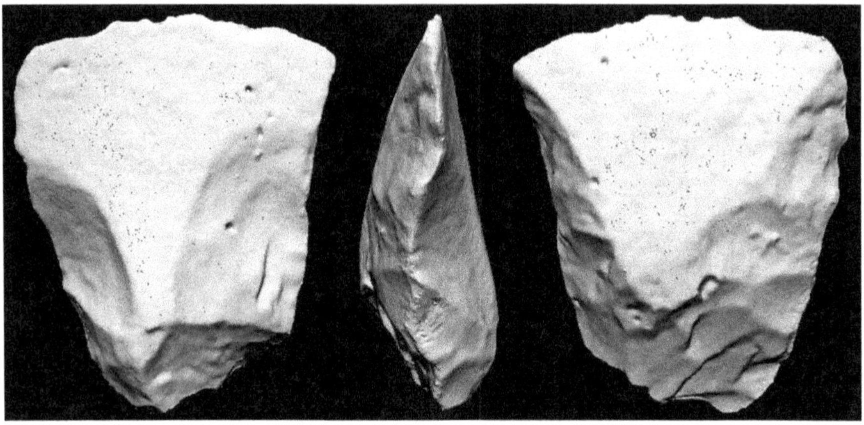

Figure 16.5. Cleaver from Morgaon with two ventral surfaces, indicating that the flake blank was struck from the ventral surface of a very large flake. The piece is 143 mm long. The termination of the flake where the two ventral surfaces meet each other had been used as the cleaver bit. Image by the author.

The site of Morgaon has an age estimate of over 780,000 years (Sangode et al., 2007). Here biface production was nearly exclusively focused on cleavers made on large flakes of basalt. The most common way of creating a flake blank was to strike a large flake from the proximal end of another, even larger flake, creating a biconvex flake with two ventral surfaces, sometimes referred to as a "Janus" flake (Deo, Mishra, Rajaguru, & Ghate, 2007; Mishra et al., 2009). The lateral edge of the flake where the two ventral surfaces met was used as the cleaver bit (Figure 16.5). Out of a random sample of 31 Morgaon cleavers analyzed for this chapter, 11 preserved traces of two ventral surfaces, which is one of the highest proportions in any Acheulean assemblage (cf. Sharon, 2007). In this technique, there is no a priori reason why the finished cleaver should be oriented one way or the other in relation to the axis of the core flake, as the gentle convexity used to make the bit covers its entire ventral surface. It was possible to tell the strike direction of the flake blank on 14 of the 31 Morgaon cleavers sampled for this study. Of these 14 cleavers, 12 were struck from the left (when looking at the ventral surface), and only 2 had been struck from the right (Figure 16.6).

The Victoria West technique is a particular way of making cleaver and handaxe blanks, known from southern Africa (Li, Kuman, Lotter, Leader, & Gibbon, 2017; Sharon & Beaumont, 2006). In the Victoria West technique, bifacial cores were prepared around much of their perimeter, with a flatter upper surface and a steeply domed lower surface (similar to Levallois). The cores were struck with a preferential blow on the side (unlike Levallois); the resulting large flake, which removed most of the upper surface, was used as a biface blank. The cores were teardrop shaped in plan, with the narrower end corresponding to the butt of the finished tool. Interestingly, the cores were invariably struck from the same side (Leader, 2014; Li et al., 2017; Sharon & Beaumont, 2006; also see Figure 16.7). Across four different assemblages, this corresponded to the dominant direction of strikes evident on the bifaces themselves (Sharon & Beaumont, 2006). Out of 226 flake bifaces from Victoria West sites where it was possible to tell blank strike direction, 180 were from the left, and 46 were

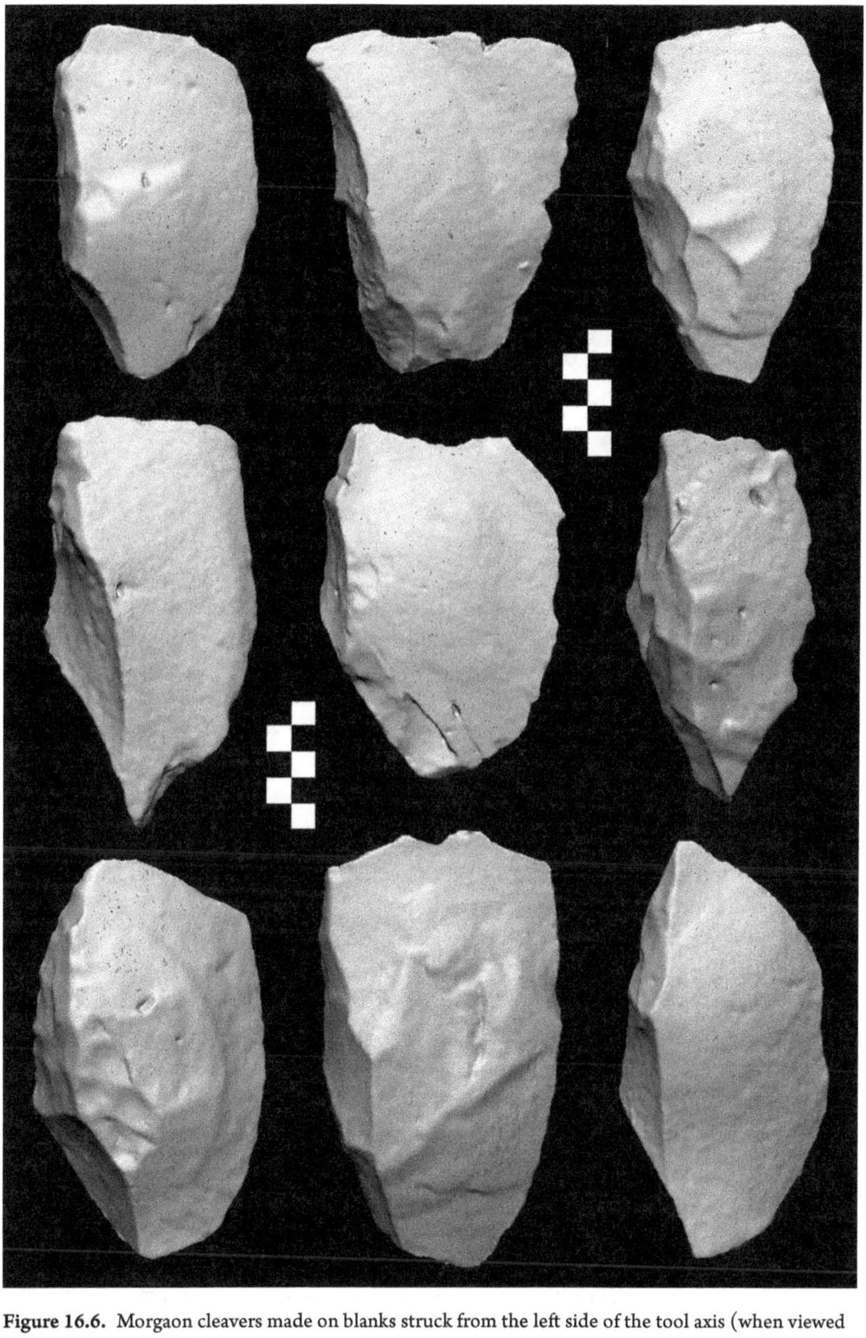

Figure 16.6. Morgaon cleavers made on blanks struck from the left side of the tool axis (when viewed from the ventral surface). Image by the author.

Figure 16.7. Victoria West core from Vingerfontein showing the platform from which the preferential cleaver flake blank was struck with the scar left by that flake blank outlined in black. Note that the cleaver flake blanks were always struck from this same side of the Victoria West cores, even though the cores appear to have been symmetrical before the preferential flakes were struck. Reproduced with the permission of the University of Cambridge Museum of Archaeology and Anthropology. Accession no. Z.36167.

from the right (Sharon, 2007). Since Victoria West cores are symmetrical about their long axis prior to the cleaver blank being struck, there seems to be no technological reason to strike them on one side rather than the other.

The Tabelbala-Tachengit method of northwestern Africa is another complex cleaver blank production strategy, quite similar to that of Victoria West (Alimen, 1978; Tixier, 1956; also see Figure 16.8). One of the principal differences between the two methods is the platform isolation on the Tabelbala-Tachengit cores, which were struck from the left corner rather than the right side as on Victoria West cores (Sharon, 2007) (Figure 16.9). As the cores were approximately symmetrical prior to the isolation of the platform (Figure 16.8C), there again seems to be no a priori reason why the left rather than the right corner should have been chosen.

Notably, Acheulean assemblages relying on less formal flake blank production strategies, such as Hunsgi V and Yediyapur IV in India, do not display the same dominance of a particular blow direction (Sharon, 2007). It is hypothesized that the dominance of flake blanks struck from a particular direction at Chirki, Morgaon, and in the Victoria West and Tabelbala-Tachengit industries is due to the increased importance of social transmission in reproducing these relatively complicated knapping strategies. At the site of Canteen Kopje, the Victoria West industry is dated to around 1 Mya and appears to be the culmination of three phases of increasingly complex core preparation (Leader, 2014; Li et al., 2017). Over-imitation may then have emerged during the Acheulean in response to increasing technological complexity.

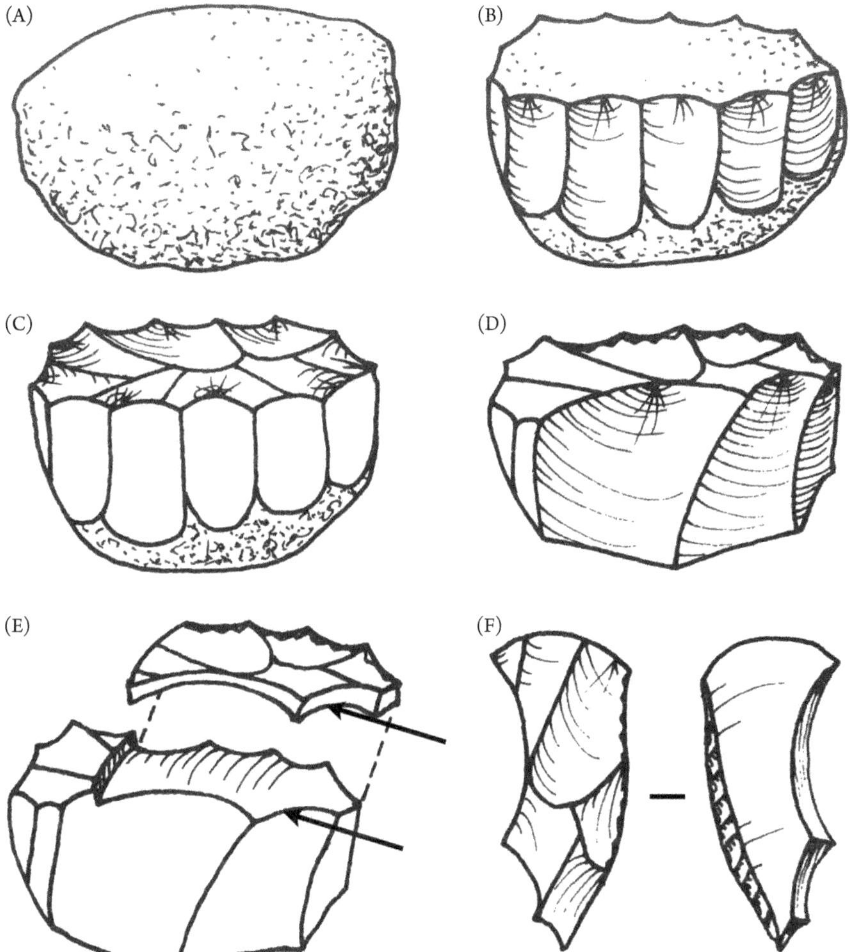

Figure 16.8. The Tabelbala-Tachengit method of cleaver blank production. Key: A = the unmodified cobble; B = creating facets around the cobble perimeter; C = shaping the upper surface; D = three large flake scars to isolate the platform; E = striking of the cleaver blank; F = both sides of the finished cleaver with marginal trimming on the ventral. Adapted from Clark (1992), The earlier Stone Age/Lower Palaeolithic in North Africa and the Sahara, *New Light on the Northeast African Past*, Heinrich-Barth-Institut, and Tixier (1956), *Le hachereau dans l'Acheuléen nord-africain*, Société préhistorique française.

LATE ACHEULEAN TEACHING

Over-imitation is sufficient for copying complex tasks where the actions are bold and clear. The platform isolation in the Tabelbala-Tachengit method, for example, involved the removal of large flakes on the left side and proximal end of the core (Alimen, 1978; Tixier, 1956; also see Figure 16.8). However, by the Late Acheulean (less than

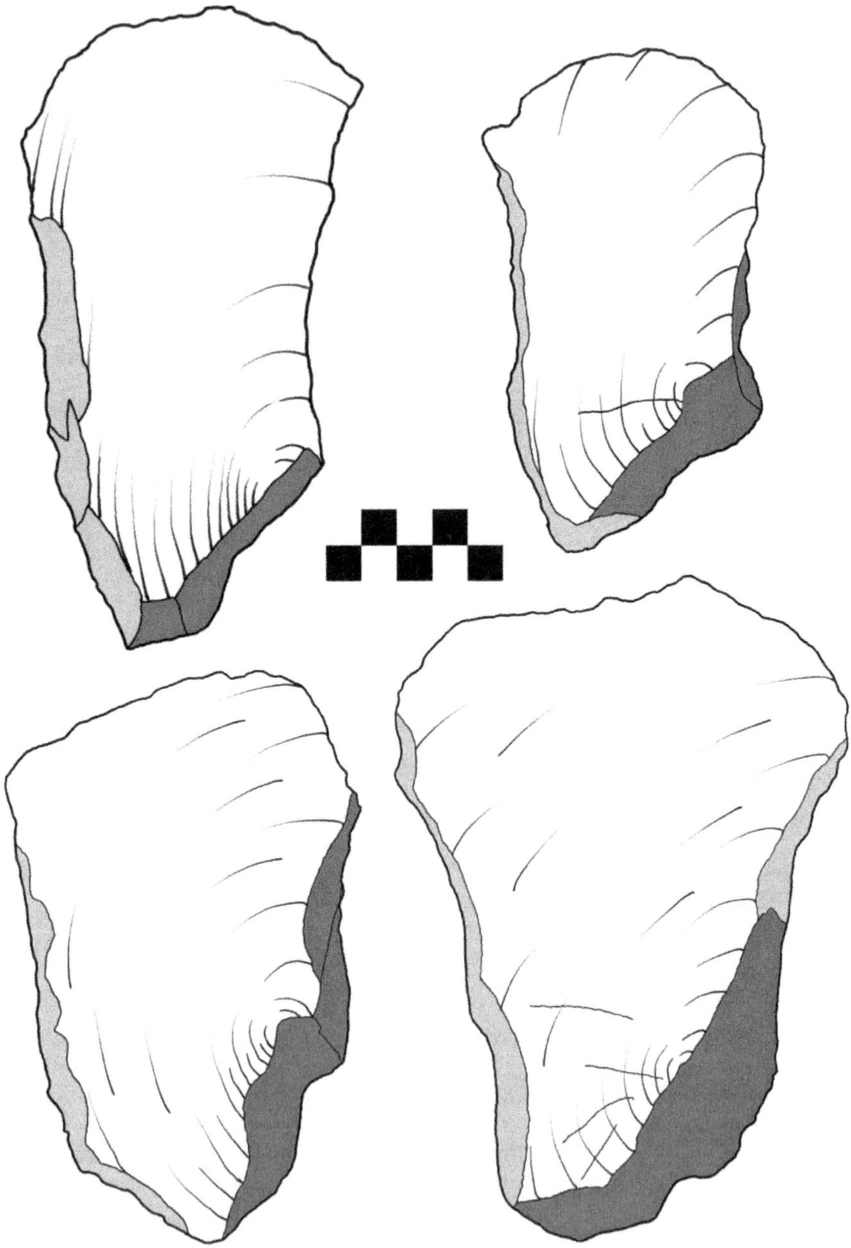

Figure 16.9. Four cleavers from Tachengit. Dark gray denotes the platform, and light gray denotes retouch. Note that they are all struck from the bottom right corner when viewed from the ventral surface. Image by the author.

600,000 years ago), more subtle platform preparation is apparent, and this is something that would probably be difficult to replicate by observation alone.

At the roughly 500,000-year-old site of Boxgrove in Britain (Pitts & Roberts, 1997), the technique of platform faceting is evident (Stout, Apel, Commander, & Roberts, 2014). Platform faceting involves the removal of small flakes, typically less than 0.25 cm in length, from bifacial edges. This serves to strengthen the edge by blunting it and steepening the angle between the two surfaces of the biface (Callahan, 1979), with the multiple small facets of the scars giving further strength. This edge then can serve as a platform that will withstand a greater strike force, so that larger flakes may propagate from it. Related to this platform preparation technique, and something often done in the process of faceting, is moving the plane of intersection toward one surface of the biface (sometimes called "turning the edge"). This results in the force of the main strike being directed across the surface of the biface rather than at an angle to it, resulting in invasive flakes that travel to the midline of the piece or beyond, thereby thinning it (Callahan, 1979). This technique has been documented at the terminal Acheulean site of Patpara in India (Shipton, 2016), and it is also evident at Boxgrove (Figures 16.10). Once the invasive flakes are struck, they tend to remove the faceted platform and restore the plane of intersection to a more medial position. Invasive flaking allows a biface to be thinned and underpins the more refined bifaces evident at Late Acheulean sites such as Boxgrove (Coolidge & Wynn, 2009; also see Figures 16.10).

The extra steps involved in invasively flaking a biface require more hierarchically organized levels of sub-goals than basic biface knapping does (Muller, Clarkson, & Shipton, 2017). Higher levels of sub-goals are more difficult to perceive directly and their purpose, more difficult to comprehend (Gärdenfors & Högberg, 2017). Simple forms of teaching such as evaluative feedback, drawing attention, and demonstration may not have been sufficient to convey invasive flaking techniques, with the teaching of concepts becoming a necessity (Gärdenfors & Högberg, 2017).

Given the small size of the flakes removed in the process of platform faceting and raising the plane of intersection, these techniques involve only very subtle movements of the hammer. They might easily be confused with other techniques, such as grinding the edge, or even disregarded as missed blows. Furthermore, the intermediate effects of these techniques are also very subtle, to the extent that they only become visible through close inspection of the tool. It is hypothesized that imitation and even over-imitation would not be sufficient to transmit these techniques between hominins, as their subtlety requires explanation as to what is being done and why.

The experience of the knapper Chris Clarkson is interesting in this regard, as he states that despite knowing these techniques in principle, he was not able to fully utilize them until he was actively taught to do so: "[An expert knapper] showed me some valuable stuff about isolating platforms and turning edges that I knew in theory, but didn't really know too well in practice (personal communication, 2017)."

An experiment explicitly designed to test the effects of teaching on handaxe knapping divided naïve participants into two groups, with one group having only simple gestural instruction (pointing to salient parts of the stones) and the other full verbal teaching (Putt, Woods, & Franciscus, 2014). While the handaxes produced by both groups were of similar quality, the principal difference between them was that only in the verbal-teaching group did participants copy their instructor to the extent that

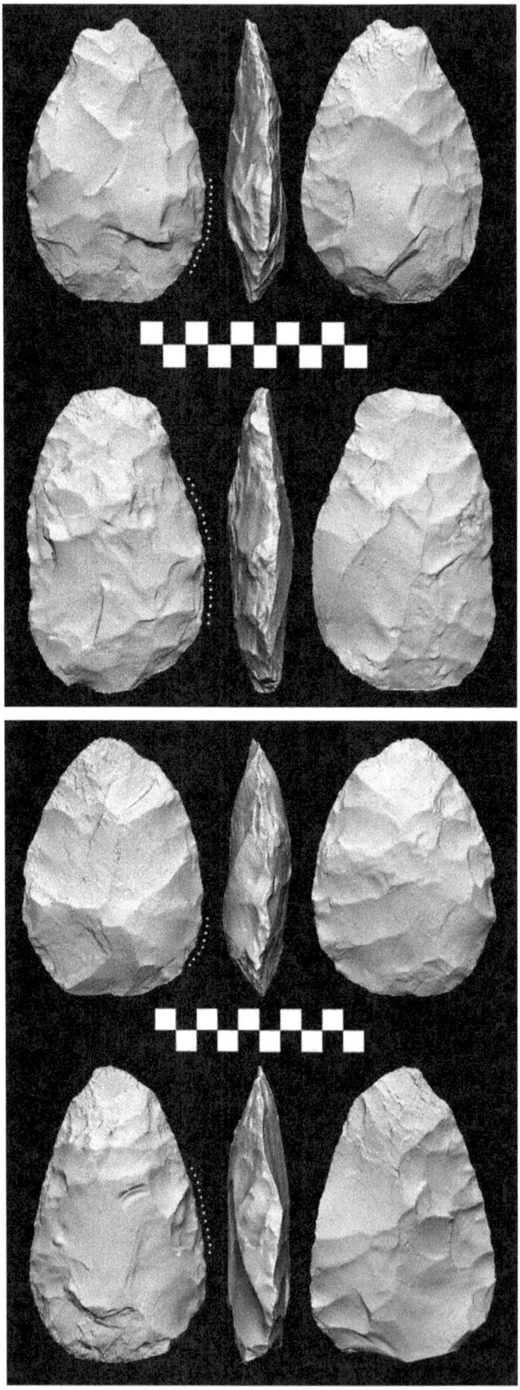

Figure 16.10. Four Boxgrove handaxes with probable faceting (denoted by dots) and planes of intersection offset to one surface. Note the highly invasive flaking and the thinness and symmetry of the pieces, particularly in comparison to the earlier Acheulean bifaces shown in Figures 16.1 and 16.4–16.6. Image by the author.

they were preparing platforms. As a result, the verbal-teaching group was able to produce large flakes and used fewer removals to achieve similar results. In other respects, such as the number of failed attempts, the verbal-teaching group was in fact less efficient, possibly because they were over-ambitiously attempting platform preparation at a very early stage of knapping competency (Putt et al., 2014).

Some of the techniques used to make the refined bifaces of the Late Acheulean—in particular, platform preparation and raising the plane of intersection—are so subtle and difficult to master that even over-imitation would likely have been insufficient for their transmission. It seems likely that transmitting the concepts and methods underpinning such techniques requires symbolic instruction, in which specific details of the stone tool can be referentially isolated and their relationship to one another explained.

CUMULATIVE CULTURAL EVOLUTION IN THE ACHEULEAN?

If the Acheulean is characterized by varying levels of high-fidelity social transmission, then should it also attest to cumulative cultural evolution? The addition of shaping in the early Acheulean to the knapping repertoire of the Oldowan (Gowlett, 1986; Roche, 2005) may have been an early instance of cumulative cultural evolution. The classic Acheulean, with its techniques for standardized large flake blank production and more symmetrical handaxes (Gallotti & Mussi, 2017; Li et al., 2017; Saragusti, Sharon, Katzenelson, & Avnir, 1998; Texier, 1996), represented further advances in knapping ability. The exquisite handaxes of the Late Acheulean, made with soft hammers (Stout et al., 2014) and the advanced knapping techniques described here, took knapping skill very close to the level displayed today by expert *H. sapiens* knappers (Edwards, 2001). However, these changes occurred over the course of more than a million years, so while they may have represented cumulative cultural evolution, its glacial pace was still quite unlike that of our own species.

High-fidelity social transmission is necessary for cumulative cultural evolution, but it is not the only ingredient. Dean and colleagues (2014) have identified demography, social structure, and cognition as key factors. It has also been suggested that small population size and shorter life histories may have contributed to the cultural stasis of the Acheulean (Nowell & White, 2010). There are aspects of cognition that are likely pertinent to cumulative culture, but which seem to have been deficient in the Acheulean in comparison to later cultural periods (Shipton et al., 2013). These include *hierarchical organization*, the ability to structure multiple behaviors into an ordered sequence; *recursion*, the ability to embed discrete behavioral routines within a larger sequence; and *generativity*, the ability to combine two existing behaviors into something new. The Acheulean-to-Middle Paleolithic transition was characterized by stone tool reduction sequences with more stages (Moncel, Moigne, Sam, & Combier, 2011; Shipton, 2016), recursive flaking strategies such as recurrent Levallois (Tryon, McBrearty, & Texier, 2005), and the combination of existing flaking concepts to create new ones (White & Ashton, 2003).

CONCLUSION

Social transmission allows both the horizontal sharing of knowledge and skills and their vertical continuity across generations. As such, it is an integral part of our cultural adaptation. Non-human great apes display some of the most robust social transmission seen in the animal kingdom, including foraging techniques using hammers. But unlike human children, other extant apes do not copy intentional yet seemingly redundant actions.

The complex percussive technology of the Oldowan and early Acheulean may have placed a selective premium on higher-fidelity social transmission. Modeling suggests higher-fidelity social transmission will only evolve when the behavioral repertoire of a species is large, and the search space does not provide information that can be exploited by individual learning (Acerbi, Jacquet, & Tennie, 2012). The Oldowan-to-Acheulean transition was characterized by a variety of new stone tool technologies, such as the Karari industry with its core-scrapers (Harris & Isaac, 1976), indicating a diverse repertoire of knapping behavior. Experimenting with modern novice knappers has failed to produce handaxes or even competency in basic freehand percussion without social observation of the process (Geribàs, Mosquera, & Vergès, 2010), indicating that individual learning could not have been enough to sustain the Early Stone Age. Oldowan and Acheulean knapping would thus have created the selective conditions for higher-fidelity imitation.

By the classic phase of the Acheulean, beginning sometime before 1 Mya, standardized methods for large flake production became widespread, including some methods whose steps were so causally opaque that they must have been transmitted through over-imitation. The four examples described in this chapter all had a bias toward striking flake blanks from a particular side of the core, something for which there is no obvious technological explanation. The over-imitation of complex knapping sequences with causally opaque steps may have resulted in distinct biases toward striking large flake blanks from a particular, seemingly arbitrary side of the cores.

Many bifaces of the Late Acheulean are remarkably thin and symmetrical. Underpinning this finesse was the ability to strike invasive flakes from faceted platforms on raised planes of intersection. Such platform preparation is physically subtle and conceptually remote, so that the techniques probably could not have been transmitted by observation alone; and indeed, experiments indicate that even simple gestural teaching is not enough (Putt et al., 2014). Symbolic instruction may be essential for transmitting these advanced knapping techniques. A key developmental function of linguistic interaction in children today is offering causal explanations, which allow for generalizations that they can subsequently use (Legare & Lombrozo, 2014). The unique childhood stage of human life history is critical to both human language and teaching (Locke & Bogin, 2006; Nielsen & Shipton, 2015), and there are some suggestions that childhood was emerging as a new life history stage by the time of the Late Acheulean (De Castro, Rozzi, Martinón-Torres, Pérez, & Rosas, 2003; Robson & Wood, 2008). Despite the evidence for very high–fidelity social transmission in the Acheulean, cumulative culture appears to have been meagre. In comparison to ourselves, Acheulean hominins were deficient in three cognitive traits that support the ability to intermingle ideas and are probably critical to the generation of cumulative culture: hierarchical organization, recursion, and generativity. High-fidelity social transmission without the modern ability to recombine ideas was likely a major factor in the extraordinary temporal and geographic span of the Acheulean.

ACKNOWLEDGMENTS

Sushma Deo and Sheila Mishra of the Deccan College generously granted me permission to study the Morgaon and Chirki assemblages. Thanks are also due to Imogen Gunn at the Cambridge Museum of Archaeology and Anthropology for setting up the Vingerfontein Victoria West core for photography. The British Museum is gratefully acknowledged for access to the Boxgrove biface collections.

REFERENCES

Acerbi, A., Jacquet, P. O., & Tennie, C. (2012). Behavioral constraints and the evolution of faithful social learning. *Current Zoology, 58*(2), 307–318.

Alimen, M. H. (1978). *L'évolution de l'Acheuléen au Sahara nord-occidental: Saoura, Ougarta, Tabelbala*. Paris: Éditions CNRS.

Baena, J., Lordkipanidze, D., Cuartero, F., Ferring, R., Zhvania, D., Martín, D., . . . Rubio, D. (2010). Technical and technological complexity in the beginning: The study of Dmanisi lithic assemblage. *Quaternary International, 223*, 45–53.

Beyene, Y., Katoh, S., WoldeGabriel, G., Hart, W. K., Uto, K., Sudo, M., . . . Asfaw, B. (2013). The characteristics and chronology of the earliest Acheulean at Konso, Ethiopia. *Proceedings of the National Academy of Sciences of the United States of America, 110*(5), 1584–1591.

Blackwell, B. A. B., Fevrier, S., Blickstein, J. I. B., Paddayya, K., Perraglia, M., Jhaldiyal, R., & Skinner, A. R. (2001). ESR dating of an Acheulean quarry site at Isampur, India. *Journal of Human Evolution, 40*(3), A3.

Boesch, C. (1991). Teaching among wild chimpanzees. *Animal Behaviour, 41*(3), 530–532.

Callahan, E. (1979). *The basics of biface knapping in the eastern fluted point tradition: A manual for flintknappers and lithic analysts*. Bethlehem, CT: Eastern States Archeological Federation.

Clark, J. D. (1992). The Earlier Stone Age/Lower Palaeolithic in North Africa and the Sahara. In F. Klees & R. Kuper (Eds.), *New light on the Northeast African past: Current prehistoric research: Contributions to a symposium, Cologne 1990* (pp. 17–37). Köln, Germany: Heinrich-Barth-Institut.

Coolidge, F. L., & Wynn, T. (2009). *The rise of* Homo sapiens: *The evolution of modern thinking*. Oxford, UK: Wiley-Blackwell.

Corbey, R., Jagich, A., Vaesen, K., & Collard, M. (2016). The Acheulean handaxe: More like a bird's song than a Beatles' tune? *Evolutionary Anthropology: Issues, News, and Reviews, 25*(1), 6–19.

Corvinus, G. (1983). *A survey of the Pravara River system in western Maharashtra, India: Volume 2, The excavations of the Acheulian site of Chirki-on-Pravara, India*. Tübingen, Germany: Institut für Urgeschichte.

Dean, L. G., Kendal, R. L., Schapiro, S. J., Thierry, B., & Laland, K. N. (2012). Identification of the social and cognitive processes underlying human cumulative culture. *Science, 335*(6072), 1114–1118.

Dean, L. G., Vale, G. L., Laland, K. N., Flynn, E. G., & Kendal, R. L. (2014). Human cumulative culture: A comparative perspective. *Biological Reviews, 89*(2), 284–301.

De Castro, J. M. B., Rozzi, F. R., Martinón-Torres, M., Pérez, S. S., & Rosas, A. (2003). Patterns of dental development in Lower and Middle Pleistocene hominins from Atapuerca (Spain). In J. L. Thompson, G. E. Krovitz, & A. J. Nelson (Eds.), *Patterns of growth and development in the genus* Homo (pp. 246–270). Cambridge, UK: Cambridge University Press.

Delagnes, A., & Roche, H. (2005). Late Pliocene hominid knapping skills: The case of Lokalalei 2C, West Turkana, Kenya. *Journal of Human Evolution, 48*(5), 435–472.

de la Torre, I. (2004). Omo revisited: Evaluating the technological skills of Pliocene hominids. *Current Anthropology, 45*(4), 439–465.

Deo, S. G., Mishra, S., Rajaguru, S. N., & Ghate, S. (2007). Antiquity of Acheulian culture in upland Maharasthra: A geoarchaeological approach. In A. R. Sankjhyan & V. R. Rao (Eds.), *Human origins, genome, & people of India: Genomic, palaeontological & archaeological perspectives* (pp. 292–308). Calcutta: Concept Publishers.

Diez-Martín, F., Yustos, P. S., Uribelarrea, D., Baquedano, E., Mark, D. F., Mabulla, A. Z. P., . . . Domínguez-Rodrigo, M. (2015). The origin of the Acheulean: The 1.7 million-year-old site of FLK West, Olduvai Gorge (Tanzania). *Scientific Reports, 5*(17839), 1–9.

Edwards, S. W. (2001). A modern knapper's assessment of the technical skills of the Late Acheulean biface workers at Kalambo Falls. In J. D. Clark (Ed.), *Kalambo Falls prehistoric site* (Vol. 3, pp. 605–611). Cambridge, UK: Cambridge University Press.

Feibel, C. S. (2004). Quaternary lake margins of the Levant Rift Valley. In N. Goren-Inbar & J. D. Speth (Eds.), *Human paleoecology in the Levantine Corridor* (pp. 21–36). Oxford, UK: Oxbow Books.

Gallotti, R., & Mussi, M. (2017). Two Acheuleans, two humankinds: From 1.5 to 0.85 Ma at Melka Kunture (Upper Awash, Ethiopian highlands). *Journal of Anthropological Sciences, 95*, 1–46.

Gärdenfors, P., & Högberg, A. (2017). The archaeology of teaching and the evolution of *Homo docens*. *Current Anthropology, 58*(2), 188–208.

Geribàs, N., Mosquera, M., & Vergès, J. M. (2010). What novice knappers have to learn to become expert stone toolmakers. *Journal of Archaeological Science, 37*(11), 2857–2870.

Goldman-Neuman, T., & Hovers, E. (2009). Methodological considerations in the study of Oldowan raw material selectivity: Insights from AL 894 (Hadar, Ethiopia). In E. Hovers & D. R. Braun (Eds.), *Interdisciplinary approaches to the Oldowan* (pp. 71–84). Dordrecht, The Netherlands: Springer Science + Business Media.

Gowlett, J. A. J. (1986). Culture and conceptualisation: The Oldowan-Acheulian gradient. In G. N. Bailey & P. Callow (Eds.), *Stone Age prehistory: Studies in memory of Charles McBurney* (pp. 243–260). Cambridge, UK: Cambridge University Press.

Harmand, S., Lewis, J. E., Feibel, C. S., Lepre, C. J., Prat, S., Lenoble, A., . . . Roche, H. (2015). 3.3-million-year-old stone tools from Lomekwi 3, West Turkana, Kenya. *Nature, 521*(7552), 310–315.

Harris, J. W. K., & Isaac, G. L. (1976). The Karari industry: Early Pleistocene archaeological evidence from the terrain east of Lake Turkana, Kenya. *Nature, 262*(5564), 102–107.

Horner, V., & Whiten, A. (2005). Causal knowledge and imitation/emulation switching in chimpanzees (*Pan troglodytes*) and children (*Homo sapiens*). *Animal Cognition, 8*(3), 164–181.

Horner, V., Whiten, A., Flynn, E. G., & De Waal, F. B. M. (2006). Faithful replication of foraging techniques along cultural transmission chains by chimpanzees and children. *Proceedings of the National Academy of Sciences of the United States of America, 103*(37), 13878–13883.

Hovers, E. (2009). Learning from mistakes: Flaking accidents and knapping skills in the assemblage of AL 894 (Hadar, Ethiopia). In K. D. Schick & N. P. Toth (Eds.), *The cutting edge: New approaches to the archaeology of human origins* (Vol. 3, pp. 137–148). Gosport, IN: Stone Age Institute Press.

Kale, V. S. (1990). Time-scale of landscape development in the Deccan Plateau. *Tropical Geomorphology Newsletter, 10*, 4–5.

Key, A. J. M., & Dunmore, C. J. (2015). The evolution of the hominin thumb and the influence exerted by the non-dominant hand during stone tool production. *Journal of Human Evolution, 78*, 60–69.

Kivell, T. L. (2015). Evidence in hand: Recent discoveries and the early evolution of human manual manipulation. *Philosophical Transactions of the Royal Society of London. Series B, Biological Sciences, 370*(1682), 20150105.

Koops, K., Schöning, C., Isaji, M., & Hashimoto, C. (2015). Cultural differences in ant-dipping tool length between neighbouring chimpanzee communities in Kalinzu, Uganda. *Scientific Reports, 5*(12456), 1–8.

Leader, G. M. (2014). *New excavations at Canteen Kopje, Northern Cape Province, South Africa: A technological comparison of three earlier Acheulean assemblages with new interpretations on the Victoria West phenomenon.* Doctoral thesis, University of the Witwatersrand, Johannesburg, South Africa.

Legare, C. H., & Lombrozo, T. (2014). Selective effects of explanation on learning during early childhood. *Journal of Experimental Child Psychology, 126*, 198–212.

Lepre, C. J., Roche, H., Kent, D. V, Harmand, S., Quinn, R. L., Brugal, J.-P., . . . Feibel, C. S. (2011). An earlier origin for the Acheulian. *Nature, 477*(7362), 82–85.

Lewis, H. M., & Laland, K. N. (2012). Transmission fidelity is the key to the build-up of cumulative culture. *Philosophical Transactions of the Royal Society of London. Series B, Biological Sciences, 367*(1599), 2171–2180.

Lewis, J. E., & Harmand, S. (2016). An earlier origin for stone tool making: Implications for cognitive evolution and the transition to *Homo*. *Philosophical Transactions of the Royal Society of London. Series B, Biological Sciences, 371*(1698), 1–8.

Li, H., Kuman, K., Lotter, M. G., Leader, G. M., & Gibbon, R. J. (2017). The Victoria West: Earliest prepared core technology in the Acheulean at Canteen Kopje and implications for the cognitive evolution of early hominids. *Royal Society Open Science, 4*(6), 170288.

Locke, J. L., & Bogin, B. (2006). Language and life history: A new perspective on the development and evolution of human language. *Behavioral and Brain Sciences, 29*(3), 259–280.

Luncz, L. V, Mundry, R., & Boesch, C. (2012). Evidence for cultural differences between neighbouring chimpanzee communities. *Current Biology, 22*(10), 922–926.

Lyons, D. E., Damrosch, D. H., Lin, J. K., Macris, D. M., & Keil, F. C. (2011). The scope and limits of overimitation in the transmission of artefact culture. *Philosophical Transactions of the Royal Society of London. Series B, Biological Sciences, 366*(1567), 1158–1167.

Marshall-Pescini, S., & Whiten, A. (2008a). Chimpanzees (*Pan troglodytes*) and the question of cumulative culture: An experimental approach. *Animal Cognition, 11*(3), 449–456.

Marshall-Pescini, S., & Whiten, A. (2008b). Social learning of nut-cracking behavior in East African sanctuary-living chimpanzees (*Pan troglodytes schweinfurthii*). *Journal of Comparative Psychology, 122*(2), 186–194.

Marzke, M. W. (2013). Tool making, hand morphology and fossil hominins. *Philosophical Transactions of the Royal Society of London. Series B, Biological Sciences, 368*(1630), 20120414.

Mgeladze, A., Lordkipanidze, D., Moncel, M.-H., Despriée, J., Chagelishvili, R., Nioradze, M., & Nioradze, G. (2011). Hominin occupations at the Dmanisi site, Georgia, Southern Caucasus: Raw materials and technical behaviors of Europe's first hominins. *Journal of Human Evolution, 60*(5), 571–596.

Mishra, S., Deo, S. G., Abbas, R., Naik, S., Shete, G., Agrawal, N., & Rajaguru, S. N. (2009). Excavations at the early Acheulian site of Morgaon, Maharashtra (2000–2007). In *Recent research trends in South Asian archaeology* (pp. 121–137). Pune: Deccan College Post Graduate and Research Institute.

Moncel, M.-H., Moigne, A.-M., Sam, Y., & Combier, J. (2011). The emergence of Neanderthal technical behavior: New evidence from Orgnac 3 (Level 1, MIS 8), southeastern France. *Current Anthropology, 52*(1), 37–75.

Moore, M. W. (2003). Australian aboriginal biface reduction techniques on the Georgina River, Camooweal, Queensland. *Australian Archaeology, 56*(1), 22–34.

Morgan, T. J. H., Uomini, N. T., Rendell, L. E., Chouinard-Thuly, L., Street, S. E., Lewis, H. M., . . . Laland, K. N. (2015). Experimental evidence for the co-evolution of hominin tool-making teaching and language. *Nature Communications, 6*(6029), 1–18.

Muller, A., Clarkson, C., & Shipton, C. (2017). Measuring behavioral and cognitive complexity in lithic technology throughout human evolution. *Journal of Anthropological Archaeology, 48*, 166–180.

Nielsen, M., Mushin, I., Tomaselli, K., & Whiten, A. (2014). Where culture takes hold: "Overimitation" and its flexible deployment in Western, Aboriginal, and Bushmen children. *Child Development, 85*(6), 2169–2184.

Nielsen, M., & Shipton, C. (2015). Childhood and the evolution of higher-effort teaching. *Behavioral and Brain Sciences, 38*, 34–35.

Nowell, A., & White, M. J. (2010). Growing up in the Middle Pleistocene: Life history strategies and their relationship to Acheulian industries. In A. Nowell & I. Davidson (Eds.), *Stone tools and the evolution of human cognition* (pp. 67–82). Boulder, CO: University Press of Colorado.

Paddayya, K. (2007). Evolution within the Acheulian in India: A case study from the Hunsgi and Baichbal Valleys, Karnataka. *Bulletin of the Deccan College Research Institute, 66*, 95–111.

Paddayya, K., Jhaldiyal, R., & Petraglia, M. D. (2006). The Acheulian quarry at Isampur, Lower Deccan, India. In N. Goren-Inbar & G. Sharon (Eds.), *Axe age: Acheulian tool-making from quarry to discard* (pp. 45–73). London: Equinox.

Parry, R., Dietrich, G., & Bril, B. (2014). Tool use ability depends on understanding of functional dynamics and not specific joint contribution profiles. *Frontiers in Psychology, 5*, 1–15.

Petraglia, M. D., LaPorta, P., & Paddayya, K. (1999). The first Acheulian quarry in India: Stone tool manufacture, biface morphology, and behaviors. *Journal of Anthropological Research, 55*(1), 39–70.

Pitts, M. W., & Roberts, M. B. (1997). *Fairweather Eden: Life in Britain half a million years ago as revealed by the excavations at Boxgrove.* London: Random House.

Putt, S. S., Woods, A. D., & Franciscus, R. G. (2014). The role of verbal interaction during experimental bifacial stone tool manufacture. *Lithic Technology, 39*(2), 96–112.

Rein, R., Nonaka, T., & Bril, B. (2014). Movement pattern variability in stone knapping: Implications for the development of percussive traditions. *PloS One, 9*(11), e113567.

Robson, S. L., & Wood, B. A. (2008). Hominin life history: Reconstruction and evolution. *Journal of Anatomy, 212*(4), 394–425.

Roche, H. (2005). From simple flaking to shaping: Stone knapping evolution among early hominins. In V. Roux & B. Bril (Eds.), *Stone knapping: The necessary conditions for a uniquely hominin behaviour* (pp. 35–48). Cambridge, UK: McDonald Institute for Archaeological Research.

Rossano, M. J. (2017). Cognitive fluidity and Acheulean over-imitation. *Cambridge Archaeological Journal, 27*(3), 495–509.

Sakura, O., & Matsuzawa, T. (1991). Flexibility of wild chimpanzee nut-cracking behavior using stone hammers and anvils: An experimental analysis. *Ethology, 87*(3-4), 237–248.

Sangode, S. J. S., Mishra, S., Naik, S., & Deo, S. G. (2007). Magnetostratigraphy of the Quaternary sediments associated with some Toba tephra and Acheulian artefact bearing localities in the western and central India. *Gondwana Magazine, 10*, 111–121.

Saragusti, I., Sharon, I., Katzenelson, O., & Avnir, D. (1998). Quantitative analysis of the symmetry of artefacts: Lower Paleolithic handaxes. *Journal of Archaeological Science, 25*(8), 817–825.

Savage-Rumbaugh, E. S., & Fields, W. M. (2006). Rules and tools: Beyond anthropomorphism. In N. P. Toth & K. D. Schick (Eds.), *The Oldowan: Case studies into the earliest Stone Age* (pp. 223–241). Gosport, IN: Stone Age Institute Press.

Sharon, G. (2007). *Acheulian large flake industries: Technology, chronology, and significance.* Oxford, UK: Archaeopress.

Sharon, G., Alperson-Afil, N., & Goren-Inbar, N. (2011). Cultural conservatism and variability in the Acheulian sequence of Gesher Benot Ya'aqov. *Journal of Human Evolution, 60*(4), 387–397.

Sharon, G., & Beaumont, P. B. (2006). Victoria West: A highly standardized prepared core technology. In N. Goren-Inbar & G. Sharon (Eds.), *Axe age: Acheulian tool-making from quarry to discard* (pp. 181–199). London: Equinox.

Shipton, C. (2010). Imitation and shared intentionality in the Acheulean. *Cambridge Archaeological Journal, 20*(2), 197–210.

Shipton, C. (2013). *A million years of hominin sociality and cognition: Acheulean bifaces in the Hunsgi-Baichbal Valley, India.* Oxford, UK: Archaeopress.

Shipton, C. (2016). Hierarchical organization in the Acheulean to Middle Palaeolithic transition at Bhimbetka, India. *Cambridge Archaeological Journal, 26*(4), 601–618.

Shipton, C., Clarkson, C., Pal, J. N., Jones, S. C., Roberts, R. G., Harris, C., . . . Petraglia, M. D. (2013). Generativity, hierarchical action and recursion in the technology of the Acheulean to Middle Palaeolithic transition: A perspective from Patpara, the Son Valley, India. *Journal of Human Evolution, 65*(2), 93–108.

Shipton, C., & Nielsen, M. (2015). Before cumulative culture. *Human Nature, 26*(3), 331–345.

Shipton, C., Petraglia, M. D., & Paddayya, K. (2009). Stone tool experiments and reduction methods at the Acheulean site of Isampur Quarry, India. *Antiquity, 83*(321), 769–785.

Stout, D., Apel, J., Commander, J., & Roberts, M. B. (2014). Late Acheulean technology and cognition at Boxgrove, UK. *Journal of Archaeological Science, 41*, 576–590.

Stout, D., Quade, J., Semaw, S., Rogers, M. J., & Levin, N. E. (2005). Raw material selectivity of the earliest stone toolmakers at Gona, Afar, Ethiopia. *Journal of Human Evolution, 48*(4), 365–380.

Stout, D., & Semaw, S. (2006). Knapping skill of the earliest stone toolmakers: Insights from the study of modern human novices. In N. P. Toth & K. D. Schick (Eds.), *The Oldowan: Case studies into the earliest Stone Age* (pp. 307–320). Gosport, IN: Stone Age Institute Press.

Stout, D., Semaw, S., Rogers, M. J., & Cauche, D. (2010). Technological variation in the earliest Oldowan from Gona, Afar, Ethiopia. *Journal of Human Evolution, 58*(6), 474–491.

Tennie, C., Braun, D. R., Premo, L. S., & McPherron, S. P. (2016). The island test for cumulative culture in the Paleolithic. In M. N. Haidle, N. J. Conard, & M. Bolus (Eds.), *The nature of culture: Based on an interdisciplinary symposium "The nature of culture", Tübingen, Germany* (pp. 121–133). Dordrecht, The Netherlands: Springer Science + Business Media.

Tennie, C., Call, J., & Tomasello, M. (2009). Ratcheting up the ratchet: On the evolution of cumulative culture. *Philosophical Transactions of the Royal Society of London. Series B, Biological Sciences, 364*(1528), 2405–2415.

Tennie, C., Call, J., & Tomasello, M. (2012). Untrained chimpanzees (*Pan troglodytes schweinfurthii*) fail to imitate novel actions. *PloS One, 7*(8), e41548.

Tennie, C., Premo, L. S., Braun, D. R., & McPherron, S. P. (2017). Early stone tools and cultural transmission: Resetting the null hypothesis. *Current Anthropology, 58*(5), 652–672.

Texier, P.-J. (1996). L'Acheuléen d'Isenya (Kenya), approche expérimentale des principales chaînes opératoires lithiques. In *La vie prehistorique* (pp. 58–63). Paris: Société préhistorique française.

Tixier, J. (1956). Le hachereau dans l'Acheuléen nord-africain. Notes typologiques. In *Congrès préhistorique de France: Compte rendu de la XV session: Poitiers-Angouleme 15–22 juillet 1956* (pp. 914–923). Paris: Société préhistorique française.

Tixier, J., & Inizian, M.-L. (1983). *Préhistoire de la pierre taillée. 1. Terminologie et technologie.* Paris: Valbonne.

Tomasello, M. (1999). The human adaptation for culture. *Annual Review of Anthropology, 28*(1), 509–529.

Toth, N. P., Schick, K. D., & Semaw, S. (2006). A comparative study of the stone tool-making skills of Pan, Australopithecus, and Homo sapiens. In N. P. Toth & K. D. Schick (Eds.), *The Oldowan: case studies into the earliest Stone Age* (pp. 155–222). Gosport IN: The Stone Age Institute Press.

Tryon, C. A., McBrearty, S., & Texier, P.-J. (2005). Levallois lithic technology from the Kapthurin Formation, Kenya: Acheulian origin and Middle Stone Age diversity. *African Archaeological Review, 22*(4), 199–229.

Westergaard, G. C., & Suomi, S. J. (1994). A simple stone-tool technology in monkeys. *Journal of Human Evolution, 27*(5), 399–404.

White, M. J., & Ashton, N. M. (2003). Lower Palaeolithic core technology and the origins of the Levallois method in North-Western Europe. *Current Anthropology, 44*(4), 598–609.

Whiten, A., Horner, V., Litchfield, C. A., & Marshall-Pescini, S. (2004). How do apes ape? *Animal Learning & Behavior, 32*(1), 36–52.

Whiten, A., McGuigan, N., Marshall-Pescini, S., & Hopper, L. M. (2009). Emulation, imitation, over-imitation and the scope of culture for child and chimpanzee. *Philosophical Transactions of the Royal Society of London. Series B, Biological Sciences, 364*(1528), 2417–2428.

Williamson, R. A., & Markman, E. M. (2006). Precision of imitation as a function of preschoolers' understanding of the goal of the demonstration. *Developmental Psychology, 42*(4), 723–731.

Wynn, T. (2000). Symmetry and the evolution of the modular linguistic mind. In P. Carruthers & A. Chamberlain (Eds.), *Evolution and the human mind: Modularity, language and meta-cognition* (pp. 113–139). Cambridge, UK: Cambridge University Press.

17

KNAPPING IN THE DARK
STONE TOOLS AND A THEORY OF MIND

James Cole

INTRODUCTION

The Paleolithic archaeological record remains the primary vehicle by which researchers are able to assess the behavioral and cognitive abilities of our hominin ancestors. The Paleolithic covers a period of time that stretches from roughly 3.3 million (Mya) to 10 thousand (Kya) years ago (Harmand et al., 2015). Much of this Stone Age is, by definition, focused on lithic tools because non-lithic artifacts generally preserve poorly, though there are a few exceptions (Thieme, 1997; Warren, 1911). In comparison, the record of the last 120,000 years has proved extraordinary for organic artifacts made or modified by a range of hominin species (e.g., d'Errico, Henshilwood, Vanhaeren, & van Niekerk, 2005; Radovčić, Sršen, Radovčić, & Frayer, 2015; Vanhaeren et al., 2006; Zilhão et al., 2010). Researchers interested in the deep-time origins of our species are united in the desire to understand the way the hominin makers of Paleolithic artifacts thought, behaved, and engaged with the social and physical landscapes that framed their existence.

As such, a long history of theoretical and empirical studies (a small sample of this literature includes Davidson & Noble, 1993; Gowlett, 1979, 1996; Gowlett, Gamble, & Dunbar, 2012; Stout, 2011; Wynn, 1979, 1981, 1985, 1991, 2002) has sought to model how artifact manufacture can illuminate the workings of the prehistoric human mind (but see Stout, Hecht, Khreisheh, Bradley, & Chaminade, 2015, for a useful counter-perspective). Such models often focus on the *chaîne opératoire* of artifact manufacture, including raw material selection, management, and movement; the complexity of knapping reduction sequences and degrees of shaping and refinement; use; and discard. In addition, studies often show a general progression toward increasing technological and cognitive complexity in the *chaîne opératoire* as technology changed from a mode 1 simple core-and-flake-based toolkit to a mode 4 blade-and-prepared-core toolkit (Clark, 1961; Gamble, 2013; Lycett & Norton, 2010). Some of these cognitive models were recently formalized in a volume edited by Wynn and Coolidge (2017), which should prove most informative in advancing our way of thinking about the minds of past hominins.

PREDICTING HOMININ COGNITION

Beyond focusing on the details of artifact manufacture, researchers have sought to access past hominin cognition through brain size and predicted social group structure using the social brain hypothesis as a primary framework (e.g., Dunbar, 1998a, 2003, Dunbar, Gamble, & Gowlett, 2010, 2014; Gamble, Gowlett, & Dunbar, 2014). This approach has proven useful for understanding when language may have evolved. Certainly, it has facilitated discussion of whether markers can be detected archaeologically prior to the more obvious signposts of beads and art (Aiello & Dunbar, 1993; Cole, 2015a, 2015b; Davidson & Noble, 1989, 1993; Deacon, 1997; d'Errico et al., 2003; Dunbar, 1996, 1998b, 2003, 2007; Gamble, 2012; McNabb, 2012; Shultz, Nelson, & Dunbar, 2012). With regard to the development of language, it is important to note that in order to achieve an understanding of language (visual or verbal), the ability to mentalize is essential (Dunbar, 1998b; Gamble et al., 2014; Origgi & Sperber, 2000).

Mentalizing is the capability to understand or infer what another individual is thinking (Gamble et al., 2014, p. 18), and it includes a suite of skills referred to by philosophers of mind as "orders of intentionality" (Premack & Woodruff, 1978). Orders of intentionality are a series of self-reflective mental states that form a recursive hierarchy, yielding an ordinal scale of cognitive complexity as more mental states are added to the sequence. For example, from Shakespeare's *Othello*, Dunbar (2004, p. 162) illustrates five orders of intentionality: Shakespeare *intended* (1) that the audience *realize* (2) that the eponymous Moor *believed* (3) that his servant Iago was being honest when he claimed to *know* (4) that Desdemona *loved* (5) Cassio. The orders of intentionality and a theory of mind (ToM) have been correlated, with a ToM requiring an individual to imagine two mental states, their own and that of someone else. Therefore, a ToM is equivalent to a second order of intentionality.

A ToM is one of the most important and fundamental cognitive abilities; it underpins all of the key components that make us human, including language, symbolism, culture, and social organization (Gamble et al., 2014). A ToM is often defined as the ability to comprehend the mental state of one's own mind, as well as the mental state of an "other," and recognize that the other's mental state may differ from one's own (Baron-Cohen, 2001). Therefore, a ToM is essential to the ability to attribute different mental states, desires, and beliefs to others (Krupenye, Kano, Hirata, Call, & Tomasello, 2016; Premack & Woodruff, 1978) and must consequently underpin the ability to attain a second order of intentionality or higher. Given that orders of intentionality are an ordinal scale, a great deal of cognitive and social complexity can be realized if only a small number of mental states are linked together (Premack & Woodruff, 1978). Modern humans tend to operate at a fifth order of intentionality (occasionally, six for some individuals) (Gamble et al., 2014), so when the orders are applied to the evolutionary record (see later discussion), fifth order is generally taken as the maximum.

Before we can discuss the evolutionary and archaeological record in regard to orders of intentionality and ToM, we must first understand which species may have acquired these abilities. There has been much recent debate about whether non-human primates and animals like elephants, dolphins, and corvids have access to a ToM. As can be imagined, this is a contentious issue with no straightforward answer. Some suggest that mirror self-recognition equates to a ToM (Plotnik, De Waal, & Reiss, 2006; Povinelli et al., 1997; Reiss & Marino, 2001; Savanah, 2013), though there are valid arguments against such behavior genuinely representing a true understanding of the contents of another's mind (de Veer & van den Bos, 1999; Nielsen & Dissanayake, 2004). As I have summarized these positions elsewhere (Cole, 2015b, 2015c), I will not reiterate them here. However, it is worth noting that ToM experiments tend to be conducted with trained, hand-reared, or captive primates with extensive human contact (Davidson & Noble, 1989), and this human exposure may inherently bias the results of such experiments (see Call & Tomasello, 2008; van der Vaart & Hemelrijk, 2012, for useful summaries of these arguments). I would emphasize that in order to truly assess the cognitive capabilities of our closest living hominid cousins and other intelligent animals, wild populations should remain the focus of such research.

Research from Crockford and colleagues (Crockford, Wittig, Mundry, & Zuberbühler, 2012) suggests that wild chimpanzees genuinely understand when other members of their group are ignorant of knowledge that they themselves possess (in this instance, the presence of a potential predator); presumably, it is on this basis that they communicate a warning to modify the behavior of the rest of the group. This clearly needs further investigation, but there are strong parallels and implications for a ToM being present in these primates. A more recent study by Krupenye and colleagues (2016) further suggests that great apes have an implicit understanding of false-belief states that promote their cognitive abilities beyond merely being good predictors of behavior based on external cues (Call & Tomasello, 2008). False-belief state experiments are used with human infants to ascertain a ToM (typically, children are able to pass these tests around 4 to 5 years of age) (Gamble et al., 2014), implying that perhaps the great apes in the Krupenye et al. study have an implicit ToM. This new work certainly raises intriguing possibilities of shared cognitive behaviors between humans and other extant primates, which have also been suggested elsewhere (e.g., Cheney & Seyfarth, 1990; Seyfarth & Cheney, 2011, 2013, 2017a, 2017b). However, I think Dunbar (2007) is still on the right track when he states that great apes are only just able to achieve second-order intentionality—they do it, but not very well, and not all the time. Therefore, we see ToM in great apes as primates hovering on the border of a cognitive barrier, poised to break through. The difference with modern humans, and presumably, some of our evolutionary ancestors, is that we are consciously self-aware that we have a ToM. It is the conscious realization of this mental state that in turn acts as a springboard to the higher orders of intentionality.

Without a ToM, the ability for abstract thought, language, and symbolic construction remains elusive, and therefore, a lack of a ToM must at least partly explain why animal communication is iconic and indexical rather than symbolic (Barbieri, 2010; Deacon, 1997; Peirce, 1974; Wynn, 1995). Though it is fully acknowledged that human communication can be iconic and indexical, the key difference is that we can also easily incorporate symbolism, something other animals struggle to do, if at all. I have provided examples of iconic, indexical, and symbolic communications elsewhere (Cole, 2015c; Cole 2017). In general, however, in the semiotic system offered by Peirce (1974), icons represent through their resemblance to an object, and indexes represent by pointing in some fashion (e.g., physical, temporal, causative, etc.) to an object. In contrast, symbols represent through socially or culturally agreed-on arbitrary conventions. ToM is therefore a critical factor of the human condition, essential to the cognitive separation of primate communication from the complexity that can be applied to the hominin fossil and archaeological record.

Orders of intentionality have been projected back onto the hominin evolutionary record using modern-day primate brain sizes and the known brain sizes of fossil hominins. These have been interpolated onto the hominin fossil record based on correlations between frontal lobe volumes, predicted group size, and achievable levels of intentionality as a predictive exercise in estimating hominin cognitive levels (Dunbar, 1992, 2004, 2007; Gamble et al., 2014, p. 146; also see Figure 17.1). Of note, this represents a fairly broad-brush approach to estimating levels of hominin cognition, as there is very little information regarding population variability within species, given the sparse nature of the fossil record. However, there is still some within-species variability in brain size, as can be seen in Figure 17.1; the implications of this for cognitive abilities are discussed further later in the chapter.

Table 17.1 summarizes the schema of Gamble and colleagues (2014, p. 146) for assigning hominins to orders of intentionality and the archaeological record through modes of technology (see Clark, 1961; Gamble, 2013, p. 64, Box 2.3, and Figure 17.1).

From Figure 17.1 and Table 17.1, it is clear that previous attempts to assign orders of intentionality to the hominin record have been difficult, and brain size variability suggests that some individuals within the same species may have attained different orders of intentionality. This is particularly clear for the Sima de los Huesos hominins (designated as Early Neandertals in Figure 17.1, based on Meyer et al., 2015, 2016). They are assigned to a third order of intentionality based on their brain size, yet they have been associated with complex behaviors interpreted as possibly ritual in nature (Carbonell & Mosquera, 2006); these are perhaps more indicative of a fourth or fifth order of intentionality, such as has been assigned to later Neandertals. In addition, the resolution of the social brain hypothesis and its application of orders of intentionality do not explicitly take into account the relation between behavioral and cognitive plasticity in relation to brain size (cf. van Schaik, 2013).

However, this is not necessarily a problem; even within our own species, we do not all operate at fifth-order or sixth-order intentionality. Having this mix of behavioral and cognitive variation within the hominin record is a truer reflection of the complexities of our own evolutionary story. McNabb and Cole (2015) describe the within-species variability in absolute brain sizes as "variable-equilibrium" and see hominin encephalization as a staircase with punctuated increases in brain size (Cole, 2012; McNabb, 2012; Shultz et al., 2012). At a general scale, however, according to the predictions of the social brain hypothesis, it appears that the australopithecines, paranthropines, and early members of the *Homo* genus (*H. rudolfensis* and *habilis*) would have a ToM or second order of intentionality. Once *H. ergaster* and *H. erectus* appear, a third order of intentionality has been attained, which seemingly corresponds to the appearance of mode 2 or bifacial handaxes in the archaeological record. A fourth order of intentionality remains the preserve of *H. heidelbergensis* and the Neandertals, broadly corresponding to the emergence of mode 3 prepared-core technology, with *H. sapiens* attaining a fifth order and mode 4 bladed prepared core technology.

Returning briefly to the language question, it seems unlikely that a fully developed ability for grammatical language emerged in hominin cognition at the same time a ToM was realized. I have previously proposed (Cole, 2015c) that a developed ability for language based on symbolic interaction incorporating material culture not only

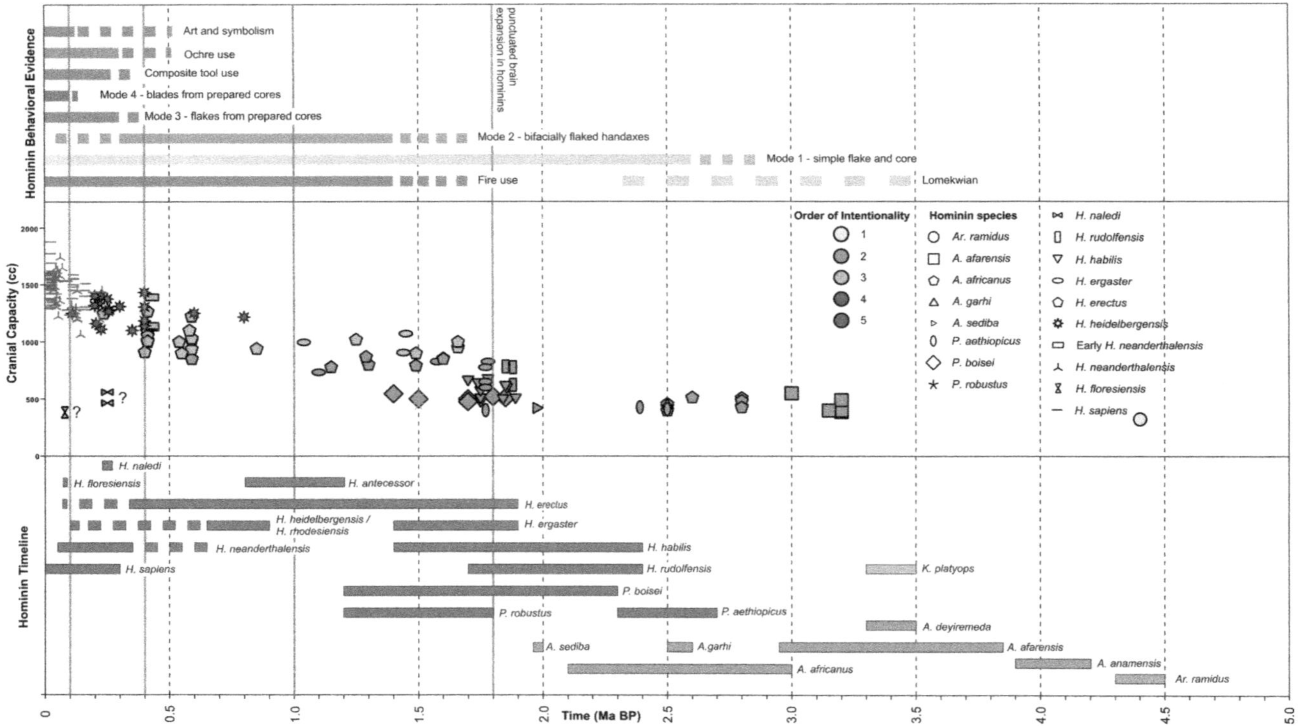

Figure 17.1. The fossil hominin timeline against hominin brain size, orders of intentionality, the first appearance datum (FAD), and last appearance datum (LAD) for the hominin behavioral record. The brown lines illustrate the punctuated changes in hominin brain size at 100 thousand years ago (Kya), 400 Kya, 1 million years ago (Mya), and 1.8 Mya (Shultz et al., 2012). Cranial capacities and dates are from Shultz et al. (2012) and supplemented by additional information from Berger et al. (2015), Brown et al. (2004), and Dirks et al. (2017). The uncertainty surrounding the presence of *H. heidelbergensis* and the beginning of *H. neanderthalensis* reflects recent publications (e.g., Meyer et al., 2015, 2016); note the assignation of early *H. neanderthalensis* for the Atapuerca hominins (ca. 430 Kya) as a result. Orders of intentionality and their application to the fossil record after Dunbar (1992, 2004), Dunbar and Shultz (2007), and Gamble, Gowlett, and Dunbar (2014). Technological mode descriptions and dates adapted from Box 2.3 (pp. 64–65) in Gamble (2013), *Settling the Earth: The Archaeology of Deep Human History*, Cambridge University Press. Art and symbolism include beads, cave art, portable art, and the Trinil clam shell (Joordens et al., 2015). The extended use of ochre includes recent evidence from South Africa (Watts, Chazan, & Wilkins, 2016). Timeline compiled by the author.

requires a second order of intentionality but may only be truly attainable with third-order intentionality. Under the schema of Gamble and colleagues (2014), this suggests that *H. ergaster* and *H. erectus* would have had a language system (Gowlett et al., 2012), probably based on nonverbal communication and visual display (McNabb, 2012). This capability would certainly make sense for a group of hominins that managed to disperse across most of the Old World and successfully adapt to a range of (albeit similar; Dennell, 2004) environmental conditions. However, a system of communication based on directed gestures and vocal punctuation is almost certainly possible with a second order of intentionality. It is only with a fifth order of intentionality that a full comprehension of the symbolic abstract occurs, and grammatical language or speech subsequently develops as a selective advantage to allow the expression of the symbolic abstract. Grammatical language is required to explain complex notions between individuals and groups in a way that facilitates an equal understanding. Nonverbal visual display utilizing the body or material culture is simply not expressive or plastic enough to convey the full meaning of a totally abstract notion such as, for example, the supernatural. Recent work by Shultz and colleagues (2012) suggests that grammatical language may have developed close to 100 Kya, corresponding with a punctuated (as opposed to gradual) increase in brain size and near the emergence of a fifth order of intentionality in the hominin record (Figure 17.1).

There have, of course, been criticisms of the social brain hypothesis (e.g., Barrett, Henzi, & Rendall, 2007; de Ruiter, Weston, & Lyon, 2011) and the application of orders of intentionality to the hominin behavioral record. Dunbar (2007) himself argues that there is no real need for the social brain hypothesis to correspond to the archaeological record, as only limited insight is gained from behavioral evidence. The social brain hypothesis explicitly deals with the mental processes underlying social behavior, rather than overt behavior or aspects of cognition that focus on instrumental skills like tool-making. Tools, in effect, become a red herring, as the mindsets that lie at the core of the social brain hypothesis are unlikely to leave a visible trace in the fossil record that archaeologists may relate to tools (Dunbar, 2007). Therefore, assigning orders of intentionality to the hominin fossil record based on brain size holds true as an estimate of cognitive ability.

However, extensive archaeological studies have identified material culture as an active participant in maintaining and structuring social relations (Barham, 2010; Gamble, 1999, 2007; Gosden & Marshall, 1999; Ingold, 2007). These results are supported by ethnographic studies that illustrate tools mediate social relations, beliefs, and social practices (Killick, 2004). Even if it is often unclear which hominin species definitively produced which different tool types, tool-making and material culture creation are intrinsically social and cognitive acts related to problem-solving and learning, however they were achieved (e.g., through imitation, observation, or demonstration) (Bamforth & Finlay, 2008; Barham, 2010; Stout et al., 2002; Stout & Chaminade, 2012). Therefore, tools have great potential to provide insight into the behavioral and achieved cognitive complexities of their hominin creators.

STONE TOOLS AND ORDERS OF INTENTIONALITY

Anthropogenically modified stone artifacts from the Paleolithic encompass a number of lithic technologies, geographical regions, and descriptive terminologies. The

Table 17.1. Orders of Intentionality, Hominins Species, and Technological Modes

Order of Intentionality	*Achieved By*	*Hominin Species*	*Technological Mode*
Fifth	Modern humans with language as we know it	*H. sapiens*, some *H. neanderthalensis* (?)	Mode 4—prepared core technology (blades)
Fourth	Last common ancestor with/and Neandertals	*H. heidelbergensis, H. neanderthalensis*	Some mode 2—bifacial handaxes, mostly mode 3—prepared core technology (flakes)
Third	All large-brained hominins (>900 cc)	*H. ergaster, H. erectus, H. antecessor*, some *H. heidelbergensis*	Some mode 1—simple flake and core, mostly mode 2—bifacial handaxes
Second (ToM)	5-year-old children (*H. sapiens*), all small-brained hominins (400–900 cc), and possibly great apes	*A. afarensis, A. africanus, A. garhi, A. sediba, P. boisei, P. robustus, P. aethiopicus, H. rudolfensis, H. naledi, H. habilis*, some *H. ergaster*, some *H. erectus*	Lomekwian and mode 1—simple flake and core, some mode 2—bifacial handaxes
First	Monkeys, lesser apes, and some mammals such as elephants and dolphins, small-brained hominins (<400 cc)	*A. ramidus*, some *A. afarensis*	Unknown, but perhaps similar to flakes produced by capuchins (cf. Proffitt et al., 2016)

Note: Adapted from Table 5.2 (p. 146) in Gamble et al. (2014), *Thinking Big: How the Evolution of Social Life Shaped the Human Mind*, Thames & Hudson; also see Box 2.3 (pp. 64–65) in Gamble (2013), *Settling the Earth: The Archaeology of Deep Human History*, Cambridge University Press; and Clark (1961), *World Prehistory: In New Perspective*, Cambridge University Press.

Table 17.2. Relations between Technology, Behaviors, and Orders of Intentionality

Technological Mode	*Material Culture Description*	*Behavioral Implication*	*Order of Intentionality*
1 (Early Stone Age/Lower Paleolithic: includes the Lomekwian in this schema) ca. 3.3 Mya to ca. <10 Kya	• Deliberate lithic tool production to create edges for use • No standard form imposition; tool shape largely governed by raw material size, shape, and mechanical flaking properties • Consists of pebble tool industries dominated by small flake removals (<10 cm) and chopping tools (Oldowan), or large flake removals (Lomekwian) • Possible bone or wood tools that have limited evidence for anthropogenic modification	• Hominins have a realized sense of self that compliments the egocentric, goal-directed behavior reflected in the strategies of tool production. • Evidence for some forward planning in raw material procurement • Social communications governed by egocentric, dyadic, gestural, and attention-directed auditory signals with a presumably greater repertoire than extant primates • Imitative learning present	First order
2 (Early Stone Age/Lower Paleolithic: Acheulean) ca. 1.7 Mya to ca. 100–60 Kya	• Lithic tools predominantly based on large flakes (>10 cm) or bifacially reduced cores • Consists mostly of bifacially knapped handaxes and cleavers (large cutting tools [LCT]), although flakes, flake tools, and cores still being produced • Regional variation in shape and form primarily affected by raw material	• Hominins have a consciously realized ToM that marks the beginning of abstract thought; this in turn is reflected in the imposition of deliberate shape and form on handaxes that can only have been knapped through the knapper having a mental construct (no matter how fluid) of the artifact before the process started.	Second order (ToM)

	• Deliberate imposition of shape and form to LCTs evidenced through the presence of a mental construct in regard to LCT form with a degree of conceptual standardization; final LCT form remains a fluid concept with no evidence for an increase in artifact symmetry or standardized form through time. • Organic artifacts may be in use (e.g., wooden spears for hunting).	• Evidence for goal-directed behavior associated with greater planning capabilities and complex imitative and active social learning, organized hunting, and controlled use of fire • Group organization reflecting complex social communications/language grounded in visual display	
Toward the end of Mode 2 as the dominant technological expression ca. 400 Kya to 200 Kya	• Individual groups may produce artifacts of extraordinary design, such as giant handaxes, S-twist handaxes, handaxe pairs, symmetrical handaxes. • An element of prepared core technology (mode 3) may enter the behavioral record, although there is still a strong emphasis on large flake production and bifacially reduced cores.	• If assemblages have a definite bias toward "true" symmetry or contain artifacts of "extraordinary design" (e.g., giant handaxes), then it may be that such artifacts have an implication beyond the purely functional and may hold some social or cultural significance. • If there is the presence of mode 3 and composite tools within mode 2 assemblages, then perhaps there is a more sustained cognitive breakthrough beyond a ToM.	Second to third order

(*continued*)

Table 17.2. Continued

Technological Mode	*Material Culture Description*	*Behavioral Implication*	*Order of Intentionality*
3 (Prepared Core, Middle Stone Age/Middle Paleolithic: Levallois) ca. 300 Kya to ca. <40 Kya	• A shift from producing lithic tools from cores and flakes to preparing cores to extract flakes of a particular form and size • Prepared core technology (e.g., Levallois) focuses on producing standardized flakes with the potential for later modification (e.g., into points or handaxes). • This type of lithic production also indicates the presence of composite tools. • Regional variation possibly driven by cultural influences rather than raw material, although raw material may still govern shape and size of artifact to a certain degree • Use of organic material culture for composite tool creation • Use of ochre evident	• Hominins have a commonality of understanding (cultural affinities) and a clear sense of shape and form that begin to play a role beyond the purely functional. • The capability to produce composite tools displays an ability for abstract thought beyond a functional level, which may manifest itself in the beginnings of cultural signaling seen within the archaeological record, such as the use of pigments. • Artifacts maintain a predominantly functional significance but may carry social meaning in regard to the creator or group. • Social communication is centered around complex gesture and utterance incorporated within visual display.	Third to fourth order

4 (Later Stone Age/Upper Paleolithic Blade and bladelet dominated assemblages, e.g., Aurignacian) ca. 120 Kya to ca. <10 Kya	• Continued emphasis on flake production with a predetermined shape and form • Flake blanks within this category are primarily concerned with composite tool production with limited secondary shaping. • Use of organic material culture for composite tool creation. This category includes an expanded repertoire of complex organic tools (such as harpoon heads). • In addition, material culture with a purely non-utilitarian design enters the record in the form of ornamentation (beads), art (cave and portable), and figurines (animal, humanoid, and anthropomorphic). • Clear evidence for regional variation in material culture production on a cultural basis	• Hominins have a commonality of understanding, a clear sense of shape and form, and the capacity for fully symbolic and functional abstract thought evidenced through the presence of non-utilitarian and composite material culture (decoration) and behaviors (e.g., symbolic burial). • Social communication is centered around visual display, gesture, and fully grammatical language. • Artifacts carry social meaning in relation to the creator and user (individual and group) and are now fully complicit in identity propagation of the individual and the group.	Fifth order (occasionally sixth order)

Note: Summary of how orders of intentionality map onto the archaeological record. The shades of gray correspond to those used in Figure 17.1. Adapted from Table 8.4 (pp. 182–187) in Cole (2017), Accessing hominin cognition: Language and social signalling in the Lower to Middle Paleolithic, *Cognitive Models in Palaeolithic Archaeology*, Oxford University Press. Timing of technological modes adapted from Box 2.3 (pp. 64–65) in Gamble (2013), *Settling the Earth: The Archaeology of Deep Human History*, Cambridge University Press.

technological variability is unsurprising, given the span of more than 3.3 million years. Diverse approaches to this technological variability have emphasized its different aspects, including flake production, typological form, metrical variation, core reduction, and microwear analysis (Foley & Lahr, 2003; Gowlett, 2009).

Clark's (1961) technological modes (Figure 17.1; Table 17.1) have provided a usefully broad comparative scale, if one that perhaps lacks the fine granularity needed to tackle the versatility of hominin behavior. Although there are valid issues with Clark's modes and their application by archaeologists to the Paleolithic record (Bar-Yosef & Belfer-Cohen, 2001), the technological modes do provide a useful framework for comparing data throughout the span of time (Gamble, 2013), though they tend to ignore the often more nuanced local or regional records and raw material availability. The modes also allow for broad comparisons of lithic technology that sidestep the rigid homotaxial sequence adopted by many researchers, and they take into account the continuities between technologies and time divisions. Such a big-picture approach is needed to deal with the equally big picture of hominin cognitive abilities, where the modes reflect a broad trajectory of increasing technological complexity, greater control of knapping techniques, and raw material utilization (Foley & Lahr, 2003). Although the use of modes is by no means ideal, they do serve as a useful heuristic for the purposes of this discussion. Future work will focus on producing a cognitive model that incorporates the local small-scale complexities and variability within the broad designations of hominin cognition.

I have previously used the modes of technology in conjunction with the identity model to relate the orders of intentionality to the archaeological record based on the minimum levels of cognitive complexity required to incorporate material culture into active social signaling at the individual and group levels (Cole, 2017). The relationships for the modes of technology, their behavioral implications, and orders of intentionality under the identity model schema are summarized in Table 17.2.

Here it is important to note that although the boundaries between material culture categories are based on Clark's technological modes, in reality we know that the archaeological record, hominin behavior, and hominin cognition are rarely (if ever) so neat and tidy. Therefore, the application of the orders of intentionality to the archaeological record should focus on derived (or developed) artifacts and what they imply for hominin cognition. Figure 17.1 highlights the fact that, even within the same species, brain size varies, and so too would the order of intentionality according to the assignments under the social brain hypothesis. It is strongly advocated here that we seek to acknowledge the complexity of the behavioral record and apply the same reasoning: Perhaps different cultural groups within the same species attained different orders of intentionality and expressed them differently in their behavioral record. Therefore, if a derived element from a higher category of material culture (focus here on the Material Culture Description column in Table 17.2, rather than the more familiar modes) is securely provenanced within an assemblage, then it may be cautiously inferred that the creating hominins may have attained the higher order of intentionality. For example, if a predominantly mode 3 lithic assemblage (generally third- to fourth-order intentionality) contains an element of ornamentation such as beads (fifth-order intentionality), then that group (perhaps only at a local level) may have actively breached the fifth-order intentionality barrier. This in turn would suggest that the rest of the species may have had the potential to do the same within the variable-equilibrium framework. Indeed, much of our difficulty in modeling the cognitive abilities of our hominin ancestors comes from the fact that we tend to approach this at a broad species level, ignoring the nuanced potential for within-species cognitive variation as expressed through cultural variation in the artifactual record. Further difficulties lie in a general lack of understanding of population size and networks within and between Paleolithic hominin groups (discussed further later).

When Figure 17.1 and Tables 17.1 and 17.2 are compared, there is an apparent mismatch in the order of intentionality assigned to the species and associated behavioral record. Perhaps the biggest difference is the suggestion that a second order of intentionality is only breached by those hominins producing mode 2 bifaces, starting with *H. ergaster* and *H. erectus*, rather than early *Homo* or the paranthropines. It has been proposed elsewhere (Cole, 2015c, 2017; McNabb & Cole, 2015) that these differences can be partially explained by differences in biological change (e.g., brain size increases, which must come first) versus behavioral change (which follows). The social brain predictions for hominin intentionality based on brain size are therefore a good measure of maximum hominin cognitive *potential*, whereas the order of intentionality as seen through the archaeological record and the identity model illustrate the *realized* cognitive level. In reality, hominins may well have lived somewhere between the two prediction ranges, much the same way modern humans fluctuate

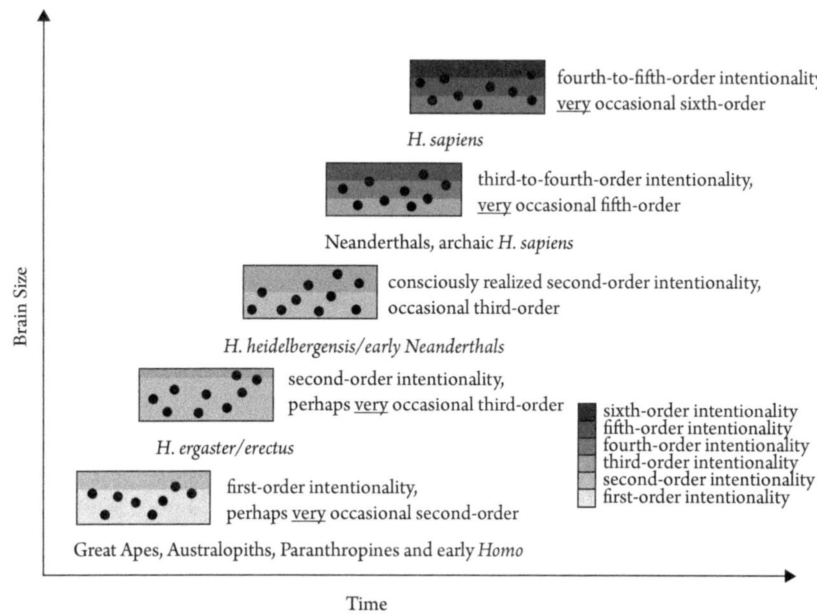

Figure 17.2. Schematic of the variable-equilibria model. The black dots indicate variability of brain size within species, not only between individuals but also within populations and groups. The species are suggested examples of those that have attained the corresponding orders of intentionality. However, the examples are not limited to those named. Adapted from Figures 5 and 6 (p. 109) in McNabb and Cole (2015), The mirror cracked: Symmetry and refinement in the Acheulean handaxe, *Journal of Archaeological Science: Reports*, and republished with permission from Elsevier.

between fifth- and sixth-order intentionality and great apes fluctuate between first- and second-order intentionality.

The key component for Table 17.2 versus Figure 17.1 and Table 17.1 is the suggestion that the biface makers of Acheulean (or mode 2) artifacts are firmly grounded within a consciously realized second order of intentionality or a ToM. The reasoning is explained in more detail in McNabb and Cole (2015) but centers on the notion that mode 2 bifaces share a conceptual standardization (McNabb, Binyon, & Hazelwood, 2004), even though the exact final form remains a fluid concept, allowing its adaptation to raw material quality, knapper skill, and desired function and aesthetic. In order to make a mode 2 biface, a knapper would have needed to hold an abstract concept of the handaxe in the mind's eye (Ashton & McNabb, 1994) before the manufacture process could proceed. That is, randomly bashing a nodule will not produce a mode 2 biface (but see McPherron, 2000). This implies that in order to knap a handaxe, the knapper must have had the ability for abstract thought—to conceive of the artifact and knapping strategy before removing the first flake from the nodule. A ToM, in turn, is an example of abstract thought. Hominins learned how to make handaxes by recognizing the intention of others' handaxes within a social and cultural framework that influenced their own knapping. The act of handaxe-making in a social context therefore implies and requires a ToM (McNabb & Cole, 2015).

Mode 2 biface manufacture does not, however, require a third order of intentionality. What does require third-order intentionality is the creation of composite tools (mode 3) and the use of material culture as an active agent in mediating social relationships and creating symbols, although this also fits well within a fourth-order bracket, and a language system grounded in visual display and complex gestures (Cole, 2017). It is not until a fifth order of intentionality that we would expect to see a full-blown ability for grammatical language and the creation of material culture (e.g., beads, art, and figurines) representing a complex imaginary mythology that can only be explained through spoken language (Cole, 2017).

As mentioned earlier, the archaeological record rarely (if ever) fits into such neat categories, however much modern researchers would like it to. Sites often suggest that hominins may have been operating at a higher (or lower) cognitive level than would generally be predicted. Surprising findings include Neandertal symbolism and jewelry (Radovčić et al., 2015; Welker et al., 2016; Zilhão et al., 2010) and giant handaxes, S-twists, and pairs perhaps used as vehicles for social signaling in the Lower Paleolithic (Hopkinson & White, 2005; Pope, Russel, & Watson, 2006; Wenban-Smith, 2004; White, 1998; White & Plunkett, 2004). The explanatory link here between expected level of cognition and behavioral output is the variable-equilibrium between the maximum cognitive potential of species (as predicted by the social brain hypothesis; see Figure 17.1) and the obtained behavioral threshold (as summarized in Figure 17.2 and Table 17.2). The variable-equilibria of species cognition within a staircase framework of punctuated increasing brain size (Shultz et al., 2012) mean that if social and environmental conditions allow, then individual communities may break through their previously realized behavioral threshold and engage in behaviors that match a higher level of cognitive potential. Within the Lower Paleolithic, the archaeological patterning suggests that small, isolated populations were occasionally able to innovate beyond the broader species cognitive level and engage with material culture production at social and symbolic levels associated with higher orders of intentionality.

These innovations do not often feed into the broader species level, since small, isolated populations by definition are associated with poor or limited social networks between groups, which inhibits cultural and cognitive transmission to the wider population (Cole, 2015a, 2017).

Therefore, when we examine the archaeological record, we must take into account the full range of complexity present between the cognitive potential of a species, their normal expected behavioral output, and those moments when groups or communities extend themselves to their full cognitive and behavioral potential within the context of population dynamics and social networks. We must not only take a broad-brush approach to understanding the past but also drill down to site-specific details, in order to populate the canvas with details. Fortunately, Wynn has long been a proponent of such an approach (Wynn, 2002, 2009; Wynn & Coolidge, 2004; Wynn, Hernandez-Aguilar, Marchant, & McGrew, 2011), and we should endeavor to incorporate such detail in our analyses. To ignore this degree of complexity and attempt to reduce behavioral and cognitive complexity to a general species level means that we will never be able to identify the sparks that drove the cognitive steps between each order of intentionality.

CONCLUSION

This chapter has explored the relationships between a ToM, orders of intentionality, and the Paleolithic record. Such a correlation has highlighted the disparities between cognitive predictions based on brain size, such as the social brain hypothesis, and those based on examining archaeological artifacts viewed as the behavioral results of cognitive processes. This disconnect, however, is not necessarily the result of incompatible methods of analysis or the inability of modern researchers to access the minds of our hominin ancestors. Rather, the differences may more truly represent hominin behavior and cognition. Modern humans have a range of complex behaviors and cognitive states that cannot be applied universally across our species, and it is perhaps therefore unfair for us to assume that we are the only human species to have such variability. If we embrace the complexity of the archaeological and fossil records and view them through the lens of the variable-equilibria model, we can begin to see that the variability within species at the individual and group levels produces a richer interpretation of the past than has previously been attained.

This holistic interpretation of varied hominin inter- and intra-species complexity may be supported by recent Paleogenetic evidence suggesting that *H. sapiens* is a mosaic species, incorporating genetic inputs from several archaic hominin species. It is not unreasonable to assume, therefore, that the hominin species we interacted with (e.g., the Neandertals and Denisovans) may have been more like us than we have previously thought and acknowledged. Indeed, the Neandertals and Denisovans must have been recognizably human, not only in a biological sense, but behaviorally as well. This would include complex stone tools and organic technologies, complex language systems, and symbolic material culture. Otherwise, we must question what drove the desire to interbreed (even if it was only at a limited scale?); after all, practically and figuratively speaking, you can't start a fire without a spark.

No doubt, future discoveries regarding human species and the origins of lithic technology will change the boundaries proposed here for relating the archaeological record to orders of intentionality. However, we should embrace these changes and welcome the increasing degrees of complexity such discoveries suggest. After nearly 150 years of studying the Paleolithic, we are entering an era of unprecedented methodological rigor and scientific study, still just barely scratching the surface of what it means to be human and understanding our hominin ancestors.

ACKNOWLEDGMENTS

I would like to thank a number of colleagues for their input through various discussions on this topic, including John McNabb, Clive Gamble, John Gowlett, and Robin Dunbar. In addition, the reviewers' comments for this chapter were extremely helpful and supportive; any remaining errors are my own. I would also like to thank Tom, Fred, and Leee for the invitation to contribute to this volume. I will always be thankful for the many conversations I have had with Tom regarding handaxes and hominin cognition; they were not only highly informative but also extremely enjoyable. Tom has certainly defined the discipline of cognitive archaeology and inspired many of us to follow in his footsteps.

REFERENCES

Aiello, L. C., & Dunbar, R. I. M. (1993). Neocortex size, group size and the evolution of language. *Current Anthropology, 34*(2), 184–193.

Ashton, N. M., & McNabb, J. (1994). Bifaces in perspective. In N. Ashton & A. David (Eds.), *Stories in stone: Proceedings of the anniversary conference at St. Hilda's College Oxford, April 1994* (pp. 182–191). London: Lithic Studies Society.

Bamforth, D. B., & Finlay, N. (2008). Introduction: Archaeological approaches to lithic production skill and craft learning. *Journal of Archaeological Method and Theory, 15*(1), 1–27.

Barbieri, M. (2010). On the origin of language: A bridge between biolinguistics and biosemiotics. *Biosemiotics, 3*(2), 201–223.

Barham, L. S. (2010). A technological fix for Dunbar's dilemma? In R. I. M. Dunbar, C. Gamble, & J. A. J. Gowlett (Eds.), *Social brain, distributed mind* (pp. 367–389). London: British Academy.

Baron-Cohen, S. (2001). Theory of mind in normal development and autism. *Prisme, 34*(1), 174–183.

Barrett, L., Henzi, P., & Rendall, D. (2007). Social brains, simple minds: Does social complexity really require cognitive complexity? *Philosophical Transactions of the Royal Society of London. Series B, Biological Sciences, 362*(1480), 561–575.

Bar-Yosef, O., & Belfer-Cohen, A. (2001). From Africa to Eurasia: Early dispersals. *Quaternary International, 75*(1), 19–28.

Berger, L. R., Hawks, J., de Ruiter, D. J., Churchill, S. E., Schmid, P., Delezene, L. K., . . . Zipfel, B. (2015). *Homo naledi*, a new species of the genus *Homo* from the Dinaledi Chamber, South Africa. *Elife, 4*, 1–35.

Brown, P., Sutikna, T., Morwood, M. J., Soejono, R. P., Saptomo, E. W., & Due, R. A. (2004). A new small-bodied hominin from the Late Pleistocene of Flores, Indonesia. *Nature, 431*(7012), 1055–1061.

Call, J., & Tomasello, M. (2008). Does the chimpanzee have a theory of mind? 30 years later. *Trends in Cognitive Sciences, 12*(5), 187–192.

Carbonell, E., & Mosquera, M. (2006). The emergence of a symbolic behaviour: the sepulchral pit of Sima de los Huesos, Sierra de Atapuerca, Burgos, Spain. *Comptes Rendus Palevol, 5*(1–2), 155–160.

Cheney, D. L., & Seyfarth, R. M. (1990). *How monkeys see the world: Inside the mind of another species.* Chicago, IL: University of Chicago Press.

Clark, G. (1961). *World prehistory: In new perspective.* Cambridge, UK: Cambridge University Press.

Cole, J. (2012). The identity model: A method to access visual display and hominin cognition within the Palaeolithic. *Human Origins, 1*, 24–40.

Cole, J. (2015a). Examining the presence of symmetry within Acheulean handaxes: A case study in the British Palaeolithic. *Cambridge Archaeological Journal, 24*(4), 713–732.

Cole, J. (2015b). Handaxe symmetry in the Lower and Middle Palaeolithic: Implications for the Acheulean gaze. In F. Coward, R. Hosfield, M. I. Pope, & F. Wenban-Smith (Eds.), *Settlement, society and cognition in human evolution: Landscapes in mind* (pp. 234–257). Cambridge, UK: Cambridge University Press.

Cole, J. (2015c). Hominin language development: A new method of archaeological assessment. *Biosemiotics, 8*(1), 67–90.

Cole, J. (2017). Accessing hominin cognition: Language and social signalling in the Lower to Middle Paleolithic. In T. Wynn & F. L. Coolidge (Eds.), *Cognitive models in Palaeolithic archaeology* (pp. 157–195). New York: Oxford University Press.

Crockford, C., Wittig, R. M., Mundry, R., & Zuberbühler, K. (2012). Wild chimpanzees inform ignorant group members of danger. *Current Biology, 22*(2), 142–146.

Davidson, I., & Noble, W. (1989). The archaeology of perception: Traces of depiction and language. *Current Anthropology, 30*(2), 125–155.

Davidson, I., & Noble, W. (1993). Tools and language in human evolution. In K. R. Gibson & T. Ingold (Eds.), *Tools, language and cognition in human evolution* (pp. 363–388). Cambridge, UK: Cambridge University Press.

Deacon, T. W. (1997). *The symbolic species: The co-evolution of language and the brain.* New York: W.W. Norton & Co.

Dennell, R. W. (2004). Hominid dispersals and Asian biogeography during the Lower and Early Middle Pleistocene, c. 2.0–0.5 Mya. *Asian Perspectives, 43*(2), 205–226.

d'Errico, F., Henshilwood, C. S., Lawson, G., Vanhaeren, M., Tillier, A.-M., Soressi, M., . . . Julien, M. (2003). Archaeological evidence for the emergence of language, symbolism, and music—An alternative multidisciplinary perspective. *Journal of World Prehistory, 17*(1), 1–70.

d'Errico, F., Henshilwood, C. S., Vanhaeren, M., & van Niekerk, K. L. (2005). *Nassarius kraussianus* shell beads from Blombos Cave: Evidence for symbolic behaviour in the Middle Stone Age. *Journal of Human Evolution, 48*(1), 3–24.

de Ruiter, J., Weston, G., & Lyon, S. M. (2011). Dunbar's number: Group size and brain physiology in humans reexamined. *American Anthropologist, 113*(4), 557–568.

de Veer, M. W., & van den Bos, R. (1999). A critical review of methodology and interpretation of mirror self-recognition research in nonhuman primates. *Animal Behaviour, 58*(3), 459–468.

Dirks, P. H. G. M., Roberts, E. M., Hilbert-Wolf, H., Kramers, J. D., Hawks, J., Dosseto, A., . . . Berger, L. R. (2017). The age of *Homo naledi* and associated sediments in the Rising Star Cave, South Africa. *eLife, 6*, e24231.

Dunbar, R. I. M. (1992). Neocortex size as a constraint on group size in primates. *Journal of Human Evolution, 22*(6), 469–493.

Dunbar, R. I. M. (1996). *Grooming, gossip and the evolution of language*. London: Faber and Faber.

Dunbar, R. I. M. (1998a). The social brain hypothesis. *Evolutionary Anthropology: Issues, News, and Reviews, 6*(5), 178–190.

Dunbar, R. I. M. (1998b). Theory of mind and the evolution of language. In J. R. Hurford, M. Studdert-Kennedy, & C. Knight (Eds.), *Approaches to the evolution of language* (pp. 92–110). Cambridge, UK: Cambridge University Press.

Dunbar, R. I. M. (2003). The social brain: Mind, language, and society in evolutionary perspective. *Annual Review of Anthropology, 32*(1), 163–181.

Dunbar, R. I. M. (2004). *The human story: A new history of mankind's evolution*. London: Faber and Faber.

Dunbar, R. I. M. (2007). The social brain and the cultural explosion of the human revolution. In P. Mellars, K. Boyle, O. Bar-Yosef, & C. Stringer (Eds.), *Rethinking the human revolution: New behavioural and biological perspectives on the origin and dispersal of modern humans* (pp. 91–98). Cambridge, UK: McDonald Institute for Archaeological Research.

Dunbar, R. I. M., Gamble, C., & Gowlett, J. A. J. (Eds.). (2010). *Social brain, distributed mind*. London: British Academy.

Dunbar, R. I. M., Gamble, C., & Gowlett, J. A. J. (Eds.). (2014). *Lucy to language: The benchmark papers*. Oxford, UK: Oxford University Press.

Dunbar, R. I. M., & Shultz, S. (2007). Evolution in the social brain. *Science, 317*(5843), 1344–1347.

Foley, R. A., & Lahr, M. M. (2003). On stony ground: Lithic technology, human evolution and the emergence of culture. *Evolutionary Anthropology: Issues, News, and Reviews, 12*(3), 109–122.

Gamble, C. (1999). *The Palaeolithic societies of Europe*. Cambridge, UK: Cambridge University Press.

Gamble, C. (2007). *Origins and revolutions: Human identity in earliest prehistory*. Cambridge, UK: Cambridge University Press.

Gamble, C. (2012). When the words dry up: Music and material metaphors half a million years ago. In N. Bannan (Ed.), *Music, language, and human evolution* (pp. 81–106). Oxford, UK: Oxford University Press.

Gamble, C. (2013). *Settling the earth: The archaeology of deep human history*. New York: Cambridge University Press.

Gamble, C., Gowlett, J. A. J., & Dunbar, R. I. M. (2014). *Thinking big: How the evolution of social life shaped the human mind*. London: Thames & Hudson.

Gosden, C., & Marshall, Y. (1999). The cultural biography of objects. *World Archaeology, 31*(2), 169–178.

Gowlett, J. A. J. (1979). Complexities of cultural evidence in the Lower and Middle Pleistocene. *Nature, 278*(5699), 14–17.

Gowlett, J. A. J. (1996). Mental abilities of early *Homo*: Elements of constrain and choice in rule systems. In P. Mellars & K. R. Gibson (Eds.), *Modelling the early human mind* (pp. 191–215). Oxford, UK: Oxbow Books.

Gowlett, J. A. J. (2009). The longest transition or multiple revolutions? Curves and steps in the record of human origin. In M. Camps & P. R. Chauhan (Eds.), *Sourcebook of Paleolithic transitions: Methods, theories and interpretations* (pp. 65–78). Berlin: Springer Verlag.

Gowlett, J. A. J., Gamble, C., & Dunbar, R. I. M. (2012). Human evolution and the archaeology of the social brain. *Current Anthropology, 53*(6), 693–722.

Harmand, S., Lewis, J. E., Feibel, C. S., Lepre, C. J., Prat, S., Lenoble, A., . . . Roche, H. (2015). 3.3-Million-year-old stone tools from Lomekwi 3, West Turkana, Kenya. *Nature, 521*(7552), 310–315.

Hopkinson, T., & White, M. J. (2005). The Acheulean and the handaxe: Structure and agency in the Palaeolithic. In C. Gamble & M. Porr (Eds.), *The hominid individual in context: Archaeological investigations of Lower and Middle Palaeolithic landscapes, locales and artefacts* (pp. 13–28). London: Routledge.

Ingold, T. (2007). Materials against materiality. *Archaeological Dialogues, 14*(1), 1–16.

Joordens, J. C. A., d'Errico, F., Wesselingh, F. P., Munro, S., de Vos, J., Wallinga, J., . . . Roebroeks, W. (2015). *Homo erectus* at Trinil on Java used shells for tool production and engraving. *Nature, 518*(7538), 228–231.

Killick, D. (2004). Social constructionist approaches to the study of technology. *World Archaeology, 36*(4), 571–578.

Krupenye, C., Kano, F., Hirata, S., Call, J., & Tomasello, M. (2016). Great apes anticipate that other individuals will act according to false beliefs. *Science, 354*(6308), 110–114.

Lycett, S. J., & Norton, C. J. (2010). A demographic model for Palaeolithic technological evolution: The case of East Asia and the Movius Line. *Quaternary International, 211*(1), 55–65.

McNabb, J. (2012). The importance of conveying visual information in Acheulean society: The background to the visual display hypothesis. *Human Origins, 1*, 1–23.

McNabb, J., Binyon, F., & Hazelwood, L. (2004). The large cutting tools from the South African Acheulean and the question of social traditions. *Current Anthropology, 45*(5), 653–677.

McNabb, J., & Cole, J. (2015). The mirror cracked: Symmetry and refinement in the Acheulean handaxe. *Journal of Archaeological Science: Reports, 3*, 100–111.

McPherron, S. P. (2000). Handaxes as a measure of the mental capabilities of early hominids. *Journal of Archaeological Science, 27*(8), 655–663.

Meyer, M., Arsuaga, J. L., de Filippo, C., Nagel, S., Aximu-Petri, A., Nickel, B., . . . Pääbo, S. (2016). Nuclear DNA sequences from the Middle Pleistocene Sima de los Huesos hominins. *Nature, 531*(7595), 504–507.

Meyer, M., Arsuaga, J.-L., Nagel, S., Martínez, I., Gracia, A., de Castro, J. M. B., . . . Pääbo, S. (2015). Nuclear DNA sequences from the hominin remains of Sima de los Huesos, Atapuerca, Spain. In *Proceedings of the 5th annual meeting of the European Society for the study of Human Evolution, 10–12 September 2005, London, UK* (Vol. 4, p. 162). Leipzig, Germany: Max Planck Institute for Evolutionary Anthropology.

Nielsen, M., & Dissanayake, C. (2004). Pretend play, mirror self-recognition and imitation: A longitudinal investigation through the second year. *Infant Behavior and Development, 27*(3), 342–365.

Origgi, G., & Sperber, D. (2000). Evolution, communication and the proper function of language. In P. Carruthers & A. Chamberlain (Eds.), *Evolution and the human mind: Modularity, language and meta-cognition* (pp. 140–169). Cambridge, UK: Cambridge University Press.

Peirce, C. S. (1974). Basis of pragmatism. In C. Hartshorne & P. Weiss (Eds.), *The collected papers of Charles Sanders Peirce* (Vol. 1, pp. 1931–1935). Cambridge, MA: Harvard University Press.

Plotnik, J. M., De Waal, F. B. M., & Reiss, D. (2006). Self-recognition in an Asian elephant. *Proceedings of the National Academy of Sciences of the United States of America, 103*(45), 17053–17057.

Pope, M. I., Russel, K., & Watson, K. (2006). Biface form and structured behaviour in the Acheulean. *Lithics: The Journal of the Lithic Studies Society, 27*, 44–57.

Povinelli, D. J., Gallup, G. G., Eddy, T. J., Bierschwale, D. T., Engstrom, M. C., Perilloux, H. K., & Toxopeus, I. B. (1997). Chimpanzees recognize themselves in mirrors. *Animal Behaviour, 53*(5), 1083–1088.

Premack, D., & Woodruff, G. (1978). Does the chimpanzee have a theory of mind? *Behavioral and Brain Sciences, 4*, 515–526.

Proffitt, T., Luncz, L. V, Falótico, T., Ottoni, E. B., de la Torre, I., & Haslam, M. (2016). Wild monkeys flake stone tools. *Nature, 539*(7627), 85–88.

Radovčić, D., Sršen, A. O., Radovčić, J., & Frayer, D. W. (2015). Evidence for Neandertal jewelry: Modified white-tailed eagle claws at Krapina. *PLoS One, 10*(3), 1–14.

Reiss, D., & Marino, L. (2001). Mirror self-recognition in the bottlenose dolphin: A case of cognitive convergence. *Proceedings of the National Academy of Sciences of the United States of America, 98*(10), 5937–5942.

Savanah, S. (2013). Mirror self-recognition and symbol-mindedness. *Biology and Philosophy, 28*(4), 657–673.

Seyfarth, R. M., & Cheney, D. L. (2011). Animal cognition: Chimpanzee alarm calls depend on what other know. *Current Biology, 22*(2), R51–R52.

Seyfarth, R. M., & Cheney, D. L. (2013). Social cognition: The primate mind before tools, language, and culture. In G. Hatfield & H. Pittman (Eds.), *Evolution of mind, brain, and culture* (pp. 105–122). Philadelphia: University of Pennsylvania Press.

Seyfarth, R. M., & Cheney, D. L. (2017a). Precursors to language: Social cognition and pragmatic inference in primates. *Psychonomic Bulletin & Review, 24*(1), 79–84.

Seyfarth, R. M., & Cheney, D. L. (2017b). The origin of meaning in animal signals. *Animal Behaviour, 124*, 339–346.

Shultz, S., Nelson, E., & Dunbar, R. I. M. (2012). Hominin cognitive evolution: Identifying patterns and processes in the fossil and archaeological record. *Philosophical Transactions of the Royal Society of London. Series B, Biological Sciences, 367*(1599), 2130–2140.

Stout, D. (2011). Stone toolmaking and the evolution of human culture and cognition. *Philosophical Transactions of the Royal Society of London. Series B, Biological Sciences, 366*(1567), 1050–1059.

Stout, D., Bril, B., Roux, V., De Beaune, S. A., Gowlett, J. A. J., Keller, C. M., . . . Stout, D. (2002). Skill and cognition in stone tool production: An ethnographic case study from Irian Jaya 1 [and comments and reply]. *Current Anthropology, 43*(5), 693–722.

Stout, D., & Chaminade, T. (2012). Stone tools, language and the brain in human evolution. *Philosophical Transactions of the Royal Society of London. Series B, Biological Sciences, 367*(1585), 75–87.

Stout, D., Hecht, E. E., Khreisheh, N., Bradley, B. A., & Chaminade, T. (2015). Cognitive demands of Lower Paleolithic toolmaking. *PLoS One, 10*(4), 1–18.

Thieme, H. (1997). Lower Palaeolithic hunting spears from Germany. *Nature, 385*(6619), 807–810.

van der Vaart, E., & Hemelrijk, C. K. (2012). "Theory of mind" in animals: Ways to make progress. *Synthese, 191*(3), 335–354.

Vanhaeren, M., d'Errico, F., Stringer, C. B., James, S. L., Todd, J. A., & Mienis, H. K. (2006). Middle Paleolithic shell beads in Israel and Algeria. *Science, 312*(5781), 1785–1788.

van Schaik, C. P. (2013). The costs and benefits of flexibility as an expression of behavioural plasticity: A primate perspective. *Philosophical Transactions of the Royal Society of London. Series B, Biological Sciences, 368*(1618), 1–9.

Warren, S. H. (1911). On a Palaeolithic (?) wooden spear. *Quarterly Journal of the Geological Society of London, 67*, xcix.

Watts, I., Chazan, M., & Wilkins, J. (2016). Early evidence for brilliant ritualized display: Specularite use in the Northern Cape (South Africa) between ~500 and ~300 Ka. *Current Anthropology, 57*(3), 287–310.

Welker, F., Hajdinjak, M., Talamo, S., Jaouen, K., Dannemann, M., David, F., . . . Hublin, J.-J. (2016). Palaeoproteomic evidence identifies archaic hominins associated with the Châtelperronian at the Grotte du Renne. *Proceedings of the National Academy of Sciences of the United States of America, 113*(40), 11162–11167.

Wenban-Smith, F. F. (2004). Handaxe typology and Lower Paleolithic cultural development: Ficrons, cleavers and two giant handaxes from Cuxton. *Lithics: The Journal of the Lithic Studies Society, 25*, 11–21.

White, M. J. (1998). Twisted ovate bifaces in the British Lower Palaeolithic: Some observations and implications. In N. Ashton, F. Healy, & P. Pettitt (Eds.), *Stone Age archaeology: Essays in honour of John Wymer* (Vol. 6, pp. 98–104). Oxford, UK: Lithic Studies Society.

White, M. J., & Plunkett, S. J. (2004). *Miss Layard excavates: A Palaeolithic site at Foxhall Road, Ipswich, 1903–1905*. Liverpool: Western Academic & Specialist Press.

Wynn, T. (1979). The intelligence of later Acheulean hominids. *Man, 14*, 371–391.

Wynn, T. (1981). The intelligence of Oldowan hominids. *Journal of Human Evolution, 10*(7), 529–541.

Wynn, T. (1985). Piaget, stone tools and the evolution of human intelligence. *World Archaeology, 17*(1), 32–43.

Wynn, T. (1991). Tools, grammar and the archaeology of cognition. *Cambridge Archaeological Journal, 1*(2), 191–206.

Wynn, T. (1995). Handaxe enigmas. *World Archaeology, 27*(1), 10–24.

Wynn, T. (2002). Archaeology and cognitive evolution. *Behavioral and Brain Sciences, 25*(3), 389–402.

Wynn, T. (2009). Hafted spears and the archaeology of mind. *Proceedings of the National Academy of Sciences of the United States of America, 106*(24), 9544–9545.

Wynn, T., & Coolidge, F. L. (2004). The expert Neandertal mind. *Journal of Human Evolution, 46*(4), 467–487.

Wynn, T., & Coolidge, F. L. (Eds.). (2017). *Cognitive models in Palaeolithic archaeology*. New York: Oxford University Press.

Wynn, T., Hernandez-Aguilar, R. A., Marchant, L. F., & McGrew, W. C. (2011). "An ape's view of the Oldowan" revisited. *Evolutionary Anthropology, 20*(5), 181–197.

Zilhão, J., Angelucci, D. E., Badal-García, E., d'Errico, F., Daniel, F., Dayet, L., . . . Zapata, J. (2010). Symbolic use of marine shells and mineral pigments by Iberian Neandertals. *Proceedings of the National Academy of Sciences of the United States of America, 107*(3), 1023–1028.

18

A CRITICAL ANALYSIS OF THE EVIDENCE FOR SEXUAL DIVISION OF TASKS IN THE EUROPEAN UPPER PALEOLITHIC

Sophie A. de Beaune

INTRODUCTION

Currently, we are able to provide an accurate picture of the lifeways of human groups in Europe during the Upper Paleolithic (UP), based on analyses and interpretations of remains left behind. To facilitate the analysis of activities we can break down their different stages, which are represented by tasks. For example, the activity of hide working includes several distinct tasks—flesh removal, tanning, and currying. The distribution of tasks within a group is directly linked to its social organization, since even if the entire group can participate in a single activity, the task can vary for each individual. The tasks, in turn, can be analyzed and subdivided into technical actions and gestures; however, we will not address these subdivisions in detail here (see de Beaune, 2000).

Despite our reliable knowledge of the material life of these populations, we know very little about the distribution of tasks according to age and sex. Division of labor, by gender or simply by individual, is a central component of just about all human societies. It is also a precursor to development of social inequality and conflict. While researchers have periodically addressed these topics for the Paleolithic, no major synthesis has been provided for some time. Moreover, we will see here that many interpretations are based on naïve or inaccurate stereotypes. Three types of explanations have been proposed:

(1) Simple suppositions based on ethnographic comparisons, particularly concerning the sexual division of tasks, through reference to what is most often observed among living or recent populations
(2) Explanations based on indirect archaeological evidence
(3) Explanations based on direct evidence, which, as we will see, are rare and difficult to interpret

The analysis in this chapter is focused on the UP and its margins because the division of tasks for the very early periods of the Paleolithic concern pre-human species for which we have very little information. It is possible that the social organization of the first representatives of the genus *Homo* and the australopithecines was closer to that of the modern great apes than to that of *H. sapiens*. In addition, it is generally accepted that European UP sites were occupied by *H. sapiens*, though this is sometimes debated for the earliest sites (e.g., archaic Aurignacian); some researchers attribute it to Neandertals. This is not highly significant relative to the ideas presented here, since lifeways do not appear to have changed considerably from 40,000 to 10,000 BC.

Nevertheless, it is of interest to consider what happened before the UP (i.e., during the Lower and the Middle Paleolithic [LP/MP]) or just after (during the Epipaleolithic and the Mesolithic periods) if the available data can provide insight regarding what happened during the UP.

I am aware that gender archaeology is a common topic in Anglo-American archaeology. The volume of literature on it is tremendous, so for my discussion here I have selected only the most relevant references, without claiming completeness (Bolger, 2013).[1] Here I consider that, even when discussing male and female tasks, it is possible that technical specialization is not related only to sex.

"SIMPLE" SUPPOSITIONS BASED ON ETHNOGRAPHIC COMPARISONS

It is generally accepted that, until recently, current hunting and gathering peoples—the Arctic Inuit, Aboriginal Australians, Kalahari San, and similar groups—lived in relatively small groups, without centralized authority, standing armies, or bureaucratic systems (Ingold, 2004; Lee & Daly, 2004). These societies were relatively egalitarian and did not include members whose only task was to produce a specific type of object (Woodburn, 1982). Morgan's evolutionist ideas inspired Marx and Engels to characterize the original state of human society as one of "primitive communism" (Lee, 1979, 1988; Testart, 1985). The "communism" of hunter-gatherer societies was embedded in relations of immediate kinship, essentially "familistic" (Service, 1966), what Marshall Sahlins called the "domestic mode of production" (Sahlins, 1972). Of course, some authors distinguish between simple and complex hunter-gatherers, arguing that not all the hunters and gatherers lived in small mobile bands; some of them lived in semi-sedentary settlements, had chiefs, and stored some resources (Hayden, 2013; Price & Brown, 1985; Testart, 1982, 2005, 2012).

Regarding the equality of UP societies, several authors have shown that the term *equality* needs qualification and that nascent forms of social hierarchy existed during this period in some ecological conditions, such as in the southwest of France or the central Russian plain (Conkey, 1985; Mellars, 1985; Soffer, 1985). Brian Hayden has widely developed this idea, comparing some UP regional groups with northwestern North America (Hayden, 2001, 2013). I will come back to this point later in the chapter.

The majority of researchers nonetheless suppose that, as in extant or recent "simple" hunter-gatherer societies, Paleolithic populations must have had some distribution of tasks, which in its simplest form consisted of a sexual division of labor

[1] I will not address here the thesis that there are cognitive differences between men and women (e.g., that women might have more gray matter (Luders, Gaser, Narr, & Toga, 2009), emphasize better (Baron-Cohen, Knickmeyer, & Belmonte, 2005), or be better at remembering the locations of objects (Eals & Silverman, 1994), relative to men). Ethnographical counter-examples, as well as our daily life experience, are enough to show that these competences depend on social context.

(Lee, 1979, 1982; Testart, 1982). Most tasks were gendered, and big-game collective hunting appears to have been in general, though not exclusively, a male task (Knüsel & Smith, 2013; Murdock & Provost, 1973; Testart, 1986). According to Lee (1979), women in hunter-gatherer societies contribute to most of the family's daily calories through gathering and sharing activities, but this is a somewhat oversimplified generalization based on a single case (i.e., the Kalahari San). Contributions to diet from different sources of food and by different genders vary tremendously across the globe. We would expect a very different dietary mix in glacial Europe than in Holocene southern Africa. Furthermore, in an archaeological context, the organic objects traditionally associated with female activities (e.g., digging sticks, skin bags, etc.) would leave no trace. This may be the reason why the "man-the-hunter" model was widely promoted after the 1965 anthropological symposium of that name, organized by Irven DeVore and Richard Lee, as reflected in the following quotes: "Hunting is so universal and is so consistently a male activity that it must have been a basic part of the early cultural adaptation, even if it provided only a modest proportion of the food supplies"; "It is also likely that early woman would not have remained idle during the Pleistocene and that plant foods, which are so important in the diet of inland hunter-gatherers today, would have played a similar role in the diet of early peoples" (Lee & DeVore, 1968b). Thus hunting is viewed as central to explaining human evolution (Tiger, 1969; Washburn & Lancaster, 1968; Zihlman, 2013). Gender and the importance of women's work was a topic much discussed during the conference "Man the Hunter" and after the publication of the proceedings (Lee & Daly, 2004; Lee & DeVore, 1968a; Sterling, 2014). The gathering or "woman the gatherer" hypothesis was developed as a corrective to the omission of women and women's activities in such reconstructions (Dahlberg, 1981; Zihlman, 1998).

In any case, the trope of man = tool-maker = hunter versus woman = mother = gatherer, which is highly speculative when the social organization of UP populations is considered, has been widely accepted for over a century (Clottes, 2015; Guthrie, 1984, 2005). Following the data published in the *Ethnographic Atlas* (Murdock, 1967), stone-working was basically deemed a male activity (Murdock & Provost, 1973),[2] and women were implicitly identified with gathering plant resources and "soft" technologies such as hide working and the production of clothes, textiles, cords, lines, and (for later prehistory) ceramics (Finlay, 2013; Zihlman, 2013). Among extant hunter-gatherer populations, all members of the community participate in daily tasks depending on their mobility. The elderly and the women usually gather and hunt small animals near the camp, while the young mothers take care of the younger children. A schematic engraving found at the Magdalenian Gönnersdorf site, representing a woman who appears to be carrying a baby on her back, tends to support the idea that the same was true during the UP (Bosinski, 2007; Bosinski & Fischer, 1974). It is thus assumed that children collected firewood near the camp, helped women gather vegetal resources, carried water, and collected the small animals captured in traps. Adult

[2] In 1962, George Murdock founded *Ethnology: An International Journal of Cultural and Social Anthropology*, published by the University of Pittsburgh, in which he published installments of his *Ethnographic Atlas* between 1962 and 1980—a data set eventually containing 1,167 cultures coded for over 100 variables.

men hunted large animals and traveled further to acquire specific resources that were not available near the camp. This scenario was commonly spread in popular literature, cinema, and comics, as can be seen, for instance, in Zdenek Burian's pictures and countless other reconstructions of Paleolithic life (Gifford-Gonzalez, 1993).

Following the first interpretations from the beginning of the twentieth century (Bégouën, 1924, 1939; Breuil, 1952; Reinach, 1903), Guthrie argued that Paleolithic art was a hunter's art, and as a consequence of the above-mentioned schematic vision of gendered activities, he came to the conclusion that it was a male activity (Guthrie, 1984, 2005). Today's researchers have consistently demonstrated that the animals most often represented in Paleolithic caves are among the ones least frequently consumed as food (Delporte, 1990; Iakovleva & Pinçon, 1997). Even if confirmed that the pictured animals were hunted animals, it does not follow that this was a male activity. In a provocative manner, Guthrie likened female figurines to the images of women in *Playboy* and suggested that the artists were men with erotic preoccupations. Of course, specialists of Paleolithic art criticized this simplistic interpretation, which underestimated the cultural context of UP art (Hays-Gilpin, 2004; White, 2006).

The ethnographic basis for these reconstructions is far from evident, however. It is founded on the supposition that there are universally male activities (e.g., large-animal hunting) and universally female activities (small-animal hunting, gathering, collecting) (Endicott, 2004). Some scholars have gone as far as to attempt to identify universal behavioral laws. They argue that there are features of human society that are universal or nearly so (Brown, 1991), and that the broad division of labor into hunters and gatherers by sex is one of these universals. Then the fact that such disparate societies should share these traits suggests that they have deep evolutionary ancestry (Codding, Bird, & Bird, 2011; Kaplan, Hill, Lancaster, & Hurtado, 2000; Marlowe, 2007). According to Alain Testart (1986), some tasks appear to have never been performed by women, such as killing an animal with a cutting weapon, but there are two arguments against this hypothesis. First, the theory that women could not draw blood in hunter-gatherer societies assumes that these societies were specific entities that can be distinguished from other types of societies. What we define as hunter-gather societies, however, results from an artificial classification made by Western researchers, who distinguish them only by the absence of agriculture. The absence of some elements, however, is a criterion that is clearly insufficient to define a society. This category of societies that survive through hunting and gathering (what about fishing?) is in reality very heterogeneous. François Sigaut (2009) has made the analogy between this classification and the one made by zoologists between vertebrates and invertebrates, recognized as insufficient as early as the late eighteenth century when scientists realized that the absence of vertebrates was not sufficient to define a homogeneous group. Furthermore, Testart's thesis implies that, by some miracle, the sexual division of labor in nonagricultural societies was governed by the same ideological principles from Australia to the Kalahari, and from the Amazon to Siberia. So far, nobody has demonstrated the existence of "universals" in human societies (Lenclud, 2013). On the contrary, Robert L. Kelly (1995) has documented the tremendous diversity in hunting and gathering societies around the world, and Susan Kent (1996) has observed similar diversity in Africa. Testart himself had to recognize that there are some exceptions to the supposed "law," following which women could not draw blood: Agta women regularly hunt with machetes or bows and arrows (Estioko-Griffin & Griffin, 1981;

Hayden, 1992; Testart, 1986). Besides the Agta, there are many cases of women acting as hunters, whether temporarily or permanently, especially in high-latitude locations where most of the diet comes from hunted foods. Testart's hypothesis also ignores the documented participation of female Native American hunter-gatherers in the mistreatment of captives (Donohoe, 2009). This "blood ideology" seems to be more that of Testart than of hunter-gatherers.

In addition, the ethnographic data must be contextualized, since some domestic activities, such as tending fires, working hides or wood, and making rope, are realized by both men and women, depending on the society, as a function of the global distribution of activities in groups. Many hunter-gatherer men gather vegetable foods when the need or opportunity arises, even if it is not a usual daily activity. In most foraging societies, if a man's primary task is hunting, women often hunt as well (Endicott, 2004). Ethnographically, women often play an important role in the acquisition of animal protein, but they serve predominantly as trappers and as beaters to drive game and perform supporting roles as butchers, participants in drives, and supernatural support. Hunting, after all, is more than the simple act of killing an animal, and the ethnographic accounts drawn by Wadley (1998) demonstrate the wide variety of ancillary activities considered by various groups as part of hunting, from sewing to meat sharing.

Stone flaking, by contrast, appears to be practiced by men in nearly all of the societies in which it is known, although there are exceptions (Casey, 1998; Kelly, 1995; Sassaman, 1998). For instance, women flaking stone is found among the Konso of South Ethiopia, the Arawe of New-Guinea, and the Tiwi and the Jerramungup of North and Western Australia (Arthur, 2010; Bird, 1993; Goodale, 1971; Weedman, 2006). Moreover, if stone flaking appears as a mainly male activity, it might be because the ethnographic data are biased: The few observations we have were made by male observers, often in societies with strong gendered division in the use of space, and almost always in places where people no longer relied on chipped stone, as highlighted by Kathryn Weedman's (2005) work on hide workers in Ethiopia. Linda Owen (2005) points out that comparative studies of male and female activities in extant or recent hunter-gatherer societies have contributed to creating stereotypes of their roles by reducing the number of variables in order to simplify statistical analyses. Today, a more critical attitude toward the use of ethnographic analogy in archaeological interpretation is expressed by many researchers (Bolger, 2013).

However, it might be the case in many societies, even when women and men ostensibly perform the "same" tasks, that distinctions are maintained in technology, materials, type of product, and/or consumers. For instance—and Testart (1986) is correct on this point—men and women's hunting techniques and weapons often (but not always) differ (Endicott, 2004). In parts of West Africa, women and men use different kinds of looms and weave distinct types of cloth that are explicitly called "women's cloth" and "men's cloth" (Kent, 1971; Kriger, 2006). Among the Wola in Highland Papua New Guinea, on the few occasions women knap, they use a bipolar technique that men consider inappropriate when they knap (Sillitoe & Hardy, 2003). The ethnohistorical evidence from Sydney Harbor (Australia) indicates that both men and women fish from shore or in canoes. However, men invariably used a two- or four-pronged, barbed fishing spear, whereas women used hook and line, with a shell fishhook that they manufacture themselves and vegetable fiber line (McDonald, 1992).

Also, all members of a group may participate in the same activity but do not perform the same tasks. Hunting by driving large herds of herbivores, for example, is an activity that requires many participants and solicits all members of the group, but not to perform the same task. In Mbuti bow-and-arrow hunts, women act as beaters to drive game within range of the armed men (Turnbull, 1965). Netsilik women and children help men with seal hunting, and everyone fishes (Balikci, 1970). It is also possible that the different phases of a manufacturing sequence might not all be carried out by the same individuals. In textile production, in some cases, different tasks within the production sequence are performed by different genders. For example, in royal textile workshops of eighteenth-century Benin, men spun and women wove (Kriger, 2006). At times, the distinctions in gendered task assignments are so subtle that the ethnographer might miss them (Costin, 2013). In the Paleolithic, animal carcasses, for example, might have been cut into pieces by some individuals, while the meat was removed by others responsible for preparing meals.

Furthermore, even when an archaeological site appears to yield indices of task specialization, we cannot presume the identities of those who performed these tasks. For instance, when Kenneth Sassaman (1998) claims—without specific evidence—that in the Eastern Woodlands of North America, men hunted game using a stone tool technology, namely hafted bifaces, distinct from that made and used by women for other activities, he seems to apply stereotypes to archaeological data. It is not necessarily one sex or the other, but perhaps a specific group, as is the case in some modern hunting societies, that stores animal products, and a particular group is responsible for providing the meat reserves for the entire community. In any case, such indices are difficult to observe in archaeological data: The absence of an activity at a site does not prove anything, since it could have been performed in a part of the site that was not excavated or, in the case of nomadic groups practicing seasonal movements, in a different camp or at a different time of year.

EXPLANATIONS BASED ON INDIRECT ARCHAEOLOGICAL EVIDENCE

Several authors have attempted to model the social organization of UP populations on the basis of indirect data. Gendered tasks are not directly visible, but some archaeological remains are interpreted to demonstrate a sexual division of tasks. Let us examine a few of these attempts.

Emergence of Task Diversification

For Steven L. Kuhn and Mary C. Stiner (2006), the sexual division of tasks among *H. sapiens* is linked with dietary diversification and, in particular, the emergence of small-prey hunting and gathering. These authors start from the classic hypothesis that men hunted large prey, while the tasks of women and children were limited to hunting small prey and gathering. To support their demonstration, they compared the diet of anatomically modern humans (AMH) and Neandertals.

According to Kuhn and Stiner, Neandertals hunted only large prey. However, this activity would have required the cooperation of all members of the group to secure a sufficient amount of meat to feed everyone. In contrast, they portray UP

AMH groups as having had a much more pronounced sexual division of labor, which enabled women to exploit plant resources, thus widening the subsistence breadth and adapting groups more effectively to the environment and resulting in higher population densities (Kuhn & Stiner, 2006). Modern counter-examples do exist, such as that of the Inuits, who had a diet consisting only of meat and for whom the sexual division of tasks was as strong as among other groups: The women worked hides, made clothes, cooked, and tended fires, while the men constructed dwellings and hunted and fished (Hayden, 2012).

Although the idea that labor division appeared at the same time as task diversification (and thus resource diversification) is interesting, it is not certain that this "innovation" really occurred between the MP and UP. We have recently learned that the Neandertal diet was more diverse than previously thought. Use-wear analyses of stone tools and the organic residues on their surfaces have shown that Neandertals hunted small prey and fished. Moreover, these analyses were made on artifacts from sites distant enough from each other—in France (Payre, Ardèche), Spain (Bolomor Cave), and Italy (Fumane)—to eliminate the hypothesis that this could be a local phenomenon (Hardy & Moncel, 2011). Other analyses of the teeth of Neandertals from Spy (Belgium) and Shanidar (Iraq) showed that they consumed a variety of vegetal resources (Henry, Brooks, & Piperno, 2010). It therefore appears that the diets of AMH and Neandertals were not very different from each other. The question of how important such foods were to the Neandertal diet still remains. Was it as relatively important as it is among the San, who eat some meat but also a lot of fruits and vegetables? Or was it as episodic as it is among the Inuit, who eat much more meat than vegetables? The available Neandertal data are still too scattered to allow us to answer with certainty. The interesting fact is that Neandertals could and did use vegetable foods and quick, small game, as well as big-game meat, whatever the proportions might have been.

Furthermore, we now clearly know that some tasks, such as the manufacturing of handaxes, required a high skill level and thus would have been associated with some form of specialization already in the MP. I will come back to this point later.

As seen from this discussion, the field of archaeology has changed dramatically in the past two decades, as women archaeologists have challenged their male colleagues' exclusive focus on hard artifacts such as spear points rather than tougher-to-find evidence of women's work. Through the study of the perishable artifacts such as basketry, cordage, and weaving, Jim Adovasio and Olga Soffer (Adovasio, Soffer, & Page, 2007; Soffer, Adovasio, & Hyland, 2002) argue that, even if women did not participate in hunting activities, they invented all kinds of critical materials, including the clothing necessary for life in colder climates, the ropes used to make the rafts that enabled long-distance travel by water, and the nets used for communal hunting. Even more importantly, women played a central role in the development of language and social life—in short, in our becoming human. Even if we don't know exactly who did what, Adovasio and Soffer show that activities were much more diversified than previously inferred from artifacts of bone, antler, ivory, and stone.

Activity Zone Locations in Occupation Sites

The presence of zones reserved for specific activities in occupation sites has been interpreted as a topographic reflection of a division of gendered tasks. This type of study is based on the hypothesis that some artifacts reflect activities associated with a particular sex. Weapons, tools, and flaking by-products would thus be associated with hunting, butchery, and stone flaking, which are strictly men's tasks, while fireplaces, needles, and scrapers would reflect women's tasks like hide working and food preparation. Such assumptions have been largely discounted by Linda Owen (2005), who showed that these stereotypes are simplistic (and sexist). It is in fact possible for a location to be dedicated to a specific activity while still being frequented by both men and women.

Let us examine a few examples from the literature. At Mal'ta in Siberia, Mikhail Gerasimov thought that he could distinguish two sectors of the site. One, which he interpreted as masculine, was associated with bifaces, bone daggers, and bird figurines, while the other, interpreted as feminine, contained scrapers, needles, awls, necklaces, and female figurines (Gerasimov, 1958, cited by Baffier, Julien, & Karlin, 1981). We can see that this interpretation made in the late 1950s was highly ethnocentric.

Another more recent case concerns the Epipaleolithic site of Ohalo II, located on the banks of Lake Tiberias (or the Sea of Galilee) in Israel. This site, dated to between 23,000 and 22,000 BC (calibrated), was occupied by hunter-gatherer populations with no agriculture or domesticated animals. Nonetheless, the site contains at least 90,000 remains from 142 plant species that were consumed or used in their wild state. In particular, 19,000 grass seeds were well preserved because they were charred in fireplaces before the site was submerged by the lake, thus protecting them. Among these grasses, the ancient wild ancestors of grains, in particular emmer wheat and barley, are present (Nadel et al., 2004).

A basalt stone found in one of the huts (number 1) appears to have served as a grindstone to process these grains, a hypothesis supported by the presence of 150 grain fragments encrusted in its upper face, 127 of which belong to the barley and wheat families (*Hordeum, Triticum, Aegilops*). Other grains were scattered around the stone. In addition, an alignment of stones covered with ash suggests the presence of ovens. From here, it is not a giant leap to imagine that an ancestor of bread made from wild grains was cooked here (Piperno, Weiss, Holst, & Nadel, 2004).

Meanwhile, Dani Nadel and Ehud Weiss (Weiss, Kislev, Simchoni, Nadel, & Tschauner, 2008) recognized that this hut had two very distinct activity zones separated by a passage area between them. One zone in the back of the northern part of the hut appears to have been dimly lit; it contained the grindstone and the scattered wild cereal grains. The second zone, with the stone flaking by-products, was probably better lit, since it was located near the entrance. The authors suggest that the first zone, linked to food preparation, was associated with women, while the second zone, dedicated to tool and weapon manufacturing, would have been associated with men. Although this interpretation is tempting, we must remain cautious and cannot eliminate the hypothesis of a physical separation of tasks related to hygienic concerns rather than gendered division.

Similarly, Françoise Audouze (2010) has suggested that the distribution of backed bladelets (here used only as projectile weapon elements) around fireplaces at the Magdalenian site of Verberie (Oise, France) could correspond to a male activity zone, while the presence of scrapers furthest from the fireplaces could indicate a hide-working zone and thus be associated with women. She nonetheless tempers this interpretation, noting that there could be technical reasons for such a spatial distribution: Weapons could have been manufactured near the fireplaces because heat is needed to melt the adhesives, while hide working must be done far from the fire to avoid the risk of hides being damaged by flying sparks (Audouze & Beyries, 2007). We must thus be wary of assuming that the position of such activity zones is linked to social norms.

At Pincevent, Grégory Debout (2007) went even further: He identified a zone that appeared to be dedicated to working hide with characteristic flint tools, the presence of ochre, and a one-meter^2 accumulation of ash that might correspond to a drying zone. Three skill levels in blade manufacturing were also distinguished at this site. The most successful productions appear to have been carried outed by true experts, others by flakers with an average skill level (perhaps apprentices in training). Finally, the worst-quality productions were perhaps the work of young individuals in the very early stages of their apprenticeship. However, these poor-quality productions are located near the area where there is evidence for hide working—leading to the interpretation that this was a female activity zone with children playing and imitating the adults nearby.

The problem raised by these different interpretations is that they are based on the suppositions described earlier concerning the distribution of tasks among men and women in non-industrial societies. We can clearly see, however, that the specialization of space does not necessarily reflect that of its occupants. This is why some authors have begun to look for more direct evidence of task distribution.

EXPLANATIONS BASED ON DIRECT EVIDENCE

One parameter is sufficiently visible in archaeological contexts to allow the identification of technical specialization: skill level. It has been used to evaluate the degree of competence of individuals and therefore their possible specialization. Let us see if this evidence tells us something about task gendering.

Evidence for Different Skill Levels

Theoretically, specialization can be identified on the basis of technical criteria, such as the different skill levels required to carry out various different tasks or the standardization of certain products. The second argument can be eliminated right away because standardization can result both from the work of a specialist or a workshop or from strict technical or social constraints imposed by a single operational procedure (Le Dosseur, 2005).

Archeological experiments to knap stones indicate that high levels of expertise are acquired over long periods of training and through processes of interactive learning under the supervision of a skilled practitioner (Eren, Bradley, & Sampson, 2011). Such experiments give us an idea of the effort involved in obtaining knapping expertise that would enable a knapper to efficiently and regularly produce standardized and elaborate forms.

The presences of different skill levels has been confirmed at Étiolles and Pincevent, where Nicole Pigeot and Monique Olive (Olive & Morgenstern, 2004; Pigeot, 1987) demonstrated that the knappers of Magdalenian artifacts did not all have the same stone-flaking talents. The proficient flakers were able to manufacture long blades and tools of an excellent quality. Other individuals had a more mediocre skill level that was nonetheless sufficient to manufacture the occasional tools necessary for daily activities. Flakes that were probably never used were also found, likely made by young children, who had even less skill. It was also proposed that beginning flakers, probably children given their low level of skill, were also present at Verberie (Janny, 2010). It is easy to imagine young children observing the work of adults and imprinting the movements and actions that had to be accomplished. They would then imitate these actions with their clumsy hands, while sometimes getting advice or being reprimanded. In any case, the stability of stone tools over thousands of years indicates that knowledge and know-how were transmitted from one generation to the next. We once again encounter the reconstruction of Gregory Debout here, though without any allegation of the sexual identity of the actor.

We do find this allegation in Nicole Pigeot's (2010) study of blade manufacturing at Étiolles. In addition to two skill levels, perhaps corresponding to novices and proficient flakers, she argues that she was able to detect productions that were well mastered but simplified. This intermediary category, which she attributes to adults, could be associated with women who occasionally flaked to satisfy immediate needs, but for whom this was not their primary activity. Why women and not some not yet fully training teenagers? She does not answer this question. She implicitly links female lithic production to a more expedient technical strategy (Bamforth & Finlay, 2008). Women are still seen by many archaeologists as stone-tool users, not makers; and stoneworkers as fathers teaching their sons to knap (Johansen & Stapert, 2008). In the same vein, Schick and Clark (2003) suggest that small flakes in East African Acheulean sites could have been made by elderly, female, or juvenile individuals, who might have needed to have such material ready to hand in case they opportunistically encountered scavenging occasions while gathering, whereas localities with large bifaces made from materials obtained from distant sources reflected planned hunting by males.

Peter Hiscock (2014) claims that making stone artifacts requires highly skilled individuals who have been taught and have practiced for extensive periods, something that occurred as early as the LP. These specialized competencies were not limited to knapping itself but also included procuring material, which demands an understanding of the inorganic materials being worked and an ability to track geological resources across the landscape. Competency and expertise could develop specialization through forms of social learning and teaching over multiples generations, and even could stimulate competition between knappers. Peter Hiscock does not hesitate to talk of "craft specialization in which a fraction of the group focused their energies on knapping" in the Oldowan and Acheulean industries (Hiscock, 2014).

Jacques Pelegrin (2007) does not agree with the hypothesis of true flaking specialists who would have been the only ones able to fulfill a function essential to the community's survival. This is why he distinguishes between a "Paleolithic expert" and a "Neolithic specialist." The former had a definite skill and was able to carry out specific technical operations but did so in a context that remained within the range of ordinary productions, while the latter, who had exceptional skills, made only certain types of products to fulfill an external demand. The pertinence of this distinction based on competence levels is debatable since we do know that specialization does not always imply an exceptional skill level. This is the case, for example, for Neolithic pottery, whose realization required only an average skill level but which nonetheless was manufactured in a context of specialization (Gomart & Burnez-Lanotte, 2012; Vieugué, 2012). One can be a specialist and not be very skilled, and activities requiring little skill can be practiced by specialists (e.g., many traditional caste occupations in India). Specialization has a lot to do with the time demands of an activity and whether it interferes with other activities. It may also be an option for individuals who are unable to participate in other activities.

On the gender of knappers, it was commonly accepted that it was a male activity, following the values attributed to male and female roles until the "Man The Hunter" conference in 1965 (Lee & DeVore, 1968a). Another argument regarding the repartition of tasks by sex appeared in the *Ethnographic Atlas* (Murdock & Provost, 1973). Stone-working was identified as a male activity in 67 societies and in only 6 as a male or female activity. Despite many biases and flawed calculations, this evidence is still used uncritically by many researchers. In fact, stone tools made and used by women are known from a diverse range of ethnographic contexts, as we saw earlier (Finlay, 2013). But it is very difficult to extrapolate from this the sex of UP knappers.

Evidence for Specialization

What then do these different skill levels indicate? Are we looking at a simple division of labor according to age and sex with domestic production, or a true specialization in the sense that the activity was performed by only a limited number of individuals (or groups) who redistributed their products among a larger community? This type of specialization is even more difficult to identify as it may concern only part of the *chaîne opératoire*. Gaëlle Le Dosseur (2005, p. 127) proposes addressing the question by evaluating the time required to carry out certain activities, which could sometimes be long enough to suggest a "full-time" activity, at least on specific occasions. This is the case, for example, for the manufacturing of bone beads, hundreds of which have been found at several Natufian sites. How can we explain the approximately 13,300 ivory beads found in the three graves at Sunghir (Russia), dated between 25,000 and 27,500 ^{14}C before present (BP), and which would have required 30 minutes per bead, or 3.33 years at 40 hours per week—totaling 6,650 hours of work (Trinkaus, Buzhilova, Mednikova, & Dobrovolskaya, 2015)! After various experimentations, Randall White (1999) has even suggested that each bead required 45 minutes, which implies many more hours of work. It is difficult in this case not to imagine an artisan specialized in the manufacturing of personal ornaments. But the vast amount of time represented by the Sunghir burial beads could also represent part-time (seasonal or evening) work by many people. Traditional folk are rarely idle—when they are not doing things related directly to survival, they are often busy making and mending. Without knowing the number of individuals involved in the activity, however, it is impossible to evaluate the amount of time devoted to it, relative to food procurement activities, let alone whether the activity required dedicated specialists or was gendered.

We can also consider the possibility of specialization when we see that a specific raw material was used far from the place where it was collected and/or worked—though we must be certain that these different tasks were not performed by the same people who traveled far away to procure raw materials that they then worked and brought back to their usual residence site. This supposes that we are able to show the existence of exchange networks organized by individuals other than the users of the final products. Such networks and the specialization that they imply have been identified in the Early Neolithic. For example, the circulation of obsidians from central and eastern Anatolia over more than 900 km toward the Near East by land and toward Cyprus by sea indicates specialized productions that were introduced into settlements in the form of finished objects (Astruc, 2005). For the UP, François Bon (2005) sought to learn whether the same could have occurred in an earlier period by analyzing the lithic objects found at several occupation sites in southwestern France and dated to the Early Aurignacian period. The raw materials from which these objects were made had both local and distant origins. We could consider the possibility of task specialization according to the raw material reduction phase, with some workshops located at or near the flint sources and others near the occupation sites. However, even if there were two types of workshops, the skills employed were the same, since an analysis of the flaking techniques showed that they were the same for all the raw materials. He infers that there were neither exchanges nor specialized flakers and that the objects that circulated were manufactured and transported by the same people, or at least the same groups. This result is interesting and well founded, but can we imagine that two individuals, or two different groups, had the same flaking skills? In any case, we have no basis for extending Bon's conclusions to the entire UP. And, once again, nothing can be said about the sex of these individuals.

Another possibility was explored by Maria-Anne Dobres (1995) based on a study of the bone tools from several Magdalenian sites. Dobres hypothesizes that the variability of technical "strategies" from one site to another could be due to a different organization of task distribution. For example, the percentage of repaired tools is highly variable, attaining 20% in some cases (e.g., at La Vache, where three-quarters of the tools repaired are harpoons). She hypothesizes that the composition of groups is not organized in the same manner at all the sites (e.g., sites with many repaired tools; sites where tools were made, etc.). At the sites where tools were manufactured, activities were planned for the long term, whereas at the sites where they were repaired, activities responded to an immediate need. According to Dobres, the less variability we observe in technical strategies, the more tool production activities were formalized. She interprets the variability observed among sites as evidence that specialization was low, since even if only some people had the skills and access to raw materials necessary to manufacture the objects (in stone and bone materials), everyone could repair them. She also suggests, however, that technical skills could vary according to age, sex, and social status.

The Evidence of Grave Goods

There are abundant ethnographic and historic examples of men buried with their weapons and tools and of women buried with their tools. It is thus logical to think that grave goods interred with the dead are informative regarding the specific activities of different individuals. Unfortunately, although we have some examples of this kind of evidence for the hunter-gatherer populations of the Mesolithic periods, it is more difficult to find for the UP. For the later periods of prehistory, in the first Mesolithic necropolises, skeletal remains are sufficiently abundant to enable statistical analyses. It is indeed easier to determine the sex of inhumed individuals in a larger population, since the features of sexual dimorphism, such as height and skeletal gracility, are easier to observe by comparing individuals within the same population. Skeletal data sometimes enable determinations of whether grave goods indeed differ between male or female tombs. In the Hoëdic necropolis, for example, personal ornaments buried with the dead appear to differ according to sex, but technical objects have greater utility for addressing the question of task distribution.

One of the most remarkable examples is provided by the Bogebakken necropolis at Vedbaek, a Danish coastal site excavated in 1975 that yielded 22 tombs dated to approximately 4,800 BC. All the bodies were deposited in an extended position and impregnated with red ocher. Grave goods were composed of animal teeth and red deer antler. Only men, however, were also accompanied by one or two flint blades attached to their waist. In one of the tombs (No. 8), a young woman around 18 years old, who perhaps died during childbirth, was accompanied by an infant, possibly stillborn. Both were adorned with numerous ornaments made of red deer and boar teeth, as well as perforated *Neritina fluviatilis* shells, and the child was laid out on a swan wing. But what is most intriguing is that the child was also accompanied by a flint blade, like the adult males in the necropolis (Albrethsen & Petersen, 1976; Petersen, 1975). This suggests that this was a little boy and that the blades—perhaps dagger blades used for hunting or combat—constituted a male attribute rather than a tool used by the woman during her lifetime.

Mesolithic tombs would thus be the oldest to provide evidence for a gendered division of tasks, even if this practice may have existed well before then. Of course, more recent tombs dated to the Neolithic and Metal Ages also show significant differences in their grave goods, the most spectacular being the abundant warrior tombs of the Bronze Age. Even for these periods, however, we must remain cautious in our conclusions, since while some objects have a high symbolic value, they provide little information on their relationship to daily activities (Péré-Noguès, 2008).

What about older burials? The first difficulty is that there are very few such burials, and they are scattered through time and space. There are really not many UP burials: just over three per thousand years for the whole Eurasia (Riel-Salvatore & Gravel-Miguel, 2013). It is thus difficult to detect recurrences, and we observe, on the contrary, a very high level of variability in funerary practices, whether in terms of the position of the body or the presence or absence of ornaments and grave goods. Two other difficulties add to this dispersion. The first concerns the association of grave goods with the deceased. Because many Paleolithic burials were found long ago, we often do not know if the flint or osseous tools found in the tomb were part of the funerary deposit or were accidentally introduced into the sediment filling the grave. There are nonetheless a few known cases of tools that were clearly associated with the deceased person. One of the oldest is the burial of an adolescent, dated to 23,440 ± 190 BP and attributed to the Gravettian, excavated in 1942 in the cave of Arene Candide in Liguria (Henry-Gambier, 2005). The young man, decorated with hundreds of perforated shells and laid on a bed of red ochre, held a 25-cm-long flint blade in his right hand, and four elk antler pierced batons had been placed by his sides. Similarly, the two children of the Sunghir burial, dated between 25,000 and 27,500 ^{14}C BP (Trinkaus et al., 2015), now believed to be a boy 11 to 13 years old and a girl perhaps 9 to 11 years old,[3] were accompanied by numerous spear points and lances.

[3] The sex assessment of these individuals is unreliable, and Trinkaus urges treating these data cautiously.

These weapons were clearly associated with the bodies, apparently without consideration for their age or sex. Consequently, they do not enable us to determine the role of the buried individuals (Henry-Gambier, 2005; Trinkaus, Buzhilova, Mednikova, & Dobrovolskaya, 2014).

The second difficulty with older burials is that given the poor state of preservation, the age of the deceased (pre-puberty), or the unreliability of the methods used, it is often difficult to determine the sex of the inhumed individual (Henry-Gambier, 2005). In previous decades, the sex of buried skeletons was attributed on the basis of robustness of the skull or the long bones, or the grave goods associated with them, which were usually interpreted in a sexist manner, resulting in an under-representation of females. More recent paleoanthropological studies have shown that these attributions were sometimes erroneous (Henry-Gambier, 2001). This is the case for the skeleton known as "the man of Menton," discovered in the cave of Cavillon (or La Barma de Caviglione, Italy) and dated to the Gravettian. The individual, deposited in a semi-flexed position, was abundantly covered with ochre and wore an impressive headdress composed of 200 perforated shells that appear to have been attached to a hairnet. But it is based on the presence of numerous tools, including a bone dagger and two large flint blades, that this individual was identified as a male. However, one analysis of the coxal bone indicates that it is in fact an adult female (Brůžek, 1991; De Lumley, Stalens, & Brůžek, 1992). This shows the "pressing need to re-examine existing collections of human remains in order to place the current study of gender, the body, and mortuary ritual on a more secure footing. Equally important in current research programs is the use of DNA, stable isotope analysis, and other types of scientific investigation in order to bring the study of the gendered body and the analysis of mortuary remains fully into the twenty-first century" (Bolger, 2013). Sébastien Villotte and his colleagues (Villotte, Brůžek, & Henry-Gambier, 2011) have begun to perform such re-analysis on fossils from the European Gravettian.

We can conclude from these few examples that there are no clearly significant differences in grave goods associated with UP male and female tombs. We must thus be cautious in our interpretations of the nature of the grave goods accompanying the dead, as they can reveal things other than the distribution of tasks. We have to admit that grave goods are imperfect indicators of gender identity, that gender categories were not always the same, and that scholars can "play with" the rules. We have to keep in mind, and a few ethnographic examples could confirm it, that the variations in funerary practices within a single population can depend on a combination of information related to the association of the deceased with distinct categories (sex, age, class, origin), but also on the circumstances of their death.

The Evidence of Osteological Analysis

Interpretations of gender in mortuary contexts must be based on skeletal material that has been sexed by osteological analysis, rather than through associated grave goods. The skeletal information provides data on health, nutrition, activity, physical stress or workload, risk, and mortality. Information in large quantity from skeletal analysis is relatively new. Mark Nathan Cohen and Sharon Bennett (1998) have presented a sample based on a review of the literature of applications of skeletal analysis to issues of gender inequality. But the major difficulty with much of archaeological data is that

Paleolithic remains are isolated individuals. There is no evidence of large necropolises before the Mesolithic period. So there is some danger of facile over-interpretation of one individual skeletal to a whole population.

Concerning the division of tasks, one of the few ways to provide direct evidence on male and female tasks in the past has been the study of the dental remains and the osteological markers of activity. Dental wear patterns have been used to reconstruct activities in which the teeth are used as a tool or are used to act as a "third hand." Osteoarthritis, or other bone changes, results from regular adult activities such as climbing, kneeling, grinding grain, and carrying heavy loads. But once we consider the influence of age and genes, there is very little remaining evidence to suggest that activity is the main cause of arthritic bone changes. There are, however, a few exceptions, including Inuit skeletons with elbow arthritis related to extreme activities. Another type of alteration concerns the areas where muscles are attached to bones. Repeated uses of certain muscles produce changes to points of attachment. But once again, the causes of such bone changes are numerous and include age and genetic influences. These stress markers have been linked by some authors with subsistence change or sexual division of labor (Hawkey & Merbs, 1995; Jurmain, 1999; Molleson, 1994; Molnar, 2006; Robb, 1998; Villotte, Churchill, Dutour, & Henry-Gambier, 2010).

This kind of analysis is more common for the Neolithic and the Bronze and Iron Ages because of the presence of cemeteries with series of skeletal remains permitting statistical studies. For example, Alison Macintosh and colleagues (Macintosh, Pinhasi, & Stock, 2014) presented an overview of analysis on bone adaptation among agriculturalist and metallurgist populations. She particularly studies variation of the upper limb asymmetry among males and females in Central Europe at the transition between the Neolithic and Early Bronze Age. Marked lateralization has also been documented, which may be the result of unilateral loading associated with the manufacture and use of many stone, bone, and metal tools and weapons. Similarly, Aline Thomas (2014) hypothesized an activity of archery in 36 male skeletons associated with arrowheads from European Middle Neolithic cemeteries. She observed a functional adaptation in the forearm bones and the clavicle in response to mechanical loads, as well as enthesopathies (or musculoskeletal stress, i.e., lesions of the tendon attachments) suggesting repeated forceful use of upper limb muscles, both findings supporting her hypothesis.

For the beginning of the Neolithic, the study of the cemeteries of Vedrovice (Moravia, Czech Republic) and Nitra Horné Krškany (western Slovakia) showed dental evidence of people having used their teeth as tools, possibly for the working of plant fibers and the production of cord or rope (Frayer, 2004; Jarašová & Dočkalová, 2008). Another example concerns Molleson's (1994, 2007) observations of Abu Hureyra's Neolithic (Syria) female skeletons. She links the grooving of the chewing surfaces of front teeth to obtaining fibers for the production of baskets and sieves by some women. From the observation of metatarsals, phalanges, and other skeletal indications, she suggests that the daily chore of grinding grain fell predominantly to women, who kneeled with their toes curled under their feet.

Concerning hunter-gatherer populations, some studies are dedicated to Mesolithic necropolises. In Mesolithic southern Scandinavia, tooth wear in adult women reflects working hides by chewing, while wear on the teeth of men indicates holding materials.

Although tooth wear increases with age for both sexes, it appears earlier among women (Alexandersen, 1993; Blankholm, 2008).

Brigitte M. Holt (2003) examined the evolution of mobility during the UP and Mesolithic through analysis of 81 UP and Mesolithic European femora and tibiae. The relationship between levels of mobility and lower-limb diaphyseal structure was used to test (and support) the hypothesized decrease in mobility during the Mesolithic. Among other results, she observed that sexual dimorphism levels in diaphyseal strength remain low throughout the three time periods (Early and Late UP and the Mesolithic), suggesting that there were no sex differences in behavior for UP and Mesolithic foragers. However, other analysis of human lower-limb bone structures from Neolithic and MP/UP samples indicates that sexual division of labor decreased over time, in particular in the relative mobility of males and females from hunting-gathering to agricultural subsistence (Ruff, 1987).

In the case of subsistence strategies, unilateral enthesopathy of the medial epicondyle has been recognized previously as a good marker of spear/harpoons throwing by hand (Dutour, 1986). Sébastien Villotte and Christopher J. Knüsel (2014) analyzed enthesopathies of the elbow in prehistoric, pre-industrial, and modern European populations. The increased prevalence of pathological changes of the right medial epicondyle suggests lateralized limb use that corresponds with "thrower's elbow." This suggests that males, but not females, preferentially employed movements involving throwing motions in these hunter-gatherer and early farming groups. Based on this evidence, these authors postulate the existence of a persistent sexual division of labor in these prehistoric European populations, involving one or several strenuous activities linked to unilateral limb use.

In a previous analysis, Sébastien Villotte and colleagues (2010) assessed 16 males and 21 females from UP or Mesolithic European populations. Males exhibited lesions that can be confidently associated with throwing activities, while no females had such lesions. A sexual division of tasks concerning hunting seems most likely, based on the presence of injuries specific to spear throwing in male individuals and their absence among females. This division of labor would have begun during childhood or adolescence. Villotte and his team note in another publication that humeral asymmetry through the late Pleistocene is especially high among the males relative to the females, and consider the possibility of a division of labor between unimanual tasks (mostly male) and bimanual tasks (mostly female). At the same time, there is a general pattern of increased asymmetry with larger body size, but it remains unclear to what extent it reflects body size or sexual effects on bilateral humeral loading (Sparacello, Villotte, Shackelford, & Trinkaus, 2016).

Marked levels of upper limb lateralization are not uncommon in UP samples, reflecting in most cases bone remodeling related to limb dominance and in other cases the effect of pathologically induced changes. The adult male of La Barma Grande also shows an unusually high degree of upper limb bilateral asymmetry, with the right side well above the mean values of other UP male specimens. However, comparisons involving normal and pathological Paleolithic remains, as well as recent skeletal samples, suggest that the phenomenon may be a secondary effect of pathological conditions such as muscular trauma (Churchill, 1994; Churchill & Formicola, 1997).

Another study based on musculoskeletal stress markers of the upper limb on skeletal remains from the Natufian and the early farming populations in the Levant indicates a gender-based division of labor in both the Natufian and Neolithic (Eshed, Gopher, Galili, & Hershkovitz, 2004). Sládek and colleagues (2016) compared upper limb bone bilateral asymmetry in a large ($n > 1200$) European sample distributed among 11 archaeological periods from the early UP through the twentieth century. They intended to show that the trajectory from UP to Neolithic was different for males and females. For instance, changes in manipulative behavior were sex-specific with a probable higher impact of changes in hunting behavior on male asymmetry (e.g., shift from unimanual throwing to use of the bow and arrow) and food grain processing in females—specifically, use of two-handed saddle querns in the early agricultural periods and one-handed rotary querns in later agricultural periods (Sládek et al., 2016).

Although this chapter is specially focused on the UP, it is nonetheless interesting to note that an analysis of activity-related dental wear patterns in 19 Neandertal fossils from three different sites reveals that all individuals have cultural striations, but those detected on the adult females are longer than the striations found in adult males. Regarding the distribution of dental chipping, in males this trait is more prevalent in the maxillary dentition, whereas in females most of the dental chipping is on their mandibular teeth. The differences detected in the overall activity-related dental wear pattern denote a difference or a division of labor by age and sex in Neandertals when the mouth is used as a third hand (i.e., in activities other than the provisioning of food). These findings provide new evidence for the lifestyle of this Pleistocene fossil human species, indicating subtle differences that suggest individual roles within the groups (Estalrrich & Rosas, 2015). This analysis contradicts that by Kuhn and Stiner presented earlier.

The literature regarding the markers of activities on bones and dental remains is too large to review here. However, Robert Jurmain and colleagues (Jurmain, Cardoso, Henderson, & Villotte, 2012) have comprehensively reviewed the topic and show the limits of these analyses. The fact that some examples are self-contradictory may indicate that mobility and subsistence gender tasks varied in time and space in different hunter and gatherer populations. In any case, such analyses are surely one of the best ways to reconstruct activities of the past populations and obtain direct evidence of the sexual division of tasks, provided one avoids overgeneralization.

Male and Female Handprints

Research in Paleolithic rock art has suggested gendered explanations, but most of them are interpretations influenced by a specific ideology (Hays-Gilpin, 2004, 2013). Few of these studies have tried to support their assertions with archaeological evidence. One of these hypotheses, presented next, is interesting because the author proposes to prove that the artists were women—which is not so common—and deduces his theory from archaeological evidence and not just from firm conviction.

According to Dean Snow (2013), the artists who decorated caves were women. He bases this interpretation on a study of negative hand stencils (i.e., stencils made by blowing paint against hands placed on the cave wall, which outline the hand while the hand itself leaves an unpainted or negative space) in a few decorated caves. To establish the sex of the authors of the hand stencils, he used an index developed by John Manning (2002) based on the length of the index and annular fingers. According to this index, these fingers are the same length for women, whereas for men the index finger is shorter than the annular finger. Snow was not the first to use this index; it was previously tested on the negative handprints in the cave of Masri II east of Borneo (Indonesia). Snow tested this index on 32 hands from eight different caves (including 16 from El Castillo, 6 from Gargas, and 5 from Pech Merle) and deduced that 24 of them were female. Based on a comparison with the hands of male and female American students, he also concluded that sexual dimorphism was greater during the UP than it is now. A few years earlier, R. Dale Guthrie (2005) performed a similar study based on the width of the palm and the thumb and concluded that the majority of painted hands belonged to adolescents.

The samples used in these studies are small because most of the handprints were incomplete (i.e., intentionally represented using only two or three fingers). We can thus question the statistical validity of such a study, and even more so because negative handprints are found only within a short time range, corresponding to the Gravettian period. Nonetheless, in the few caves concerned, Guthrie and Snow agree that adult male hands correspond to only around 10% of the total number of hands represented. Men were therefore not the only ones to participate in artistic activities.

CONCLUDING PROPOSITIONS

Based on the information presented here, I propose that there was a reasoned distribution of activities within groups, accompanied by an emerging social hierarchy, but that it is very difficult to make conclusions about the gendering of activities because the sex of artisans is unknown.

A Distribution of "Workshop" Activities

It appears that a form of task distribution existed, for at least two reasons. First, life within a society implies a capacity to manage the equilibrium of all the means employed to ensure the survival of the group. François Sigaut often discussed this idea[4] with reference to the "workshop" concept arising from the school of Frédéric Le Play and applied to "wild peoples" by Paul Descamps (1923). Here, the notion of a workshop designated the manner in which a given activity was distributed among a social group in the form of simultaneous or successive, but connected, tasks. This implies that a prehistorian (and an anthropologist of techniques) is not interested in a particular activity, but rather in the entire repertoire of activities of a social group and in the different tasks composing these activities, in order to understand the balance of the whole ensemble.[5] Thanks to this notion, the problem of task distribution between the sexes can be seen in a new light—that is, there are no specifically male or female tasks. To understand the nature of this distribution, we consider all of the tasks practiced by all of the members of a social group and their organization within workshops. Along these lines, based on the data gathered from 71 hunter-gatherer populations,

[4] Particularly in the context of seminars at the École des Hautes Études en Sciences Sociales, Paris (see de Beaune, 2013).

[5] This notion of "workshop" explains how small changes can have unexpectedly significant repercussions.

Nicole M. Waguespack (2005) showed that the amount of time spent by women to collect vegetal resources varied depending on the degree of dependence on the meat of large herbivores hunted by men. The most time-consuming task was obtaining and sharing meat for daily consumption, which influenced the time spent collecting wild plants. The more they abandoned the collection of wild plants, the more time would be devoted to other technical and non-food-related activities. Another example comes from the Inuit: If an Inuit family has no son, the father may train one or two daughters to hunt. Similarly, families with few daughters may teach sons sewing and other female skills, and orphans learn both men's and women's skills (Briggs, 1970; Endicott, 2004). Hetty Jo Brumbach and Robert Jarvenpa observed that different activities among the Chipewyan require greater or lesser degrees of logistical organization, and this balance can change over time as external factors change. For instance, as part of the population became more sedentary, another part (men) needed to "become increasingly mobile and logistically organized" (Brumbach & Jarvenpa, 1997). Also, as prey size increases, the tasks of butchering, processing, storing, and distributing food tend to fall on women rather than men, at least in these particular circumpolar societies (Jarvenpa & Brumbach, 2009). This suffices to show that the dichotomous view of women collecting vegetal resources and men hunting is simplistic and does not take into account the balance between all of the tasks carried out by the members of a community.

The second reason is that regardless of the task or activity considered, we can reasonably suppose that it was preferentially assigned to the individual, male or female, who was best equipped to succeed at it. However, while some activities can be performed by any member of the group, others—particularly in the technical and artistic domains—require a certain degree of competence. Knowing that individual aptitudes vary from one individual to another, we can infer that there are specialists in bone, flint, hide, and other material working to whom the manufacturing and maintenance of artifacts was entrusted. Without going so far as to suggest craft specialization, we can imagine that even if the entire group had the same global knowledge, some adults focused on performing the tasks for which they were best suited. To take a contemporary example, all Tuareg blacksmiths know more or less how to make a great variety of artifacts, but they often choose to specialize in making whatever kind of article they feel most skilled at doing; some make camel saddles, others wood spoons, or spades, or jewels (Casajus, 1987). Also relevant here is the case of a 35- to 40-year-old man of the Fish Creek group from Australia: He was a skillful craftsman, able to repair spear points and spear throwers and produce pipes and bullroarers, who was exempted from all food tasks (McCarthy & McArthur, 1960). As another example, Deborah Olausson (2008) explored the flint technology in southern Scandinavia during the Late Neolithic, especially craft expertise and its potential as a factor in aggrandizer strategies. She argues that there are elements of natural aptitude that enabled certain individuals to excel at flintknapping, allowing them to create objects of exceptional size and beauty. If native ability in a particular domain is a rare commodity, then harnessing it and developing it through practice provides an opportunity for a potential aggrandizer to control prestige goods and accrue social capital. We should not exclude the likelihood that the same would have occurred in the European UP.

We have seen that there is clear evidence for different competence levels in knapping stone during the UP, implying that adults transmitted their knowledge and skills to younger individuals through apprenticeship modalities (see, for instance, Tostevin, 2012, on the transmission of flintknapping behaviors). The skill level necessary to manufacture handaxes indicates that this specialization already existed tens or even hundreds of thousands of years previously, in the MP. These groups required social structures with which to facilitate the learning process. Peter Hiscock (2014) argues that the development of apprentice–master relationships would have enhanced differentiated labor roles and the growth of specialization within small groups.

This must have been true in the artistic domain as well, since the art in Lascaux and elsewhere could not have been done by just anybody. The techniques used throughout the UP—engravings, sculpture and painting—do not appear to have significantly evolved during the 30,000 years of portable and parietal art; on the contrary, they appear to have been fully developed from the beginning of the UP (de Beaune, 2019). These artists had not only specific capabilities but also the privilege to enter into the depths of the cavern and to leave their mark, characteristics that would have distinguished them from their peers. Independent of a particular status (shaman, sorcerer, priest, etc.) granted to the artist by at least some members of the group, he or she was an individual whose skills were recognized. This suggests that the artist would have played a specific role in the group, perhaps comparable to that of the carvers of totem poles in a Northwest Coast native community, whose talents justified their maintenance by the rest of the group (Leroi-Gourhan, 1982). Of note, these carvers were never full-time artists; they were also engaged in food-collecting activities such as fishing. Apprentices were trained by artists from their communities (artists were usually high-status individuals; Marie Mauzé, personal communication). Like skilled stone workers, the artists were probably accompanied by young apprentices, since one would never venture alone into a cavern, for basic security reasons, and because help from an assistant to hold the lamp or prepare the colorants was indispensable.

Furthermore, simple logic suggests that, even if women were restricted to domestic tasks, they must also have been able to carry out the technical tasks traditionally assigned to men. For instance, if they were able to cook, they must have been able to make a fire. If they cut carcasses into pieces and worked hides, they used stone and bone tools; as these tools wear out quickly, they must have been able to make or at least resharpen them and thus to work stone and bone. All of these techniques are usually considered to be reserved for men.

Possible Evidence for an Emerging Social Hierarchy

If we accept that there was a dense population during the UP, that groups periodically gathered into large residential units, or that they were relatively sedentary (Conkey, 1980, 1985; Hayden, 1981; Mellars, 1985), then we can suppose that individuals with a particular status emerged. They would have been responsible for things such as organizing and coordinating the activities of the group, ensuring the distribution of food and other resources, and perhaps serving as mediators in conflicts.

The existence of such persons is difficult to prove on the basis of archaeological evidence. The fact that only some individuals were buried during the UP indicates potential social differentiation. However, distinct treatments of the inhumed bodies make it difficult to determine whether any specific treatment existed for men or for women, or the old versus the young. UP burials differ widely in terms of their elaboration, even within single regions, and sometimes even within a single site. There is no such thing as a "typical" UP burial. Gender ratios changed over time and are difficult to interpret (Riel-Salvatore & Gravel-Miguel, 2013), and there appear to be no significant differences between male and female ornaments (Marian Vanhaeren, personal communication). Brigitte M. Holt and Vincenzo Formicola (2008) have noted that for the Late UP there was an intriguing and unexpected incidence of individuals affected by congenital disorders, which probably indicates selective burial practices for abnormal individuals. We therefore cannot eliminate the hypothesis that inhumation was reserved for certain members of the group, and perhaps their families, related to a particular social status.

Brian Hayden (2001, 2013) has proposed a revision of the traditional model of these societies as largely egalitarian, like the hunter-gatherer societies observed in Africa (Childe, 1954). In his study, the groups occupying the most favorable environments, such as in southwestern France, would have belonged to the same category of relatively complex and hierarchical hunter-gatherer societies as those of the Pacific Northwest Coast of North America. These societies had significant socioeconomic inequalities, were more sedentary, and had higher demographic densities. Moreover, they openly fought among themselves in rivalries based on wealth, possessed prestige objects, obtained certain resources and wealth in an exclusive manner, and shared very little. Some archaeological evidence at Magdalenian sites in southwestern France suggests similarities with these societies, which could be due to an underlying surplus production resulting from the massive exploitation of migratory and other species: salmon on the Pacific Northwest Coast plateau, reindeer and other herd animals in France. Based on an analysis of the geographic distribution of sites and their surface area, as well as of some remains such as burials and personal ornaments, Hayden even suggests that there was a specialization of sites and thus of tasks, and that different social statuses among individuals already existed in Neandertal societies. To him it is evident that the sexual distribution of tasks was highly developed, as it is in all known hunter-gatherer societies (Hayden, 2012).

Technological studies of the grave goods associated with burials and their comparison with ornaments and faunal assemblages from contemporary sites and burials reveal the exceptional character of some of the inhumations, like those of Saint-Germain-La Rivière, La Madeleine, La Grotte des Enfants, and Sunghir (e.g., Vanhaeren, 2002; Vanhaeren & d'Errico, 2001, 2003; White, 1999). As observed in a number of hunter-gatherer populations, and contrary to the supposed egalitarian character of UP societies, these items may indicate that an individual was a member of a privileged social group (Vanhaeren & d'Errico, 2003).

In conclusion, regarding the division of labor along gender lines in the Paleolithic, there is evidence for emerging activity specialization, but it is difficult or impossible to reliably assign it to specific genders. I will end this chapter by insisting again that we must be cautious in our interpretations, as what we seek to understand is the identity of our Paleolithic ancestors, identity that forcibly evolved throughout millennia of prehistory and was subject to the ebbs and flows that shaped the history of Paleolithic societies (de Beaune, 2016; Lenclud, 2003).

ACKNOWLEDGMENTS

I thank two anonymous reviewers for their very valuable comments, which helped me improve this chapter.

REFERENCES

Adovasio, J. M., Soffer, O., & Page, J. (2007). *The invisible sex: Uncovering the true roles of women in prehistory*. Walnut Creek, CA: Left Coast Press.

Albrethsen, S. E., & Petersen, E. B. (1976). Excavation of a Mesolithic cemetery at Vedbæk, Denmark. *Acta Archeologica, 47*, 1–28.

Alexandersen, V. (1993). Teeth: Health, disease and use. In S. Hvass & B. Storgaard (Eds.), *Digging into the past: 25 years of archaeology in Denmark* (p. 81). Copenhagen: Aarhus Universitetsforlag.

Arthur, K. W. (2010). Feminine knowledge and skill reconsidered: Women and flaked stone tools. *American Anthropologist, 112*(2), 228–243.

Astruc, L. (2005). Approches multi-échelles de la spécialisation technique: Les industries lithiques taillées du Néolithique pré-céramique au Proche-Orient. *Techniques & Culture, 46/47*, 79–104.

Audouze, F. (2010). Domesticity and spatial organization at Verberie. In E. B. W. Zubrow, F. Audouze, & J. G. Enloe (Eds.), *The Magdalenian household: Unraveling domesticity* (pp. 145–175). Albany, NY: State University of New York Press.

Audouze, F., & Beyries, S. (2007). Chasseurs de rennes d'hier et d'aujourd'hui. In S. Beyries & V. Vaté (Eds.), *Les Civilisations du Renne d'hier et d'aujourd'hui: Approches ethnohistoriques, archéologiques et anthropologiques* (pp. 185–208). Antibes: Éditions APDCA.

Baffier, D., Julien, M., & Karlin, C. (1981). La femme et la préhistoire. *Pénélope, 4*, 95–99.

Balikci, A. (1970). *The Netsilik Eskimo*. Prospect Heights, IL: Waveland Press.

Bamforth, D. B., & Finlay, N. (2008). Introduction: Archaeological approaches to lithic production skill and craft learning. *Journal of Archaeological Method and Theory, 15*(1), 1–27.

Baron-Cohen, S., Knickmeyer, R. C., & Belmonte, M. K. (2005). Sex differences in the brain: Implications for explaining autism. *Science, 310*(5749), 819–823.

Bégouën, H. (1924). La magie aux temps préhistoriques. *Mémoires de l'Académie des sciences, 12*(II), 417–432.

Bégouën, H. (1939). Les bases magiques de l'art préhistorique. *Scientia, 4*(33), 202–216.

Bird, C. F. M. (1993). Woman the tool maker: Evidence for women's use and manufacture of flaked stone-tools in Australia and New Guinea. In H. du Cros & L. Smith (Eds.), *Women in archaeology: A feminist critique* (pp. 22–30). Canberra: Australian National University.

Blankholm, H. P. (2008). Southern Scandinavia. In G. Bailey & P. Spikins (Eds.), *Mesolithic Europe* (pp. 107–131). Cambridge, UK: Cambridge University Press.

Bolger, D. (2013). Introduction: Gender prehistory—The story so far. In D. Bolger (Ed.), *A companion to gender prehistory* (pp. 1–19). Chichester, UK: Wiley-Blackwell.

Bon, F. (2005). Être un ou plusieurs: Quelles différences pour l'Aurignacien? *Techniques & Culture, 46/47*, 35–49.

Bosinski, G. (2007). *Gönnersdorf und Andernach-Martinsberg: Späteiszeitliche Siedlungsplätze am Mittelrhein*. Koblenz, Germany: Gesellschaft für Archäologie am Mittelrhein und Mosel e.V.

Bosinski, G., & Fischer, G. (1974). *Die Menschendarstellungen von Gönnersdorf: Der Ausgrabung 1968. Der Magdalénien—Fundplatz Gönnersdorf 1*. Wiesbaden, Germany: Steiner.

Breuil, H. (1952). *Quatre cents siècles d'art pariétal: Les cavernes ornées de l'Âge du Renne*. Montignac: Centre d'Études et de Documentation préhistoriques.

Briggs, J. L. (1970). *Never in anger: Portrait of an Eskimo family*. Cambridge, MA: Harvard University Press.

Brown, D. E. (1991). *Human universals*. New York: McGraw-Hill.

Brumbach, H. J., & Jarvenpa, R. (1997). Ethnoarchaeology of subsistence space and gender: A subarctic Dene case. *American Antiquity*, *62*(3), 414–436.

Brůžek, J. (1991). *Fiabilité des procédés de détermination du sexe à partir de l'os coxal: Implications à l'étude du dimorphisme sexuel de l'homme fossile*. Doctoral thesis, Institut de paleontologie humaine, Museum national d'Histoire naturelle, Paris.

Casajus, D. (1987). Craft and ceremonies: The Inadan in Tuareg society. In A. Rao (Ed.), *The other nomads: Peripatetic minorities in cross-cultural perspective* (pp. 291–310). Köln-Wien: Böhlau Verlag.

Casey, J. (1998). Just a formality: The presence of fancy projectile points in a basic tool assemblage. In S. Kent (Ed.), *Gender in African prehistory* (pp. 83–103). London: SAGE.

Childe, V. G. (1954). Early forms of societies. In C. Singer, E. J. Holmyard, A. R. Hall, & T. I. Williams (Eds.), *A history of technology* (Vol. 1, pp. 38–57). Oxford, UK: Clarendon Press.

Churchill, S. E. (1994). *Human upper body evolution in the Eurasian Later Pleistocene*. Doctoral thesis, University of New Mexico.

Churchill, S. E., & Formicola, V. (1997). A case of marked bilateral asymmetry in the upper limbs of an Upper Paleolithic male from Barma Grande (Liguria), Italy. *International Journal of Osteoarchaeology*, *7*(1), 18–38.

Clottes, J. (2015). *Pourquoi l'art préhistorique?* Paris: Gallimard.

Codding, B. F., Bird, R. B., & Bird, D. W. (2011). Provisioning offspring and others: Risk–energy trade-offs and gender differences in hunter–gatherer foraging strategies. *Proceedings of the Royal Society of London B: Biological Sciences*, *278*(1717), 2502–2509.

Cohen, M. N., & Bennett, S. (1998). Skeletal evidence for sex roles and gender hierarchies in prehistory. In B. D. Miller (Ed.), *Sex roles and gender hierarchies* (pp. 273–296). Cambridge, UK: Cambridge University Press.

Conkey, M. W. (1980). The identification of prehistoric hunter-gatherer aggregation sites: The case of Altamira. *Current Anthropology*, *21*(5), 609–630.

Conkey, M. W. (1985). Ritual communication, social elaboration, and the variable trajectories of Paleolithic material culture. In T. D. Price & J. A. Brown (Eds.), *Prehistoric hunter-gatherers: The emergence of cultural complexity* (pp. 299–323). New York: Academic Press.

Costin, C. L. (2013). Gender and textile production in prehistory. In D. Bolger (Ed.), *A companion to gender prehistory* (pp. 180–202). Chichester, UK: Wiley-Blackwell.

Dahlberg, F. (Ed.). (1981). *Woman the gatherer*. New Haven, CT: Yale University Press.

de Beaune, S. A. (2000). *Pour une archéologie du geste: Broyer, moudre, piler, des premiers chasseurs aux premiers agriculteurs*. Paris: CNRS Éditions.

de Beaune, S. A. (2013). François Sigaut (1940–2012), agronome et technologue. *Revue de synthèse*, *134*(1), 115–124.

de Beaune, S. A. (2016). *Qu'est-ce que la Préhistoire?* Paris: Gallimard.

de Beaune, S. A. (2019). Vertical transfer of knowledge on the *Longue durée*: Technical perspective on the development of Upper Paleolithic art in Europe. In S. Hansen, F. Klimscha, &

J. Renn (Eds.), *Proceedings of the international conference: "Prehistoric networks in the Longue durée: Palaeolithic innovations enabling the Neolithic revolution."* Berlin: Topoi.

Debout, G. (2007). Un atelier de peausserie il y a 12 000 ans à Pincevent? In S. Beyries & V. Vaté (Eds.), *Les Civilisations du Renne d'hier et d'aujourd'hui: Approches ethnohistoriques, archéologiques et anthropologiques* (pp. 439–455). Antibes: Éditions APDCA.

Delporte, H. (1990). *L'image des animaux dans l'art préhistorique.* Paris: Picard.

de Lumley, H., Stalens, H., & Brůžek, J. (1992). Nouvelles données sur la sépulture de l'homme de Menton, Grotte du Cavillon (Baoussé-Roussé, Grimaldi, Italie). *Communication à la réunion de la Société préhistorique française, séance de janvier 1992.*

Descamps, P. (1923). L'atelier chez les sauvages. *Revue de l'Institut de sociologie, 1*(3), 352–378.

Dobres, M. (1995). Gender and prehistoric technology: On the social agency of technical strategies. *World Archaeology, 27*(1), 25–49.

Donohoe, F. (2009). "Hand him over to me and I shall know very well what to do with him": The gender map and ritual native female violence in early America. In K. Williams, R. Jones, H. Van Koten, & C. Murray (Eds.), *Debating the difference: Gender, representation and self-representation.* Dundee, Scotland: University of Dundee.

Dutour, O. J. (1986). Enthesopathies (lesions of muscular insertions) as indicators of the activities of Neolithic Saharan populations. *American Journal of Physical Anthropology, 71*(2), 221–224.

Eals, M., & Silverman, I. (1994). The hunter-gatherer theory of spatial sex differences: Proximate factors mediating the female advantage in recall of object arrays. *Ethology and Sociobiology, 15*(2), 95–105.

Endicott, K. L. (2004). Gender relations in hunter-gatherer societies. In R. B. Lee & R. H. Daly (Eds.), *The Cambridge encyclopedia of hunters and gatherers* (Reprint ed., pp. 411–418). Cambridge, UK: Cambridge University Press.

Eren, M. I., Bradley, B. A., & Sampson, C. G. (2011). Middle Paleolithic skill level and the individual knapper: An experiment. *American Antiquity, 76*(2), 229–251.

Eshed, V., Gopher, A., Galili, E., & Hershkovitz, I. (2004). Musculoskeletal stress markers in Natufian hunter-gatherers and Neolithic farmers in the Levant: The upper limb. *American Journal of Physical Anthropology, 123*(4), 303–315.

Estalrrich, A., & Rosas, A. (2015). Division of labor by sex and age in Neandertals: An approach through the study of activity-related dental wear. *Journal of Human Evolution, 80*, 51–63.

Estioko-Griffin, A., & Griffin, P. B. (1981). Woman the hunter: The Agta. In F. Dahlberg (Ed.), *Woman the gatherer* (pp. 121–151). New Haven, CT: Yale University Press.

Finlay, N. (2013). Gender and lithic studies in prehistoric archaeology. In D. Bolger (Ed.), *A companion to gender prehistory* (pp. 142–160). Chichester, UK: Wiley-Blackwell.

Frayer, D. W. (2004). The dental remains from Krškany (Slovakia) and Vedrovice (Czech Republic). *Anthropologie, 42*(1), 71–103.

Gerasimov, M. M. (1958). Paleoliticheskaya stoyanka Mal'ta (Raskopki 1956–1957 gg.). *Sovetskaya Etnografiya, 3*, 28–52.

Gifford-Gonzalez, D. (1993). You can hide, but you can't run: Representation of women's work in illustrations of Palaeolithic life. *Visual Anthropology Review, 9*(1), 3–21.

Gomart, L., & Burnez-Lanotte, L. (2012). Techniques de façonnage, production céramique et identité de potiers: Une approche technologique de la céramique de style non rubané du site du Staberg à Rosmeer (Limbourg, Belgique). *Bulletin de la Société préhistorique française, 2012*, 231–250.

Goodale, J. C. (1971). *Tiwi wives: A study of the women of Melville Island, North Australia.* Seattle, WA: University of Washington Press.

Guthrie, R. D. (1984). Ethological observations from Paleolithic art. In H.-G. Bandi, W. Huber, M.-R. Sauter, & B. Sitter (Eds.), *La contribution de la zoologie et de l'éthologie à l'interprétation de l'art des peuples chasseurs préhistoriques* (pp. 35–74). Fribourg, Switzerland: Éditions Universitaires.

Guthrie, R. D. (2005). *The nature of Paleolithic art.* Chicago, IL: University of Chicago Press.

Hardy, B. L., & Moncel, M.-H. (2011). Neanderthal use of fish, mammals, birds, starchy plants and wood 125–250,000 years ago. *PLoS One, 6*(8), 1–10.

Hawkey, D. E., & Merbs, C. F. (1995). Activity induced musculoskeletal stress markers (MSK) and subsistence strategy changes among ancient Hudson Bay Eskimos. *International Journal of Osteoarchaeology, 5*(4), 324–338.

Hayden, B. (1981). Research and development in the Stone Age. *Current Anthropology, 22*(5), 519–548.

Hayden, B. (1992). Observing prehistoric women. In C. Claasen (Ed.), *Exploring gender through archaeology: Selected papers from the 1991 Boone conference* (pp. 33–47). Madison, WI: Prehistoric Press.

Hayden, B. (2001). Richman, poorman, beggarman, chief: The dynamics for social inequality. In G. M. Feinman & T. D. Price (Eds.), *Archaeology at the millennium: A sourcebook* (pp. 231–271). New York: Kluwer/Plenum.

Hayden, B. (2012). Neandertal social structure? *Oxford Journal of Archaeology, 31*(1), 1–26.

Hayden, B. (2013). *Naissance de l'inégalité. L'invention de la hiérarchie durant la préhistoire.* Paris: CNRS Éditions.

Hays-Gilpin, K. A. (2004). *Ambiguous images: Gender and rock art.* Walnut Creek, CA: Altamira Press.

Hays-Gilpin, K. A. (2013). Gender and prehistoric rock art. In D. Bolger (Ed.), *A companion to gender prehistory* (pp. 122–141). Chichester, UK: Wiley-Blackwell.

Henry, A. G., Brooks, A. S., & Piperno, D. R. (2010). Microfossils in calculus demonstrate consumption of plants and cooked foods in Neanderthal diets (Shanidar III, Iraq; Spy I and II, Belgium). *Proceedings of the National Academy of Sciences of the United States of America, 108*(2), 486–491.

Henry-Gambier, D. (2001). *La sépulture des enfants de Grimaldi (Baoussé-Roussé, Italie): Anthropologie et palethnologie funéraire des populations de la fin du Paléolithique supérieur.* Paris: Comité des Travaux historiques et scientifiques.

Henry-Gambier, D. (2005). Évolution des pratiques funéraires en Italie au Paléolithique supérieur. In D. Vialou, J. Renault-Miskovsky, & M. Patou (Eds.), *Comportements des hommes du Paléolithique moyen et supérieur en Europe: Territoires et milieux* (pp. 213–229). Liège: ERAUL 111.

Hiscock, P. (2014). Learning in lithic landscapes: A reconsideration of the hominid "toolmaking" niche. *Biological Theory, 9*(1), 27–41.

Holt, B. M. (2003). Mobility in Upper Paleolithic and Mesolithic Europe: Evidence from the lower limb. *American Journal of Physical Anthropology, 122*(3), 200–215.

Holt, B. M., & Formicola, V. (2008). Hunters of the Ice Age: The biology of Upper Paleolithic people. *American Journal of Physical Anthropology, 137*(S47), 70–99.

Iakovleva, L., & Pinçon, G. (1997). *La frise sculptée du Roc-aux-Sorciers: Angles-sur-l'Anglin (Vienne).* Paris: Comité des Travaux historiques et scientifiques.

Ingold, T. (2004). On the social relations of the hunter-gatherer band. In R. B. Lee & R. H. Daly (Eds.), *The Cambridge encyclopedia of hunters and gatherers* (Reprint ed., pp. 399–410). Cambridge, UK: Cambridge University Press.

Janny, F. (2010). Technologie lithique à Verberie: Production domestique et apprentissage. In E. B. W. Zubrow, F. Audouze, & J. G. Enloe (Eds.), *The Magdalenian household: Unraveling domesticity* (pp. 51–75). Albany, NY: State University of New York Press.

Jarašová, I., & Dočkalová, M. (2008). Dental remains from the Neolithic settlements in Moravia, Czech Republic. *Anthropology, 46*(1), 77–101.

Jarvenpa, R., & Brumbach, H. J. (2009). Fun with Dick and Jane: Ethnoarchaeology, circumpolar toolkits, and gender "inequality." *Ethnoarchaeology, 1*(1), 57–78.

Johansen, L., & Stapert, D. (2008). Stone age kids and their stones. In M. Sørensen & P. Desrosiers (Eds.), *Technology in archaeology: Proceedings of the SILA Workshop: The study of technology as a method for gaining insight into aocial and cultural aspects of prehistory* (pp. 15–39). Copenhagen: National Museum of Denmark.

Jurmain, R. (1999). *Stories from the skeleton: Behavioral reconstruction in human osteology*. London: Taylor and Francis.

Jurmain, R., Cardoso, F. A., Henderson, C., & Villotte, S. (2012). Bioarchaeology's Holy Grail: The reconstruction of activity. In A. L. Grauer (Ed.), *A companion to paleopathology* (pp. 531–542). New York: Wiley-Blackwell.

Kaplan, H., Hill, K., Lancaster, J., & Hurtado, A. M. (2000). A theory of human life history evolution: Diet, intelligence, and longevity. *Evolutionary Anthropology: Issues, News, and Reviews*, 9(4), 156–185.

Kelly, R. L. (1995). *The foraging spectrum: Diversity in hunter-gatherer lifeways*. Washington, DC: Smithsonian Institution Press.

Kent, K. P. (1971). *Introducing West African cloth*. Denver, CO: Denver Museum of Natural History.

Kent, S. (Ed.). (1996). *Cultural diversity among twentieth-century Foragers: An African perspective*. Cambridge, UK: Cambridge University Press.

Knüsel, C. J., & Smith, M. J. (2013). The osteology of conflict: What does it all mean? In C. J. Knüsel & M. J. Smith (Eds.), *The Routledge handbook of the bioarchaeology of human conflict* (pp. 656–694). London: Routledge.

Kriger, C. E. (2006). *Cloth in West African history*. Lanham, MD: Altamira Press.

Kuhn, S. L., & Stiner, M. C. (2006). What's a mother to do? The division of labor among Neandertals and modern humans in Eurasia. *Current Anthropology*, 47(6), 953–981.

Le Dosseur, G. (2005). La spécialisation des tâches au début de la néolithisation au Levant: L'exploitation des matières osseuses au Natoufien. *Techniques & Culture, 46/47*, 105–138.

Lee, R. B. (1979). *The !Kung San: Men, women and work in a foraging society*. Cambridge, UK: Cambridge University Press.

Lee, R. B. (1982). Politics, sexual and non-sexual, in an egalitarian society. In E. Leacock & R. B. Lee (Eds.), *Politics and history in band societies* (pp. 37–59). Cambridge, UK: Cambridge University Press.

Lee, R. B. (1988). Reflections on primitive communism. In T. Ingold, D. Riches, & J. Woodburn (Eds.), *Hunters and gatherers, Volume 1: History, evolution and social change* (pp. 252–268). Oxford, UK: Berg.

Lee, R. B., & Daly, R. H. (2004). Introduction: Foragers and others. In R. B. Lee & R. H. Daly (Eds.), *The Cambridge encyclopedia of hunters and gatherers* (Reprint ed., pp. 1–19). Cambridge, UK: Cambridge University Press.

Lee, R. B., & DeVore, I. (Eds.). (1968a). *Man the hunter: The first intensive survey of a single, crucial stage of human development—Man's once universal hunting way of life.* Chicago, IL: Aldine Transaction.

Lee, R. B., & DeVore, I. (1968b). Problems in the study of hunters and gatherers. In R. B. Lee & I. DeVore (Eds.), *Man the hunter: The first intensive survey of a single, crucial stage of human development—Man's once universal hunting way of life* (pp. 3–12). Chicago, IL: Aldine Transaction.

Lenclud, G. (2003). Identité et changement sont-ils compatibles? In M. Bley (Ed.), *Grand dictionnaire de la philosophie* (pp. 512–516). Paris: Éditions Larousse-CNRS Éditions.

Lenclud, G. (2013). *L'universalisme ou le pari de la raison: Anthropologie, histoire, psychologie.* Paris: EHESS-Gallimard-Le Seuil.

Leroi-Gourhan, A. (1982). *Les racines du monde: Entretiens avec Claude-Henri Rocquet.* Paris: Pierre Belfond.

Luders, E., Gaser, C., Narr, K. L., & Toga, A. W. (2009). Why sex matters: Brain size independent differences in gray matter distributions between men and women. *Journal of Neuroscience, 29*(45), 14265–14270.

Macintosh, A. A., Pinhasi, R., & Stock, J. T. (2014). Divergence in male and female manipulative behaviors with the intensification of metallurgy in Central Europe. *PLoS One, 9*(11), 1–17.

Manning, J. T. (2002). *Digit ratio: A pointer to fertility, behavior, and health.* New Brunswick, NJ: Rutgers University Press.

Marlowe, F. W. (2007). Hunting and gathering: The human sexual division of foraging labor. *Cross-Cultural Research, 41*(2), 170–195.

McCarthy, F. D., & McArthur, M. (1960). Food consumption and dietary levels of groups of Aborigines living on naturally occurring foods. In C. P. Mountford (Ed.), *Records of the Australian-American scientific expedition to Arnhem Land, 1948. Vol. II. Anthropology and nutrition* (pp. 90–135). Melbourne: Melbourne University Press.

McDonald, J. (1992). The Great Mackerel rockshelter excavation: Women in the archaeological record? *Australian Archaeology, 35*(1), 32–50.

Mellars, P. (1985). The ecological basis of social complexity in the Upper Paleolithic of Southern France. In T. D. Price & J. A. Brown (Eds.), *Prehistoric hunter-gatherers: The emergence of cultural complexity* (pp. 271–297). New York: Academic Press.

Molleson, T. I. (1994). The eloquent bones of Abu Hureyra. *Scientific American, 271*(2), 70–75.

Molleson, T. I. (2007). Bones of work at the origins of labour. In S. Hamilton, R. D. Whitehouse, & K. I. Wright (Eds.), *Archaeology and women: Ancient and modern issue* (pp. 185–198). Walnut Creek, CA: Left Coast Press.

Molnar, P. (2006). Tracing prehistoric activities: Musculoskeletal stress marker analysis of a stone-age population on the Island of Gotland in the Baltic Sea. *American Journal of Physical Anthropology, 129*(1), 12–23.

Morgan, L. J. (1877). *Ancient society, or researches in the lines of human progress, from savagery through barbarism to civilization.* LL. D. New York: Henry Holt & Co.

Murdock, G. P. (1967). Ethnographic atlas: A summary. *Ethnology, 6*(2), 109–236.

Murdock, G. P., & Provost, C. (1973). Factors in the division of labor by sex: A cross-cultural analysis. *Ethnology, 12*(2), 203–225.

Nadel, D., Weiss, E., Simchoni, O., Tsatskin, A., Danin, A., & Kislev, M. E. (2004). Stone Age hut in Israel yields world's oldest evidence of bedding. *Proceedings of the National Academy of Sciences of the United States of America, 101*(17), 6821–6826.

Olausson, D. J. (2008). Does practice make perfect? Craft expertise as a factor in aggrandizer strategies. *Journal of Archaeological Method and Theory, 15*(1), 28–50.

Olive, M., & Morgenstern, M. (2004). L'organisation de l'espace habité. In N. Pigeot (Ed.), *Les derniers Magdaléniens d'Étiolles: Perspectives culturelles et paléohistoriques, l'unité d'habitation Q31* (pp. 181–220). Paris: CNRS Éditions.

Owen, L. R. (2005). *Distorting the past: Gender and the division of labor in the European Palaeolithic*. Tübingen, Germany: Kerns Verlag.

Pelegrin, J. (2007). Réflexion sur la notion de "spécialiste" dans la taille de la pierre au Paléolithique. In R. Desbrosse & A. Thévenin (Eds.), *Des grottes de Châtelperron au musée des Antiquités nationales: Hommage à Henri Delporte* (pp. 315–318). Paris: Comité des Travaux historiques et scientifiques.

Péré-Noguès, S. (2008). Recherches autour des "marqueurs funéraires" à travers l'exemple de quelques sépultures féminines de la nécropole du Fusco (Syracuse). *Pallas: Revue d'histoire antique, 2008*, 151–171.

Petersen, E. B. (1975). A survey of the late Paleolithic and the Mesolithic of Denmark. In S. K. Kozlowsk (Ed.), *The Mesolithic in Europe* (pp. 77–129). Warsaw: University Press.

Pigeot, N. (1987). Magdaléniens d'Étiolles: Économie du débitage et organisation sociale (l'unité d'habitation U5). *Supplément À Gallia préhistoire, 25*, 9–157.

Pigeot, N. (2010). Éléments d'une organisation sociale magdalénienne à Étiolles: Du savoir-faire au statut social des personnes. In E. B. W. Zubrow, F. Audouze, & J. G. Enloe (Eds.), *The Magdalenian household: Unraveling domesticity* (pp. 198–212). Albany, NY: State University of New York Press.

Piperno, D. R., Weiss, E., Holst, I., & Nadel, D. (2004). Processing of wild cereal grains in the Upper Palaeolithic revealed by starch grain analysis. *Nature, 430*(7000), 670–673.

Price, T. D., & Brown, J. A. (Eds.). (1985). *Prehistoric hunter-gatherers: The emergence of cultural complexity*. New York: Academic Press.

Reinach, S. (1903). L'art et la magie à propos des peintures et des gravures de l'Âge du Renne. *Anthropologie, XIV*, 257–266.

Riel-Salvatore, J., & Gravel-Miguel, C. (2013). Upper Paleolithic mortuary pratices in Eurasia: A critical look at the burial record. In S. Tarlow & L. N. Stutz (Eds.), *The Oxford handbook of the archaeology of death and burial* (pp. 303–346). Oxford, UK: Oxford University Press.

Robb, J. E. (1998). The interpretation of skeletal muscle sites: A statistical approach. *International Journal of Osteoarchaeology, 8*(5), 363–377.

Ruff, C. B. (1987). Sexual dimorphism in human lower limb bone structure: Relationship to subsistence strategy and sexual division of labor. *Journal of Human Evolution, 16*(5), 391–416.

Sahlins, M. (1972). *Stone Age economics*. London: Tavistock.

Sassaman, K. E. (1998). Lithic technology and the hunter-gatherer sexual division of labour. In K. A. Hays-Gilpin & D. S. Whitley (Eds.), *Reader in gender archaeology* (pp. 159–171). London: Routledge.

Schick, K. D., & Clark, J. D. (2003). Biface technological development and variability in the Acheulean industrial complex in the Middle Awash region of the Afar Rift, Ethiopia. In M. Soressi & H. L. Dibble (Eds.), *Multiple approaches to the study of bifacial technologies* (pp. 1–29). Philadelphia, PA: University of Pennsylvania, Museum of Archaeology and Anthropology.

Service, E. R. (1966). *The hunters*. Englewood Cliffs, NJ: Prentice-Hall.

Sigaut, F. (2009). Propos contre-révolutionnaires sur le Néolithique, l'agriculture, etc. In J.-P. Demoule (Ed.), *La révolution néolithique dans le monde* (pp. 181–196). Paris: CNRS Éditions.

Sillitoe, P., & Hardy, K. (2003). Living lithics: Ethnoarchaeology in Highland Papua New Guinea. *Antiquity*, *77*(297), 555–566.

Sládek, V., Ruff, C. B., Berner, M., Holt, B. M., Niskanen, M., Schuplerová, E., & Hora, M. (2016). The impact of subsistence changes on humeral bilateral asymmetry in Terminal Pleistocene and Holocene Europe. *Journal of Human Evolution*, *92*, 37–49.

Snow, D. R. (2013). Sexual dimorphism in European Upper Paleolithic cave art. *American Antiquity*, *78*(4), 746–761.

Soffer, O. (1985). Patterns of intensification as seen from the Upper Paleolithic of the Central Russian plain. In T. D. Price & J. A. Brown (Eds.), *Prehistoric hunter-gatherers: The emergence of cultural complexity* (pp. 235–270). New York: Academic Press.

Soffer, O., Adovasio, J. M., & Hyland, D. C. (2002). Perishable technologies and invisible people: Nets, baskets, and "Venus" wear ca. 26,000 BP. In B. A. Purdy (Ed.), *Enduring records: The environmental and cultural heritage of wetlands* (pp. 233–245). Oxford, UK: Oxbow Press.

Sparacello, V. S., Villotte, S., Shackelford, L. L., & Trinkaus, E. (2016). Patterns of humeral asymmetry among Late Pleistocene humans. *Comptes rendus Palevol*, *16*(5/6), 680–689.

Sterling, K. (2014). Man the hunter, woman the gatherer? The impact of gender studies on hunter-gatherer research (a retrospective). In V. Cummings, P. Jordan, & M. Zvelebil (Eds.), *The Oxford handbook of the archaeology and anthropology of hunter-gatherers* (pp. 151–173). Oxford, UK: Oxford University Press.

Testart, A. (1982). *Les chasseurs-cueilleurs ou l'origine des inégalités*. Paris: Société d'Ethnologie.

Testart, A. (1985). *Le communisme primitif*. Paris: Éditions de la Maison des Sciences de l'Homme.

Testart, A. (1986). *Essai sur les fondements de la division sexuelle du travail chez les chasseurs-cueilleurs*. Paris: Éditions de l'EHESS.

Testart, A. (2005). *Éléments de classification des sociétés*. Paris: Errance.

Testart, A. (2012). *Avant l'histoire: L'évolution des sociétés, de Lascaux à Carnac*. Paris: Gallimard.

Thomas, A. (2014). Bioarchaeology of the Middle Neolithic: Evidence for archery among early European farmers. *American Journal of Physical Anthropology*, *154*(2), 279–290.

Tiger, L. (1969). *Men in groups*. New York: Random House.

Tostevin, G. B. (2012). *Seeing lithics: A middle-range theory for testing for cultural transmission in the Pleistocene*. Oxford, UK: Oxbow Books.

Trinkaus, E., Buzhilova, A. P., Mednikova, M. B., & Dobrovolskaya, M. V. (2014). *The people of Sunghir: Burials, bodies and behavior in the Earlier Upper Paleolithic*. New York: Oxford University Press.

Trinkaus, E., Buzhilova, A. P., Mednikova, M. B., & Dobrovolskaya, M. V. (2015). The age of the Sunghir Upper Paleolithic human burials. *Anthropologie (Brno)*, *53*(1–2), 221–231.

Turnbull, C. M. (1965). *Wayward servants: The two worlds of the African Pygmies*. Garden City, NY: Natural History Press.

Vanhaeren, M. (2002). *Les fonctions de la parure au Paléolithique supérieur: De l'individu à l'unité culturelle*. Doctoral thesis, Université Bordeaux I, Bordeaux, France.

Vanhaeren, M., & d'Errico, F. (2001). La parure de l'enfant de La Madeleine (fouilles Peyrony): Un nouveau regard sur l'enfance au Paléolithique supérieur. *Paléo*, *13*, 201–237.

Vanhaeren, M., & d'Errico, F. (2003). Le mobilier funéraire de la Dame de Saint-Germain-la-Rivière (Gironde) et l'origine paléolithique des inégalités. *Paléo, 15*, 195–238.

Vieugué, J. (2012). Spécialisation fonctionnelle des premières productions céramiques dans les Balkans (6100–5500 av. J.-C.). *Bulletin de la Société préhistorique française, 2012*, 251–265.

Villotte, S., Brůžek, J., & Henry-Gambier, D. (2011). Caractéristiques biologiques des sujets adultes gravettiens: Révision de l'âge au décès et du sexe. In N. Goutas, L. Klaric, D. Pesesse, & P. Guillermin (Eds.), *À la recherche des identités gravettiennes: Actualités, questionnements et perspectives: Actes de la table ronde sur le Gravettien en France et dans les pays limitrophes* (pp. 209–216). Paris: Mémoires de la Société préhistorique française.

Villotte, S., Churchill, S. E., Dutour, O. J., & Henry-Gambier, D. (2010). Subsistence activities and the sexual division of labor in the European Upper Paleolithic and Mesolithic: Evidence from upper limb enthesopathies. *Journal of Human Evolution, 59*(1), 35–43.

Villotte, S., & Knüsel, C. J. (2014). "I sing of arms and of a man . . .": Medial epicondylosis and the sexual division of labour in prehistoric Europe. *Journal of Archaeological Science, 43*, 168–174.

Wadley, L. (1998). The invisible meat providers: Women in the Stone Age of South Africa. In S. Kent (Ed.), *Gender in African prehistory* (pp. 69–81). London: SAGE.

Waguespack, N. M. (2005). The organization of male and female labor in foraging societies: Implications for Early Paleoindian archaeology. *American Anthropologist, 107*(4), 666–676.

Washburn, S. L., & Lancaster, C. S. (1968). The evolution of hunting. In R. B. Lee & I. DeVore (Eds.), *Man the hunter: The first intensive survey of a single, crucial stage of human development—Man's once universal hunting way of life* (pp. 293–303). Chicago, IL: Aldine Transaction.

Weedman, K. (2005). Gender and stone tools: An ethnographic study of the Konso and Gamo hideworkers of southern Ethiopia. In L. Frink & K. Weedman (Eds.), *Gender and hide production* (pp. 175–196). Walnut Creek, CA: Altamira Press.

Weedman, K. (2006). Gender and ethnoarchaeology. In S. M. Nelson (Ed.), *Handbook of gender archaeology* (pp. 247–294). Walnut Creek, CA: Altamira Press.

Weiss, E., Kislev, M. E., Simchoni, O., Nadel, D., & Tschauner, H. (2008). Plant-food preparation area on an Upper Paleolithic brush hut floor at Ohalo II, Israel. *Journal of Archaeological Science, 35*(7), 2400–2414.

White, R. (1999). Intégrer la complexité sociale et opérationnelle: La construction matérielle de l'identité sociale à Sungir. In H. Camps-Fabrer & M. Julien (Eds.), *Préhistoire d'os: Recueil d'études sur l'industrie osseuse préhistorique* (pp. 319–331). Aix-en-Provence: Publications de l'Université de Provence.

White, R. (2006). Looking for biological meaning in cave art. *American Scientist, 94*(4), 371–372.

Woodburn, J. (1982). Egaliarian societies. *Man, New Series, 17*(3), 431–451.

Zihlman, A. L. (1998). Woman the gatherer: The role of women in early hominid evolution. In K. A. Hays-Gilpin & D. S. Whitley (Eds.), *Reader in gender archaeology* (pp. 91–105). London: Routledge.

Zihlman, A. L. (2013). Engendering human evolution. In D. Bolger (Ed.), *A companion to gender prehistory* (pp. 23–44). Chichester, UK: Wiley-Blackwell.

19

THE ENHANCED WORKING MEMORY MODEL
ITS ORIGIN AND DEVELOPMENT

Frederick L. Coolidge

INTRODUCTION

Nearly two decades ago, I was pleasantly engaged reading an article in *Scientific American* (Tattersall, 2000), which claimed that about 500 thousand years ago (Kya), there may have been five or six species of human types living nearby one another in Africa. About 100 Kya, Tattersall claimed, there were only two remaining types, *Homo sapiens* and Neandertals. Finally, he stated that about 30 Kya, there was only us, as Neandertals had gone extinct. At the end of the article, Tattersall speculated that the cause of the Neandertal extinction was their lack of language. *Wait a minute,* I thought: If modern vervet monkeys teach their young at least three different sounds for three different kinds of predators, which elicits three different kinds of behavior (Price et al., 2015; Seyfarth, Cheney, & Marler, 1980), then surely Neandertals had at least that capability. Interestingly, modern putty-nosed monkeys are thought to have two different warning sounds for two different predators (eagles and leopards), which also elicit two distinct behaviors, either climbing up trees away from leopards or down into bushes for eagles. But even more interesting is that if the two sounds are combined, it seems to signal that the group should move to another location (Schlenker, Chemla, Arnold, & Zuberbühler, 2016). While this combinatoriality of sounds is very rare in non-human primate communication, it nonetheless made the claim that Neandertals were *alinguistic,* especially with their much larger brains, all the more improbable. However, if monkeys had different sounds for different predators that elicited different behaviors, then surely Neandertals had language or at the very least communicative abilities. But this raises a dilemma: What *was* the cause of the Neandertal extinction? Though I had degrees in psychology and a postdoctoral fellowship in clinical neuropsychology, I had neither in anthropology or archaeology. I would need to collaborate with someone who had the training I lacked, and that someone would turn out to be Thomas Wynn. And thus was my introduction to the exciting and still relatively new field that is cognitive archaeology.

EXECUTIVE FUNCTIONS

In 2000, Coolidge, Thede, and Young investigated the heritability of executive functions of the frontal lobes in a sample of twin children and adolescents. Executive functions consist of the ability to make decisions and plan, strategize, organize, inhibit, and temporally sequence events. They were brought to the world's attention by Russian neuropsychologist Alexander Luria in 1962. In the 2000 study, executive functions were found to be a highly heritable trait (approximately 77% of the variance associated with them can be attributed to genetic influences) with a polygenic basis (additive genetic influence). This information applies to Neandertals as follows: If they had successfully survived for 200,000 years or longer in Europe and Asia, at least until *H. sapiens* entered those same regions about 40 Kya, then they had to have possessed a reasonable array of executive functions, including decision-making, planning, organizing, inhibition, and sequencing. Further, even the popular literature said that there was little evidence that warfare between the two human types had caused the Neandertal extinction. The idea that *H. sapiens* had perhaps simply out-competed Neandertals suggested that perhaps there were differences in the executive functioning of the two human species. Executive functioning heritability, in turn, implied the possibility that a genetic event had occurred in *H. sapiens*—but not Neandertals—to enhance their executive functions. That is, perhaps a beneficent genetic mutation had occurred in the many genes responsible for executive functions, enhancing them well beyond the standard functional array. In this scenario, *H. sapiens* would survive because of an enhanced ability to reason, while Neandertals would go extinct. An enhanced ability to reason might have directly meant that *H. sapiens* was able to extract more food and other resources than Neandertals from the same environments. Thus, neither war between the two human types nor a lack of language in Neandertals would have been required, rather, simply an enhanced ability to extract environmental resources in *H. sapiens*.

In a fortuitous situation, our Anthropology faculty contained someone well versed in applying cognitive models to archaeology: Thomas Wynn. In one of his classic papers, "An Ape's View of the Oldowan" (1989), he and primatologist William McGrew argued that the cognitive prerequisites of Oldowan stone tools were essentially the same as the tool-fashioning culture of extant chimpanzees. His response to the idea that executive functions may have been at the root of the Neandertal extinction was that the idea was interesting and akin to the same reasoning used by Steven Mithen in his 1996 book *The Prehistory of the Mind*, except that Mithen had used the term *cognitive fluidity* (but not from the cognitive literature) and a cathedral metaphor for the workings of the human mind. We agreed that he would supply archaeological evidence for executive functions, while I would supply psychological evidence for them gleaned from the clinical neuropsychological literature (research that was primarily based on brain-damaged patients), and we would co-publish the results. Wynn liked the idea that the contentious issues of language and symbolic thinking could be circumvented in the Neandertal extinction debate and that the concept of executive functions was firmly rooted in the cognitive sciences. Thus, in 2001 in the *Cambridge Archaeological Journal*, we published our first article together, "Executive Functions of the Frontal Lobes and the Evolutionary Ascendancy of *Homo sapiens*." In it, we examined the archaeological evidence for three major executive functions: temporal sequencing, inhibition, and planning/organization.

The executive function of temporal sequencing is well known to require Broca's area in the prefrontal cortex, as well as cerebellar input and other brain regions. Cognitively, it requires a complex linkage of steps, with each preceding step successfully completed before a subsequent step. Neolithic examples abound: loom-weaving, ceramic manufacturing, and metallurgy (e.g., iron and tin production), with the latter probably heralding the end of the Neolithic period. Identifying Paleolithic examples of sequential memory was much harder, and even sophisticated stone tool techniques such as Levallois are not *prima facie* evidence for temporal sequencing. Wynn chose barbed points from Katanda, the Democratic Republic of Congo, that date to about 88–77 Kya, as evidence of a final product requiring a tight set of sequenced actions. Bow-and-arrow technology dating to about the same time period was also a task requiring a complex sequence of actions (e.g., Coolidge, Haidle, Lombard, & Wynn, 2016).

With regard to tasks of inhibition, evidence for agriculture seemed to be a prime example for the delay of gratification required for cultivating, planting, and storing agricultural products successfully. Remote game traps would also be evidence for inhibition, but we noted that there was little or no evidence from Middle Paleolithic foraging activities on the part of either Neandertals or *H. sapiens*, and remote traps do not appear to predate agricultural activities.

Finally, we addressed the archaeological evidence for organizing and planning. Here we cited the colonization of the Sahul, which took place somewhere between 40 and 60 Kya, as evidence of not only sophisticated organization and planning but also temporal sequencing, as preparing a boat and people for a successful voyage to a land mass that could not be seen from the point of origin would absolutely require modern executive functions (or, conversely but extremely unlikely, simply a huge amount of luck). Noble and Davidson (1991) noted that such a journey required modern behavior and modern language, and Klein (2000) also claimed that it would have required sophisticated technologies, although none highlighted the cognitive perquisites nor interpreted them as evidence of essentially fully modern executive functions.

In summary, we argued for enhanced executive functions in *H. sapiens*, but those that came relatively late in the archaeological record. We also noted that Mithen (1996) had earlier proposed a "big bang" of human culture about 60–30 Kya when the full accessibility among his proposed mental modules arose as function of some complex social and intellectual interactions. Klein and Edgar (2002) also viewed the Upper Paleolithic period as a cultural explosion and reasoned that it might have been due to a genetic event in *H. sapiens*. And so our idea had some earlier support, both for the idea of a genetic event giving *H. sapiens* fully modern thinking and for a cultural explosion that might have occurred as a function of that event.

Working Memory

A generic definition of working memory might be the amount of information one can hold or process in one's mind despite interference. Originally, the term was virtually synonymous with short-term memory, and the latter was almost always tested in a verbal acoustic paradigm, in which subjects would be asked to repeat increasing amounts of verbal information such as strings of numbers. In 1974, experimental psychologists Alan Baddeley and Graham Hitch published their classic paper, "Working Memory," where they envisioned working memory as a multicomponent model that would encompass a short-term verbal store, a visuospatial component, and, importantly, an overarching component (the central executive) that attended to the information provided by the other two subsystems. We were attracted to this more circumscribed cognitive model of working memory, primarily because we saw that the concept of executive functions we used from the clinical neuropsychological literature was nearly synonymous with Baddeley and Hitch's central executive's functions, especially its attention to tasks and goals. In addition, their two subsystems, the phonological loop and visuospatial sketchpad, provided additional heuristic value for interpreting the archaeological record.

In 2000, Baddeley added a fourth component, the episodic buffer, proposed to serve as a multimodal integrative memory store in the direct service of the central executive. He hypothesized that the episodic buffer temporarily blended information from the two subsystems, and accessed relevant information from long-term memory stores in order to complete short- and long-term goals. An analogous concept, episodic memory, was proposed by experimental psychologist Endel Tulving in 1972. Baddeley viewed the concepts as roughly similar, but emphasized that his conception was not simply the recall of episodes from long-term explicit memory, but rather an important multimodal integrative mechanism that could create "an ice-hockey-playing elephant" (Baddeley, 2001, p. 857). In reality, Tulving's original concept of episodic memory was more akin to a memory system that intentionally recalled previous experiences from long-term memory, and he clearly differentiated this ability from semantic memory, the stored knowledge of facts and meanings. Later, however, Tulving (2002) imbued episodic memory with far more complexity and power, and he elaborated on its relationships to the concept of self, the subjectiveness of time, and conscious awareness of this subjectivity, which he labeled *autonoetic consciousness*.

Further, just as the concept of executive functions from the clinical neuropsychological literature may have heralded Baddeley's central executive, and Tulving's concept of episodic memory may have preceded Baddeley's episodic buffer, so too Crowder and Morton's (1969) concept of precategorical acoustic storage (PAS) predated Baddeley's phonological loop subsystem. PAS was viewed as an initial temporary memory store for acoustic information, and it was labeled *precategorical* because the information was held in an acoustic store while "meanings" were sought for the sounds in long-term memory. Baddeley further differentiated phonological storage from PAS by proposing that the phonological loop was not only a temporary storage place for sounds but also contained an articulatory rehearsal system, whereby the sounds could be consciously and willfully repeated, either vocally or subvocally, until they gained more permanent storage in explicit long-term memory.

Heritability of Working Memory

As noted earlier, Coolidge and colleagues (2000) found a high heritability (77%) and polygenic basis for executive functions in a twin sample, where an array of executive functions were measured by parental reports. A subsequent study, in which executive functions in twins were measured by laboratory assessments, directly supported these findings. Friedman and colleagues (2008) found all of their measures of executive functions were highly intercorrelated because they were all influenced by a highly heritable (99%) common factor, although the individual measures varied in heritability. Other genetic studies supported the high heritability of the two major subsystems of working memory. Thus, the working memory model and the genetic heritability of working memory nicely fit with our suspicion that some relatively recent (200 Kya, more or less) genetic event in working memory, working memory capacity, or one of its components may have given *H. sapiens* essentially modern thinking and, in all likelihood, fully modern language. We labeled the result of that event *enhanced working memory* (Coolidge & Wynn, 2005).

Regardless of the specific nature of enhanced working memory, it appears that something extraordinary happened in cognitive abilities, attested by dramatic change in the archaeological record more recently than 100 Kya. Although there are cultural gradualists (e.g., McBrearty & Brooks, 2000), others (e.g., Klein & Edgar, 2002; Mithen, 1996) proposed that there was a cultural explosion around 50 Kya, and Mithen and Klein and Edgar hypothesized that it might have been due to a genetic mutation. Neither, however, was more specific at that time about the nature of the mutation, nor its specific cognitive or neuronal sequelae. However, many archaeologists recognized that personal ornamentation, depictive and sometimes mystical cave art, highly ritualized burials, bow-and-arrow technology, and enigmatic figurines like the Hohlenstein-Stadel lion-man began appearing about 80 or so Kya. In our (Wynn & Coolidge) 2010 article, we reasoned then that fully modern minds (with fully modern syntactical language) must have been in place by at least 40 Kya or earlier.

"What I Will Not Address," said Thomas Wynn

Immediately after our 2001 publication, Tom said there were two contentious issues he preferred to avoid: Neandertals and language. At the time, I was "sitting in" on a course Tom taught on *H. erectus*: specifically, Nariokotome (KNM-WT 15000), also known as Turkana Boy. The skeleton was discovered in 1984 by Kamoya Kimeu, a member of a research team at Nariokotome, Kenya, near Lake Turkana with Richard Leakey. Amazingly, this skeleton, estimated to be about 1.6 million years old, was nearly complete. It had been identified as that of a boy about 8 to 11 years old who may have died from a spinal injury. He had already attained a height of 160 centimeters (63 inches) tall, and like all *H. erectus*, he would have been even taller as an adult (perhaps 168 to 170 centimeters [64 to 67 inches]). Our text in the course was written by Richard Leakey and anthropologist Alan Walker, both of whom devoted major parts of their careers to studying Nariokotome. I admit my mind drifted sometimes as I was learning more detail than I wanted to know about skeletons, skeletal growth, teeth, and other physical qualities. At one point, I distinctly remember coming into full focus when Tom said, "And then hominins started living on the ground." My immediate thought was, *So they were living and* sleeping *on the ground*. And so I raised my hand to confirm this suspicion. Tom said, "Yes." So I asked, "Were there any archaeological changes at that time?" "Indeed," Tom said, "Their stone tools became bifacial, symmetrical, and overall much more sophisticated than previous stone tools." Since I had completed both my MA thesis and PhD dissertation on learning during sleep and the differential effects of sleep stages on memory, my mind churned with possibilities. We then combed through the anthropological and psychological literature for the possibility that sleeping on the ground in large groups (for protection) might produce quantitatively and qualitatively different and better sleep than sleeping in a nest in a tree as the australopithecines, like Lucy, had done.

We wrote a rather long treatise on the effects of the tree-to-ground sleep transition in early *Homo*, with a tangential section about how the major cognitive leap from the australopithecines to *H. erectus* may have ultimately influenced the cognitive abilities of later *Homo* species like *H. sapiens* and Neandertals. We reasoned that sleeping on the ground may have allowed for a single integral sleep period with increased periods of rapid eye movement (REM) sleep and perhaps even more slow-wave sleep. We proposed that the content of these REM periods may have allowed for threat simulation and social rehearsal, making these hominids more prepared for subsequent waking threats and interactions. Dreams are associated with creativity and innovation by a myriad of evidence: literature (Coleridge's poem "Kubla Khan; or, A Vision in a Dream: A Fragment"), scientific discoveries (Mendeleev's periodic table), mathematical theorems (Ramanujan's work in mathematical theory), and musical compositions (Tartini's "Trillo del Diavolo"). Based on this evidence, we speculated that the design for these new, more complex stone tools could have come to an expert stone knapper in a dream. And finally, experimental evidence supported the idea that procedural memories like stone knapping and visuospatial memories like wayfinding are not only consolidated during REM and slow-wave sleep but are also enhanced during sleep (Coolidge & Wynn, 2006). The only objection Tom had was the section on Neandertals. He suggested (to my chagrin) that we eliminate it. "But all is not lost," he assured me. "I think it could be the basis for another paper."

THE EXPERT NEANDERTAL MIND

Easily one of the greatest mysteries in anthropology is the extinction of Neandertals. We began our (Wynn & Coolidge) 2004 article, "The Expert Neandertal Mind," with four basic premises:

(1) Modern humans replaced Neandertals, which meant that after the arrival of anatomically and behaviorally modern *H. sapiens* in Europe about 45 Kya, Neandertals went extinct. Yes, many modern humans carry some Neandertal genes (about 1% to 4%), but some modern human populations have none. So it is fair to say Neandertals went extinct as a distinct population, but it is not correct to say they live on distinctly within us. Some relatively small amount of unique DNA is carried by some modern humans, but not all.

(2) We posited a cognitive difference between us and Neandertals, with those cognitive differences based on the gross neuroanatomical differences between the respective brains, something assessed from skulls and endocasts. Subsequent studies (e.g., Bastir et al., 2011; Bruner, 2004, 2010; Bruner & Iriki, 2016; Weaver, 2005, 2010) have substantiated these differences (e.g., parietal expansion, and larger cerebella and olfactory bulbs in *H. sapiens*). While there are many cultural similarities between the two human types, especially in their hunting and gathering activities, it appears that Neandertals never fully acquired the trappings of modern *H. sapiens* culture. We proposed that the reason was a cognitive difference—not a dramatic one, perhaps, but a small and very significant difference in thinking and reasoning. What were these trappings? Cave art, figurines, highly ritualized burials, and a greater variety of stone and bone tools. Neandertal apologists (e.g., Rodríguez-Vidal et al., 2014) counter-argue that a single and simple tic-tac-toe etching in a cave in Gibraltar is the cognitive equivalent of the cave paintings in Lascaux or Chauvet, but clearly, it is not. They argue that the

etching is evidence for symbolic and abstract reasoning, but again, that reasoning seems a bit hasty to us. Others argue that the Neandertal flower burial in Shanidar Cave in Iraq is the cognitive equivalent of the elaborate and highly ritualized burial in Sungir, Russia, but once more, it is not. Yes, Neandertals may have intentionally buried their dead, but the unquestionable qualitative differences between just these examples imply to us that the fundamental difference between Neandertal and *H. sapiens* cultures was cognitive.

(3) This cognitive difference need not have been dramatic. Neandertals may have had nearly the full complement of cognitive faculties of modern humans. Undoubtedly, they could reason, make decisions, create goals and attain them, and dream. We avoided overemphasizing differences, but we also avoided overemphasizing the two human types' similarities. Instead, we posited a small cognitive difference with profound long-term consequences. This is, as we elucidated later (Wynn, Overmann, & Coolidge, 2016), the excluded middle position between "the two species were indistinguishable" (e.g., Villa & Roebroeks, 2014) and "we were completely different" (a strawman position first articulated in the late nineteenth century).

(4) Modern thinking is largely based on cognitive abilities that evolved long ago in hominins, primates, and even mammals. Thus, we postulated that this small positive cognitive difference between Neandertals and *H. sapiens* was the result of some genetic event that occurred within the last 200,000 years or so. We also have stated repeatedly that it was not a *single* dominant gene. We do think it was a genetic or epigenetic event, but most likely a mutated but beneficent gene that joined many others (polygenic influence) already controlling some cognitive function, enhancing that function.

The Nature of the Cognitive Difference: Enhanced Working Memory

In 2007, Baddeley stated he was "intrigued" by our speculations that "possession of working memory was the crucial advantage held by *H. sapiens* over Neanderthal man" (Baddeley, 2007, p. 336). However, he also said that he found it difficult to see how our hypothesis could be tested. It is important to note at the outset of this discussion that we did *not* limit the possession of working memory to *H. sapiens*. How could we? If extant primates make decisions (executive functions), store sounds for processing (phonological store), process and store visual images (visuospatial sketchpad), and recall past scenarios with a what, where, and when signature (episodic buffer/memory), then Neandertals without a doubt possessed all of the multiple components of Baddeley's working memory model. Again, what we proposed is that *H. sapiens* in the past 200,000 years or more recently possessed *enhanced* working memory (Wynn & Coolidge, 2010). Baddeley is correct, however, that standard laboratory measures of working memory capacity cannot be applied to prehistoric people, so we needed to detect evidence for enhanced working memory in the archaeological record, and perhaps, even assess differences in varying levels of working memory capacity.

The Phonological Loop

Admittedly, we have remained somewhat hesitant to declare the exact nature of the enhancement to working memory, although we have speculated on a number of possibilities. Baddeley and his colleagues proposed what we think is a strong candidate: They recognized the evolutionary advantage of the phonological loop as a language acquisition device (e.g., Baddeley, Gathercole, & Papagno, 1998). Further, Baddeley and Logie (1999) called the phonological loop a "major bottleneck for language comprehension" (p. 41). Interestingly, they did not see its primary purpose as matching sounds to known meanings in long-term memory, although that must be a critical evolutionary component, but they saw its chief evolutionary function as a mechanism for learning new words. And why would the phonological loop be viewed as perhaps the strongest candidate for the specific nature of enhanced working memory? As we wrote in the second edition of our book *The Rise of* Homo sapiens (Coolidge & Wynn, 2018), modern human brains were indeed profoundly influenced by the development of the mammalian lineage about 200 million years ago (Mya). The overarching surface of mammals' brains, the cortex, was much more highly expanded, gyrified, and convoluted than reptilian or fish brains, to the extent that it was labeled the *neocortex*. The elaborated and more complex mammalian neocortex gave mammals much greater flexibility in responding to their environments and more complicated behavioral repertoires than other animals. However, we noted that it was the primate lineage, which developed about 65 Mya from the mammalian lineage, that *most* profoundly influenced the modern human brain and its functions.

The first primates were small, nocturnal (probably to avoid daytime predation), tree-dwelling animals who competed with reptiles, birds, and other predatory animals for foodstuffs like grass, leaves, and fruit. Human brains take up only about 2% of a body's mass but require about 20–25% of the total caloric intake to function properly; for this reason, brains are considered an expensive metabolic tissue. As fruit has far more calories than grass or leaves, it would allow for an increase in brain size relative to body size over grass or leaves. So, in order to compete with other animals for nutritious fruits, the first primates had to coordinate their foraging activities with one another, and they did so through vocalizations. The result was a social brain, and its neuronal foundations were naturally selected for in order to use sounds to communicate and coordinate group activities. Aboitiz, García, Bosman, and Brunetti (2006) noted that phonological storage capacity can be phylogenetically tracked to earlier homologues in hominid evolution and to current primate brain systems. They believed that an expanding memory system allowed more complex memories representing multiple items to be combinatorially manipulated, permitting "the maintenance of a 'state of mind' that captures attentional and memory resources" (p. 51). Aboitiz, Aboitiz, and García (2010) hypothesized that an increase in acoustic short-term memory capacity allowed for increasing more complex utterances, which in turn made for stronger social bonds and facilitated more complex messaging among group members. Further, they proposed that this increased capacity to maintain sounds for processing allowed the development of long-distance dependencies within a complex vocalization, such that linguistic elements could interact to create additional meanings (i.e., recursive linguistic structuring).

Chomsky, Recursion, and the Pragmatics of Speech

Chomsky and his colleagues (e.g., Bolhuis, Tattersall, Chomsky, & Berwick, 2014; Fitch, Hauser, & Chomsky, 2005; Hauser, Chomsky, & Fitch, 2002) declared that (1) the hallmark of modern language is *recursion*; (2) fully modern language appeared suddenly and completely about 100 Kya in one person because of a single gene, and then spread rapidly; (3) animal communication is useless in studying human language, and so are vocalization and speech studies because they are not synonymous with language; (4) language was not subject to natural selection; and (5) the original purpose of language was *not* communication. Their second claim has no support from genetics research whatsoever, their third claim is highly contentious and a strong majority of linguists do not support the position, and their fourth and fifth claims remain unelaborated and unexplained in any of Chomsky or his colleagues' writings to date, and thus they resemble the arguments of creationists. However, their first claim is an intriguing one, even if it lacks any depth of explanation in their subsequent writings.

We wrote an extensive article on the nature of recursion, "Recursion: what it is, who has it, and how did it evolve?" (Coolidge, Overmann, & Wynn, 2011). In it, we acknowledged the computational power of recursion, and we pondered what we had written about much earlier (in e.g., Wynn & Coolidge, 2006), the purposes of speech (i.e., its pragmatics). Although complex utterances can be expressed though recursion, we hypothesized that it was not their *complexity* that was evolutionarily adaptive per se but, rather, the evolutionary value of the thoughts and their ultimate consequences that was released through recursion. Of course, this presupposes that recursion is defined as embedding a phrase within a phrase, a definition we identified in our 2010 paper as "strong." The weaker and less common definition of recursion is that it is both an instruction and an output, giving rise to combinations of discrete smaller units into increasingly larger units. According to the weaker definition, if recursion calls an instruction into the episodic buffer of working memory and the instruction is presented first, then no great burden is placed on working memory. Alternatively, if the instruction occurs at the end of phrase, there is an added burden on working memory. However, we were interested in the evolutionary implications of the stronger definition (even though we had argued against it).

In 2006, we proposed that recursion may have affected the range of speech acts or the pragmatics of speech. A *speech act* refers to the act that is done or performed by speaking (e.g., Adams, 2002). There is far from a general consensus on a single taxonomy assessing the intent of communication, although speech act analyses typically list exclamatives (shouts of pain, pleasure, or surprise), imperatives (commands), declaratives (statements of fact, greetings, denials), interrogatives (questions or requests), and subjunctives (expressions of subjective statements, such as wishes, possibilities, hypotheticals, and counterfactuals). Interestingly, the first four can be expressed by simple, even single-syllable utterances (at least in English): exclamations ("Ouch!"), imperatives ("Move!"), declaratives ("Nice!"), and interrogatives ("Where?"). Thus, recursion does not appear to be a necessary precondition for these speech acts. However, the subjunctive mode of speech, "what if" thinking, does appear to require recursion or longer canonical utterances. Thus, recursion may have allowed the formation and release of subjunctive thinking, but recursion, in turn, would have required an expanded phonological loop. It is also interesting to note that the first four speech acts do not require complex thoughts about what the receiver of the speech is thinking. Properly estimating the reactions, thoughts, or feelings of the receiver of speech by the speaker is called *theory of mind* (e.g., Baron-Cohen, 2000; Kinderman, Dunbar, & Bentall, 1998).

The Cognitive Prerequisites for Diplomatic Speech

Ambrose (2010) noted that Middle Stone Age and Middle Paleolithic humans were not overly effective hunters, rarely left the boundaries of their home range, did not schedule an effective strategic use of land, and may have suffered intense interpersonal violence. This confluence of behaviors suggested to Ambrose a limited capacity to plan, as well as limited and problematic interactions with other groups. He hypothesized that the neurological capacities for planning could have been in place, but that social and linguistic innovations, prompted by a catastrophic environmental change like the explosion of the Toba volcano in Indonesia about 77 Kya, may have been necessary to release this capacity. Ambrose and others have hypothesized that the Toba explosion resulted in a human genetic bottleneck in Africa, reducing the effective population (those reproducing) to as few as 2,000 ancestors of *all* extant *H. sapiens*. Regardless of the cause of the population bottleneck, a volcanic winter, about 6 years in length, occurred about the time of the Toba explosion, and this winter prompted a sudden ice age in the northern hemisphere that lasted about 1,800 years (Ambrose, 2003).

Ambrose (2010) noted that some behavioral sequelae of this bottleneck appear to have been a transition from restricted landscapes and limited interactions with others to expanded geographic and social landscapes, particularly in the form of material exchanges, trading, and other forms of cooperation. Ambrose believed that one of these innovations may have been a "language of diplomacy." This was based on the concept of *indirect speech*, by Pinker, Nowak, and Lee (2008), who noted that people frequently insinuate their intentions. Pinker and colleagues also proposed that human conversations are often the result of a target person anticipating what another person in the interaction may be thinking about the interaction, what the other person thinks that the target person is thinking about the interaction, what other humans think about the interaction and relationship, and so on. Ambrose suggested that diplomatic speech may have facilitated the mutual trust required to maintain reciprocal cooperation, presumably for interpersonal interactions within a group and/or interactions with others in distant groups. Obviously, such types of thinking require an ability to read the intentions of others (theory of mind). Ambrose further speculated that humans using diplomatic speech, which probably fostered mutually rewarding relationships, would be able to survive in new, harsher, and more unpredictable environments.

In 2009, we again addressed a purported limitation in Neandertal cognition, at an evolution conference at the Wellcome Trust Genome Campus, Hinxton, UK. We speculated on the cognitive prerequisites of diplomatic speech, and we hypothesized that one possibility for Neandertal extinction may have been the inability to speak indirectly. Social dominance hierarchies appear to be an endemic and pervasive characteristic of most primate social interactions, and it is highly probable that dominance hierarchies have been characteristic throughout primate evolution. Dunbar (2004) notes that physical grooming fosters social bonding and reduces interpersonal aggression in non-human primates, but that the evolution of the genus *Homo* required another grooming mechanism: language. According to Dunbar, language serves the same initial purposes as physical grooming, but could be conducted on much larger scales, thereby allowing group size to expand. We reasoned that in a highly structured social dominance hierarchy, reliance on direct forms of speech such as imperatives might suffice. However, even within a single group, indirect forms of speech might have had survival value. "Get out of my way!" might suffice for those higher in the dominance hierarchy, but for those of lower status, "Would you mind please moving over" might engender greater cooperation, trust, formation of alliances, and positive feelings, and perhaps, even greater reciprocal cooperation in future interactions. Furthermore, successful social interactions between or among groups would have undoubtedly benefitted from indirect speech or a language of diplomacy.

We reasoned that diplomatic speech would require several cognitive prerequisites, the first of which was an expanded phonological loop capacity. As noted earlier, phonological storage has been thought to play a critical role in language production and comprehension. Adults who have greater phonological storage capacity have also been found to score better on verbal tests of intelligence and verbal fluency, and better on retroactive and proactive interference tasks (Kane & Engle, 2002). Children with greater phonological storage capacity have larger vocabularies, produce longer utterances, and demonstrate a greater range of syntactic construction (Adams & Gathercole, 2000). Our second cognitive prerequisite was recursive thinking and recursive speech (e.g., "Violet said that Sunny wants to escape"). However, the *sine qua non* cognitive prerequisite for recursion had to have been expanded phonological storage capacity. Modern speech and thought are replete with recursive phrases, and theoretically, modern language is capable of infinite generative recursion (e.g., "Klaus said that Violet said that Sunny wants to escape"). Linguist Eric Reuland (2010) has emphasized the combinability property of recursion—that is, two expressions can be merged, and intrinsically, one of them may then determine the further combinatorial properties of the result. In this manner, combinability is a core characteristic of recursion, and its selective advantage over time is the ability to make more complex computations. This view of recursion, consisting of combinatorially manipulated complex memories, is also supported by Aboitiz and colleagues (2006), who posit that recursion is highly dependent on sufficient phonological storage capacity.

Recursion, however, is more than just combinability, and the actual structure of embedded phrases can make differential demands on working memory and phonological storage capacity. For example, direct speech (particularly imperatives) often places instructions at the beginning of a statement, which places minimal burdens on working memory. However, if the instruction occurs at the end of a phrase or a combination of phrases, there is then an added demand on working memory. We argued that indirect speech sometimes places the instruction at the end of a statement, or the full intention of the utterance is not known until its end (e.g., "Would you mind terribly, if it is not particularly inconvenient, moving over?"). Aboitiz and colleagues (2006) have also noted that one important aspect of syntax is hierarchical phrase construction, a fundamental property of recursion. Recursivity permits the manipulation of elements in the hierarchy, allowing, for example, embedded phrases to be moved into varying positions within an utterance. Importantly for our present argument, Aboitiz and colleagues note that changing the canonical order in a recursive utterance also can transform the utterance into a passive form, and we hypothesized that one form of this passivity would be the formation of more polite, indirect, and diplomatic forms of speech from more direct forms. Aboitiz and colleagues also claim that the capacity for linguistic recursion could *only* have originated in an expanding working memory system, highly dependent on short-term memory mechanisms that allow speakers and listeners to keep track of temporal and long-distance dependencies among embedded phrases and syntactic movements. Thus, a language of diplomacy can require the transformation of embedded phrases into non-threatening forms, which, as we and others have noted, is a core property of recursion.

Our third cognitive prerequisite for a language of diplomacy was theory of mind. Deficits in theory of mind have long been noted in psychological disorders such as autism and schizophrenia, and periodically the deficits have been purported to be causal and a core deficit in some disorders, like depression, anxiety, paranoia, and delusional states (e.g., Kinderman et al., 1998). Theory of mind has also been described as consisting of four independent skills: detection of the intentions of others, detection of eye-direction, shared attention, and a final component called the theory of mind module. The final component, whose onset in humans is thought to develop by the age of 4, contains a complex set of social-cognitive rules, and combined with the other three components, creates the full-fledged, adult-like theory of mind (e.g., Baron-Cohen, 1997; Gerrans, 2002).

There have been numerous speculations (and much debate) that theory of mind is based on the mirror neuron system (Borg, 2007, 2013; Hickok, 2014). Mirror neurons were originally identified in macaques, when a set of neurons in the ventral premotor cortex became active when observing conspecifics engaging in motor tasks. Strong evidence from various brain imaging techniques have confirmed the presence of a similar system in humans (e.g., Arbib & Mundhenk, 2005). Interestingly, the anterior part of the intraparietal sulcus has also received much recent attention. Whereas theory of mind has traditionally been associated with prefrontal cortices, current neuropsychology and neuroimaging evidence points to the parietal cortex as the primary locus of category construction and concept formation (e.g., Medin & Atran, 2004). It is the region where the brain integrates multiple sensory modalities into comprehensive models of the world (e.g., Lou et al., 2004). The ability to interpret correctly the actions, intentions, and goals of others may also be considered an important step in successful social interactions, including evaluations of unlikely or unpredicted actions of others. Properly interpreting the actions, intentions, and goals of others is obviously useful in planning one's own actions and forming personal goals. It now appears that the epicenter for these latter abilities is the intraparietal sulcus (e.g., Coolidge, 2014; Hamilton & Grafton, 2006) and perhaps the retrosplenial cortex (e.g., Coolidge & Hicks, 2016; Vann, Aggleton, & Maguire, 2009).

In Coolidge and Wynn (2009), we also proposed that variations in a language of diplomacy can be internally self-debated through subvocal articulation (i.e., inner speech) and posed by alternative grammatical constructions such as "What if I were to state this communication in this manner or should I state it in this other manner?" These statements appear to require expanded phonological storage capacity, recursion (or recursive thinking), and theory of mind. And yet, we argued that these are necessary but not sufficient conditions for a language of diplomacy.

Our fourth and another *sine qua non* prerequisite for a language of diplomacy was the executive functions of the frontal lobes. In our first paper together (Coolidge &

Wynn, 2001), we had traced the origin for the metaphor of executive functions to Vermont country doctor John Martyn Harlow (1848, 1868), who described a young man, Phineas Gage, who suffered a partial frontal lobectomy as a result of a railroad construction accident. Harlow noted that Gage's use of language, memory, and general intellect appeared to be intact, but he could no longer make and execute his daily plans of action. As Gage was the foreman of a railroad crew, these planning and execution skills were critical, and although he recovered from his physical head wound, he could not regain his job. What Gage lost was his command over his executive functions as described previously: temporal sequencing, planning, organizing, and inhibiting prepotent responses. We suggested that central executive functions are required for a language of diplomacy, and furthermore, that they are critical to the expression of full theory of mind. Our evidence for this contention comes from studies of children's executive functions and theory of mind. In a study of children aged 2 to 4 years old, Hughes and Ensor (2007) found stronger evidence that executive functions are a prerequisite for theory of mind than vice versa. They concluded that "children's growing competence in executive functions provides an important platform for their acquisition of a theory of mind" (p. 1457).

Carlson, Moses, and Claxton (2004), exploring the specific executive functions involved in theory of mind, tested children aged 3 and 4 with theory-of-mind tasks, inhibitory control tasks, planning tasks, and vocabulary measures. They found that the inhibition tasks were significantly related to theory of mind (while covarying the effects of age and receptive vocabulary), whereas planning tasks shared no unique variance with theory-of-mind tasks. Their study reinforced the hypothesis that inhibitory processes underpin the relationship between executive functions and theory of mind. The authors further concluded that this relationship appeared to be especially strong for tasks measuring conflict inhibition, rather than tasks measuring delay inhibition. In a provocative study of the relationship between executive functions and problem behaviors in childhood, Hughes and Ensor (2008) found that executive functions fully mediated the relationship between early verbal ability and later problem behaviors. It appeared that the salience for executive functions for problem behaviors may have been that the children were using language as a tool for self-regulation, particularly of negative emotions (like anger and agitation) and impulsive behaviors. It has been previously hypothesized that children engage in "private speech" as a means of solving problems, attaining goals, and regulating their behavior (e.g., Berk, 1992). Although this hypothesis was not tested directly by Hughes and Ensor, the notion that inhibitory executive functions help mediate inner speech in its regulation of outer speech and diplomatic speech remains an intriguing possibility. For example, the emotional regulation and inhibition of prepotent behaviors in diplomatic interpersonal communications (e.g., a tendency to fear strangers) would be of critical importance to diplomatic communicatory success. Further, debating alternative grammatical expressions through inner speech might also require the inhibition of choices that, although prepotent, would not serve the functions of diplomacy as well as some other counterintuitive choices might. Thus, inhibitory executive functions might regulate emotional expressions in the act of diplomacy and be called upon to inhibit and choose among alternative grammatical constructions in inner speech.

More recently, in a study with implications for cultural and familial effects on theory of mind, Hughes and colleagues (2014) found the theory-of-mind module to be virtually invariant across 5- and 6-year-olds in the UK, Italy, and Japan; however, they did find that in more "collectivistic" cultures (i.e., Japan) children showed a "less advanced understanding of the subjective nature of mental states" (p. 8). It is important to note that this does not mean Japanese adults may have lesser theory of mind than any worldwide peers, as the delays in theory-of-mind reasoning exhibited by 5- and 6-year-old Japanese children may quickly dissipate; however, it does indicate that culture and familial factors may impact theory of mind early in life. Further, the nature of a culture and its family interactions are also subject to a biological leash; thus, it is at least theoretically possible that Neandertal culture and family interactions could have affected the ability to infer properly the mental states of others. That hypothesis is consistent with archaeological evidence that Neandertals may have had smaller groups than *H. sapiens* living at the same time and traveled less far, and there is also less evidence for their long-distance trading (Coolidge & Wynn, 2018).

The Evolution of Syntax and Grammar

Although somewhat synonymous terms, *syntax* often refers to the rules for ordering of words in an utterance, while *grammar* more often refers to all of the rules and processes that structure a language, including syntax, inflections, punctuation, and semantics. In Coolidge (2012), I hypothesized that the limits of working memory capacity, particularly phonological storage capacity, might have promoted the development of word order (syntax/grammar) in language. While the order of "subject-verb-object" may be completely arbitrary in a language, an imposed order of words in a language, as opposed to random or capricious order of words, may help to bypass the limitations of working memory capacity, particularly the limitations of the phonological loop. As noted earlier, instructions or the subject of the utterance placed at the beginning of a statement appears to place minimal burdens on working memory. For example, "The little boy hurt himself while playing on a swing in a rural school playground" should place minimal demands on working memory, as the subject (S) and the verb (V) are mentioned at the outset of the sentence, while the object (O) of the subject and verb are placed at the end of the sentence. This S-V-O structuring appears to increase the intelligibility of the utterance, as the related syntactical elements in it are closer together rather than far apart.

However, if the S occurs at the end of the sentence, the initial information must be held in abeyance temporarily, thus placing greater cognitive demands on the domain-general working memory capacity as well as phonological storage. For example, "While playing on a swing in a rural school playground, the little boy hurt himself." In the latter construction, the initial information is essentially unmeaningful but must be held in short-term memory until it is revealed at the end of utterance who the actor was and what the action was. Consistent with my hypothesis, cross-cultural comparative studies of languages reveal that languages without an imposed word order are rare. Further, V-O-S, O-V-S, and O-S-V are very rare, as the greatest numbers of speakers in the world use S-V-O orderings, as in English, the Romance languages, Chinese, Swahili, and others. The largest numbers of distinct languages (not the number of users) use S-O-V ordering, and some languages like German use a combination of S-O-V and S-V-O, with the latter being more common. Thus, one experimentally testable hypothesis is whether S-O-V and S-V-O word orderings (as opposed to the rarer

V-O-S, O-V-S, and O-S-V word orderings) place fewer demands on working memory capacity.

THE VISUOSPATIAL SKETCHPAD, EPISODIC BUFFER, AND EPISODIC MEMORY

Paleoneurologist Emiliano Bruner (2004) demonstrated that a derived feature of *H. sapiens'* brains in the past 100,000 years appears to be expanded parietal lobes compared to Neandertals. We contacted Emiliano by email right after seeing his article. When the Wenner-Gren Foundation awarded us funding in 2008 for an international conference focused on our enhanced working memory hypothesis, Bruner was one of our first of 16 invitees. The conference culminated in a special issue of *Current Anthropology* devoted to the invitees' unique contributions to our conference theme: "Working Memory: Beyond Language and Symbolism" (Wynn & Coolidge, 2010). Bruner's subsequent research (e.g., Bruner & Iriki, 2016) established two critical areas of expansion within the medial and superior portions of the parietal lobes: the intraparietal sulcus and the precuneus.[1]

These anatomical regions play a major role in the cognitive processes of the visuospatial sketchpad, the episodic buffer, and episodic memory. The cognitive functions of the latter three make them a second prime candidate as to the specific nature of our enhanced working memory concept. Cognitive scientist Roger Shepard (1997) noted that one of the critical evolutionary advantages of recollections may have been "thought experiments." He stated that every real-world experiment to solve some problem had to have been preceded by corresponding thought experiments. However, he envisioned that they drew upon 'countless' prior real experiments, which had been recombined in mind to produce new solutions. The advantage of this internalized knowledge would be that one could avoid "trial and possibly fatal error" (p. 24). Although Shepard did not specifically mention "episodic" memory, it is evident that he appreciated one of its chief evolutionary advantages. Baddeley (2000, 2001) also proposed that greater working memory capacity would allow for reflection on and comparison of multiple past experiences. This might allow an individual to actively choose a future action or create an alternative action, rather than simply choosing the path of highest probable success. Although an individual would still be better off choosing alternatives simply based on the past (an example of an inflexible anticipatory process), compared to someone without the benefit of past experience, Baddeley proposed that greater working memory capacity would allow for the formulation of mental models more likely to be successful as future behaviors.

Tulving (2002) proposed that the ability to simulate and contemplate alternative future scenarios may have been the driving force in the evolution of the episodic memory system. The episodic memory system, which includes mental time travel,

[1] I have tried to convince Bruner that the medial and superior expansion of the parietal lobes would also have meant significant inferior displacement of the inferior parietal regions, particularly the supramarginal gyrus, angular gyrus, and the retrosplenial cortex (which sits inferiorly to the inferior parietal lobes. Admittedly, such internal brain tissue reorganizations are difficult to detect in fossils and endocasts.

allows awareness of not only the past but also awareness of alternative future scenarios. "This awareness allows autonoetic creatures to reflect on, worry about, and make plans for their own and their progeny's future in a way that those without this capability possibly could not. *H. sapiens*, taking full advantage of its awareness of its continued existence in time, has transformed the natural world into one of culture and civilization that our distant ancestors, let alone members of other species, possibly could not imagine" (Tulving, 2002, p. 20).

Episodic memory (along with classical and operant conditioning) is a subtype of associative learning, the ability to recall specific experiences with "what, where, and when" qualities. It has been argued that many animals, including birds and mammals, meet the behavioral definition for episodic memory, and thus it has a long evolutionary history (e.g., Allen & Fortin, 2013). However, the human episodic memory system is qualitatively different, as it includes the ability to place one's self in past and future events (autobiographical memory). These features appear to provide a particularly rich form of memory, one that would also allow for detailed simulations of future events. In fact, Addis, Wong, and Schacter (2007) have argued that it is not the *sole* purpose of episodic memory to recall prior episodes with complete fidelity, but its evolutionary adaptiveness was the ability to recall the *essence* of prior events in order to solve novel problems. The neural basis for human episodic memory is a complex network involving the hippocampus, parahippocampal cortex, entorhinal cortex, amygdala, precuneus, and retrosplenial cortex. In conjunction with the prefrontal cortex, this neural network creates a highly specialized system for the formation and use of episodic memories. Interestingly, this complex network also forms the neural basis for the resting-state default mode network of the brain. All hominins prior to modern *H. sapiens* undoubtedly possessed a basic form of episodic memory, as previously noted. Also, because of the paleoneurological evidence from Bruner's work (2004, 2010; Bruner & Iriki, 2016) on a distinct parietal lobe expansion in modern *H. sapiens*, it may be hypothesized that one possible discrete cognitive shift associated with this expansion was enhanced visuospatial abilities (i.e., the visuospatial sketchpad), which may have given rise to autonoetic thinking, human autobiographical memories, and the ability to construct future scenarios. Archaeology, in the guise of narrative imagery, provides strong evidence for autonoetic thinking in the Later Stone Age of Africa and Upper Paleolithic of Europe (e.g., snares, weirs, managed foraging, and parietal and mobile art). Earlier evidence is controversial (e.g., the curated Acheulean handaxe as a material scaffold of the autonoetic self).

An enhanced visuospatial sketchpad may have also aided basic navigation, which would have been useful, of course, once *H. erectus* had made the full transition to ground life. The landscape of *H. erectus* changed dramatically, not only by an expansion of territory about 10-fold over the australopithecines, but also by exploring new and disrupted environments, to the extent that *H. erectus* has been labeled a "weed species" (e.g., Cachel & Harris, 1995; Wells & Stock, 2007). For example, it appears that *H. erectus* evolved in Africa but left it in a series of migrations to the Middle East, Europe, Asia, India, and Southeast Asia beginning at least about 1.5 Mya. The "weed" part of the analogy is that *H. erectus* never appears to have had substantial or large populations, but kept "popping up" "in distant places and disrupted environments. Further, it may have been that *H. erectus*' enhanced visuospatial abilities (relative to those of the australopithecines and habilines) are what aided these very distant migrations. Further, it is possible that an enhanced ability to "see" one's place in the environment might also have repercussions for one's phenomenological world, as will be discussed next.

Parietal Lobes: Egocentric and Allocentric Perception

The superior, medial, and inferior portions of the parietal lobe (e.g., intraparietal sulcus, precuneus, supramarginal gyrus, angular gyrus) may be particularly involved in self- and other perceptions (e.g., Lou et al., 2004). In addition, the retrosplenial cortex (RSC), which has been implicated in cognitive functions like episodic memory, navigation, imagination, and planning (Vann et al., 2009), is located inferiorly to the parietal lobes and posterior to the splenium (the latter is the thickest and most posterior part of the corpus callosum). In a neural network conjunction with the parietal lobes, hippocampus, and thalamus, the RSC appears to play a major synergistic role in navigation, especially with novel environments and spatial memories. This network has also been shown to have a major role in episodic and autobiographical memories. Its most critical cognitive function, however, may reside in its translational role between an egocentric viewpoint (a view from one's own self, known primarily to be a medial and/or posterior parietal lobe function) and an allocentric one (an independent viewpoint, or a view from another person's or place's perspective). According to empirical work by Vann, Aggleton, and Maguire (2009), the place and grid cells of the hippocampus index locations contained within episodic or autobiographical memories, which the RSC translates into egocentric information such that a location in a memory may be viewed from other viewpoints (i.e., allocentrism). It is also suspected that the RSC may also act as a short-term storage buffer for the visuospatial information being translated. Many human neurophysiological studies have confirmed that the RSC is significantly activated by many kinds of spatial navigational tasks, including passive viewing of scenery, virtual-interactive spatial navigation, and active navigation of both new environments and highly familiar environments.

The RSC is also highly active when topographical information needs updating or for use of one's own motion (ideothetic) to plan routes. Human and animal studies involving brain damage in these regions confirms the loss or major degradation of the aforementioned spatial abilities. Thus, the "worldly" success of *H. erectus* may have depended on the natural selection of a network of brain regions that could reciprocally translate egocentric and allocentric viewpoints. Finally, Burgess, Becker, King, and O'Keefe (2001) have proposed that this translational RSC model may be related to imaginative or creative thinking for its basic ability to reconstruct scenes or imagine alternative scenes, and so the "thought experiments" of Shepard (1997), Baddeley's (2001) claim that the episodic buffer may create alternative future behavioral options, and Tulving's (2002) hypothesis that episodic memory played a major role in hominin cognitive evolution may all neurologically reside in this critical neural network of parietal lobes, hippocampus, thalamus, and RSC. It is also interesting to speculate further that the dramatic changes in technology from mode 1 to mode 2 stone tools may be behaviorally and neurologically tied to the enhanced navigation abilities of *H. erectus* because they share a common neurological substrate. In other words, mode 2 bifacial handaxes and enhanced navigational abilities may have been co-evolved behaviors, and the expansion of parietal lobes as described by Bruner and colleagues could only have meant that these cognitive abilities must have advanced even further in *H. sapiens*.

ULTIMATE ORIGINS OF LEARNING AND MEMORY

Through our collaboration, I have gained insight into the benefits of remaining unentangled in controversies like language and symbolism, and why our enhanced working memory hypothesis has managed to bypass the standard explanations of how fully modern minds came into place, that is, when *H. sapiens* had modern language and symbolic thinking. One major problem with the latter argument is that archaeological evidence for language and symbolic thinking rests in artifacts that may or may not actually reflect the cognitive processes underlying symbolic thinking. For example, Henshilwood, d'Errico, Vanhaeren, Van Niekerk, and Jacobs (2004) claimed that the 75,000-year-old bead ornaments and ochre engravings discovered in Blombos Cave, South Africa were "unambiguous" markers of symbolically mediated behavior and "a proxy for modern syntactical language" (p. 404). Perhaps an even more egregious example of this rush to imbue archaeological artifacts with fully symbolic thinking is Rodríguez-Vidal and colleagues' (2014) discovery of a simple rock engraving of four simple cross-hatched lines in a cave in Gibraltar, attributed to Neandertals, which they interpreted as a "symbolic code," an "abstract pattern," and evidence for "abstract thought and expression" (p. 13301). However, as Botha (2008, 2010) noted, Henshilwood and colleagues' final argument requires several inferential steps or bridging arguments, and he warned of the dangers of each of the inferential leaps. The first, of course, is whether the beads were intentionally made and worn as personal ornaments. As we have noted (Coolidge & Wynn, 2011), it is acceptable to us, given the archaeological evidence, that the shells were intentionally modified. However, can it be confidently ascertained that they were personal ornaments? Could they have been used as a counting or tallying device? Could they have been used for navigational purposes, just as modern US Army Rangers use small, rosary-like beads on a string to measure distances by the length of their stride (e.g., Wynn, Overmann, Coolidge, & Janulis, 2017)? The next inference, that the beads have symbolic meaning, appears unwarranted. As others have noted, in the Saussurian meaning of a symbol, usually a word, there is an arbitrary link between the sound (or written sign) and the thing it represents. It must also be learned and transmitted collectively, and there must be agreement within a community about its meaning. However, it is possible that the shell beads were highly individualistic, private, idiosyncratic productions, whose meaning resided only in their makers. They may not have had any referential meaning, other than perhaps, "I make beads. I wear beads. I use beads." In this respect, we noted that Henshilwood and colleagues made a very curious observation. They stated that the appearance and disappearance of such artifacts from about 80 to 60 Kya may imply that something might have served as a barrier to their transmission. We argued that the absence of collectively shared meanings might have been a likely barrier. The latter reinforces the notion that the beads might have been highly individualistic and, at best, had a group-delimited meaning. Further, rather than symbols, perhaps the beads were simply indices with group-delimited meanings. For example, they might have meant, "We wear beads. We are different from the people who do not wear beads"; in other words, they might have been simple indices of group membership or social roles without any explicit or implicit meaning beyond the association of appearance and social role. As we further argued, the latter *might* require some self-awareness and perhaps even recognition of others' perspectives, but it certainly would not be an indication of higher levels of theory of mind, as they argued.

THE EVOLUTIONARY FOUNDATIONS OF LEARNING AND MEMORY

The most common dichotomy in cognitive psychology for learning and memory systems is a differentiation between associative and non-associative learning. *Associative learning* consists of classical conditioning, operant conditioning, and role modeling. In classical conditioning, a novel stimulus is paired with an unconditioned stimulus and comes to elicit a conditioned response. In operant conditioning, the behavior may increase or decrease in frequency, depending on its consequences. Reinforcers increase the probability that a behavior will reoccur, and punishment decreases the probability that a behavior will reoccur. Role modeling consists of consciously or unconsciously copying the behaviors of others. Behaviorists have long argued that elaborate behavior repertoires may have evolved from just these three types of conditioning. Aye, but here's the rub: As has been previously noted, there is a tendency to imbue more cognitive complexity than is warranted when interpreting archaeological artifacts. Because Pavlov's dog would eventually salivate to the mere sound of a bell without the actual presence of food, we might be tempted to assume that the bell came to "symbolize" food to the dog. Classical conditioning has been demonstrated in very simple organisms like planaria, which lack a definitive brain and possess only a rudimentary nervous system. Thus, the underlying cognitive mechanisms must be more carefully delineated in anthropology, and any interpretations for "symbolic thinking" must first be judged on whether the behavior in question can be explained in simpler terms (i.e., associative learning).

Non-associative learning consists of sensitization (an organism responds to relevant stimuli with increasing strength or reliability) and habituation (irrelevant stimuli are ignored). Habituation has often been touted as the simplest form of learning (i.e., learning not to respond to repeated stimuli). Evolutionarily, non-associative learning appeared even before the advent of life, as molecules have affinities (sensitization) or disaffinities (habituation) for other elements and molecules based on chemical and atomic predilections for elements to coagulate in the first place to create stable energy states (minimizing free energy and forming stable subatomic ratios). A homeostatic balance between sensitization and habituation has been proposed by Eisenstein and Eisenstein (2006), who purported that organisms have an innate tendency to be either sensitizers or habituators. This explains the chicken in the system (an evolutionary tendency to sensitize or habituate) but ignores the egg (how did an organism develop either tendency in the first place?). According to the "selfish metabolism" hypothesis of molecular biologist Victor de Lorenzo (2014), the survival and replication of the earliest molecules before four billion years ago was predicated on their ability to exploit new chemical landscapes in the interest of their own metabolism. Only subsequently, with metabolic and autocatalytic abilities, were these molecules more successful than competing forms at replication and diversification in the gene pool. Thus, even in prebiotic conditions, the affinity of a molecule for another molecule that enhances its survival (metabolism) is a protoform of sensitization, and the disaffinity of a molecule for another molecule (which does not enhance its metabolism) is a protoform of habituation (Coolidge, in press). These initial molecular coagulations were invariably subject to Darwinian natural selection, likely based on the selfish metabolism hypothesis. It is possible that other molecular forms failed to perpetuate and expand because they were not able to enhance their own metabolic processes and subsequently replicate.

The bacteria *Escherichia coli* (*E. coli*) is used as a model for studying the first walled-cells, prokaryotes, which appeared about 3.9 billion years ago (although the first prokaryotes contained only a single strand of RNA and not double-stranded DNA like modern bacteria). *E. Coli* possess flagella, which when coordinated makes them approach relevant stimuli (sensitization) and when uncoordinated makes them avoid irrelevant stimuli (habituation). Flatworms like planaria and amphioxus are used as models for the first bilaterians (multicellular eukaryotes, which have a nucleus and DNA). Studies have shown that these modern flatworms are capable of associative and non-associative learning, which may indicate that the first simple flatworms, which appeared about 545 Mya, may have been capable of both forms of learning. Further, Ginsburg and Jablonka (2010) proposed that the explosion in the numbers and diversity of life during the Cambrian period (545–525 Mya) occurred specifically because of the bilaterians' ability to exhibit unlimited associative learning. The ability to use previously memorized experiences may have allowed bilaterians to anticipate future events and rewards and to discriminate among various stimuli. Further, the first beginnings of the animal kingdom (metazoan phylum) are now attributed to jellyfish (ctenophores: comb jellies) rather than sponges, according to a recent phylogenomic study (Shen, Hittinger, & Rokas, 2017). Their appearance about 600 Mya serves as a prototype of the first multicellular eukaryotes. However, sponges are sessile, and comb jellies move but often inadvertently (wind, waves, currents, etc.). With the advent of the first simple flatworms at about 545 Mya, movement became intentional.

Notice also that the various cells of flatworms at this time had already become highly specialized. There were cells devoted to sensations (perception of light, vibrations, chemicals, etc.) and cells devoted to movement. And most importantly, there were cells devoted to the interpretation of those sensations to guide appropriate movements, and those were the first rudimentary brain cells (neurons). This Cambrian explosion of life forms also may have given rise to speciatic war. As animals could move, they could also become better predators, and there was also an explosion of defenses against predation—for example, hard body parts, shells, teeth, fangs, nails, thorns, and toxins. Further, particular niches also served to protect against predation (e.g., nocturnal or diurnal activities, mud burrowing, flight, etc.). However, to return to the origins of memory, this explosion in the numbers and diversity of life, as Ginsburg and Jablonka have argued, occurred because organisms became capable of virtually unlimited associative learning, in addition to their already long-standing non-associative learning abilities.

Three final observations on the origins of learning and memory are in order. First, it is curious that cognitive psychologists always list associative learning first, then non-associative learning (as I did, too, at the outset of this discussion), when clearly non-associative learning was evident at the very advent of life about 4 billion years ago and associative learning, in all likelihood, did not arise until much later (i.e., at the beginning of the metazoan phylum at about 600 Mya). The second observation is a question: Why is it called *non*-associative learning if there is a clear and stable *association* between a stimulus and a response, as is the case in sensitization and habituation? The final observation is actually a hypothesis: Because non-associative learning involves the ability of a molecule or organism to form a stable relationship between a stimulus and a response, associative learning may have been an *exaptation* of this core principle of non-associative learning (i.e., forming stable relationships between stimuli and responses). In summary, the ultimate origin of learning and memory systems may reside in the chemical and atomic affinities and disaffinities of basic elements for one another. These basic properties serve as the prototypic basis for non-associative learning, with associative learning being an exaptation of the basic processes underlying non-associative learning.

ACKNOWLEDGMENTS

It has been my great pleasure and honor to work with Thomas Wynn for almost two decades. He is truly a gentleman and a scholar, and after his research work in Stirling, Scotland, he is also a very good judge of fine whiskey. I thank my lucky stars that I walked into his office nearly 20 years ago to say, "I think Tattersall's got this wrong." I also thank my lucky stars that Leee Overmann walked into my office in 2008 and said, "I think I'll add a (fourth) major," to which I replied, "Why don't you get a master's degree with me and Tom?" The highlight of all the graduations I have ever attended in 42 years of academic life was seeing her receive a doctorate from Oxford University (Keble College) in 2016.

REFERENCES

Aboitiz, F., Aboitiz, S., & García, R. R. (2010). The phonological loop. *Current Anthropology, 51*(S1), S55–S65.

Aboitiz, F., García, R. R., Bosman, C., & Brunetti, E. (2006). Cortical memory mechanisms and language origins. *Brain and Language, 98*(1), 40–56.

Adams, A.-M., & Gathercole, S. E. (2000). Limitations in working memory: Implications for language development. *International Journal of Language & Communication Disorders, 35*(1), 95–116.

Adams, C. (2002). Practitioner review: The assessment of language pragmatics. *Journal of Child Psychology and Psychiatry, 43*(8), 973–987.

Addis, D. R., Wong, A. T., & Schacter, D. L. (2007). Remembering the past and imagining the future: Common and distinct neural substrates during event construction and elaboration. *Neuropsychologia, 45*(7), 1363–1377.

Allen, T. A., & Fortin, N. J. (2013). The evolution of episodic memory. *Proceedings of the National Academy of Sciences of the United States of America, 110*(S2), 10379–10386.

Ambrose, S. H. (2003). Did the super-eruption of Toba cause a human population bottleneck? Reply to Gathorne-Hardy and Harcourt-Smith. *Journal of Human Evolution, 45*(3), 231–237.

Ambrose, S. H. (2010). Coevolution of composite-tool technology, constructive memory, and language: Implications for the evolution of modern human behavior. *Current Anthropology, 51*(S1), S135–S147.

Arbib, M. A., & Mundhenk, T. N. (2005). Schizophrenia and the mirror system: An essay. *Neuropsychologia, 43*(2), 268–280.

Baddeley, A. D. (2000). The episodic buffer: A new component of working memory? *Trends in Cognitive Sciences, 4*(11), 417–423.

Baddeley, A. D. (2001). Is working memory still working? *American Psychologist, 56*(11), 851–864.

Baddeley, A. D. (2007). *Working memory, thought, and action*. New York: Oxford University Press.

Baddeley, A. D., Gathercole, S. E., & Papagno, C. (1998). The phonological loop as a language learning device. *Psychological Review, 105*(1), 158–173.

Baddeley, A. D., & Hitch, G. J. (1974). Working memory. In G. H. Bower (Ed.), *The psychology of learning and motivation: Advances in research and theory* (Vol. 8, pp. 47–89). New York: Academic Press.

Baddeley, A. D., & Logie, R. H. (1999). Working memory: The multi-component model. In A. Miyake & P. Shah (Eds.), *Models of working memory: Mechanisms of active maintenance and executive control* (pp. 28–61). New York: Cambridge University Press.

Baron-Cohen, S. (1997). *Mindblindness: An essay on autism and theory of mind*. Cambridge, MA: MIT Press.

Baron-Cohen, S. (2000). Theory of mind and autism: A review. *International Review of Research in Mental Retardation, 23*, 169–184.

Bastir, M., Rosas, A., Gunz, P., Peña-Melian, A., Manzi, G., Harvati, K., . . . Hublin, J.-J. (2011). Evolution of the base of the brain in highly encephalized human species. *Nature Communications, 2*, 1–8.

Berk, L. E. (1992). Children's private speech: An overview of theory and the status of research. In R. M. Diaz & Laura E. Berk (Eds.), *Private speech: From social interaction to self-regulation* (pp. 17–53). Hillsdale, NJ: Lawrence Erlbaum Associates.

Bolhuis, J. J., Tattersall, I., Chomsky, N., & Berwick, R. C. (2014). How could language have evolved? *PLoS Biology, 12*(8), 1–6.

Borg, E. (2007). If mirror neurons are the answer, what was the question? *Journal of Consciousness Studies, 14*(8), 5–19.

Borg, E. (2013). More questions for mirror neurons. *Consciousness and Cognition, 22*(3), 1122–1131.

Botha, R. (2008). Prehistoric shell beads as a window on language evolution. *Language & Communication, 28*(3), 197–212.

Botha, R. (2010). On the soundness of inferring modern language from symbolic behaviour. *Cambridge Archaeological Journal, 20*(3), 345–356.

Bruner, E. (2004). Geometric morphometrics and paleoneurology: Brain shape evolution in the genus *Homo*. *Journal of Human Evolution, 47*(5), 279–303.

Bruner, E. (2010). Morphological differences in the parietal lobes with the human genus: A neurofunctional perspective. *Current Anthropology, 51*(1), S77–S88.

Bruner, E., & Iriki, A. (2016). Extending mind, visuospatial integration, and the evolution of the parietal lobes in the human genus. *Quaternary International, 405*, 98–110.

Burgess, N., Becker, S., King, J. A., & O'Keefe, J. (2001). Memory for events and their spatial context: Models and experiments. *Philosophical Transactions of the Royal Society of London. Series B, Biological Sciences, 356*(1413), 1493–1503.

Cachel, S., & Harris, J. W. K. (1995). Ranging patterns, land-use and subsistence in *Homo erectus* from the perspective of evolutionary biology. In J. Bowe & S. Sartono (Eds.), *Human evolution in its ecological context. Paleo-anthropology: Evolution and ecology of Homo erectus* (pp. 51–66). Leiden, The Netherlands: Pithecanthropus Centennial Foundation.

Carlson, S. M., Moses, L. J., & Claxton, L. J. (2004). Individual differences in executive functioning and theory of mind: An investigation of inhibitory control and planning ability. *Journal of Experimental Child Psychology, 87*(4), 299–319.

Coolidge, F. L. (2012). On the emergence of grammatical language as a means of bypassing the limitations of working memory capacity: Comment on “Interaction between lexical and grammatical language systems in the brain” by Alfredo Ardila. *Physics of Life Reviews, 9*(2), 217–218.

Coolidge, F. L. (2014). The exaptation of the parietal lobes in *Homo sapiens. Journal of Anthropological Sciences, 92*, 295–298.

Coolidge, F. L. (in press). The origins of learning and memory systems. *Human Evolution.*

Coolidge, F. L., Haidle, M. N., Lombard, M., & Wynn, T. (2016). Bridging theory and bow hunting: Human cognitive evolution and archaeology. *Antiquity, 90*(349), 219–228.

Coolidge, F. L., & Hicks, J. M. (2016). Evolutionary implications of the neuropsychological functions of the retrosplenial cortex. Poster presented at the meeting of the European Society for the study of Human Evolution, Madrid, Spain.

Coolidge, F. L., Overmann, K. A., & Wynn, T. (2011). Recursion: What is it, who has it, and how did it evolve? *Wiley Interdisciplinary Reviews: Cognitive Science, 2*(5), 547–554.

Coolidge, F. L., Thede, L. L., & Young, S. E. (2000). Heritability and the comorbidity of attention deficit hyperactivity disorder with behavioral disorders and executive function deficits: A preliminary investigation. *Developmental Neuropsychology, 17*(3), 273–287.

Coolidge, F. L., & Wynn, T. (2001). Executive functions of the frontal lobes and the evolutionary ascendancy of *Homo sapiens. Cambridge Archaeological Journal, 11*(3), 255–260.

Coolidge, F. L., & Wynn, T. (2005). Working memory, its executive functions, and the emergence of modern thinking. *Cambridge Archaeological Journal, 15*(1), 5–26.

Coolidge, F. L., & Wynn, T. (2006). The effects of the tree-to-ground sleep transition in the evolution of cognition in early *Homo. Before Farming, 4*, 1–18.

Coolidge, F. L., & Wynn, T. (2009). Did Neandertals lack diplomatic speech? Cognitive prerequisites for speaking indirectly. In *Evolution of brain, behavior, & intelligence.* Hinxton, UK: Wellcome Trust Centre.

Coolidge, F. L., & Wynn, T. (2011). Comment on “The Still Bay and Howiesons Poort, 77–59 ka: Symbolic material culture and the evolution of the mind during the African Middle Stone Age.” *Current Anthropology, 52*(3), 380–382.

Coolidge, F. L., & Wynn, T. (2018). *The rise of* Homo sapiens: *The evolution of modern thinking* (2nd ed.). Oxford, UK: Oxford University Press.

Crowder, R. G., & Morton, J. (1969). Precategorical acoustic storage (PAS). *Attention, Perception, & Psychophysics, 5*(6), 365–373.

de Lorenzo, V. (2014). From the selfish gene to selfish metabolism: Revisiting the central dogma. *BioEssays, 36*(3), 226–235.

Dunbar, R. I. M. (2004). Gossip in evolutionary perspective. *Review of General Psychology, 8*(2), 100–110.

Eisenstein, E. M., & Eisenstein, D. (2006). A behavioral homeostasis theory of habituation and sensitization: II. Further developments and predictions. *Reviews in the Neurosciences, 17*(5), 533–558.

Fitch, W. T., Hauser, M. D., & Chomsky, N. (2005). The evolution of the language faculty: Clarifications and implications. *Cognition, 97*(2), 179–210.

Friedman, N. P., Miyake, A., Young, S. E., DeFries, J. C., Corley, R. P., & Hewitt, J. K. (2008). Individual differences in executive functions are almost entirely genetic in origin. *Journal of Experimental Psychology: General, 137*(2), 201–225.

Gerrans, P. (2002). The theory of mind module in evolutionary psychology. *Biology and Philosophy, 17*(3), 305–321.

Ginsburg, S., & Jablonka, E. (2010). The evolution of associative learning: A factor in the Cambrian explosion. *Journal of Theoretical Biology, 266*(1), 11–20.

Hamilton, A. F. de C., & Grafton, S. T. (2006). Goal representation in human anterior intraparietal sulcus. *Journal of Neuroscience, 26*, 1133–1137.

Harlow, J. M. (1848). Passage of an iron rod through the head. *Boston Medical and Surgical Journal, 39*, 389–393.

Harlow, J. M. (1868). Recovery from the passage of an iron bar through the head. *Publications of the Massachusetts Medical Society, 2*(3), 327–346.

Hauser, M. D., Chomsky, N., & Fitch, W. T. (2002). The faculty of language: What is it, who has it, and how did it evolve? *Science, 298*(5598), 1569–1579.

Henshilwood, C., d'Errico, F., Vanhaeren, M., Van Niekerk, K., & Jacobs, Z. (2004). Middle stone age shell beads from South Africa. *Science, 304*(5669), 404.

Hickok, G. (2014). *The myth of mirror neurons: The real neuroscience of communication and cognition*. New York: W. W. Norton & Company.

Hughes, C., Devine, R. T., Ensor, R., Koyasu, M., Mizokawa, A., & Lecce, S. (2014). Lost in translation? Comparing British, Japanese, and Italian children's theory-of-mind performance. *Child Development Research, 2014*, 1–10.

Hughes, C., & Ensor, R. (2007). Executive function and theory of mind: Predictive relations from ages 2 to 4. *Developmental Psychology, 43*(6), 1447–1459.

Hughes, C., & Ensor, R. (2008). Does executive function matter for preschoolers' problem behaviors? *Journal of Abnormal Child Psychology, 36*(1), 1–14.

Kane, M. J., & Engle, R. W. (2002). The role of prefrontal cortex in working-memory capacity, executive attention, and general fluid intelligence: An individual-differences perspective. *Psychonomic Bulletin & Review, 9*(4), 637–671.

Kinderman, P., Dunbar, R. I. M., & Bentall, R. P. (1998). Theory-of-mind deficits and causal attributions. *British Journal of Psychology, 89*(2), 191–204.

Klein, R. G. (2000). *The human career: Human biological and cultural origins* (2nd ed.). Chicago, IL: University of Chicago Press.

Klein, R. G., & Edgar, B. (2002). *The dawn of human culture: A bold new theory on what sparked the "Big Bang" of human consciousness*. New York: John Wiley & Sons.

Lou, H. C., Luber, B., Crupain, M., Keenan, J. P., Nowak, M., Kjaer, T. W., . . . Lisanby, S. H. (2004). Parietal cortex and representation of the mental self. *Proceedings of the National Academy of Sciences of the United States of America, 101*(17), 6827–6832.

Luria, A. R. (1962). *Higher cortical functions in man*. Berlin: Springer Science & Business Media.

McBrearty, S., & Brooks, A. S. (2000). The revolution that wasn't: A new interpretation of the origin of modern human behavior. *Journal of Human Evolution, 39*(5), 453–563.

Medin, D. L., & Atran, S. (2004). The native mind: Biological categorization and reasoning in development and across cultures. *Psychological Review, 111*(4), 960–983.

Mithen, S. J. (1996). *The prehistory of mind: The cognitive origins of art, religion and science.* London: Thames & Hudson.

Noble, W., & Davidson, I. (1991). The evolutionary emergence of modern human behaviour: Language and its archaeology. *Man, 26*(2), 223–253.

Pinker, S., Nowak, M. A., & Lee, J. J. (2008). The logic of indirect speech. *Proceedings of the National Academy of Sciences of the United States of America, 105*(3), 833–838.

Price, T., Wadewitz, P., Cheney, D., Seyfarth, R., Hammerschmidt, K., & Fischer, J. (2015). Vervets revisited: A quantitative analysis of alarm call structure and context specificity. *Scientific Reports, 5*, 1–11.

Reuland, E. (2010). Imagination, planning, and working memory: The emergence of language. *Current Anthropology, 51*(S1), S99–S110.

Rodríguez-Vidal, J., D'Errico, F., Pacheco, F. G., Blasco, R., Rosell, J., Jennings, R. P., . . . Finlayson, C. (2014). A rock engraving made by Neanderthals in Gibraltar. *Proceedings of the National Academy of Sciences of the United States of America, 111*(37), 13301–13306.

Schlenker, P., Chemla, E., Arnold, K., & Zuberbühler, K. (2016). *Pyow-hack* revisited: Two analyses of putty-nosed monkey alarm calls. *Lingua, 171*, 1–23.

Seyfarth, R. M., Cheney, D. L., & Marler, P. (1980). Vervet monkey alarm calls: Semantic communication in a free-ranging primate. *Animal Behaviour, 28*(4), 1070–1094.

Shen, X.-X., Hittinger, C. T., & Rokas, A. (2017). Contentious relationships in phylogenomic studies can be driven by a handful of genes. *Nature Ecology & Evolution, 1*(5), 126–148.

Shepard, R. (1997). The genetic basis of human scientific knowledge. In G. R. Bock & G. Cardew (Eds.), *Characterizing human psychological adaptations* (pp. 4–13). Chichester, UK: John Wiley & Sons.

Tattersall, I. (2000). Once we were not alone. *Scientific American, 282*(1), 56–62.

Tulving, E. (1972). Episodic and semantic memory. In E. Tulving & W. Donaldson (Eds.), *Organization of memory* (pp. 382–423). Oxford, UK: Academic Press.

Tulving, E. (2002). Episodic memory: From mind to brain. *Annual Review of Psychology, 53*(1), 1–25.

Vann, S. D., Aggleton, J. P., & Maguire, E. A. (2009). What does the retrosplenial cortex do? *Nature Reviews Neuroscience, 10*(11), 792–802.

Villa, P., & Roebroeks, W. (2014). Neandertal demise: An archaeological analysis of the modern human superiority complex. *PLoS One, 9*(4), 1–10.

Weaver, A. H. (2005). Reciprocal evolution of the cerebellum and neocortex in fossil humans. *Proceedings of the National Academy of Sciences of the United States of America, 102*(10), 3576–3580.

Weaver, A. H. (2010). Cerebellum and brain evolution in Holocene humans. In D. Broadfield, M. Yuan, K. D. Schick, & N. P. Toth (Eds.), *The human brain evolving: Paleoneurological studies in honor of Ralph L. Holloway* (pp. 97–106). Bloomington, IN: Stone Age Institute Press.

Wells, J. C. K., & Stock, J. T. (2007). The biology of the colonizing ape. *American Journal of Physical Anthropology, 134*(S45), 191–222.

Wynn, T., & Coolidge, F. L. (2004). The expert Neandertal mind. *Journal of Human Evolution, 46*(4), 467–487.

Wynn, T., & Coolidge, F. L. (2006). The effect of enhanced working memory on language. *Journal of Human Evolution, 50*(2), 230–231.

Wynn, T., & Coolidge, F. L. (2010). Beyond symbolism and language: An introduction to Supplement 1, Working Memory. *Current Anthropology, 51*(S1), S5–S16.

Wynn, T., & McGrew, W. C. (1989). An ape's view of the Oldowan. *Man, 24*(3), 383–398.

Wynn, T., Overmann, K. A., & Coolidge, F. L. (2016). The false dichotomy: A refutation of the Neandertal indistinguishability claim. *Journal of Anthropological Sciences, 94*, 1–21.

Wynn, T., Overmann, K. A., Coolidge, F. L., & Janulis, K. (2017). Bootstrapping ordinal thinking. In T. Wynn & F. L. Coolidge (Eds.), *Cognitive models in Palaeolithic archaeology* (pp. 197–213). Oxford, UK: Oxford University Press.

20

MATERIALITY AND THE PREHISTORY OF NUMBER

Karenleigh A. Overmann

INTRODUCTION

Materiality is important to numbers: Consider the relative difficulty of computing with and without material forms—fingers, pencil-and-paper, or calculator. For almost everyone, the former is far easier and permits greater accuracy, speed, and complexity of calculation than the latter. Visualizing and remembering numbers are similarly facilitated, as material devices (aside from the fingers) present and retain numerical information over longer spans of time and in greater volumes. But material devices for counting and computing are more than external, ancillary support to the brain: When we manipulate numbers, whether on our fingers, as beads on an abacus, or in the form of equations with written notations, thinking and doing with numbers are intertwined to a degree that it is difficult, and perhaps meaningless, to try to separate them.

The centrality of material forms to numbers makes them components of a cognitive system in which brains, bodies, and materiality interact to produce systemic outcomes greater than what any part can achieve in isolation (Malafouris, 2013). The ability to recruit and incorporate materiality is the trait that particularly distinguishes human cognition from that of other species. In this view, cognition is *extended* (i.e., includes materiality as a constitutive component) and *enacted* (our interactions with materiality are "transformational and not merely informational"; di Paolo, Rhohde, & de Jaegher, 2010, p. 39). As a cognitive component, materiality has agency and semiotic function (Malafouris, 2013): As we engage them or the world through them, material forms shape our behavioral and psychological responses, and over generations of such interactivity, our material forms become increasingly fine-tuned to elicit specific psychological and behavioral responses. Materiality acquires semiotic value through this interactivity: What it is and what we do with it give it meaning (Malafouris, 2013). The latter is particularly important in numerical cognition, as materially based meaning precedes language for numbers, and change in material form can alter how numbers are conceptualized and structured.

Since numbers involve material forms and such artifacts and behavioral traces may be all there is for reconstructing past ways of thinking, archaeology, as the science of materiality past and present, is the discipline best equipped to discern where and when numbers originated. Certainly, the material records of Middle Stone Age (MSA) Africa and Upper Paleolithic (UP) Europe provide tantalizing hints of possible counting devices: stringed beads from Blombos Cave, South Africa (75,000 years ago; d'Errico, Henshilwood, Vanhaeren, & van Niekerk, 2005), a notched bone from

Border Cave in the Lebombo Mountains, South Africa (42,000 years; d'Errico et al., 2012), and stenciled handprints in Cosquer Cave, France (Clottes, Courtin, & Vanrell, 2005; Overmann, 2014; Rouillon, 2006). Determining whether such artifacts might have represented counting, however, can be difficult to achieve through archaeological analysis.

Prehistoric devices may look like modern ones for counting, but resemblance is an unreliable criterion, subject to typological imposition. Prehistoric devices are often isolated in time and space, so they cannot signify unambiguous meaning or change over time. Manufactural characteristics (e.g., surface disposition and the use of same or different tools in creating marks; d'Errico, 1991, 1998) and use-wear patterns are often ambiguous regarding the possible use of prehistoric devices for counting, especially when there are other, quite plausible social purposes (e.g., beads are ornaments, notches can decorate, and handprints have ritualistic intentions we may never fully grasp). Further, counting in traditional societies generally involves materials unlikely to leave an archaeological trace—fingers, gestures, notched sticks, knotted strings, torn leaves, body paint, marks on the ground, pebbles, grains, and so on—implying that the archaeological record may underestimate the time depth for counting, perhaps significantly so, especially on timescales and in environments where materials readily and reliably decay.

EVOLUTIONARY PRECURSORS OF THE HUMAN CAPACITY FOR NUMBERS

Besides the material record, we can consider the other components in the cognitive system for numbers (e.g., brain function and form, physical anatomy, and behaviors). As material forms are recruited and incorporated, the other components interact with them to produce number systems with striking cross-cultural similarities (e.g., they start with subitizable quantities and often use the hands to represent numbers; behaviors like pairing and one-to-one correspondence can scaffold numerical concepts, thus obviating the need for pre-existing numerical lexicons). The ability to recruit and incorporate materiality into numerical cognition has several evolutionary precursors, including quantity perception, language, concept formation and manipulation, parietal encephalization, and categorization. Many of these precursors have implications for the timeline on which the human capacity for numbers likely developed. Others bear on what numbers are as concepts, or they inform the materiality, behaviors, and psychological functioning used in numbers, including those of the past. Still others are integral to the way humans think about numbers, and influence the material forms used to represent and manipulate them, just as the material forms influence structure and other characteristics in numerical concepts.[1]

[1] While the discussion here focuses on how number systems elaborate, it is not assumed that they necessarily do so. Absent the requisite conditions (e.g., size and internal complexity of groups; contact between groups) that motivate elaboration, number systems may be stable over long periods of time or possibly decline through disuse, rather than grow and elaborate.

Quantity Perception

Objects have quantity, and organisms have the ability to appreciate quantity, the perceptual ability known as *numerosity* (Piazza & Izard, 2009). Besides humans, a range of other species has demonstrated the ability to appreciate quantity—not just other primates but mammals generally, as well as birds, amphibians, fish, and perhaps even insects. Thus, it is reasonable to assume the hominin species ancestral to humans appreciated quantity too, and that the ability would have been a function of the parietal lobe, as it is in contemporary primate species (Orban et al., 2006). Yet despite demonstrating the ability to appreciate quantity, no other species has, or is suspected of having, number concepts. Something other than the ability to subitize (i.e., rapidly and unambiguously distinguish small quantities up to three or four) and appreciate magnitude (discern bigger from smaller in quantities above the subitizing range if the difference is above a noticeability threshold) is needed to yield numbers from numerosity.

Language

Linguists have speculated that the ability needed to crystalize concepts of discrete quantity from the perceptual experience of quantity is language, or a computational capability that underlies both language and numbers to which language provides access (Chomsky, 2004; Hurford, 1987, 2007). However, cross-linguistic patterning in numerical language suggests the influence of language-external mechanisms, rather than rules inherent to language alone. For example, the first lexical numbers to emerge across languages share a characteristic "one, two, three, many" patterning (e.g., Menninger, 1992)[2], as does grammatical number (the differentiation of singulars from plurals) if and when languages develop this subsequent feature (Overmann, 2015). As this patterning is consistent with the perceptual experience of quantity, it suggests that the structure of language is influenced by the perceptual experience of quantity. Another language-external mechanism is suggested by the way words for numbers are compounded. Grouping emerges because it becomes increasingly difficult to distinguish quantities above about four, known as the subitization constraint. Such grouping is often done by amounts that imply the fingers, "ten-ness" that language absorbs in producing lexical terms for higher quantities. That is, there is no inherent "ten-ness" to language; rather, grouping reflects the use of material forms (i.e., the fingers), implying linguistic structure is influenced by the material forms used for numbers.

Language is undoubtedly key in explicating, expressing, and conventionalizing numerical concepts (e.g., Saxe, 2012). However, the assumption that humans have

[2] Menninger's work, originally published in 1934, is dated but provides an example of the myriad observations that the first numbers to emerge are consistent with the perceptual experience of quantity (e.g., individuated names for subitizable quantities and "many" for unindividuable non-subitizable quantities). There is in fact significant variability in how these emerge: for example, in whether "one" emerges before "two," "two" before "one," or both together as defined against one another (e.g., Closs, 1993), and, if there is a "three," whether it is formed as "two and one," "one and two," "one and one and one," or is uniquely named. See discussion in Epps (2006) and the reply to Coolidge's commentary on Overmann (2015).

numbers because they have language ignores the semiotic meaningfulness of material quantity to perceptual experience. Further, while it is necessary, language is also insufficient for developing numbers, since all human societies have language but some lack numbers (most famously, the Pirahã of Amazonian Brazil). Numerical cognition is also highly dependent on manuovisual access to the material forms that make numerical concepts tangible, manipulable, and representable (Malafouris, 2010). Indeed, language has been called a "slow and hesitant" means of accessing numerical intuitions (Sfard & Linchevski, 1994, p. 198). Further, the prevalence of alinguistic numerical representation by means of fingers, gesture, and material devices (attested ethnographically) suggests that number concepts develop through material engagement in advance of the development and availability of an associated vocabulary. This in turn suggests the ability to manipulate materiality—indeed, the ability to incorporate materiality into the cognitive system for numbers as a constitutive component (Malafouris, 2013)—is the critical piece that yields number concepts from numerosity.

Concept Formation and Manipulation

When extant brains perform mathematical tasks, they recruit neurological circuits involved in planning and executing motor movements (Andres, Seron, & Olivier, 2007; Heimann, Umilta, & Gallese, 2013; Penner-Wilger et al., 2007; Tschentscher, Hauk, Fischer, & Pulvermüller, 2012). The temporal roots of recruiting motor circuits to manipulate concepts may be quite deep. About 2 million years ago, the habilines demonstrated the earliest unambiguous sign of a new relationship with materiality: They began to retain and reuse the stones they had fashioned and used, rather than discarding them and creating new ones on subsequent occasions. This separation of an object from the process for which it had been made and used created what may have been the first concept of "tool" as an object with an existence separate from both process and maker, a concept that likely recruited the motor circuits involved in tool manipulation (Coolidge & Wynn, 2018).

The use of "neural muscles" in manipulating both objects and numbers is demonstrated by the mental abacus, a physical procedure in which practitioners use their fingers to operate an imaginary device to perform mathematical calculations with remarkable complexity, accuracy, and speed. Mental abacus performance is uninterrupted when practitioners keep their hands motionless but disrupted if the hands perform a different task, implying that motor circuit planning but not execution is essential (Brooks, Barner, Frank, & Goldin-Meadow, 2014; Frank & Barner, 2012). However, we need not suspect the habilines of developing number concepts, but merely recognize that the ability to leverage motor circuit planning to manipulate both tools and concepts likely began with them, which highlights the temporally deep roots of incorporating materiality into human cognition.

Parietal Encephalization

Anatomically modern humans (AMH) appeared in the fossil record about 200,000 years ago (Shea, Fleagle, & Assefa, 2007).[3] One of the traits distinguishing

[3] More recent evidence appears to push this date back to 300,000 years ago (Richter et al., 2017).

AMH from earlier species is the characteristic globular shape of the skull, influenced by the encephalization of the parietal lobe (Bruner, 2004, 2010, 2014; Bruner, de la Cuétara, & Holloway, 2011; Bruner et al., 2004; Bruner, Manzi, & Arsuaga, 2003). This volumetric increase is pertinent to numbers because the parietal lobe is the region of the brain most closely associated with them, including quantity perception (Piazza & Izard, 2009); numerical reasoning (Amalric & Dehaene, 2016); finger-counting (Kringinger et al., 2011); losing the abilities to finger-count and reason numerically (Roux, Boetto, Sacko, Chollet, & Trémoulet, 2003); manipulating numbers in verbal form (Dehaene, Piazza, Pinel, & Cohen, 2003); retrieving arithmetic facts from memory (Grabner, Ansari, et al., 2009); mathematical competence (Grabner, Ischebeck, et al., 2009); abstraction (Harvey, Klein, Petridou, & Dumoulin, 2013); metaphorizing to express abstract concepts like numbers (Lakoff & Núñez, 2000); inner speech (Geva et al., 2011); and the recursive "merge" function responsible for building compound words for higher numbers (Hung et al., 2015).

Globularity would have maximized inter- and intra-regional connectivity and avoided some of the evolutionary trade-offs associated with encephalization (Gibson, 1991; Kaas, 2000; Rilling & Insel, 1999), while parietal expansion would have co-located regions involved in number with those supporting control of the fingers, tool use, and language functions like metaphorizing and inner speech, facilitating their interaction, which seems critical to expressing numerical intuitions linguistically (Coolidge & Overmann, 2012). The general relation between a brain region's size and the adaptive value of its psychological function (Rehkämper, Frahm, & Mann, 1995) relates parietal expansion to important aspects of the human ecological niche. However, the extent to which gross morphological change can index concomitant or subsequent neural reorganization or change in psychological function is unclear. Certainly, the idea that there may have been a lag, possibly as much as tens of thousands of years, between the appearance of the modern skull and the onset of behavioral modernity suggests that anatomical modernity may do little more than signal neuroanatomical developments later supporting parietal functions important to numbers. The lag certainly reflects matters like the limits on radiocarbon dating materials older than 50,000 years and has been strongly challenged on definitional and preconceptual grounds (e.g., McBrearty & Brooks, 2000). Minimally, however, the change in gross morphology associated with AMH suggests the requisite neuroanatomy for numbers may have been available as early as 200,000 years, with the requisite functionality for numbers likely being realized at some subsequent time.

Categorization

As concepts, numbers involve categorical judgments of *relations,* or sameness and difference in multiple elements (e.g., AA and BB share the quality of containing identical elements, while AB and AD do not; Christie & Gentner, 2007). Numbers start with the judgment that two sets of objects share or differ in quantity (the property of *cardinality,* the number of objects in a set), subsequently represented by one of the sets in the comparison (often the hands, setting up the basic patterning by fives and tens). Such comparisons take the form of pairing and one-to-one correspondence, behaviors that manipulate sets of objects into arrangements whereby their cardinality can be compared (and for which language for numbers is not required). Like humans, other species can form categorical judgments of *identity* (i.e., A is identical to A but differs from B). Non-human great apes can also form categories of relations, but for them it is neither intuitive nor effortless (like it is for human children), requires specific training (while it is spontaneous in humans), and even with training never appears to become as powerful (Christie & Gentner, 2007; Langer, 1986; Thompson & Oden, 2000; Thompson, Oden, & Boysen, 1997). Non-human great apes also apparently find it difficult to distinguish cross-dimensional relations (e.g., recognizing that aaaaaaaa and BB are both composed of identical elements, aaaaaaaa and cde share small letter size, and BBB and %^&, which share neither elements nor size or kind, are both trios; Gentner & Colhoun, 2010).

Human categorization exceeds that of even closely related species, partly because of our larger brains (cortical volume, adjusted for body size, correlates with intelligence; Falk, 2011) and partly because of our ability to exploit materiality, behavioral complexity that is increasingly attested after 100,000 years. Our use of material culture is not merely motivating (i.e., things to count and reasons to count them); rather, our embodied interactivity with materiality has the potential to generate concepts (e.g., as categorizations of shared quantity are opportunized by behaviors that rearrange sets of objects). Material structures provide anchoring and stability for categorizations (Hutchins, 2005), a continuity that the brain cannot achieve on its own (Fauconnier & Turner, 1998), and a means of representing and manipulating them (Malafouris, 2010). Thus, incorporating material culture into the human cognitive system may be why our categorization has become more complex than that of other primates. Timeline-wise, in signaling social identity, the Blombos beads almost certainly index categorical judgments of social relations, along with the potential for forming other types of relational categories at 75,000 years.

Ordinality

Given concepts of *how many*, their ordering by increasing magnitude is critically important, giving numbers "most of their mathematical properties" (Russell, 1920, p. 29). Sequence, the ability to order, informs a wide variety of cognitive functions, including language, perception, executive functioning, organization and timing, learning, and movement (Bapi, Pammi, Miyapuram, & Ahmed, 2005; Salthouse, 2005). In numbers, sequence informs the ordering of counting numbers from low to high. There is evidence suggesting that sequencing involves dedicated neural circuitry: The condition known as synesthesia implies "cross-talk" between neural networks (e.g., an ordinal category experienced in color implies that neural activity related to the latter intrudes into neural activity related to the former) (Cohen Kadosh, Gertner, & Terhune, 2012; Gertner, Henik, Reznik, & Cohen Kadosh, 2013). Given possible dedicated neural circuitry and the fact that what humans sequence are cultural items (e.g., beads, days of the week, alphabets), the ability is likely to have co-evolved with and through our engagement of material culture (Wynn, Overmann, Coolidge, & Janulis, 2017).

Demographic Density

However, just as we did not suspect the habilines of numbers, we need not assume archaic humans were necessarily capable of using even restricted numbers[4] as recently as 75,000 years ago, despite their modern anatomy and behavior. In addition to numerosity, "neural muscles" for concept formation and manipulation, enhanced parietal functionality, and the expanded categorizing associated with material culture, yet another critical piece was needed: demographic density. There is an oft-noted (though poorly explained) correlation between the size of and contact between social groups and number system complexity (e.g., Epps, Bowerin, Hansen, Hill, & Zentz, 2012). Small groups living in relative isolation are more likely to have restricted numbers. Currently classified as restricted, for example, is the Pirahã number system, which has been variously described as having no numbers at all or counting only to one or two (Dryer & Haspelmath, 2013; Everett, 2005; Gordon, 2004). Simply, small or isolated groups may be able to manage their resources quite well without counting; groups that become larger and/or less isolated have increased resource management needs that they respond to by developing or extending counting.

For prehistoric peoples, the archaeological evidence suggests that for much of prehistory (i.e., prior to the Neolithic[5]), small groups lived in contact situations ranging from relative isolation to at least some contact with other groups, and which could also include more extensive contact, suggested by exotic materials being found at significant distances from their origin points. This broad scenario, were we talking about extant rather than prehistoric peoples, would be consistent with restricted numbers while admitting the possibility of some systems with numbers higher than 20 (i.e., in contexts with greater contact, etc.). If the "but archaic and extant brains aren't identical" objection is set aside, a reasonable inference based on the correlation between number system complexity and social demographics is that with the evolutionary prerequisites in place, prehistoric peoples could have had restricted numbers 75,000 years ago, well before the Neolithic. In fact, this inference fits with the artifactual evidence: The African beads, African bone, and Cosquer Cave handprints accumulated quantities consistent with restricted numbers.

INSIGHT INTO RECENT NUMERICAL PREHISTORY

For additional insight into the prehistory of numbers, we can come at the problem from the other direction, starting not with the earliest possible artifacts but the first unambiguous numbers and working backward to their likely precursors. The earliest archaeologically attested numbers come from the Ancient Near East (ANE), southern Mesopotamia in the Neolithic (8300–4500 BC). The cultural groups most closely

[4] Restricted numbers count no higher than 20 (Comrie, 2013) but often no higher than much smaller numbers, like 2 or 3.

[5] During the Neolithic, group sizes significantly increased, from dispersed hunting and gathering groups to small settlements and villages, to larger settlements, and ultimately to urban city-states. While this demographic change occurred at different times in different parts of the world, it relates directly to the requisite conditions for numerical elaboration and was presumably attended by it.

associated with these earliest numbers, and with the invention of writing as well, are the Sumerians (the first urban civilization in what is now southern Iraq), Akkadians (the first Semitic-speaking Mesopotamian empire in northern Iraq and Syria), and Elamites (an ancient cultural group in southwestern Iran). By the mid-fourth millennium BC, Mesopotamia was characterized by unprecedentedly large populations living in dense urban centers with intensive agriculture, massive construction projects, and intense periodic warfare, matters requiring extensive workforces and bureaucracies. The earliest unambiguous numbers took the form of numerical impressions on clay tablets. These are understood as an intermediate form between the clay tokens used for numbers in the Neolithic and the proto-cuneiform and cuneiform numbers used in later periods (into the Old Babylonian period, 1900–1600 BC): They are temporally intermediate, and they also bridge the earlier and later technologies through correspondences of shapes and sizes (Amiet, 1972; Schmandt-Besserat, 1992a). The correspondences warrant interpreting at least some of the Neolithic clay tokens as having numerical purpose (Englund, 2004a; Høyrup, 2002).

Objects purported to have been numerical tokens have been found as early as the 10th millennium BC (Moore, 2000). Various problems challenge interpreting such objects as numerical earlier than the fourth millennium (i.e., when their shapes and sizes unambiguously resemble numerical impressions). For example, numerical purpose is difficult to construe from the dimensions of clay objects; archaeological context is often ambiguous or undocumented; tokens may quite plausibly may have been games, toys, or funerary or ritualistic offerings; and there are large disparities between the archaeological prevalence of tokens and the commodities they are purported to have represented (Englund, 1998a, 1998b; Friberg, 1994; Shendge, 1983; Zimansky, 1993). Further, even at the point where their numerical interpretation is most secure, tokens did not comprise a single system for representing numbers but several, with shape and size conventions varying by the commodity enumerated, city of use, and time (Friberg, 1994; Nissen, Damerow, & Englund, 1993). Conservatively, numerical notations and corresponding tokens are unambiguous in the mid-to-late fourth millennium; numerical representation prior to notations (i.e., by means of tokens only) grows increasingly ambiguous the further back in time being considered, despite the archaeological record of token finds.

In whatever period they are reasonably construed to have represented counting, tokens are often discussed as if they were the first material technology used for ANE counting. However, there are several reasons to think they were not. First, token shapes and sizes encoded metrological relations (i.e., relations between things like containers of grain or jars of oil, similar to the idea that four quarts equals a gallon; Nissen et al., 1993). This "bundling" meant that one token of a higher value was equivalent to some number of tokens of a lesser value: as few as 2 but, more often, as many as 6 or 10. Importantly, quantities like 6 and 10 are not subitizable. Non-subitizable quantities must be counted to be meaningful, implying that the people who used them would have been able to count beyond the subitizable range.[6] Indeed, when numerical tokens

[6] Alternatives include either an inefficient procedure to place tokens into one-to-one correspondence with a known standard or an inexact estimation of the (non-subitizable) quantity represented by tokens. Either would have represented a pressure toward achieving greater efficiency and accuracy.

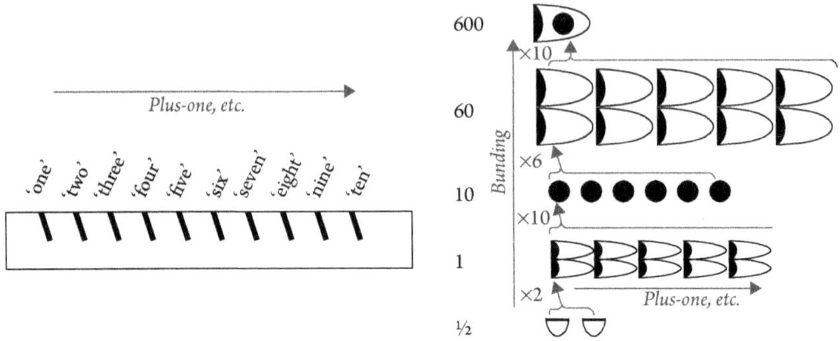

Figure 20.1. Exponential dimensions of counting devices. (Left) In a tally, the potential relations between numbers are one-dimensional (intra-exponential only) and include greater than, less than, not equal to, falls between, and plus-one. Total value is achieved by accumulating value along the single dimension. (Right) In the archaic ANE notational system for counting objects (shown: sexagesimal system S for counting most objects; Nissen et al., 1993), the relations between numbers are explicit and include those of an ordinal counting sequence (intra-exponential, the horizontal axis), as well as the relations implicit in bundling (inter-exponential, the vertical axis). Total value is achieved by accumulating value along both dimensions. The two-dimensional complexity of tokens implies one or more one-dimensional precursor (e.g., tallies, fingers). Image by the author. Developed and included as Figure 8.10 in Overmann (2016b), *Materiality in numerical cognition: Material engagement theory and the counting technologies of the Ancient Near East*, doctoral thesis, University of Oxford.

and impressions overlap in the late fourth millennium, they attest to a number system that had become capable of representing not only small non-subitizable numbers but ones in the hundreds and thousands. For example, the Cuneiform Digital Library documents at least nine clay artifacts with impressions that include a sign interpreted as 3,600 ("Cuneiform Digital Library," 2015), implying a number system that had become fairly well elaborated, as consistent with demographic factors of group size (extremely large) and inter-group contact (extensive).[7] Bundling also meant that tokens were related not only to the commodity they counted (a single dimension, which fingers and tallies instantiate) but to each other as well (a second exponential dimension), making tokens a relatively complex technology (Chrisomalis, 2010 and Figure 20.1). Two-dimensionality suggests that at least one and possibly several one-dimensional technologies necessarily preceded the two-dimensional tokens. And indeed, there is evidence of one-dimensional precursors in the region: tallies and finger-counting.

Worked bones that may have been used as tallies have been found in the Levant and dated to the Late UP (Davis, 1974; Reese, 2002; Tixier, 1974). Few similar artifacts have been found in Mesopotamia in the Neolithic, Chalcolithic, or later Bronze Age, a circumstance consistent with both non-use and failure to preserve. The latter possibility is quite likely, as organic materials like bone are known to preserve

[7] This elaboration would not have occurred in the absence of numerical concepts or words, even if the words themselves remain unknown because there was no writing system capable of expressing them at the time.

poorly in the region's climatic conditions (Coinman, 1996). Tallies are mentioned in second-millennium texts (e.g., "The debate between grain and sheep," 2005, lines 130–133). Broadly, however, tallies and tokens were used by distinct cultural groups inhabiting neighboring regions at different times. There is strong genetic differentiation (Lazaridis et al., 2016) between the Levantine hunter-gatherers of the Late UP and the Zagros Mountain groups who expanded into southern Mesopotamia during the Neolithic transition. These groups presumably represented separate numerical traditions, suggested by use of distinct bases and lexical terms by the Sumerians, Akkadians, and Elamites.

Two circumstances support the possibility of contact diffusion between these groups. First, in rejecting the idea that farmers displaced the earlier Levantine and Iranian hunter-gatherers during the Neolithic transition, "the spread of ideas and farming technology moved faster than the spread of people, as . . . determine[d] from the fact that the population structure of the Near East was maintained throughout the transition to agriculture" (Lazaridis et al., 2016). As agriculture spread through contact, numerical ideas were likely to as well, since numbers are extremely likely to diffuse between groups (attested by the prevalence of numerical loan words and similar phenomena in extant languages and number systems; Comrie, 1999, 2005). Even infrequent inter-group contact has the potential to spread numerical words, concepts, and technologies, especially given the impetus of interactions like trade; further, it need not be assumed that spreading was the only effect, since contact would have helped motivate numerical development (i.e., for purposes like ensuring reciprocity in trade).

Second, the correlation between social demographics and numerical elaboration suggests that the small, nomadic groups in periodic contact with one another in Late UP Levant could well have counted within the range of restricted numbers, given that possible tallies bearing notches within the restricted range have been found (Reese, 2002). Further, whether they were influenced by contact with Levantine groups or developed an independent numerical tradition, the Zagros populations expanding into southern Mesopotamia would have been under pressure to develop and elaborate higher numbers to support the resource management needs associated with increased group size and sedentism (15,000–10,000 BC), animal husbandry and small-scale cereal cultivation and storage (by 7000 BC), and cattle domestication (after 6500 BC) (Robson, 2008). Given the level of numerical elaboration attained by the late fourth millennium, a numerical prehistory concurrent and consistent with the demographic changes of the Neolithic transition is more plausible than the idea that numbers emerged suddenly and with two-dimensional complexity in the late fourth millennium.

Fingers are another possible one-dimensional precursor for the Neolithic tokens. Finger-counting is attested by a characteristic organization of numerical representations in groupings of 5's, 10's, and 20's, both as notations and words. For example, when the Elamites adopted the written Sumerian numerical representations whose cycles were based on tokens and metrological relations, they reorganized them as a decimal system (Englund, 2004b). Then, as cuneiform became capable of representing not just semantic value but phonetic information by the mid-third millennium, it showed that Sumerian had number-words organized by 5's, 10's, and 20's: The words for "six" through "nine" were "five-plus" compounds (e.g., the word for "six" was "five-plus-one"); the word for "30" meant "three 10's" and "40" meant "two 20's"; Blažek, 1999). Further, Akkadian (a Semitic language) was a decimal system, as numbers in Semitic languages tend to be (Friberg, 2007).

Textual evidence[8] of finger-counting has several important implications for ANE numerical cognition. First, it shows that the Mesopotamian peoples used their fingers for counting, just like extant peoples do (the latter is attested by the prevalence of decimal, quinary, and vigesimal bases and cycles in known number systems, and the persistence of "five- and ten-ness" and finger-counting despite significant levels of numerical elaboration). Second, the presence of finger-counting in modern emerging number systems implies that finger-counting may have emerged early in ANE counting (i.e., earlier than the later metrologically based systems of numerical tokens, impressions, and notations and perhaps earlier than the possible tallies found in Levantine contexts). An early emergence would be consistent with the neurological integration between the parts of the brain controlling the fingers and appreciating quantity, the use of the fingers in manipulating enumerated objects, and the ready availability of the fingers as a material device for representing quantity[9] (Overmann, 2016c). Finally, the region contained at least three indigenous numerical traditions in contact with one another.[10] Sumerian, Elamite, and Akkadian were distinct languages,[11] and differed in the details of their numerical representations (e.g., methods of compounding larger numbers from smaller ones). The interaction between these separate numerical traditions (demonstrated by notational adoption and shared linguistic forms for higher numbers) implies opportunties for numerical elaboration as a response to increased needs for numbers emerging from demographic factors, social purposes like resource management and trade, and the illuminating contrast of numerical differences.

The idea that ANE peoples realized their numbers in similar fashion to extant peoples is further supported by linguistic evidence. In contemporary languages, both the earliest lexical numbers (Menninger, 1992) and grammatical number (Corbett, 2000) take the general form "one, two, three" and (big and small) "many" across

[8] Though textual evidence of numerical language occurs much later than archaeological evidence of numerical impressions, tallies, and tokens, this is likely an artifact of lacking a writing system capable of preserving both semantic and phonetic components of language ("glottographic" writing), rather than evidence of a complex, two-dimensional number system springing into existence with unprecedented and unique rapidity.

[9] Because Mesopotamian people were characterized by anatomical and behavioral modernity, there is no reason to exclude them from sharing these traits.

[10] There is also evidence of other number systems in the region, a ternal counting system and an emesal or Sumerian dialect spoken by women that contained numbers (Lambert, 1969; Whittaker, 2002). Ternal numbers occur in scribal and priestly (i.e., male) contexts (Lambert, 1969). The emesal was possibly a literary convention rather than a woman's dialect (Whittaker, 2002), but this seems implausible because an invented counting system would be a curiously specific detail. Alternatively, either may have represented emergent number systems from a minority cultures in contact with Sumerian, Akkadian, and Elamite populations.

[11] Sumerian and Elamite are linguistic isolates, unrelated to any known language including each other, and Akkadian is an East Semitic language.

languages and cultures. This patterning conforms to and thus likely reflects the perceptual experience of quantity—subitizing and magnitude appreciation. While there is no linguistic evidence of the earliest form of their lexical numbers, "one-two-three-many" patterning does characterize grammatical number in Sumerian, Akkadian, and Elamite. Grammatical number, which distinguishes singulars from plurals, became expressible in written Sumerian, Akkadian, and Elamite during the third millennium (Dahl, 2015). As grammatical number emerges subsequent to lexical numbers (Corbett, 2000; Overmann, 2015), the presence of grammatical number in the three ancient languages implies an earlier emergence for lexical number words, though this is a relative (not absolute) chronology. Further, slight but significant differences in grammatical number suggest that the three languages developed the feature independently; this in turn suggests that the numerical traditions they reflect developed independently as well.

In total, the linguistic and archaeological evidence suggests that the Mesopotamian peoples developed their numbers in similar fashion to the way extant peoples do—through the perceptual experience of quantity, the comparison of sets for shared (or dissimilar) cardinality, and the use of fingers for numerical comparisons and representation—and that their numerical development may have started much earlier than the fourth millennium BC—possibly as early as the Late UP (in conjunction with or earlier than the Levantine tallies).

MESOPOTAMIAN ARTIFACTS AND THEIR EFFECTS ON NUMERICAL COGNITION

Analysis of the agency and semiotic function (Malafouris, 2013) of the material technologies for counting offers the possibility of gaining additional insight into their role in numerical cognition. One aspect of material agency is the idea of *affordance*, potential utility that organisms can recognize and exploit (Gibson, 1977; Greeno, 1994; Jenkins, 2008). Affordances can be encoded in material artifacts, making them available for other individuals and generations and distributing cognitive effort over space and time (Hutchins, 1995; Smith, 2007). Different material artifacts offer different affordances for counting, and newer material forms can represent extensions of and reactions to older material affordances. Essential to such an analysis is an extended sequence of material technologies used for counting and calculating, like that of the ANE (Table 20.1), which would have spanned the realization of initial number concepts (perhaps as early as the Late UP) and their elaboration into a complex mathematical system (by the Old Babylonian period). Each material form would have provided capabilities and structure, as well as limitations that would eventually motivate the incorporation of new material forms, which in turn would have been selected on the basis of affordances they shared and capabilities that differed from those of previous forms. New material forms would have solved one or more limitations of a previous form while injecting new limitations of their own; they would also have facilitated numerical elaboration by introducing new affordances, contrasts with older affordances, and the potential for generating novel patterns.

Table 20.1. Affordances and Limitations in Artifacts Used as Material Counting Technologies in the Ancient Near East

Artifact	*Affordances*	*Limitations*
Fingers	• Neurologically integrated with quantity perception • Psychological–behavioral–material bridge • Available; manipulate the objects being counted • Linearity; stable order	• Limited capacity; ephemeral • Commodity unspecified
Tallies	• Linearity; stable order • Accumulation • More capacity; persistent	• Commodity unspecified • More than three or four notches are difficult to discriminate visually; not manipulable; one-dimensional
Tokens	• Linearity; stable order (imposed: not implicit in the form) • Accumulation; greater capacity; persistent • Commodity encoded • Grouped (more discriminable); manipulable (operations; relations); two-dimensional	• Loose (imposing the need for containment); not concise; multivalent
Notations	• Linearity; stable order (imposed) • Many operations; greater capacity; persistent • Commodity specified apart from number • Grouped; two-dimensional • Integrity of form; concise; monovalent • Handwritten (fusiform gyrus becomes trained; improved hand–eye coordination; improved character recognition and recall); conceptualization as objects • Whole–part relations (with tokens) • Tables (e.g., multiplication; reciprocals) • Greater complexity of calculation	• Fixed (must calculate with tokens or manipulate relations between numbers with new algorithms)

Note: The technologies used for counting in the Ancient Near East represented four key transitions. Fingers influenced basic structure in numerical concepts (e.g., linearity; stable order). Tallies represented the transition to material culture (i.e., as opposed to using the body for counting). Tokens represented the emergence of knowledge-based numeration. Finally, written notations enabled numbers to become conceptualized as entities. Affordances and limitations show continuity and discontinuity across technologies, representing factors that motivated and influenced the selection of new material forms. Previously published as Table 1 (p. 10) in Overmann (2017), Concepts and how they get that way, *Phenomenology and the Cognitive Sciences*, available online October 31, 2017 through Springer Science + Business Media B.V. (https://doi.org/10.1007/s11097-017-9545-8).

Fingers

The digits of the hands are often an early material form used to represent number concepts, for several reasons. First, the neural interaction between the sensorimotor control of the fingers and the perception of quantity implies that finger-counting would be cognitively prepotent and likely explains its ubiquity and persistence across significant differences of numerical elaboration. Second, fingers manipulate enumerated objects, thereby bridging the psychological, behavioral, and material components of cognition (Gallagher, 2013; Mattens, 2013). Third, fingers are simply ready to hand to aid memory and mitigate attentional limits as needed. Across cultures, finger-counting becomes habitually ordered, often from an outside finger across the hand (Overmann, 2014), imparting linearity and stable order, or *ordinality*, sequential ordering by increasing magnitude (Gelman & Gallistel, 1978). This culturally inflected patterning makes numerical information more accessible (relative to random ordering or starting with the inner fingers) by decreasing demands on attention and memory and improving visual distinguishability (the latter is particularly important in *finger-montring*, the display of finger-counting patterns to others). Ordinality emerges from the interaction of psychological, behavioral, and material components: memory and ordinal sequencing, motor efficiency and habituation, and the hand's pentadactylism. However, fingers do not specify what they count, information that must be maintained separately, either in context or in memory, where it is subject to processes like interruption and forgetting. Fingers also have a relatively limited capacity for enumeration (most typically, five per hand), and they are perishable in the sense that sooner or later, the hands are needed for something other than recording quantity. These limitations may motivate the incorporation of new material forms with greater persistence and capacity for higher quantities.

Tallies

Material forms incorporated to address the fingers' limitations are likely to be structured similarly (e.g., notched tallies; knotted strings). Attributes like linearity and stable order are reinforced by the interactivity of psychological and behavioral capabilities with the material form, thereby enhancing the accessibility of numerical information. Like fingers, tallies do not specify what they count, necessitating keeping the information in context or memory. Unlike fingers, tallies solve the problems of persistence and capacity. Further, the motor activity of making notches may help enable the realization of numerical concepts like accumulation and relations like "one more." Tallies are not manipulable: It is difficult to move or subtract a notch once it has been made. Because of the subitization constraint, notches become increasingly difficult to distinguish in quantities of more than three or four; this motivates strategies like grouping or separating notches into subitizable or recognizable amounts during the manufacture process (e.g., groups of five). Because grouping and separating relate notches to each other in addition to whatever they are counting, they add a second dimension of exponential representation (Figure 20.1). Finally, tallies represent the incorporation of materiality external to the body into the cognitive system for numbers, as well as the transition to the use of materiality as a collaborative medium.

Tokens

Artifacts containing both impressions and tokens show that numerical impressions were ordered by increasing magnitude (Nissen et al., 1993). This implies that tokens were ordered in similar fashion, much as fingers and tallies had been and as would be consistent with their efficient and effective use. However, the source of that linearity and stable order is no longer apparent in the material form itself (i.e., loose tokens have a flexibility for rearrangement that notches on tallies and fingers lack). Tokens mitigated the effects of the subitization constraint through bundling, relations between token values that were encoded through conventions of shape and size (and, as discussed earlier, non-subitizable bundling amounts like 6 and 10 imply the ability to count). The shape/size conventions additionally specified what was being enumerated, resolving another limitation inherent in both fingers and tallies. Because they were loose, tokens were manipulable to a degree not possible with tallies or fingers. Manipulability enabled the performance of bundling and debundling operations, affording opportunities to conventionalize operations like subtraction and reciprocation (a type of division), form explicit relations between token values, and develop algorithms for calculating (operations, relations, and algorithms are important because they form the basis of mathematics). However, tokens were loose, and in the mid-fourth millennium, they were placed inside containers made of clay (bullae and envelopes); within centuries, the containers' external surfaces were marked with shapes resembling tokens, obviating the need for tokens inside and today widely credited as the origin of writing and the ability to represent quantity and commodity separately (e.g., Schmandt-Besserat, 1992b, 2010).

Impressions and Notations

Early numerical impressions and proto-cuneiform notations resembled tokens in their shapes, sizes, and bundling relations, conventions used to encode both quantity and commodity. In comparison, cuneiform notations separated quantity and commodity and would ultimately develop a system of place value. Impressions and notations shared the linearity and stable order of the earlier technologies. Unlike tokens, impressions and notations were fixed, making them suitable only for recording and communicating numerical information; until suitable operations and algorithms for manipulating notations were developed, tokens were still used for calculating (Friberg, 1981; Høyrup, 2002). Use of tokens for calculating and impressions/notations for recording implies comparisons between the two material forms that would have facilitated the appreciation of whole–part relations (e.g., those contained in the notion of reciprocals, as 60 relates to 2 and 30) and thus the explication and elaboration of relations between numbers. Unlike any previous technology, cuneiform notations were concise, enabling significant volumes of numerical information to be recorded (e.g., multiplication tables, tables of reciprocals) in a non-perishable, transportable, and easily recreatable medium. Scribes practiced writing tables as part of their training and in the process, learned numerical relations. This gave scribes more options for calculating than using tokens: They could additionally use information from tables or memory, a factor in developing new, complex algorithms for manipulating relations between numbers. The effects of handwriting cuneiform notations (e.g., trained object-recognition processing) would have intensified the conception of numbers as discrete objects, while the ability to write about manipulating numbers would have facilitated the development of operational concepts (Overmann, 2016a).

Conceptual Change and Structural Persistence

As concepts of discrete quantity, numbers are initially equivalences realized through behaviors like one-to-one correspondence that match fingers and objects. As forms like tallies are incorporated, numbers become collections that share quantity with enumerated objects, and the source of productive grouping once provided by fingers is no longer apparent. As manipulable forms like tokens are incorporated, numbers begin to share quantity with each other, and the source of their linearity and stable order is no longer apparent. Finally, once numbers become entities related numerically to similar entities, the source of the manipulability needed to develop relations and algorithms is no longer apparent. Simply put, numbers acquire attributes like linearity, stable order, productive grouping, and manipulable relations from the material forms used for counting; these attributes remain implicit in both the number concepts themselves and the behaviors that make and use material technologies for counting, but may no longer be implicit in newer material forms. These phenomena will be called, respectively, *conceptual change* and *structural persistence*.

Conceptual change occurs in conjunction with change in the materiality used to manipulate and represent number concepts. Change in material form is a type of abstraction (i.e., decontextualization, the extraction of content from its original circumstances to remove their influence on its meaning). Such abstraction can effectively mask immediate affordances, allowing for the emergence of novel, less constrained behavioral possibilities (Clowes & Mendonça, 2016). An example of such change is the separation of the representation of quantity and commodity (conjoined in tokens) as numerical impressions and pictographic labels, which enabled their further elaboration as mathematical and literate traditions (Schmandt-Besserat, 1992a, 1992b). Another is the development of cuneiform notations, whose concision enabled the creation of numerical tables for representing large volumes of information about the numerical relations between numbers, which in turn helped numbers become conceptualized as discrete entities related numerically to like entities. Similarly, the incorporation of tokens into a tally-based system would have fostered a concept of number that was more manipulable than an ordinal counting sequence and which contained more relations between numbers than were realizable with tallies or fingers.

Concepts may change when material forms change, but structure can persist (e.g., as in the persistence of linearity from fingers to tallies to tokens), for several reasons. One is that the material component, considered across space and time, is really a patchwork in which older and newer forms often coexist (especially when multiple, separately developed traditions interact). Another reason that structure persists is neurological, like the interaction between the intraparietal sulcus and angular gyrus that underlies, potentializes, and perpetuates finger-counting. Third, learning and practice, which represent structured interactions with materiality, are also social behaviors that transmit such knowledge between individuals and generations. Language, used for teaching, also likely reinforces qualities like linearity, as the temporal ordering of speech demands that numbers be expressed sequentially. Additionally, when numbers are entities with linear order, productive grouping, and manipulable relations, material structure is implicit in the concepts themselves, and this implicit structure may facilitate the retention and use of older material forms. Finally, materiality has agency: What a material form is capable of enables, limits, and structures what people can do with it. The structure of older material forms also influences the structuring potential of newer forms because materiality conditions people to things working in certain ways, and this narrows the range of possible behavioral outcomes. The structuring of potential solutions is also a limitation, since it narrows the range of possibilities to those that bear affinity to previous solutions. This may be why the presentation of unrelated material forms has become an effective technique for sparking novel solutions (Kirsh, 2014).

Several points need highlighting here. First, the development of ANE number concepts is not envisioned as a universal or deterministic process, but rather one whose affinities to other number systems reflects the involvement of shared psychological, physiological, and behavioral capacities, and whose variability reflects unique material choices and combinatorial decisions. Second, behaviors with material forms must be repeated at some threshold of sustained engagement to cause training effects in the brain (e.g., neuronal repurposing or regional coordination). The required threshold is currently undefined, though a similar repurposing, developing literacy from writing behaviors, appears to emerge under the behavioral demand associated with a state-level bureaucracy (Overmann, 2016a). Behavioral sustainment in turn depends on a level of social demand (also currently undefined, through likely more than what would be associated with traditional lifestyles; see e.g., Epps et al., 2012).

UP/MSA ARTIFACTS

Working forward from the evolutionary precursors and then backward from the first unambiguous numbers and their associated material forms allowed the consideration of their likely precursors and affordance interactions. This yielded new insights into the role of materiality in numerical cognition: finger-counting emerges as a function of neurological interactivity, hand availability, and object manipulation, with linearity and stable order emerging from the interactivity between the psychological, behavioral, and material components; follow-on technologies share structure and affordances with previous technologies, while injecting new capabilities and limitations into the cognitive system for numbers; and second-dimensional relations reflect those of both previous technologies and enumerated objects. These insights have implications for interpreting artifacts possibly used for counting in even earlier prehistoric periods in terms of what material forms and their change over time might suggest about matters like agency and affordances, and the availability and extent of a numerical lexicon.

Conservatively, prehistoric artifacts like beads, bones, and handprints need not represent numerical concepts, just as modern rosaries do not; one element (bead, notch) is simply matched with one of whatever is being counted (with a rosary, prayers). However, activities like making notches and stringing beads might plausibly generate numerical concepts (e.g., "accumulation," "one more"), since they represent psychological–behavioral–material interactivity. As mentioned previously, the earliest counting may leave no archaeological trace.[12] Artifacts of non-perishable materials may therefore represent a subsequent elaboration and are likely to be one-dimensional technologies (i.e., bridging fingers and two-dimensional technologies) associated with number concepts and terms occurring in the restricted range (i.e., up to 20). One-dimensional accumulation enumerates objects or events, which may or may not be recognizable from the numbers themselves once removed from cultural context (e.g., recognizable: days in a lunar cycle; not recognizable: bride-price). Given increased social needs for numbers, material forms like tallies are predicted to change over time in two ways: first, the quantities they record will increase (i.e., numbers will go up), and second, the quantities will reflect a strategy (e.g., grouping) that facilitates visual differentiation of the marks. Such changes would predictably develop under social pressures for greater efficiency in numbers, typically associated with demographic factors like larger group size and inter-group contact.

A material form like a tally entails interactivity between the psychological, behavioral, and material dimensions of cognition and thus is capable of scaffolding numerical concepts. However, a series of notches is unlikely to generate the same concept of number as that associated with, for example, a numerical notation. That is, the concept is related to the material form used to scaffold and represent it. A tally is a collection that cumulatively represents a discrete quantity. The concept of number instantiated by a tally is accessed by counting the notches or comparing them to a standard, and it represents relatively few relations between numbers (e.g., "more than," "less than," and "as many as"). In contrast, a written numeral (e.g., Old Babylonian sexagesimal numbers) is an entity that represents a discrete quantity by itself. The concept of number instantiated by a numeral is apparent from but not reducible to the form and exists within a relational system of numerical entities.

Further, notches on a tally may be collected over time, while numerals are generally handwritten as complete units. The movements and behaviors associated with each productive process engage different psychological processes and imply different opportunities for neuronal reorganization. For example, while numerosity is thought to be repurposed for recognizing symbolic numbers (Dehaene & Cohen, 2007), the extent to which similar reorganization occurs in tally-based numerical recognition is unclear. Similarly, fostering inter-regional coordination, as in recruiting the angular gyrus for arithmetic facts (Grabner, Ansari, et al., 2009; Grabner, Ischebeck, et al., 2009), depends on the availability of arithmetic facts (e.g., relations between numbers) that may not obtain in tally-based numerical representation. Neuronal engagement and training effects also require recurrence of the instantiating behavior, something plausibly associated with social demand. Whether the creation and use of, for example, an MSA tally would fall above the behavioral recurrence threshold required to repurpose neuronal functionality (in one or more individuals) is unknown. This is further complicated by the circumstance that demand thresholds for behavioral recurrence have not been established.

[12] Early counting may leave linguistic traces (e.g., "one-two-three-many" patterning, "five-plus" compounds) that provide insight into past psychological, behavioral, and material characteristics. However, language reconstruction is limited to roughly the past 10,000 years, less or more depending on whether the focus is lexical similarity or structural features (Dunn, Terrill, Reesink, Foley, & Levinson, 2005).

UP/MSA artifacts like the beads from Blombos Cave (d'Errico et al., 2005), the notched bone from Border Cave (d'Errico et al., 2012), and the stenciled handprints in Cosquer Cave (Rouillon, 2006) are either ungrouped (notches and handprints) or ambiguous regarding grouping (beads). This implies number concepts falling within the restricted range, as well as a lack of social pressure for higher numbers and grouping (and this is consistent with established demographic factors). The lack of social pressure, in turn, suggests a limited participation in behaviors associated with counting and thus a limited amount of neural repurposing. Absent social purposes and associated behaviors for numbers related to demographic factors, the elaboration of number concepts to any significant extent may have occurred as late as the Neolithic.

Some 30,000 years separate the Blombos beads from the Lebombo bone, another 15,000 the Lebombo bone and Cosquer handprints. Though all of them are the kind of material forms predictable for early counting, none settles the question of numerical origins or can be reasonably associated with more than restricted numbers. If at least some of them represented counting, as may be the case for the tallies in particular, the temporal span suggests a relative stasis between realizing and elaborating numbers that may in turn reflect the lack of significant demographic pressure. Beyond demographic factors are cultural characteristics like resourcing and technology (e.g., subsistence, sedentism, and specialization) with implications for the elaboration of material culture that emerged in MSA Africa and UP Europe. A rich material culture may be critical to realizing and elaborating numbers: Material culture is an integral part of the cognitive system generally but also provides objects whose value motivates counting, objects whose form enables and structures counting, and objects unrelated to either that may help spark the innovation and creativity that transcend habit and entrenched material structure. And once available, numbers are a cognitive technology that enables the management of complexity, allowing for even greater complexity to emerge.

ACKNOWLEDGMENTS

I would like to thank Steven Chrisomalis and Lyn Wadley for their constructive reviews of this chapter and, of course, Thomas Wynn and Frederick Coolidge for their inspiration, mentorship, and friendship.

REFERENCES

Amalric, M., & Dehaene, S. (2016). Origins of the brain networks for advanced mathematics in expert mathematicians. *Proceedings of the National Academy of Sciences of the United States of America, 113*(18), 4909–4917.

Amiet, P. (1972). *Mémoires de la délégation archéologique en Iran, Tome XLIII, Mission de Susiane* (Vol. I). Paris: Librairie Orientaliste Paul Geuthner.

Andres, M., Seron, X., & Olivier, E. (2007). Contribution of hand motor circuits to counting. *Journal of Cognitive Neuroscience, 19*(4), 563–576.

Bapi, R. S., Pammi, V. S. C., Miyapuram, K. P., & Ahmed, A. (2005). Investigation of sequence processing: A cognitive and computational neuroscience perspective. *Current Science, 89*(10), 1690–1698.

Blažek, V. (1999). *Numerals: Comparative etymological analyses and their implications: Saharan, Nubian, Egyptian, Berber, Kartvelian, Uralic, Altaic and Indo-European languages*. Brno, Czech Republic: Masarykova Univerzita.

Brooks, N. B., Barner, D., Frank, M. C., & Goldin-Meadow, S. (2014). Abacus: Gesture in the mind, not the hands. In P. Bello, M. Guarini, M. McShane, & B. Scassellati (Eds.), *Proceedings of the Cognitive Science Society* (Vol. 36, pp. 72–73). Austin, TX: Cognitive Science Society.

Bruner, E. (2004). Geometric morphometrics and paleoneurology: Brain shape evolution in the genus *Homo*. *Journal of Human Evolution, 47*(5), 279–303.

Bruner, E. (2010). Morphological differences in the parietal lobes with the human genus: A neurofunctional perspective. *Current Anthropology, 51*(1), S77–S88.

Bruner, E. (2014). Functional craniology, human evolution, and anatomical constraints in the Neanderthal braincase. In T. Akazawa, N. Ogihara, H. C. Tanabe, & H. Terashima (Eds.), *Dynamics of learning in Neanderthals and modern humans: Cognitive and physical perspectives* (Vol. 2, pp. 121–129). Tokyo: Springer.

Bruner, E., de la Cuétara, J. M., & Holloway, R. L. (2011). A bivariate approach to the variation of the parietal curvature in the genus *Homo*. *Anatomical Record, 294*(9), 1548–1556.

Bruner, E., Manzi, G., & Arsuaga, J. L. (2003). Encephalization and allometric trajectories in the genus *Homo*: Evidence from the Neandertal and modern lineages. *Proceedings of the National Academy of Sciences of the United States of America, 100*(26), 15335–15340.

Bruner, E., Saracino, B., Ricci, F., Tafuri, M., Passarello, P., & Manzi, G. (2004). Midsagittal cranial shape variation in the genus *Homo* by geometric morphometrics. *Collegium Antropologicum, 28*(1), 99–112.

Chomsky, N. (2004). *The generative enterprise revisited: Discussions with Riny Huybregts, Henk van Riemsdijk, Naoki Fukui and Mihoko Zushi*. Berlin: Mouton de Gruyter.

Chrisomalis, S. (2010). *Numerical notation: A comparative history*. Cambridge, UK: Cambridge University Press.

Christie, S., & Gentner, D. (2007). Relational similarity in identity relation: The role of language. In S. Vosniadou, D. Kayser, & A. Protopapas (Eds.), *Proceedings of the 2nd European Cognitive Science conference* (pp. 402–406). Mahwah, NJ: Lawrence Erlbaum Associates.

Closs, M. P. (1993). Native American number systems. In M. P. Closs (Ed.), *Native American mathematics* (pp. 3–43). Austin, TX: University of Texas Press.

Clottes, J., Courtin, J., & Vanrell, L. (2005). *Cosquer redécouvert*. Paris: Seuil.

Clowes, R. W., & Mendonça, D. (2016). Representation redux: Is there still a useful role for representation to play in the context of embodied, dynamicist and situated theories of mind? *New Ideas in Psychology, 40*(Special Issue on Representation), 26–47.

Cohen Kadosh, R., Gertner, L., & Terhune, D. B. (2012). Exceptional abilities in the spatial representation of numbers and time: Insights from synesthesia. *The Neuroscientist, 18*(3), 208–215.

Coinman, N. R. (1996). Worked bone in the Levantine Upper Paleolithic: Rare examples from the Wadi al-Hasa, West-Central Jordan. *Paléorient, 22*(2), 113–121.

Comrie, B. (1999). Haruai numerals and their implications for the history and typology of numeral systems. In J. Gvozdanović (Ed.), *Numeral types and changes worldwide* (pp. 81–94). Berlin: Mouton de Gruyter.

Comrie, B. (2005). Endangered numeral systems. In J. Wohlgemuth & T. Dirksmeyer (Eds.), *Bedrohte Vielfalt: Aspekte des Sprach(en)tods* (pp. 203–230). Berlin: Weißensee-Verlag.

Comrie, B. (2013). Numeral bases. In M. S. Dryer & M. Haspelmath (Eds.), *World atlas of language structures online*. Leipzig: Max Planck Institute for Evolutionary Anthropology.

Coolidge, F. L., & Overmann, K. A. (2012). Numerosity, abstraction, and the emergence of symbolic thinking. *Current Anthropology, 53*(2), 204–225.

Coolidge, F. L., & Wynn, T. (2018). *The rise of* Homo sapiens: *The evolution of modern thinking* (2nd ed.). Oxford, UK: Oxford University Press.

Corbett, G. G. (2000). *Number*. Cambridge, UK: Cambridge University Press.

Cuneiform Digital Library. (2015). Oxford, UK: Oriental Institute, University of Oxford.

Dahl, J. L. (2015). Comment on "Numerosity structures the expression of quantity in lexical numbers and grammatical number." *Current Anthropology, 56*(5), 647–648.

Davis, S. J. M. (1974). Incised bones from the Mousterian of Kebara cave (Mount Carmel) and the Aurignacian of Ha-Yonim cave (Western Galilee), Israel. *Paléorient, 2*(1), 181–182.

Dehaene, S., & Cohen, L. (2007). Cultural recycling of cortical maps. *Neuron, 56*(2), 384–398.

Dehaene, S., Piazza, M., Pinel, P., & Cohen, L. (2003). Three parietal circuits for number processing. *Cognitive Neuropsychology, 20*(3), 487–506.

d'Errico, F. (1991). Microscopic and statistical criteria for the identification of prehistoric systems of notation. *Rock Art Research, 8*, 83–93.

d'Errico, F. (1998). Palaeolithic origins of artificial memory systems: An evolutionary perspective. In C. Renfrew & C. Scarre (Eds.), *Cognition and material culture: The archaeology of symbolic storage* (pp. 19–50). Cambridge, UK: McDonald Institute for Archaeological Research.

d'Errico, F., Backwell, L., Villa, P., Degano, I., Lucejko, J. J., Bamford, M. K., . . . Beaumont, P. B. (2012). Early evidence of San material culture represented by organic artifacts from Border Cave, South Africa. *Proceedings of the National Academy of Sciences of the United States of America, 109*(33), 13214–13219.

d'Errico, F., Henshilwood, C. S., Vanhaeren, M., & Van Niekerk, K. L. (2005). *Nassarius kraussianus* shell beads from Blombos Cave: Evidence for symbolic behaviour in the Middle Stone Age. *Journal of Human Evolution, 48*(1), 3–24.

di Paolo, E. A., Rohde, M., & de Jaegher, H. (2010). Horizons for the enactive mind: Values, social interaction, and play. In J. Stewart, O. Gapenne, & E. A. di Paolo (Eds.), *Enaction: Toward a new paradigm for cognitive science* (pp. 33–87). Cambridge, MA: MIT Press.

Dryer, M. S., & Haspelmath, M. (2013). *World atlas of language structures online*. Leipzig: Max Planck Institute for Evolutionary Anthropology.

Dunn, M., Terrill, A., Reesink, G., Foley, R. A., & Levinson, S. C. (2005). Structural phylogenetics and the reconstruction of ancient language history. *Science, 309*(5743), 2072–2075.

Englund, R. K. (1998a). Review: Denise Schmandt-Besserat, *How writing came about. Written Language and Literacy, 1*, 257–261.

Englund, R. K. (1998b). Texts from the Late Uruk period. In *Mesopotamien: Späturuk-Zeit und Frühdynastische Zeit* (pp. 15–233). Freiburg, Switzerland: Universitätsverlag.

Englund, R. K. (2004a). Proto-cuneiform account books and journals. In M. Hudson & C. Wunsch (Eds.), *Creating economic order: Record-keeping, standardization and the development of accounting in the ancient Near East* (pp. 23–46). Bethesda, MD: CDL Press.

Englund, R. K. (2004b). The state of decipherment of proto-Elamite. In S. D. Houston (Ed.), *The first writing: Script invention as history and process* (pp. 100–149). Cambridge, UK: Cambridge University Press.

Epps, P. (2006). Growing a numeral system: The historical development of numerals in an Amazonian language family. *Diachronica, 23*(2), 259–288.

Epps, P., Bowerin, C., Hansen, C. A., Hill, J. H., & Zentz, J. (2012). On numeral complexity in hunter-gatherer languages. *Linguistic Typology, 16*(1), 41–109.

Everett, D. L. (2005). Cultural constraints on grammar and cognition in Pirahã: Another look at the design features of human language. *Current Anthropology, 46*(4), 621–646.

Falk, D. (2011). Evolution of the primate brain. In W. Henke, I. Tattersall, & T. Hardt (Eds.), *Handbook of paleoanthropology* (Vol. Vol. II, pp. 1133–1162). Berlin: Springer-Verlag.

Fauconnier, G., & Turner, M. (1998). Conceptual integration networks. *Cognitive Science, 22*(2), 133–187.

Frank, M. C., & Barner, D. (2012). Representing exact number visually using mental abacus. *Journal of Experimental Psychology: General, 141*(1), 134–149.

Friberg, J. (1981). Methods and traditions of Babylonian mathematics: Plimpton 322, Pythagorean triples, and the Babylonian triangle parameter equations. *Historia Mathematica, 8*(3), 277–318.

Friberg, J. (1994). Preliterate counting and accounting in the Middle East: A constructively critical review of Schmandt-Besserat's *Before writing. Orientalistische Literaturzeitung, 89*(5–6), 477–489.

Friberg, J. (2007). *A remarkable collection of Babylonian mathematical texts.* New York: Springer Science + Business Media.

Gallagher, S. (2013). The enactive hand. In Z. Radman (Ed.), *The hand, an organ of the mind: What the manual tells the mental* (pp. 209–225). Cambridge, MA: MIT Press.

Gelman, R., & Gallistel, C. R. (1978). *The child's understanding of number.* Cambridge, MA: The President and Fellows of Harvard College.

Gentner, D., & Colhoun, J. (2010). Analogical processes in human thinking and learning. In B. M. Glatzeder, V. Goel, & A. von Müller (Eds.), *Towards a theory of thinking: Building blocks for a conceptual framework* (pp. 35–48). Heidelberg: Springer.

Gertner, L., Henik, A., Reznik, D., & Cohen Kadosh, R. (2013). Implications of number-space synesthesia on the automaticity of numerical processing. *Cortex, 49*(5), 1352–1362.

Geva, S., Jones, P. S., Crinion, J. T., Price, C. J., Baron, J.-C., & Warburton, E. A. (2011). The neural correlates of inner speech defined by voxel-based lesion symptom mapping. *Brain: A Journal of Neurology, 134*(10), 3071–3082.

Gibson, J. J. (1977). The theory of affordances. In *Perceiving, acting, and knowing: Toward an ecological psychology* (pp. 127–143). Hillsdale, NJ: Lawrence Erlbaum Associates.

Gibson, K. R. (1991). Myelination and behavioral development: A comparative perspective on questions of neoteny, altriciality and intelligence. In K. R. Gibson & A. C. Petersen (Eds.), *Brain maturation and cognitive development: Comparative and cross-cultural perspectives* (pp. 29–63). Piscataway, NJ: Aldine Transaction.

Gordon, P. (2004). Numerical cognition without words: Evidence from Amazonia. *Science, 306*(5695), 496–499.

Grabner, R. H., Ansari, D., Koschutnig, K., Reishofer, G., Ebner, F., & Neuper, C. (2009). To retrieve or to calculate? Left angular gyrus mediates the retrieval of arithmetic facts during problem solving. *Neuropsychologia, 47*(2), 604–608.

Grabner, R. H., Ischebeck, A., Reishofer, G., Koschutnig, K., Delazer, M., Ebner, F., & Neuper, C. (2009). Fact learning in complex arithmetic and figural-spatial tasks: The role of the angular gyrus and its relation to mathematical competence. *Human Brain Mapping, 30*(9), 2936–2952.

Greeno, J. G. (1994). Gibson's affordances. *Psychological Review, 101*(2), 336–342.

Harvey, B. M., Klein, B. P., Petridou, N., & Dumoulin, S. O. (2013). Topographic representation of numerosity in the human parietal cortex. *Science, 341*(6150), 1123–1126.

Heimann, K., Umilta, M. A., & Gallese, V. (2013). How the motor-cortex distinguishes among letters, unknown symbols and scribbles: A high density EEG study. *Neuropsychologia, 51*, 2833–2840.

Høyrup, J. (2002). *Lengths, widths, surfaces: A portrait of Old Babylonian algebra and its kin*. New York: Springer.

Hung, Y.-H., Pallier, C., Dehaene, S., Lin, Y.-C., Chang, A., Tzeng, O. J.-L., & Wu, D. H. (2015). Neural correlates of merging number words. *NeuroImage, 122*, 33–43.

Hurford, J. R. (1987). *Language and number: The emergence of a cognitive system*. Oxford, UK: Blackwell.

Hurford, J. R. (2007). *The origins of meaning: Language in the light of evolution*. Oxford, UK: Oxford University Press.

Hutchins, E. (1995). *Cognition in the wild*. Cambridge, MA: MIT Press.

Hutchins, E. (2005). Material anchors for conceptual blends. *Journal of Pragmatics, 37*(10), 1555–1577.

Jenkins, H. S. (2008). Gibson's "affordances": Evolution of a pivotal concept. *Journal of Scientific Psychology, 12*, 34–45.

Kaas, J. H. (2000). Why is brain size so important: Design problems and solutions as neocortex gets bigger or smaller. *Brain and Mind, 1*(1), 7–23.

Kirsh, D. (2014). The importance of change and interactivity in creativity. *Pragmatics & Cognition, 22*(1), 5–26.

Kringinger, H., Koten, J. W., Horoufchin, H., Kohn, N., Arndt, D., Sahr, K., . . . Willmes, K. (2011). The role of finger representations and saccades for number processing: An fMRI study in children. *Frontiers in Psychology, 2*, 1–12.

Lakoff, G., & Núñez, R. E. (2000). *Where mathematics comes from: How the embodied mind brings mathematics into being*. New York: Basic Books.

Lambert, W. G. (1969). Review, O. R. Gurney and P. Hulin, *The Sultantepe tablets II*. *Journal of Semitic Studies, 14*(2), 242–247.

Langer, J. (1986). *The origins of logic: One to two years*. Orlando, FL: Academic Press.

Lazaridis, I., Nadel, D., Rollefson, G., Merrett, D. C., Rohland, N., Mallick, S., . . . Reich, D. (2016). Genomic insights into the origin of farming in the ancient Near East. *Nature*, 536(7617), 419–424.

Malafouris, L. (2010). Grasping the concept of number: How did the sapient mind move beyond approximation? In C. Renfrew & I. Morley (Eds.), *The archaeology of measurement: Comprehending heaven, earth and time in ancient societies* (pp. 35–42). Cambridge, UK: Cambridge University Press.

Malafouris, L. (2013). *How things shape the mind: A theory of material engagement*. Cambridge, MA: MIT Press.

Mattens, F. (2013). Perception and representation: Mind the hand! In Z. Radman (Ed.), *The hand, an organ of the mind: What the manual tells the mental* (pp. 159–184). Cambridge, MA: MIT Press.

McBrearty, S., & Brooks, A. S. (2000). The revolution that wasn't: A new interpretation of the origin of modern human behavior. *Journal of Human Evolution, 39*(5), 453–563.

Menninger, K. (1992). *Number words and number symbols: A cultural history of numbers*. (P. Broneer, Trans.). New York: Dover Publications.

Moore, A. M. T. (2000). Stone and other artifacts. In A. M. T. Moore, G. C. Hillman, & A. J. Legge (Eds.), *Village on the Euphrates: From foraging to farming at Abu Hureyra* (pp. 165–186). Oxford, UK: Oxford University Press.

Nissen, H. J., Damerow, P., & Englund, R. K. (1993). *Archaic bookkeeping: Early writing and techniques of economic administration in the ancient Near East.* (P. Larsen, Trans.). Chicago, IL: University of Chicago Press.

Orban, G. A., Claeys, K., Nelissen, K., Smans, R., Sunaert, S., Todd, J. T., . . . Vanduffel, W. (2006). Mapping the parietal cortex of human and non-human primates. *Neuropsychologia, 44*(13), 2647–2667.

Overmann, K. A. (2014). Finger-counting in the Upper Palaeolithic. *Rock Art Research, 31*(1), 63–80.

Overmann, K. A. (2015). Numerosity structures the expression of quantity in lexical numbers and grammatical number. *Current Anthropology, 56*(5), 638–653.

Overmann, K. A. (2016a). Beyond writing: The development of literacy in the Ancient Near East. *Cambridge Archaeological Journal, 26*(2), 285–303.

Overmann, K. A. (2016b). *Materiality in numerical cognition: Material Engagement Theory and the counting technologies of the Ancient Near East.* Doctoral thesis, University of Oxford.

Overmann, K. A. (2016c). The role of materiality in numerical cognition. *Quaternary International, 405*, 42–51.

Overmann, K. A. (2017). Concepts and how they get that way. *Phenomenology and the Cognitive Sciences*, 1–16. Retrieved from https://doi.org/10.1007/s11097-017-9545-8

Penner-Wilger, M., Fast, L., LeFevre, J.-A., Smith-Chant, B. L., Skwarchuk, S.-L., Kamawar, D., & Bisanz, J. (2007). The foundations of numeracy: Subitizing, finger gnosia, and fine motor ability. In D. S. McNamara & J. G. Trafton (Eds.), *Proceedings of the Cognitive Science Society* (Vol. 29, pp. 1385–1390). Austin, TX: Cognitive Science Society.

Piazza, M., & Izard, V. (2009). How humans count: Numerosity and the parietal cortex. *Neuroscientist, 15*(3), 261–273.

Reese, D. S. (2002). On the incised cattle scapulae from the East Mediterranean and Near East. *Bonner Zoologische Beitrage, 50*, 183–198.

Rehkämper, G., Frahm, H. D., & Mann, M. D. (1995). Brain composition and ecological niches in the wild or under man-made conditions (domestication): Constraints of the evolutionary plasticity of the brain. In E. Alleva, A. Fasolo, H.-P. Lipp, L. Nadel, & L. Ricceri (Eds.), *Proceedings of the NATO Advanced Study Institute on behavioural brain research in naturalistic and semi-naturalistic settings: Possibilities and perspectives (Acquafredda di Maratea, Italy, Sept. 10–20, 1994)* (pp. 83–103). Dordrecht, The Netherlands: Kluwer Academic.

Richter, D., Grün, R., Joannes-Boyau, R., Steele, T. E., Amani, F., Rué, M., . . . McPherron, S. P. (2017). The age of the hominin fossils from Jebel Irhoud, Morocco, and the origins of the Middle Stone Age. *Nature, 546*(7657), 293–296.

Rilling, J. K., & Insel, T. R. (1999). Differential expansion of neural projection systems in primate brain evolution. *Neuroreport, 10*(7), 1453–1459.

Robson, E. (2008). *Mathematics in ancient Iraq: A social history.* Princeton, NJ: Princeton University Press.

Rouillon, A. (2006). Au Gravettien, dans la grotte Cosquer (Marseille, Bouches-du-Rhône), l'Homme a-t-il compté sur ses doigts? *Anthropologie, 110*(4), 500–509.

Roux, F., Boetto, S., Sacko, O., Chollet, F., & Trémoulet, M. (2003). Writing, calculating, and finger recognition in the region of the angular gyrus: A cortical stimulation study of Gerstmann syndrome. *Journal of Neurosurgery, 99*(4), 716–727.

Russell, B. (1920). *Introduction to mathematical philosophy* (2nd ed.). London: George Allen & Unwin.

Salthouse, T. A. (2005). Relations between cognitive abilities and measures of executive functioning. *Neuropsychology, 19*(4), 532–545.

Saxe, G. B. (2012). *Cultural development of mathematical ideas.* Cambridge, UK: Cambridge University Press.

Schmandt-Besserat, D. (1992a). *Before writing: From counting to cuneiform* (2 vols). Austin, TX: University of Texas Press.

Schmandt-Besserat, D. (1992b). *How writing came about.* Austin, TX: University of Texas Press.

Schmandt-Besserat, D. (2010). The token system of the ancient Near East: Its role in counting, writing, the economy and cognition. In C. Renfrew & I. Morley (Eds.), *The archaeology of measurement: Comprehending heaven, earth and time in ancient societies* (pp. 27–34). Cambridge, UK: Cambridge University Press.

Sfard, A., & Linchevski, L. (1994). The gains and the pitfalls of reification? The case of algebra. *Educational Studies in Mathematics, 26*(2–3), 191–228.

Shea, J. J., Fleagle, J. G., & Assefa, Z. (2007). Context and chronology of early *Homo sapiens* fossils from the Omo Kibish Formation, Ethiopia. In P. Mellars, K. Boyle, O. Bar-Yosef, & C. B. Stringer (Eds.), *Rethinking the human revolution: New behavioural and biological perspectives on the origin and dispersal of modern humans* (pp. 153–162). Cambridge, UK: McDonald Institute for Archaeological Research.

Shendge, M. J. (1983). The use of seals and the invention of writing. *Journal of the Economic and Social History of the Orient, 26*(2), 113–136.

Smith, O. F. (2007). Object artifacts, image artifacts and conceptual artifacts: Beyond the object into the event. *Artifact, 1*, 4–6.

Thompson, R. K., & Oden, D. L. (2000). Categorical perception and conceptual judgments by nonhuman primates: The paleological monkey and the analogical ape. *Cognitive Science, 24*(3), 363–396.

Thompson, R. K., Oden, D. L., & Boysen, S. T. (1997). Language-naive chimpanzees (*Pan troglodytes*) judge relations between relations in a conceptual matching-to-sample task. *Journal of Experimental Psychology: Animal Behavior Processes, 23*(1), 31–43.

Tixier, J. (1974). Poinçon décoré du Paléolithique supérieur à Ksar'Aqil (Liban). *Paléorient, 2*, 187–192.

Translation of "The debate between grain and sheep." (2005). In *Electronic Text Corpus of Sumerian Literature.* Oriental Institute, University of Oxford.

Tschentscher, N., Hauk, O., Fischer, M. H., & Pulvermüller, F. (2012). You can count on the motor cortex: Finger counting habits modulate motor cortex activation evoked by numbers. *NeuroImage, 59*(4), 3139–3148.

Whittaker, G. (2002). Linguistic anthropology and the study of emesal as (a) women's language. In S. Parpola & R. M. Whiting (Eds.), *Sex and gender in the Ancient Near East: Proceedings of the 47th Rencontre Assyriologique Internationale, Helsinki, July 2–6, 2001* (Vol. Part II, pp. 633–644). Helsinki: The Neo-Assyrian Text Corpus Project.

Wynn, T., Overmann, K. A., Coolidge, F. L., & Janulis, K. (2017). Bootstrapping ordinal thinking. In T. Wynn & F. L. Coolidge (Eds.), *Cognitive models in Palaeolithic archaeology* (pp. 197–213). Oxford, UK: Oxford University Press.

Zimansky, P. (1993). Review of Denise Schmandt-Besserat's *Before writing,* Volumes I and II. *Journal of Field Archaeology, 20*(4), 513–517.

21

ENSNARING THE MIND
COGNITIVE IMPLICATIONS OF SETTING SNARES AND TRAPS

Lyn Wadley

INTRODUCTION

Everyday tasks, especially those that involve making and using tools or various products, can provide information about complex cognition similar to ours. Sometimes the procedures needed for the tasks cannot be implemented without complex cognitive abilities, such as multitasking. An example is the manufacture and application of compound adhesive made from a combination of natural ingredients that can have variable attributes depending on the season or the place of origin (Wadley, Hodgskiss, & Grant, 2009).

Traps and snares, apparatus for remote capture, provide another excellent example of material culture that engages complex cognition. They entail delayed gratification because when assembled, they are designed to function to benefit their operators, not straightaway, but in the future. Furthermore, the capture of prey is at a distance from the hunter, out of sight. Thus, traps and snares are evidence for human ability to integrate action across space and through time (Wynn & Coolidge, 2003). They provide convincing evidence of foresight (Vaesen, 2012), not least because the hunter needs to calculate how present actions will impact unseen events in the future (Coolidge & Wynn, 2005). Premeditated acts carried out in this way, without being directly overseen by humans, involve modern executive functions of the brain, which, in turn, characterize enhanced working memory and modern cognition (Coolidge & Wynn, 2005; Wynn & Coolidge, 2003, 2007a, 2007b). The central executive is the decision-making constituent of working memory, and its roles include the prevention of peripheral thoughts or actions while an actor concentrates on achieving the aims of a specific task (Wynn & Coolidge, 2007b). Snares and traps offer a good example of complex cognition in action, and their use implies sophisticated mental concepts (Wadley, 2010) as well as technical expertise. The latter involves thorough repetition of tasks, and it is based on procedural memories with some cognitive control mechanisms of working memory, like rapid task-switching and the ability to ignore distraction (Wynn, Haidle, Lombard, & Coolidge, 2017). These cognitive attributes are requirements for setting snares, even though the technology is relatively simple. Detractors can claim that spiders create snares in the form of their webs, but such behavior is instinctive with short, simple operational sequences, "cognigrams," in which the gap between challenge and solution is much less than that revealed by human action (Haidle, 2006). In addition, the spider generally keeps watch over its web, so the action is not remote.

Here, I use some traits from the working memory model as bridging theory (Botha, 2008; Wynn, 2009) to link the mental aspects involved in trapping or snaring with modern executive functions of the brain. Further, bridging theory is then required to link the elusive archaeological data to remote-capture technology that we know from ethnography and historical records. I then examine the sort of evidence necessary for bridging the gap between archaeological fauna recovered from sites and the use of snares and traps in the past. It is a difficult topic, and my conclusions are made cautiously. My principal example of the sort of circumstantial evidence pointing to snaring in the Middle Stone Age (MSA) of southern Africa comes from the faunal record 65,000–62,000 years ago in Sibudu, South Africa, and I shall shortly describe the available data.

WHAT ARE THE TECHNICAL AND COGNITIVE REQUIREMENTS FOR SETTING A TRAP OR SNARE?

Procedural knowledge in the form of prey ecology and behavior is required before traps or snares can be set and harvested (Frison, 1991; Schmitt, Madsen, & Lupo, 2004). Nevertheless, drawing on ethology for the meat quest is not necessarily evidence of cognition solely like ours; Neandertals may also have possessed this ability (Stringer et al., 2008). Chimpanzees in Uganda practice something similar for they study human behavior; they are one step ahead of human trappers, and expediently rob either living or dead blue duiker from human-laid snares (Brand, Eguma, Zuberbühler, & Hobaiter, 2014).

When a snare is manufactured, some form of cordage is necessary, as well as a flexible sapling to act as a trigger release (Figure 21.1). The technology is unpretentious and does not involve the long action sequences necessary for creating, for example, bow-and-arrow sets (Lombard & Haidle, 2012). Notwithstanding the technical simplicity of the snare, its successful use implicates several demanding cognitive processes. A decision needs to be made about whether to create a noose for trapping the prey by its foot or neck. The size and weight of the prey and its habits will determine both the structure and the placement of the snare. Only a narrative mind can envision exactly

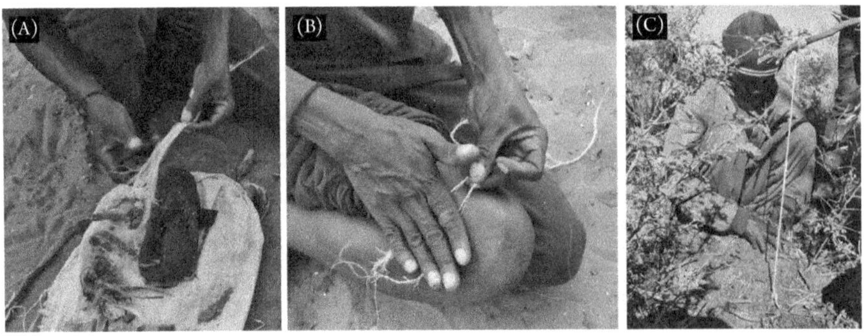

Figure 21.1. Making a simple snare in the Kalahari. (A) San hunter extracting fibers from a *Sansevieria* leaf, (B) twisting and rolling the fibers to make cord, and (C) setting a spring-loaded foot snare using a bent sapling, forked stick, and twine. Image by Gary Trower and reproduced with his permission.

where an animal's head or foot will be placed along the track, then select the correct place for an inconspicuous snare (Shaw-Williams, 2014). Furthermore, setting snares requires the ability to read an animal's tracks to understand and predict its behavior as prey. Theory of mind is a prerequisite for imagining the condition and behavior of an animal as signaled by its tracks (Shaw-Williams, 2014). Using tracks as the marker for setting a successful snare necessitates, as the minimum requirement, the cognitive capacity to concurrently hold in mind the representation of the sign (the animal tracks) and the image recalled from memory through its prior association with the sign (Stuart-Fox, 2015). In order to establish a causal relationship, both representations have to be held in what Stuart-Fox (2015) calls *incipient working memory*. Thus, causal reasoning (Lombard & Gärdenfors, 2017), through the cognitive requirements of tracking and the constant need to update the information presented by tracks, may have been one of the cognitive foundations upon which working memory was built. If this is the case, then the invention and use of the humble snare may have had far-reaching evolutionary implications. I discuss this issue again at the end of the chapter.

Over millennia, people have devised a variety of remote-capture contraptions, including pitfalls and stone-walled weirs. Pitfalls involve digging a hole in a place the prey is likely to traverse, and disguising the hole with foliage. If a small creature is to be caught, then the hole may suffice (Forbes, 2015), but a larger animal may need to be incapacitated, for example, by setting sharpened stakes in the pit. Stone walling is sometimes used to construct desert kites (Holzer, Avner, Porat, & Horwitz, 2010) and also tidal fish traps (Avery, 1975; Hine, Sealy, Halkett, & Hart, 2010; also see Figure 21.2). Traps that require people to drive game toward them do not involve delayed

Figure 21.2. Tidal fish trap in the Western Cape, South Africa. The stone walls create a weir that exploits natural pools. High tides carry fresh water and fish into the trap. When the tide retreats, the fish cannot swim past the wall and are trapped in the pool until the next high tide. Image by Graham Avery (see Avery, 1975, for a discussion of tidal fish traps) and reproduced with his permission.

gratification and are not discussed further here. Apart from the sort of traps that involve stone walling, most traps and snares have not survived, and archaeologists are obliged to use circumstantial evidence to infer their use.

TRAPS AND SNARES IN ARCHAEOLOGICAL SITES: IS THERE ANY EVIDENCE?

What credible circumstantial evidence can archaeologists use to infer the use of remote-capture-like snares in the past? Carefully examining the inventory of prey remains is one way: The use of snares should create high frequencies of faunal remains from animals otherwise difficult to obtain. Nocturnal animals or solitary forest dwellers are more easily snared than when hunted with spears or projectiles. Swift rabbits and hares and some large birds are also well suited to capture in traps or snares. Small carnivores, such as mongooses, tend to be accidentally caught in snares and traps (Lupo & Schmitt, 2002), and high frequencies of their remains most likely point to remote-capture devices rather than active hunting with weapons or nets. Small carnivores are especially prone to capture in snares, yet bigger ones may also be accidentally caught if the snare is strong enough. The reason for some of these captures is that carnivores tend to scavenge and will take bait. In Canada, for example, cougar mortalities are common when these large carnivores are accidentally caught at wolf bait stations (Knopff, Knopff, & Boyce, 2010). Remains of dangerous caracal and hyena in Klipdrift, southern Cape (Figure 21.3), may be the result of accidental snaring (Reynard, Discamps, Badenhorst, van Niekerk, & Henshilwood, 2016). Even so, the Klipdrift carnivore carcasses were used by humans because percussion and cut-marks on the bones suggest both skinning and butchery (Reynard, Discamps, Badenhorst, et al., 2016).

Small bovids lend themselves to capture by snaring (Klein, 1981), not least because many of these are non-migratory, have small home ranges, and are solitary and/or furtive. Klein reasoned that snares were used in both MSA and Later Stone Age (LSA) sites in Africa because of the similarity of the older and younger faunal collections at sites like Border Cave (Klein, 1977; also see Figure 21.3). Indeed, evidence from several African sites raises the possibility that the meat quest was similar in the MSA and more recent LSA, particularly in regard to the presence of little size class I ungulates (Brain, 1981; Marean, Abe, Frey, & Randall, 2000). Size class I bovids, such as steenbok/grysbok (*Raphicerus* spp.), blue duiker (*Philantomba monticola*), and dik-dik (*Madoqua* sp.), are well suited to being caught in snares. The mass of a blue duiker is about 5 kg, so it constitutes a really small meat package.

Large and small creatures are represented in many MSA sites. Bone in such sites is generally in the form of fragments, presumably smashed open to extract marrow. The discarded bone splinters were frequently burned; thus, it is not always possible to conduct taphonomic studies to determine whether or how the carcasses were butchered. Nevertheless, the diversity of prey in the MSA implies that people used a range of meat-getting strategies (Wadley, 2015). Active hunting with weapons like spears or arrows is suggested by the remains of animals such as zebra and wildebeest, scavenging by big, dangerous animals like rhinoceros and hippopotamus, and snaring by forest dwellers that could not readily be caught in other ways. A stone tip wedged in the vertebrae of a giant extinct buffalo at Klasies River (Figure 21.3) indicates either that

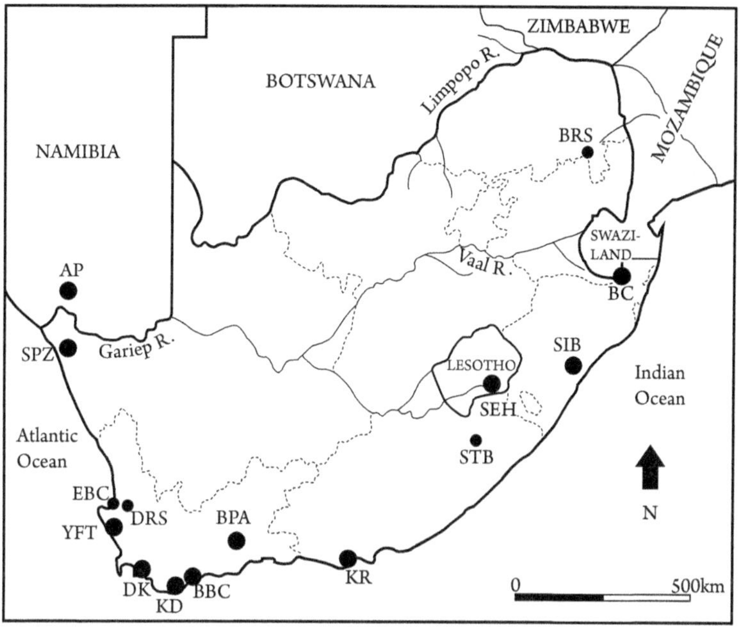

Figure 21.3. Southern African Middle Stone Age (MSA) sites referred to in the text. AP = Apollo 11; BBC = Blombos Cave; BC = Border Cave; BPA = Boomplaas; BRS = Bushman Rockshelter; DK = Die Kelders 1; DRS = Diepkloof; EBC = Elands Bay; KD = Klipdrift; KR = Klasies River; SEH = Sehonghong; SIB = Sibudu; SPZ = Spitzkloof A; STB = Strathalan B; YFT = Ysterfontein 1. Image by the author.

risky hunts took place with stone-tipped weapons or that stone-tipped stakes were hidden in a pitfall trap (Milo, 1998). Milo favors the interpretation of the pitfall trap because of the angle at which the stone entered the buffalo's vertebra.

When Clark and Kandel (2013) took data from eight southern African sites dated to between 170,000 and 40,000 years ago, they noticed that dietary breadth climaxed between about 70,000 and 60,000 years ago. This coincided with high frequencies of small, low-ranked prey (particularly browsers of size class 1). Remains of hares, rock hyraxes, grysbok/steenbok (*Raphicerus* spp.), and reedbuck (*Redunca* spp.) occur widely in southern Africa, and their distribution tends to cut across diverse environments. Thus, the remains of creatures that are clearly suitable for capture in snares are found in most Stone Age sites (Wadley, 2015).

From at least 70,000 years ago to about 55,000 years ago, faunal assemblages are principally comprised of small creatures at sites like Blombos Cave (Henshilwood et al., 2001), Diepkloof (Parkington, Poggenpoel, Rigaud, & Texier, 2005; Rigaud, Texier, Parkington, & Poggenpoel, 2006), and Sibudu (Clark & Plug, 2008; also see Figure 21.3). At Blombos, the small-creature collection pattern begins at 100,000 years ago (Badenhorst, van Niekerk, & Henshilwood, 2016). At Blombos, grysbok, steenbok, and unidentified bovids belonging to size class I, and other small creatures such as dune mole rats (*Bathyergus suillus*) and rock hyraxes were taken in both the MSA and LSA (Badenhorst et al., 2016; Henshilwood et al., 2001), suggesting that the meat-getting strategy used in the LSA may have had early origins. A study of cortical thickness of unidentifiable bone from Blombos shows that medium-size mammals were more common at the site in the MSA than is evidenced by the identifiable bone (Reynard, Henshilwood, & Badenhorst, 2014). This discovery does not, however, detract from the possibility that small mammals were obtained through remote capture. Tortoise, rock hyrax, and small-bovid remains are common at Klipdrift, near Blombos, though there is variability in the frequencies of large and small bovids between about 63,000 and 60,000 years ago (Henshilwood et al., 2014; Reynard, Discamps, Wurz, et al., 2016; Reynard, Discamps, Badenhorst, et al., 2016). Immature rock hyrax remains at Klipdrift may imply snaring (Reynard, Discamps, Badenhorst, et al., 2016). Rock hyrax was particularly well represented at Boomplaas, Western Cape, although many small creatures at this site may have been accumulated by carnivores and raptors (Faith, 2013). At Diepkloof, hyrax as well as dune mole rats, hares, and small carnivores are numerous (Parkington et al., 2005; Steele & Klein, 2013). Although the animals would have been caught most effectively in snares or traps, post-depositional damage on the bones has made it impossible to identify the agent that introduced the bones to Diepkloof (Steele & Klein, 2013). Farther north, tortoises and mammals of size class 1 were habitually eaten at Spitzkloof A Rockshelter (Dewar & Stewart, 2012), whereas rock hyrax was popular at Apollo 11 in Namibia (Thackeray, 1979; Vogelsang et al., 2010). Dune mole rats are usually present in large numbers in Cape coastal sites and were the most common prey at Die Kelders 1 (Klein & Cruz-Uribe, 2000) and Ysterfontein 1 (Avery et al., 2008; Halkett et al., 2003; Klein et al., 2004). However, small creatures like these need to be analyzed carefully to be sure they were exploited by humans; many of the Die Kelders mole rats seem to have been introduced to the site by owls (Klein & Cruz-Uribe, 2000). Tortoises were collected at Die Kelders (Steele & Klein, 2009), Blombos, Klipdrift, and Ysterfontein 1, among many other sites, but they can be collected easily by hand and require no equipment. The emphasis on small or medium-size animals in at least part of the MSA is obvious not only in southern Africa but also much farther north. In Porc-Epic Cave, Ethiopia, for example, the faunal record consists largely of animals like dik-dik, hare, hyrax, and gazelle (*Gazella* spp.) (Assefa, 2006).

CIRCUMSTANTIAL EVIDENCE FOR SNARES AT SIBUDU

Vegetation change in the Sibudu area through time has been identified using charcoal analysis (Allott, 2006). Mosaic vegetation communities are likely to have existed in Sibudu's neighborhood throughout human occupation there. Thus, forest and savanna were always side by side, but the ratios of communities varied. Some taxa appeared or disappeared through time, but when the change is reduced to its simplest terms, it seems that forest proliferated before 58,000 years ago, whereas savanna was extensive in the more recent period. In forests or closed vegetation, the use of snares is more practical than other forms of hunting, except perhaps for the use of nets, a form of capture used in the forests of central Africa (Lupo & Schmitt, 2005). Not everyone agrees with this interpretation; Churchill (1993), for example, suggests that the "surgical" nature of arrows makes them suitable for use in forests. A putative bone arrowhead was found at Sibudu (Backwell, d'Errico, & Wadley, 2008); thus, a variety of weaponry is likely to have been used at this site.

As vegetation changed, so did the proportions of fauna that were returned to Sibudu. Hunters at Sibudu seem to have preyed on animals according to their proportions in nature (Clark & Plug, 2008). Size class III (70–300 kg) and IV (300–1000 kg) ungulates increased through time at Sibudu (Clark & Plug, 2008). Remains of size class III herd animals such as blue wildebeest, hartebeest, and zebra became particularly prolific more recently than 58,000 years ago, so savanna and grassland grazers were preferentially targeted. As a result, taxonomic variability lessened in comparison with that in earlier periods (Clark & Plug, 2008). While the changes are not straightforward or completely synchronous with environmental ones, it can also be said that simplifications in technology are observable more recently than 58,000 years ago. The 58,000-year-old occupations at Sibudu provide high-resolution data, for there are multiple sedimentary horizons with overlapping ages (Jacobs, Wintle, Duller, Roberts, & Wadley, 2008). Bifacial and/or unifacial points are uncommon in the earliest 58,000-year-old occupations, and then they increase alongside the intensified hunting of herd animals. The positive correlation suggests that people were hunting the large animals with stone-tipped spears.

In contrast, the period 65,000–62,000 years ago is characterized by technological diversity in Sibudu, as well as taxonomic diversity in the form of a wide range of animals brought back to the shelter. There are no small carnivores or lagomorphs and hardly any size class I ungulates, primates, or pigeons in any of the 58,000-year-old occupations, whereas they are abundant earlier.

Sibudu has a number of strands of evidence to suggest that remote capture was practiced prior to 58,000 year ago. For the remainder of the chapter, I will concentrate on this earlier period. At the time, moist evergreen forest and riverine and savanna communities were in the vicinity of the site (Allott, 2006; Sievers, 2006). This vegetation diversity supported taxonomic diversity among animals, but between 65,000 and 62,000 years ago, animal taxa preferring forested (or closed) vegetation comprise 91.4% of the number of bones identified by species (Clark & Plug, 2008). The faunal assemblage incorporates the nocturnal Gambian giant rat (*Cricetomys gambianus*) (Clark & Plug, 2008; Glenny, 2006), which is caught most successfully in snares or traps, as well as monkeys, rock hyrax, and hares. High quantities of small mammals, including carnivores such as felids, viverrids, mustelids, canids, and mongooses, are especially characteristic of the period (Clark, 2009; Clark & Plug, 2008), but blue duiker remains eclipse all others in the faunal collection. Together with blue duiker are other small bovids like steenbok, gray duiker, red duiker, and bushbuck. The non-migratory, predictable behavior and small home ranges of all these species fulfil Lupo and Schmitt's (2002) predictions for the type of prey susceptible to capture in snares.

A mortality profile that epitomizes the age structure of a living herd is typical of natural disasters or mass killings (Klein, 1982; Klein & Cruz-Uribe, 1984), but snaring may produce this profile too, because of repeated capture events. Snares detain a taxonomically diverse range of prey of all ages (Lupo & Schmitt, 2002; Schmitt & Lupo, 2008), and thus snaring may be implied when archaeofaunal collections exhibit both age profiles that mimic those of living populations and wide species diversity (including small carnivores). When hunting with weaponry, juvenile animals were probably targeted less than adults (Speth & Clark, 2006), so higher proportions of juveniles can be expected when remote-capture techniques are employed.

At the Klasies River main site, small-bovid remains do indeed have a catastrophic age profile (Klein, 1981), and they have a taphonomy that suggests human predation (Milo, 1998), so they seem likely victims of snaring. At Sibudu, only about 11% of blue duiker remains are from juveniles (Clark, 2009), and Estes (1997) calculates that 38% of a wild blue duiker population should be under 1 year old. The Sibudu mortality profile is accordingly not characteristic of a live population of blue duiker, resembling more the live population estimate from central Africa when animals are captured in nets (Lupo & Schmitt, 2002). These figures from both places may underestimate the number of available blue duiker juveniles because of the behavior of the animals. After birth, juveniles lie still when there is danger, and newborns lie hidden for several weeks (Estes, 1997). At 3 months of age, a blue duiker walks only about 73 m, and this distance is not exceeded until it is an adult and leaves its parents' territory (Estes, 1997). In contrast, adults walk between 650 and 1770 m each day and range over almost 40% of their territory during these excursions (Estes, 1997). As they always use the same paths (Apps, 2000), people aware of their movements can catch them easily in well-located snares (Skinner & Chimimba, 2005). Juveniles are much less likely than wider-ranging adults to fall prey to a snare, and it is therefore plain that a catastrophic mortality pattern with many juveniles is unlikely for blue duikers caught in snares.

Why do I favor an interpretation of snare capture rather than net hunting or bow-and-arrow hunting for small creatures at Sibudu? The small-carnivore component among the 65,000- to 62,000-year-old remains convinces me more than the remains of blue duiker because little carnivores are likely to be unintended by-products of the hunt (and thus not victims of arrow hunting). It seems unlikely that nets were preferred over snares at Sibudu because small carnivores are more liable to be caught in snares than nets, which do not normally arrest such animals (Lupo & Schmitt, 2002). Thus the small-carnivore component at Sibudu argues against the use of nets and tends to support the interpretation of snaring before 58,000 years ago. Given that no cut-marks or other processing marks have been found on the small-carnivore bones from Sibudu (Clark, 2009), it is also not possible to be completely sure that humans brought these animals to the site. However, small carnivores are absent from Sibudu in occupations more recent than 58,000 years ago (Clark & Plug, 2008), which implies that they did not normally live and die in the rockshelter and were not consistently brought there by non-human predators.

For the period to 58,000 years ago, bush pig remains are the most prevalent type in Sibudu, after those of blue duiker (Clark & Plug, 2008). Bush pigs are belligerent, and they move nocturnally (Skinner & Chimimba, 2005). Their presence tends to support the interpretation of snaring/trapping because they are dangerous to confront face to face. They are commonly caught in snares or traps set along paths in southeastern Cameroon (Yasuoka, 2006) and in Central and West Africa (Fa, Ryan, & Bell, 2005). Since bush pigs are much bigger (weighing 69–72 kg) than blue duikers (weighing <5 kg), they require more robust snares than the flimsy ones that are adequate for duikers. Pitfalls would also be an effective method of capturing bush pigs.

Birds are an intriguing component of the Sibudu fauna. Plug and Clark (2008) suggest that most of Sibudu's bird remains were unlikely to have entered the rockshelter as human prey, except for ones like guinea fowl (*Numida meleagris*), francolins (*Pternisitis/Scleroptila* sp.), waterfowl, marine and pelagic birds, seagulls (*Larus* sp.), and avocets (*Recurvirostra avosetta*). This argument has been countered in a more recent study where bone surface modifications were examined on the Sibudu MSA bird assemblage comprising pigeons, doves, Galliformes, waders, and raptors. Some cutmarks associated with skinning and defleshing were recognized, but also human tooth marks and perforations on distal humeri produced during disarticulation of the forewing (Val, 2016; Val, de la Peña, & Wadley, 2016). Thus, at least some of the pigeons and doves were eaten by humans. Neandertals at Gorham's Cave, Gibraltar captured and ate doves at least 67,000 years ago, possibly plucking them straight from their nests or setting traps or baited nets (Blasco et al., 2014). Sibudu's inhabitants might have done the same, or perhaps they set sticky bird lime, extracted from trees such as figs (*Ficus* spp.), on branches or rock ledges frequented by the birds. The use of bird lime is a simple, effectual means of remote capture that is still used in many places today. Apart from Sibudu, bird remains have been found in the Namibian sites of Apollo 11, Fackelträger, and Haalenberg (Thackeray, 1979), in Strathalan Cave in the Eastern Cape (Opperman & Heydenrych, 1990), Sehonghong in Lesotho (Plug, 1978; Plug & Mitchell, 2008), Blombos Cave (Henshilwood et al., 2001), and Diepkloof (Steele & Klein, 2013). Coastal sites such as Klasies River (von den Driesch, 2004) and Die Kelders (Klein & Cruz-Uribe, 2000; Marean et al., 2000) have the remains of flightless birds such as the jackass penguin (*Spheniscus demersus*), but these may have been clubbed rather than caught in traps or snares.

THE BRIDGE BETWEEN ARCHAEOLOGICAL DATA, ETHOLOGY, AND COGNITIVE THEORY

Recognizing snares and traps in archaeological sites remains challenging because these devices usually leave no trace in excavated living sites. This is partly because they are made of perishable materials, but also because they are generally set at a distance from the home base. Home bases are repeatedly occupied and therefore accumulate considerable cultural debris, whereas remote-capture sites are ephemeral. Thus the absence of traps or snares in occupation sites need not imply that they were nonexistent. Although devices like tidal fish traps and desert kites have left durable traces in some parts of the world, circumstantial evidence for remote capture seems the best we can hope for at most archaeological sites. The assessment of circumstantial evidence is fraught because, as explained earlier, non-selective capture techniques do not necessarily produce a catastrophic mortality pattern. Indeed, some forms of hunting that involve face-to-face contact with prey, such as driving herd animals over a cliff, produce this mortality pattern more successfully. In principle, snaring and trapping should produce catastrophic mortality patterns, but the type of animal best caught in these devices sometimes has behavioral strategies that preclude the development of this pattern. Solitary animals like blue duikers, which are well represented before 58,000 years ago at Sibudu, tend to keep their young hidden and sedentary, thereby preventing them from wandering onto paths traversed regularly by adults. This means that juveniles will be under-represented in the record even when snaring is practiced because adults fall prey to snares set in their habitual walkways.

The low blue duiker juvenile-to-adult ratios at Sibudu provide an example of why ethological studies need to be made alongside mortality studies of archaeologically recovered faunal remains. Snares are implied by the relatively high frequencies of blue duiker and bush pig remains prior to 58,000 years ago, though it is acknowledged that these animals can be hunted by other means. The use of snares, however, is particularly supported at Sibudu by the numbers of small carnivores that are likely to be accidental victims of snares for they are prone to being caught in such devices. Most small carnivores occur in the 65,000- to 62,000-year-old period at Sibudu, suggesting that their occurrence at the site was not natural.

Raptors and carnivores can sometimes leave a faunal signature of small creatures like rock hyrax, and this gives a false impression of snaring. Barn owls and possibly other raptors brought mice, rats, and shrews to Sibudu (Glenny, 2006), and other little animals may have a similar origin in the shelter. The uneven chronological spread of small creatures such as hare, hyrax, little carnivores, and pigeons argues against important contributions of these remains by non-human predators in Sibudu. The small-animal remains are mostly found at and before 62,000 years ago, and then again in small numbers after 38,000 years ago. It seems unlikely that non-human agents would have selected these creatures early on and then overlooked them thereafter. A few blue duiker, hyrax, and small-carnivore remains reappear in the final MSA at about 38,000 years ago, but this small-creature collection strategy is not as convincing as it was in the earlier period (Clark & Plug, 2008; Collins, 2013).

I am attempting here to create a bridging argument (Botha, 2008; Wynn, 2009) between the faunal record from Sibudu and the interpretation of remote-capture technology in the MSA. Notwithstanding the suggestive evidence presented from Sibudu and supportive evidence from other MSA sites in southern Africa, methodological and taphonomic questions remain, so the interpretation of snaring must remain a hypothesis. Hopefully, the challenge can be more successfully addressed in the future, because for Thomas Wynn and Frederick Coolidge (2003), evidence for remote capture in the archaeological record is an incontrovertible indication of delayed gratification and premeditated action that results in remote capture. I agree with them that such evidence implies modern executive functions of the brain, enhanced working memory, and, ultimately, complex cognition of the kind that all people share today.

Tool-making implies several cognitive aspects, such as causal reasoning; thus tools are ideal for building a bridge to access past human cognition (Haidle, 2014). In Haidle's terminology, snares implicate "notational tool behavior" that requires abstract causal reasoning about the effects of the tool on an unseen agent (Haidle, 2014). Most discussions, including many in this book, concentrate on what tool behavior and technical innovations like snares tell us about the state of cognition of their makers. However, there are intricate recursive relationships between brains, bodies, and things (Malafouris, 2010), and it would be naïve to view tools and devices like snares as objects that passively embody information about the minds of their makers. Feedback between cognition and technology is a constant process. The period 65,000–62,000 years ago at Sibudu incorporates a range of innovations linked to hunting practices (e.g., new lithic classes including quartz arrowheads and barbs, a variety of adhesive types, and a suite of bone tools, including putative arrowheads). The innovations almost certainly acted as stimulants that inspired their makers to greater inventiveness. The recursive relationship between imagination, creation, and innovation is likely to have been an important factor in the evolution of human minds, as well as in the evolution of technology. Innovation is enabled by working memory (Wynn et al., 2017), so we should see working memory develop in concert with technology. In a modern-day example, Malafouris (2010) cites a study of London taxi drivers who are obliged to develop superior navigational skills and as a consequence possess significantly larger posterior hippocampi than control samples. Minds and things are not isolated but are intertwined (Malafouris, 2013) within their particular environments, so we can speak of "cognitive ecology" (Malafouris, 2015). Thus, I suggest that simple snaring technology embodied powerful cognitive concepts, and that the reinforcing that took place between the technical process and "cognitive ecology" would have contributed effectively to the evolution of both.

ACKNOWLEDGMENTS

I receive research funding from the National Research Foundation, which does not necessarily support the ideas represented in this chapter. I thank Graham Avery and Gary Trower for the use of images in Figures 21.1 and 21.2. I am also grateful to the anonymous reviewers for suggestions that improved this chapter.

REFERENCES

Allott, L. F. (2006). Archaeological charcoal as a window on palaeovegetation and wood use during the Middle Stone Age at Sibudu Cave. *Southern African Humanities, 18*(1), 173–201.

Apps, P. (Ed.). (2000). *Smithers' mammals of southern Africa: A field guide.* Cape Town, South Africa: Struik Publishers.

Assefa, Z. (2006). Faunal remains from Porc-Epic: Paleoecological and zooarchaeological investigations from a Middle Stone Age site in south-eastern Ethiopia. *Journal of Human Evolution, 51*(1), 50–75.

Avery, G. (1975). Discussion on the age and use of tidal fish-traps (Visvywers). *South African Archaeological Bulletin, 30*(119/120), 105–113.

Avery, G., Halkett, D., Orton, J., Steele, T. E., Tusenius, M., & Klein, R. G. (2008). The Ysterfontein 1 Middle Stone Age Rock Shelter and the evolution of coastal foraging. *South African Archaeological Society Goodwin Series, 10*, 66–89.

Backwell, L., d'Errico, F., & Wadley, L. (2008). Middle Stone Age bone tools from the Howiesons Poort layers, Sibudu Cave, South Africa. *Journal of Archaeological Science, 35*(6), 1566–1580.

Badenhorst, S., van Niekerk, K. L., & Henshilwood, C. S. (2016). Large mammal remains from the 100 ka Middle Stone Age layers of Blombos Cave, South Africa. *South African Archaeological Bulletin, 71*(203), 46–52.

Blasco, R., Finlayson, C., Rosell, J., Marco, A. S., Finlayson, S., Finlayson, G., . . . Vidal, J. R. (2014). The earliest pigeon fanciers. *Scientific Reports, 4*(5971), 1–7.

Botha, R. (2008). Prehistoric shell beads as a window on language evolution. *Language & Communication, 28*(3), 197–212.

Brain, C. K. (1981). *The hunters or the hunted? An introduction to African cave taphonomy.* Chicago, IL: University of Chicago Press.

Brand, C., Eguma, R., Zuberbühler, K., & Hobaiter, C. (2014). First report of prey capture from human laid snare-traps by wild chimpanzees. *Primates, 55*(3), 437–440.

Churchill, S. E. (1993). Weapon technology, prey size selection, and hunting methods in modern hunter-gatherers: Implications for hunting in the Paleolithic and Mesolithic. In G. L. Peterkin, H. M. Bricker, & P. A. Mellars (Eds.), *Hunting and animal exploitation in the Later*

Palaeolithic and Mesolithic of Eurasia (pp. 11–24). Arlington, VA: American Anthropological Association.

Clark, J. L. (2009). *Testing models on the emergence and nature of modern human behavior: Middle Stone Age fauna from Sibudu Cave (South Africa).* PhD Thesis, University of Michigan, Ann Arbor, MI.

Clark, J. L., & Kandel, A. W. (2013). The evolutionary implications of variation in human hunting strategies and diet breadth during the Middle Stone Age of southern Africa. *Current Anthropology, 54*(S8), S269–S287.

Clark, J. L., & Plug, I. (2008). Animal exploitation strategies during the South African Middle Stone Age: Howiesons Poort and post-Howiesons Poort fauna from Sibudu Cave. *Journal of Human Evolution, 54*(6), 886–898.

Collins, B. R. (2013). The taphonomy of the Final Middle Stone Age fauna from Sibudu Cave, South Africa. *International Journal of Osteoarchaeology, 25*(6), 805–815.

Coolidge, F. L., & Wynn, T. (2005). Working memory, its executive functions, and the emergence of modern thinking. *Cambridge Archaeological Journal, 15*(1), 5–26.

Dewar, G., & Stewart, B. A. (2012). Preliminary results of excavations at Spitzkloof Rockshelter, Richtersveld, South Africa. *Quaternary International, 270*, 30–39.

Estes, R. D. (1997). *The behavior guide to African mammals: Including hoofed mammals, carnivores, primates.* Berkeley, CA: University of California Press.

Fa, J. E., Ryan, S. F., & Bell, D. J. (2005). Hunting vulnerability, ecological characteristics and harvest rates of bushmeat species in afrotropical forests. *Biological Conservation, 121*(2), 167–176.

Faith, J. T. (2013). Taphonomic and paleoecological change in the large mammal sequence from Boomplaas Cave, Western Cape, South Africa. *Journal of Human Evolution, 65*(6), 715–730.

Forbes, V. (2015). Duck fleas as evidence for eiderdown production on archaeological sites. *Journal of Archaeological Science, 61*, 105–111.

Frison, G. C. (1991). Hunting strategies, prey behavior and mortality data. In M. C. Stiner (Ed.), *Human predators and prey mortality* (pp. 15–30). Boulder, CO: Westview Press.

Glenny, W. (2006). Report on the micromammal assemblage analysis from Sibudu Cave. *Southern African Humanities, 18*(1), 279–288.

Haidle, M. N. (2006). *Menschen–Denken–Objekte Zur problem-lösung-distanz als Kognitionsaspekt im Werkzeugverhalten von Tieren und im Laufe der Menschlichen Evolution.* PhD Thesis, University of Tübingen, Germany.

Haidle, M. N. (2014). Building a bridge—An archeologist's perspective on the evolution of causal cognition. *Frontiers in Psychology, 5*(1472), 1–15.

Halkett, D., Hart, T., Yates, R., Volman, T. P., Parkington, J. E., Orton, J., . . . Avery, G. (2003). First excavation of intact Middle Stone Age layers at Ysterfontein, Western Cape Province, South Africa: Implications for Middle Stone Age ecology. *Journal of Archaeological Science, 30*(8), 955–971.

Henshilwood, C. S., Sealy, J. C., Yates, R., Cruz-Uribe, K., Goldberg, P., Grine, F. E., . . . Watts, I. (2001). Blombos Cave, Southern Cape, South Africa: Preliminary report on the 1992–1999 excavations of the Middle Stone Age levels. *Journal of Archaeological Science, 28*(4), 421–448.

Henshilwood, C. S., Van Niekerk, K. L., Wurz, S., Delagnes, A., Armitage, S. J., Rifkin, R. F., . . . Mienies, S. S. (2014). Klipdrift Shelter, southern Cape, South Africa: Preliminary report on the Howiesons Poort layers. *Journal of Archaeological Science, 45*, 284–303.

Hine, P., Sealy, J., Halkett, D., & Hart, T. (2010). Antiquity of stone-walled tidal fish traps on the Cape coast, South Africa. *South African Archaeological Bulletin, 65*(191), 35–44.

Holzer, A., Avner, U., Porat, N., & Horwitz, L. K. (2010). Desert kites in the Negev desert and northeast Sinai: Their function, chronology and ecology. *Journal of Arid Environments, 74*(7), 806–817.

Jacobs, Z., Wintle, A. G., Duller, G. A. T., Roberts, R. G., & Wadley, L. (2008). New ages for the post-Howiesons Poort, late and final Middle Stone Age at Sibudu Cave, South Africa. *Journal of Archaeological Science, 35*(7), 1790–1807.

Klein, R. G. (1977). The mammalian fauna from the Middle and Later Stone Age (Later Pleistocene) levels of Border Cave, Natal Province, South Africa. *South African Archaeological Bulletin, 32*(125), 14–27.

Klein, R. G. (1981). Stone Age predation on small African bovids. *South African Archaeological Bulletin, 36*(134), 55–65.

Klein, R. G. (1982). Age mortality profiles as a means of distinguishing hunted species from scavenged ones in Stone Age archaeological sites. *Paleobiology, 8*(2), 151–158.

Klein, R. G., Avery, G., Cruz-Uribe, K., Halkett, D., Parkington, J. E., Steele, T. E., . . . Yates, R. (2004). The Ysterfontein 1 Middle Stone Age site, South Africa, and early human exploitation of coastal resources. *Proceedings of the National Academy of Sciences of the United States of America, 101*(16), 5708–5715.

Klein, R. G., & Cruz-Uribe, K. (1984). *The analysis of animal bones from archaeological sites.* Chicago, IL: University of Chicago Press.

Klein, R. G., & Cruz-Uribe, K. (2000). Middle and Later Stone Age large mammal and tortoise remains from Die Kelders Cave 1, Western Cape Province, South Africa. *Journal of Human Evolution, 38*(1), 169–195.

Knopff, K. H., Knopff, A. A., & Boyce, M. S. (2010). Scavenging makes cougars susceptible to snaring at wolf bait stations. *Journal of Wildlife Management, 74*(4), 644–653.

Lombard, M., & Gärdenfors, P. (2017). Tracking the evolution of causal cognition in humans. *Journal of Anthropological Sciences, 95*, 1–16.

Lombard, M., & Haidle, M. N. (2012). Thinking a bow-and-arrow set: Cognitive implications of Middle Stone Age bow and stone-tipped arrow technology. *Cambridge Archaeological Journal, 22*(2), 237–264.

Lupo, K. D., & Schmitt, D. N. (2002). Upper Paleolithic net-hunting, small prey exploitation, and women's work effort: A view from the ethnographic and ethnoarchaeological record of the Congo Basin. *Journal of Archaeological Method and Theory, 9*(2), 147–179.

Lupo, K. D., & Schmitt, D. N. (2005). Small prey hunting technology and zooarchaeological measures of taxonomic diversity and abundance: Ethnoarchaeological evidence from Central African forest foragers. *Journal of Anthropological Archaeology, 24*(4), 335–353.

Malafouris, L. (2010). The brain-artefact interface (BAI): A challenge for archaeology and cultural neuroscience. *Social Cognitive and Affective Neuroscience, 5*, 264–273.

Malafouris, L. (2013). *How things shape the mind: A theory of material engagement.* Cambridge, MA: MIT Press.

Malafouris, L. (2015). Metaplasticity and the primacy of material engagement. *Time and Mind, 8*(4), 351–371.

Marean, C. W., Abe, Y., Frey, C. J., & Randall, R. C. (2000). Zooarchaeological and taphonomic analysis of the Die Kelders Cave 1 Layers 10 and 11 Middle Stone Age larger mammal fauna. *Journal of Human Evolution, 38*(1), 197–233.

Milo, R. G. (1998). Evidence for hominid predation at Klasies River Mouth, South Africa, and its implications for the behavior of early modern humans. *Journal of Archaeological Science, 25*(2), 99–133.

Opperman, H., & Heydenrych, B. (1990). A 22,000-year-old Middle Stone Age camp site with plant food remains from the north-eastern Cape. *South African Archaeological Bulletin, 45*(152), 93–99.

Parkington, J. E., Poggenpoel, C., Rigaud, J.-P., & Texier, P.-J. (2005). From tool to symbol: The behavioral context of intentionally marked ostrich eggshell from Diepkloof, Western Cape. In F. D'Errico & L. Backwell (Eds.), *From tools to symbols: From early hominids to modern humans* (pp. 475–492). Johannesburg, South Africa: Witwatersrand University Press.

Plug, I. (1978). *Die Latere Steentydperk van die Boesman-rotsskuiling in Oos-Transvaal*. MA Thesis, University of Pretoria, South Africa.

Plug, I., & Clark, J. L. (2008). In the air: A preliminary report on the birds from Sibudu Cave, KwaZulu-Natal, South Africa. *South African Archaeological Society Goodwin Series, 10*, 133–142.

Plug, I., & Mitchell, P. J. (2008). Sehonghong: Hunter-gatherer utilization of animal resources in the Highlands of Lesotho. *Annals of the Transvaal Museum, 45*(1), 31–53.

Reynard, J. P., Discamps, E., Badenhorst, S., van Niekerk, K. L., & Henshilwood, C. S. (2016). Subsistence strategies in the southern Cape during the Howiesons Poort: Taphonomic and zooarchaeological analyses of Klipdrift Shelter, South Africa. *Quaternary International, 404*, 2–19.

Reynard, J. P., Discamps, E., Wurz, S., van Niekerk, K. L., Badenhorst, S., & Henshilwood, C. S. (2016). Occupational intensity and environmental changes during the Howiesons Poort at Klipdrift Shelter, southern Cape, South Africa. *Palaeogeography, Palaeoclimatology, Palaeoecology, 449*, 349–364.

Reynard, J. P., Henshilwood, C. S., & Badenhorst, S. (2014). Inferring animal size from the unidentified long bones from the Middle Stone Age layers at Blombos Cave, South Africa. *Annals of the Ditsong National Museum of Natural History, 4*(1), 9–25.

Rigaud, J.-P., Texier, P.-J., Parkington, J. E., & Poggenpoel, C. (2006). Le mobilier Stillbay et Howiesons Poort de l'abri Diepkloof: La chronologie du Middle Stone Age sudafricain et ses implications. *Comptes Rendus Palevol, 5*(6), 839–849.

Schmitt, D. N., & Lupo, K. D. (2008). Do faunal remains reflect socioeconomic status? An ethnoarchaeological study among Central African farmers in the northern Congo Basin. *Journal of Anthropological Archaeology, 27*(3), 315–325.

Schmitt, D. N., Madsen, D. B., & Lupo, K. D. (2004). The worst of times, the best of times: Jackrabbit hunting by middle Holocene human foragers in the Bonneville Basin of western North America. In M. Mondini, S. Muñoz, & S. Wickler (Eds.), *Colonisation, migration, and marginal areas: A zooarchaeological approach* (9th ICAZ Conference proceedings, Durham 2002, pp. 86–95). Oxford, UK: Oxford University Press.

Shaw-Williams, K. (2014). The social trackways of the evolution of human cognition. *Biological Theory, 9*(1), 16–26.

Sievers, C. (2006). Seeds from the Middle Stone Age layers at Sibudu Cave. *Southern African Humanities, 18*(1), 203–222.

Skinner, J. D., & Chimimba, C. T. (2005). *The mammals of the southern African sub-region*. Cambridge, UK: Cambridge University Press.

Speth, J. D., & Clark, J. L. (2006). Hunting and overhunting in the Levantine Late Middle Paleolithic. *Before Farming, 2006*(3), 1–42.

Steele, T. E., & Klein, R. G. (2009). Late Pleistocene subsistence strategies and resource intensification in Africa. In J.-J. Hublin & M. P. Richards (Eds.), *The evolution of hominin diets: Integrating approaches to the study of Palaeolithic subsistence* (pp. 113–126). New York: Springer.

Steele, T. E., & Klein, R. G. (2013). The Middle and Later Stone Age faunal remains from Diepkloof Rock Shelter, Western Cape, South Africa. *Journal of Archaeological Science, 40*(9), 3453–3462.

Stringer, C. B., Finlayson, J. C., Barton, N., Fernández-Jalvo, Y., Cáceres, I., Sabin, R. C., . . . Riquelme-Cantal, J. A. (2008). Neanderthal exploitation of marine mammals in Gibraltar. *Proceedings of the National Academy of Sciences of the United States of America, 105*(38), 14319–14324.

Stuart-Fox, M. (2015). The origins of causal cognition in early hominins. *Biology and Philosophy, 30*(2), 247–266.

Thackeray, J. F. (1979). An analysis of faunal remains from archaeological sites in southern South West Africa (Namibia). *South African Archaeological Bulletin, 34*(129), 18–33.

Vaesen, K. (2012). The cognitive bases of human tool use. *Behavioral and Brain Sciences, 35*(4), 203–218.

Val, A. (2016). New data on the avifauna from the Middle Stone Age layers of Sibudu Cave, South Africa: Taphonomic and palaeoenvironmental implications. *Quaternary International, 421*, 173–189.

Val, A., de la Peña, P., & Wadley, L. (2016). Direct evidence for human exploitation of birds in the Middle Stone Age of South Africa: The example of Sibudu Cave, KwaZulu-Natal. *Journal of Human Evolution, 99*, 107–123.

Vogelsang, R., Richter, J., Jacobs, Z., Eichhorn, B., Linseele, V., & Roberts, R. G. (2010). New excavations of Middle Stone Age deposits at Apollo 11 Rockshelter, Namibia: Stratigraphy, archaeology, chronology and past environments. *Journal of African Archaeology, 8*(2), 185–218.

von den Driesch, A. (2004). The Middle Stone Age fish fauna from Klasies River main site, South Africa. *Anthropozoologica, 39*(2), 33–59.

Wadley, L. (2010). Were snares and traps used in the Middle Stone Age and does it matter? A review and a case study from Sibudu, South Africa. *Journal of Human Evolution, 58*(2), 179–192.

Wadley, L. (2015). Those marvellous millennia: The Middle Stone Age of Southern Africa. *Azania: Archaeological Research in Africa, 50*(2), 155–226.

Wadley, L., Hodgskiss, T., & Grant, M. (2009). Implications for complex cognition from the hafting of tools with compound adhesives in the Middle Stone Age, South Africa. *Proceedings of the National Academy of Sciences of the United States of America, 106*(24), 9590–9594.

Wynn, T. (2009). Hafted spears and the archaeology of mind. *Proceedings of the National Academy of Sciences of the United States of America, 106*(24), 9544–9545.

Wynn, T., & Coolidge, F. L. (2003). The role of working memory in the evolution of managed foraging. *Before Farming, 2*(1), 1–16.

Wynn, T., & Coolidge, F. L. (2007a). A Stone-Age meeting of minds. *American Scientist, 96*(1), 44–51.

Wynn, T., & Coolidge, F. L. (2007b). Did a small but significant enhancement in working memory capacity power the evolution of modern thinking? In P. Mellars, K. Boyle, O. Bar-Yosef, & C. B. Stringer (Eds.), *Rethinking the human revolution: New behavioural and biological perspectives on the origin and dispersal of modern humans* (pp. 79–90). Cambridge, UK: McDonald Institute for Archaeological Research.

Wynn, T., Haidle, M. N., Lombard, M., & Coolidge, F. L. (2017). The expert cognition model in human evolutionary studies. In T. Wynn & F. L. Coolidge (Eds.), *Cognitive models in Palaeolithic archaeology* (pp. 21–43). Oxford, UK: Oxford University Press.

Yasuoka, H. (2006). The sustainability of duiker (*Cephalophus spp*) hunting for the Baka hunter-gatherers in south-eastern Cameroon. *African Study Monographs, S33*, 95–120.

22

ON THE MINDS OF BOW HUNTERS

Marlize Lombard

INTRODUCTION

The bow, as used in hunting or warfare in a symbiotic relationship with arrows, has been described as the first of five machines that changed the world (Denny, 2007). It represents *Homo sapiens*' causal understanding that the power of stored mechanical energy could overcome physical challenges, and that such energy could be harnessed to brace our subsistence and/or conflict strategies, drill holes, make fire, or even play music (Lombard, 2016). Bow hunting therefore serves as an example of the "coupling of brain with brawn" (e.g., Barham, 2013; Cotterell & Kamminga, 1990). This uniquely techno-dependent evolutionary trajectory allows us to adapt to, and effectively manage risks associated with, almost any socioeconomic environment. If interpretations of the archaeological record at the Middle Stone Age (MSA) site of Sibudu Cave in KwaZulu-Natal, South Africa, are correct, then the earliest known evidence for the application of such mechanical understanding in food-procurement strategies is represented by spring traps at about 70 thousand years ago (Kya) (Lombard & Wadley, 2016; Wadley, 2010).

Shortly thereafter (in Stone Age terms), at about 64 Kya, we have the earliest, multi-stranded evidence for the use of stone-tipped arrows at Sibudu as well as at Umhlatuzana Rock Shelter (about 90 km south of Sibudu) (Lombard, 2011; Lombard & Phillipson, 2010; Wadley & Mohapi, 2008). This interpretation is supported by the analysis of experimental and archaeological bone points from the same context at Sibudu (e.g., Backwell, d'Errico, & Wadley, 2008; Bradfield & Lombard, 2011). It is possible that we will find even earlier traces of stone-tipped bow-and-arrow technology, as suggested by a handful of backed artifacts dated to about 71 Kya from Pinnacle Point Site 5–6 on the south coast of South Africa (Brown et al., 2012). These tools, however, have not been analyzed directly for corroborating use traces, so their functionality remains speculative. More robust evidence exists for the use of bone-tipped arrows, perhaps in combination with the use of poisons, in southern Africa from about 43 to 24 Kya (e.g., d'Errico et al., 2012; Robbins, Campbell, Brook, Murphy, & Hitchcock, 2012). Thus, although bow-and-arrow technology could have been invented and reinvented in different contexts through time (e.g., Lombard & Parsons, 2010, 2011; Riede, 2008), there is strong evidence for its early use in southern Africa. Outside of Africa, the oldest known evidence for bow hunting currently comes from the cave of Potočka zijalka, Slovenia, in the form of bone points dated to 32 Kya (Odar & Maver, 2011).

This record has direct implications for theoretical models aimed at unraveling the evolution of human cognition through technology, such as the documentation of modern executive functions and working memory. For example, Coolidge and Wynn (2009) argued that technical evidence for executive functions can be found in evidence for facilities or structures designed to capture energy and/or perform tasks without human supervision. Apart from remote-capture technologies such as spring traps, fish weirs, or "desert kites" (the latter being structures that funnel gazelle into killing zones), they suggest that harpoons (projectiles with detachable heads attached by a line to the shaft or a float, which represent a complex set of separate tools functioning collectively as a hunting/fishing system) also point toward modern executive functions. Their inference is based on the fact that the individual components of harpoons must be thought through and produced well in advance, and then each component must be carefully maintained to function together optimally when needed (Coolidge & Wynn, 2009).

At the time when Coolidge and Wynn published the first edition of their book, the oldest known evidence for structures such as desert kites were dated to about 12 Kya in the Middle East, and harpoons could have been used as early as about 17 Kya in Europe. Their careful analyses of a whole range of archaeological, paleoanthropological, and neurological phenomena resulted in a parsimonious conclusion that

> modern humans evolved an enhanced working memory capacity that powered executive function ability. These enabled complex contingency planning (e.g., managed foraging), abstract reasoning with superordinate categories (e.g., Hohlenstein-Stadel), and innovation. These abilities are evidenced in the archaeological record only very late, indeed not until 32,000 years ago, and this is a puzzle. (Coolidge & Wynn, 2009, pp. 244–245)

It is therefore patent that the recent evidence for the use of spring traps and bow hunting that predates the last 30 millennia, stretching into MIS 4 at more than 60 Kya, has the potential to reset the clock regarding the contribution that such technologies could make toward the interpretation of the evolution of human cognition.

Human technology does not represent passive material culture or an epiphenomenal product of a mind already in place. Rather, over the last 3 million years, our technologies helped to co-create, shape, and transform our bodily experiences, brains, and genes in a continuous evolutionary interplay between our environments (environmental and social), behaviors, thinking, and biology (e.g., Bruner & Lozano, 2014; di Paolo, Rohde, & De Jaegher, 2010; Malafouris, 2013). Here I focus only on techno-behaviors associated with bow hunting as a potential proxy for the interpretation of enhanced levels of cognition. My focus does not imply that other technical systems do not have similar interpretative potential (e.g., Haidle, 2014; Wadley, 2013). In contrast to many other managed foraging systems such as traps or wooden objects, however, stone- and bone-tipped weaponry is traceable into deep time. Hence, such weaponry can provide direct spatiotemporal material evidence for the evolution of technology-assisted hunting. Another key reason for my focus is that archaeological evidence for bow hunting is thus far exclusive to *H. sapiens*, having never been found in association with other members of the *Homo* genus (e.g., Coolidge, Haidle, Lombard, & Wynn, 2016; Lombard & Gärdenfors, 2017; Shea & Sisk, 2010; Williams, Burke, & Lombard,

2014). Especially in the southern African context, where we currently have the earliest known evidence for bow hunting, we also know that this techno-behavior is associated exclusively with *H. sapiens*. Here, human populations are known to have been anatomically and genetically "modern" since at least 100 Kya (see, e.g., Dusseldorp, Lombard, & Wurz, 2013, for a summary of paleoanthropological record; Lombard, Schlebusch, & Soodyall, 2013, for a genetic overview), with new genetic data from the region indicating the age of about 265 to 350 Kya (Schlebusch et al., 2017).

It follows that if we are able to understand some of the cognitive mechanisms associated with a techno-behavior such as bow hunting, we might gain glimpses into what differentiates our minds from that of others in our lineage. Should we one day find evidence for bow hunting among, let's say, our Neandertal cousins, the body of work synthesized here will have equal interpretative value for understanding levels of cognitive complexity in such populations. Thus, although my impetus for exploring the cognition of bow hunting and how it could have helped shape the modern mind is the fact that it might represent our specific evolutionary path in thinking, its implications for understanding levels of cognition apply equally to any deep-time Stone Age/Paleolithic communities, regardless of spatiotemporal context. In the next section I provide an overview of the cognition of bow hunting before presenting ideas about how this technology could have interacted with the evolution of our brains.

TESTING BOW HUNTING AS A COGNITIVE NICHE ASSOCIATED WITH *H. SAPIENS*

Many researchers grasp the importance of finding early evidence for the use of bows and/or spear throwers in the archaeological record (e.g., Brooks, Nevell, Yellen, & Hartman, 2006; Iovita, Schönekeß, Gaudzinski-Windheuser, & Jäger, 2016; Pargeter, Shea, & Utting, 2016; Shea, 2006). Yet, little effort has been made to understand and explain the potential cognitive implications of bow hunting. For example, Shea and Sisk (2010) have argued that bow hunting was a key strategic innovation in the context of ecological niche broadening that aided late Pleistocene human dispersal into western Eurasia after about 50 Kya. Although they see bow hunting as enabling *H. sapiens* to overcome obstacles that constrained previous human dispersals from Africa to temperate western Eurasia, they interpret the lack of this techno-behavior in the Neandertal record not as reflecting variability in cognitive complexity but, rather, as a result of energetic constraints and time-budgeting factors associated with such complex technologies (Shea & Sisk, 2010). Similar to others (e.g., Coolidge & Wynn, 2009; Wadley, 2013), I consider technical systems a more secure footing for exploring *H. sapiens*–specific cognitive and behavioral trends—as opposed to, for example, evidence for symbolic behaviors (e.g., Donald, 1993). The potential of such technical systems to inform on human cognitive evolution should thus be thoroughly explored before rejecting their cerebral implications out of hand.

Working with Miriam Haidle (Lombard & Haidle, 2012), I tested whether a cognitive explanation for bow hunting in the Stone Age/Paleolithic record could be ruled out. We applied "cognigrams" and "effective chains" as investigative method (for full discussion of the approach and its applications, see Haidle, 2012). Our analysis confirmed that the problem–solution distance of bow hunting is much extended compared to that of spear hunting (thrusting or throwing spears), even if the manufacture of the individual components might not be more taxing cognitively than the production of the stone-tipped spears that were also wielded by Neandertals and *H. heidelbergensis* (Lombard & Haidle, 2012). The extended thought-and-action sequence or problem–solution distance of the bow-and-arrow system correlates well with Coolidge and Wynn's (2009) interpretation of harpoons as technical systems that represent modern executive functions (also see Haidle, 2014). Yet, we have also urged caution about inferring cognitive complexity from hierarchical analyses based on a *chaîne opératoire* approach, such as thought-and-action sequences, because these generally lack an explicit cognitive theory (Wynn, Haidle, Lombard, & Coolidge, 2017).

However, it is not only in contingency planning or problem–solution distance that bow hunting exceeds the cognitive requirements of spear hunting. Hunting with a bow and arrow represents a symbiotic system or machine that relies on the simultaneous, focused manipulation of both the bow and the arrow. Technological symbiosis (where neither part of the system is effective without the simultaneous manipulation of the other) enables a level of complexity and flexibility that is not possible with non-symbiotic, simple, or composite technologies (Haidle et al., 2015; Lombard & Haidle, 2012). For example, such tool sets have two different elements: (1) enhancing elements with stable capacities, such as the bow, and (2) multiple consumable elements with changing, flexible capacities, such as arrows. The application of consumable elements is actively augmented by the enhancing element—handled and controlled by the user in a way that reveals the full potential of the consumable elements only when used in unison (Haidle, 2014; Lombard & Haidle, 2012). In this scenario, cognitive flexibility regarding problem-solving, decision-making, and action-taking is amplified. For example, arrowheads and other units (such as length, weight, fletching, etc.) can be adapted instantaneously to prey type, season, situation, and/or environment, increasing the scope and potential for success (e.g., Haidle et al., 2015; Lombard, 2016). Once the concept of symbiotic technologies is understood, different elements and series of elements can be adapted and grouped in multiple ways, and in sequences of various length and complexity, to achieve diverse results. For example, bows can be

(a) Grouped with drill bits (which are sometimes hunting arrows), weights, and handling pieces to use as bow drills;
(b) Used with palm protectors, fire sticks, base wood, and tinder as fire drills;
(c) Used as simple, violin-like instruments, stroked with a stick or arrow, and applied to the mouth cavity or a gourd as sound box, as is done by Kalahari hunter-gatherers in southern Africa; or
(d) Plucked (non-symbiotically) with the fingers like a one-string guitar, also demonstrated by the Kalahari San (Lombard, 2016).

Thus, we found that a key evolutionary advantage of symbiotic technologies, such as a bow-and-arrow set, is the amplification of conceptual, technological, and behavioral modularization and flexibility (also see e.g., Carignani, 2016), enabling almost endless combinations of single elements or chains of operations, in a variety of ways, to reach single or multiple goals. Bow hunting signifies a major cognitive improvement because it offers instantaneous and spontaneous flexibility to effectively handle any one possibility or situation out of a suite of diverse foreseen (and unforeseen) scenarios (Lombard & Haidle, 2012). It allows for a range of cognitive and cultural complexity and flexibility, basic to human behavior today and currently applied in the most complex of technologies (Lombard, 2016). These outcomes demonstrate that the technologies and behaviors associated with bow hunting can be used to assess levels of cognitive complexity, and that cognition cannot be ruled out as an explanation for why we do not see bow hunting in the archaeological records of populations other than *H. sapiens*.

WAS EXPERT COGNITION ENOUGH?

It has also been suggested that much of the technical complexity we observe, today as in the past, can be explained through expert cognition, which does not require enhanced executive functioning of the human mind (e.g., Coolidge et al., 2016; Wynn & Coolidge, 2004, 2010). Coolidge and Wynn (2009), for example, juxtaposed the difference between expert cognition and executive functions in the context of Neandertal versus modern human behavior when they wrote,

> We suspect that the flexibility of Neandertals' expert cognition was not quite as successful as the executive functions of *Homo sapiens* in adjusting to changing environments and the changing response of a competing variety of *Homo*. And, in the end, the *Homo sapiens* way of life and thinking survived, and the Neandertal way of life came to an end. (Coolidge & Wynn, 2009, p. 206)

It follows that if we tested whether bow hunting could be explained through expert cognition only, then Shea and Sisk (2010) might be correct, and the choice between bow hunting and spear hunting in the Neandertal context could be explained without considering variation in cognition.

Expert cognition did not develop recently. An increase in long-term working-memory capacity probably accompanied the transition from *H. erectus* to *H. heidelbergensis*, so that expert cognition was well within the thinking range of Neandertals (Coolidge et al., 2016). In current-day performances, expertise is achieved through the chunking and chaining (modularization) of large bodies of information (i.e., complex sequences are divided into small, easily processed chunks, which are then chained together). The last component of a chunk often prompts or cues the first component of the next chunk in a hierarchical chain (thought-and-action sequence). Expertise comes with extensive repetition of chaining chunks of specific information (e.g., mastering chess or a piece of music) and then automatically reacting to physical cues (e.g., a blacksmith relying on color, radiant heat, sound, physical resistance, and muscle tensions to cue an appropriate response). An expert therefore develops vast amounts of retrieval structures, linked to almost every conceivable variation in his or her field of expertise (Coolidge et al., 2016). An expert's performance, however, is limited to a narrow field. For example, chess masters have no advantage in checkers or Go (Wynn et al., 2017), and experts do not conceive of or implement new games, a new piece of music, or a new technology. Expert cognition alone thus precludes spontaneous and/or thoughtful innovation.

From a problem-solving distance point of view, bows represent the most complex production sequence in the bow-and-arrow system (Lombard & Haidle, 2012). We found that each of the technical modules of the bow and arrow could be learned or explained through expert cognition (Coolidge et al., 2016). Some flexibility is inherent in the chunking, chaining, and cueing of information. For example, bow makers can apply at least four alternative ways to bend the bow stave (Lombard & Haidle, 2012). Variation in local conditions and personal histories cue the appropriate sequence of actions used by each bow maker. The more bows a person produces, the more routine the information-retrieval structure becomes, so that decisions require increasingly less attention, resulting in a diminished occupation of working-memory capacity. Bow production is therefore a classic example of expert cognition (Coolidge et al., 2016), and basic arrow production is no more complex than producing a stone-tipped spear (Lombard & Haidle, 2012).

In an expert system, procedures become automatic, requiring little attention during performance. These would include activities such as collecting all the necessary materials, knapping a stone tip, or shaping a bow stave or an arrow shaft. The activity of combining the different elements, however, can hardly be executed effectively without paying active attention to each element and its changing properties during the process of combination. For example, during the process of fixing a stone tip to a shaft using adhesive and binding materials, the artisan must pay simultaneous attention to the properties and position of the tip, as well as those of the shaft, and how they could best fit together. He or she needs to carefully manipulate the binding material and/or adhesive so that the arrangement of tip and shaft remains stable during binding. Evidence for such techno-behaviors might therefore be early indicators that the thought processes of associated populations were moving out of the explanatory realm of expert cognition (Wynn et al., 2017).

In the South African context, there is evidence for the use of ochre-loaded compound adhesives in association with tools of more than 60 Kya that could have functioned as arrow tips (e.g., Lombard, 2008; Wadley, Hodgskiss, & Grant, 2009). While experimenting with such adhesives, Lyn Wadley noted,

> There is no recipe that can be followed; making these glues is not like baking a cake. The technique is not routine; it entails evaluating the qualities of the ingredients and adjusting their quantities accordingly. It requires complete, undivided attention. (Wadley, 2010, p. S115)

The arrow maker needs to hold information about a range of materials, their traits, and the various effects they might have on each other in active attention and continuously process the information and adapt accordingly to achieve a successful composite technology (Wadley, 2010). We have argued that the mixing of ochre-loaded adhesives represents a symbiotic technology similar to the bow-and-arrow system itself (Lombard & Haidle, 2012). Thus, until the compound adhesive is actively mixed, the entire technical sequence can be facilitated through expert cognition (Wynn et al., 2017). The mixing process, however, demands the simultaneous, focused manipulation of several elements and the full attention of the glue maker to ensure success; the entire performance is thus no longer automatically cued. Even though the mixing event represents only a short-lived activity within a long chain of actions in stone-tipped arrow production (Lombard & Haidle, 2012), the task switching and response inhibition of the cognitive control elements of working memory are heavily engaged during such performance (Wadley, 2010). Effective chunking and chaining (i.e., expert cognition) can no longer guarantee success, although they remain an important part of the thinking process (Wynn et al., 2017).

The process just described demonstrates that extended production sequences or problem-solution distances alone cannot explain the full cognitive repertoire associated with the innovation, manipulation, and application of symbiotic technologies. We therefore see that it is often not in the collecting of materials or production of artifacts that the potential clues for cognitive complexity is captured, but rather, potential clues are sometimes captured in short-lived performances, such as the symbiotic use of a bow-and-arrow set (e.g., Lombard & Haidle, 2012) or the careful mixing of compound adhesives (Wadley, 2010). The archaeological dilemma, of course, is that we cannot excavate such moments of "neural effort." But I suggest that we can explore ways to assess whether modern minds display functional variability during simulated tasks.

A NEUROARCHAEOLOGICAL GLIMPSE INTO SHOOTING ARROWS

In a first attempt to evaluate the amount of neural effort that is required for simple spear throwing compared with arrow-shooting tasks, we used electroencephalography and event-related potential (EEG-ERP) to test directly for neuro-indicators that could assess whether shooting an arrow with a bow could be considered more cognitively challenging than throwing a spear at a static target (Williams et al., 2014). Our hypothesis was that if the act of shooting an arrow from a bow does not require more neural effort than throwing a spear, then differences in levels of cognitive complexity must be sought elsewhere in the thought-and-action sequence as described by Lombard and Haidle (2012). The excerpts from Williams and colleagues (2014) that I present in this section can only be considered preliminary, but I suggest they provide valuable indicators for future study and hypothesis testing.

During our experiment, we eliminated variables associated with invention, innovation, production, and the challenges of actualistic hunting scenarios. We focused on three cortical sites: the parietal, frontal midline, and orbitofrontal regions. These areas were selected because they each play an important role during cognitive activation states (including visuospatial or motor tasks) and have been shown to facilitate aspects of complex cognition and working memory. For example, the parietal cortex is activated when visuospatial information is processed in the mind (Rushworth & Taylor, 2006; Walter & Dassonville, 2008), which assists with visual working-memory processes (Riley, Lambert, Qi, & Constantinidis, 2013; Salazar, Dotson, Bressler, & Gray, 2012), and it might be involved in selecting and maintaining perceptual information in working memory (Xu & Chun, 2009). The frontal midline is also associated with working-memory processes (Hsieh & Ranganath, 2013), and it is activated when sustained or internalized attention is needed (Aftanas & Golocheikine, 2001). Relatively little is known about the function of the orbitofrontal cortex, but it has been shown to coordinate and integrate sensory information (Kringelbach, 2005), assist with processing incentives or motivational information (e.g., Ostlund & Balleine, 2007; Schoenbaum & Setlow, 2001), and be active in situations where decision-making is required (Larquet, Coricelli, Opolczynski, & Thibaut, 2010; Rolls & Grabenhorst, 2008).

The four electrical frequency bands used were delta, theta, alpha, and beta. Delta activity is associated with a resting state (Lu et al., 2007), some decision-making tasks (Nácher, Ledberg, Deco, & Romo, 2013), and motivational processes that are triggered by biological rewards and danger (Knyazev, 2012). Theta activity retains information in the mind during the performance of a working-memory task (Tesche & Karhu, 2000), plays an important role in declarative- and episodic-memory processing (Fell et al., 2003; Klimesch et al., 2001), memory encoding (Klimesch, Doppelmayr, Russegger, & Pachinger, 1996), sustained attention (Sauseng, Hoppe, Klimesch, Gerloff, & Hummel, 2007), and sensorimotor processing (Bland & Oddie, 2001). The alpha band is involved in highly specific attentional (Worden, Foxe, Wang, & Simpson, 2000), perceptual (e.g., Thut, Nietzel, Brandt, & Pascual-Leone, 2006) and memory-related processes (Klimesch, Schack, & Sauseng, 2005), and it plays an integral role in suppressing information that is irrelevant to the task at hand (Palva & Palva, 2007). Such inhibition improves the encoding, storage, and retrieval of information to and from memory (Jensen, Gelfand, Kounios, & Lisman, 2002; Obleser, Wöstmann, Hellbernd, Wilsch, & Maess, 2012). Beta waves are associated with mental activities that require higher-order cognitive functions (see Geethanjali, Adalarasu, & Rajsekaran, 2012; Weiss & Mueller, 2012). For example, low-amplitude beta waves reflect a strongly engaged mind where active attention is required (e.g., Gross et al., 2004).

Our results indicated that some elements of the central executive, such as inhibition and focusing on external as well as internal attention, were probably in place to a certain extent for Stone Age spear hunters (Williams et al., 2014). This interpretation suggests that spear-hunting Neandertals and *H. heidelbergensis* possessed some forms of executive functioning. In general, however, the task of throwing a spear could be executed with relatively little neural effort. It was patent, within the confinements of our method and experiment, that the task involving shooting arrows from a bow showed statistically significant higher levels of neural activity across the cortical sites and frequency bands compared with the spear-throwing task. Thus, regardless of the levels of cognitive complexity achieved by Stone Age/Paleolithic spear hunters, our results demonstrate that simply firing an arrow with a bow at a target is consistently more cognitively taxing than throwing a spear (Williams et al., 2014, p. 204). In the following discussion I present the outcomes related to the arrow-shooting task; for the detailed results relating to the spear-throwing task, see Williams and colleagues (2014, pp. 203–204).

During the initial stages of the arrow-shooting task, there was no change in parietal delta activity, followed by a slight increase that probably allowed the participants to concentrate on the visual items that were involved in or assisted with processing incoming sensory information. For example, parietal delta activity may have been involved in visual acuity, as both clarity and detail are essential for target shooting (e.g., Strydom & Ferreira, 2010). Parietal theta activity increased as the arrow-shooting task prevailed, which was perhaps needed for context updating. The arrow-shooting task required that a number of action sequences be remembered, which might reflect aspects of episodic-memory processing. There was an increase in frontal midline delta activity, followed by a decline, and a slight increase again. Increases in the delta frequency band in this region could indicate internal attention, allowing the participants to process incoming sensory information in the mind.

The statistically significant increase in frontal midline theta activity during the arrow-shooting task, compared to the spear-throwing task, demonstrates that shooting an arrow with a bow is cognitively more demanding. For example, the activity required higher levels of sustained attention and memory load. The premise that participants "mentally rehearsed" the action sequences involved in the arrow-shooting task was validated further by the slight increase in frontal midline beta activity shown during the final stages of the task and probably reflected the execution of praxis movements. Such behavior is seen as a uniquely human trait, enabling us to synchronize ideas or concepts with motor planning and task execution (Ayres, 1985; May-Benson & Cermak, 2007). Thus, when considering technologies that might reflect our cognitive and behavioral uniqueness, it is perhaps noteworthy that frontal midline beta activity was completely absent for the duration of the spear-throwing task. The significant increase in activity of this region during the arrow-shooting task suggests that sustained alertness was needed for task execution. Information irrelevant to the task may have been inhibited, allowing the participants to focus on the cognitive functions involved (e.g., internal attention, imagination, and mental imagery) (Williams et al., 2014, pp. 204–205).

During the initial stages of the arrow-shooting task, there was a slight increase in orbitofrontal delta activity, followed by a decline. This increase could have been useful for screening internal and external stimuli and increasing attention to task-relevant stimuli. To assist with encoding sensory and perceptual information into memory, there was a gradual increase in orbitofrontal theta activity, followed by a slight decline. Orbitofrontal theta activity could represent an essential frontal component in which novel stimuli are processed. The mechanistic function of orbitofrontal alpha activity could have motivated the participants to focus their attention on the arrow-shooting task. As the task progressed, orbitofrontal alpha activity continued to increase, and possibly worked in conjunction with frontal midline alpha activity. Similar to the spear-throwing task, kinesthetic sensations were probably needed during the arrow-shooting task. These stimuli would have provided participants with sensory feedback on the motor actions performed, enabling them to determine whether behavioral adjustments were needed. To improve task performance, the participants might have used positive feedback or reinforcement learning to actively recall the "correct" actions from memory. This was supported by an increase in orbitofrontal beta activity as the tasks progressed (Williams et al., 2014, pp. 204–205).

Although our observations and interpretations are presented according to cortical site, they should not be viewed as unrelated to each other. For example, parietal and frontal midline theta activity plays an important role in updating familiar sensorimotor representations, which allows an individual to execute memorized action sequences (Makeig et al., 2004). It is useful in situations when one must engage in cognitively complex tasks that require additional information to be processed. To assist with task performance, participants might have mentally rehearsed these action sequences. Theta activity generally assists with learning novel tasks that require a number of action sequences be remembered, which is associated with working-memory load and episodic-memory processing (e.g., Kawasaki & Yamaguchi, 2012). Once each of the volunteers practiced using the bow-and-arrow set as an effective unit, sustained alertness might have encouraged them to perform the task skillfully. Task practice resulted in a further increase in memory-related brain waves. For example, parietal and frontal midline theta activity increased as the arrow-shooting task prevailed. Levels of delta activity increased across all of the cortical sites recorded when the participants switched from the spear-throwing to the arrow-shooting task. This observation implies that, compared with the spear-throwing task, the arrow-shooting task required greater attention to internal processing (i.e., the selection, modulation, and maintenance of internalized information and/or the processing of information in working memory, episodic memory, and long-term memory) (Williams et al., 2014, pp. 204–205).

Changes associated with hunting technologies are seen to "imply that a gradual but substantial improvement in WM [working memory] has occurred, perhaps in WM capacity, the types of information accessible and manipulable by WM, or both" (Coolidge, Wynn, & Overmann, 2012, p. 55). Our preliminary results and interpretation of the data indicate that when participants switched from the spear-throwing task to the arrow-shooting task, several executive functions (e.g., attention, active inhibition, context updating, reinforcement learning, and memory rehearsal) were enhanced to assist with working-memory processes. Thus, we demonstrated that variation in levels of neurological activity associated with the memory systems of spear throwers and arrow shooters is a feasible hypothesis for the interpretation and explanation of the archaeological record of hunting technologies.

THE MEMORY OF A BOW HUNTER

Our exploration into the neuroarchaeology of bow hunting suggests that shooting an arrow with the purpose of hitting a target requires higher levels of working-memory load and episodic-memory processing than does throwing a spear at the same target. We also observed that indicators for mentally rehearsed praxis (i.e., conceptualization or ideation, planning, and execution of a motor act) were generally absent during the spear-throwing task but increased significantly during the arrow-shooting task (Williams et al., 2014). Praxis facilitates effective interaction with objects and environments based on "ideas" about how to manipulate both (Ayres, 1985), and these ideas stem from a creative interplay between memorized concepts and previously unthought-of applications or performances. Based on our work with cognigrams and action sequences as well as the expert-cognition model, even *H. heidelbergensis* could have learned most of the skills associated with bow-and-arrow production (Coolidge et al., 2016; Lombard & Haidle, 2012). However, based on the same body of work, we have argued that they probably could not have conceptualized, structured, and implemented the entire symbiotic technical system. Thus, for thinking through, inventing, and using such systems, something more than expert cognition was needed (Coolidge et al., 2016).

Apart from acquiring all the materials and assembling the different parts of the system, the artisan must be able to access the entire system in attention, at least occasionally—that is, ideation. The best way to achieve an extended thought-and-action sequence, in combination with the symbiotic application of a technological system and taking into consideration the variables associated with the whole sequence, is not through expert cognition but episodic memory (Coolidge et al., 2016). We therefore suggested that episodic memory is a necessary, if not sufficient, condition for bow hunting (Coolidge et al., 2016). Episodic memory implicates the recollection of events as well as event clusters (additional elements associated with the events). It involves an awareness that recalled events might vary from the original events, and that time is subjective if such events can be reimagined or modified (consciously or unconsciously) (Coolidge et al., 2016).

Although humans share basic episodic memory with most animals (e.g., Allen & Fortin, 2013), at some point "true" episodic memory (i.e., autonoetic or autobiographical memory) evolved only in hominins. Earlier hominins could probably remember and use knowledge of past events to solve "present" problems, but they were most likely not aware that they were doing so (Tulving, 2002). A clear consciousness of the "self" in a past memory, and the cognizant manipulation of such memories, is what Tulving referred to as "autonoetic awareness," which might have evolved recently and could be unique to *H. sapiens* (Coolidge, Haidle, Lombard, & Wynn, 2016; Tulving, 2002; Tulving & Kim, 2007; also see Malafouris, 2008). Schacter and Addis (2007) proposed a constructive episodic simulation hypothesis, which allows for the recombination of past details or events imagined in novel configurations for future application. These simulations can recall past events in a highly flexible manner to improve the success of immediate or future actions. They reasoned that, as the future is not an exact representation of the past, the ability to simulate future events must be inherently flexible in order to recall, extract, and recombine aspects of past events, safeguarding the success of future actions (Schacter & Addis, 2007). Their concept of constructive episodic simulation may represent the critical cognitive component for bow hunting—that is, a cognitive system able to consciously activate, deactivate, and reactivate expertise modules in new ways over spatiotemporal distances (Coolidge et al., 2016). This interpretation links to the previously suggested evolutionary advantage of symbiotic technologies, in terms of the amplification of conceptual, technological, and behavioral modularization and flexibility, which offers instantaneous and spontaneous plasticity in day-to-day problem-solving and the development of increasingly complex techno-behaviors (Lombard, 2016; Lombard & Haidle, 2012), which might have enabled our successful spread across the globe as opposed to the demise of the Neandertals (Coolidge & Wynn, 2009; Shea & Sisk, 2010).

We have therefore suggested that the cognitive requirements for conceptualizing, organizing, building, and using technical systems such as the bow and arrow effectively may have required a fully modern episodic-memory system—a cognitive system capable not only of autobiographical-memory retrieval but also of constructive episodic-memory simulations (prospection)—which could imply that early bow hunters possessed a fully autonoetic awareness (Coolidge et al., 2016). Support for our conclusion can be found in the work of d'Argembeau and Demblon (2012), who suggested that in an event-queuing paradigm, it is unlikely that future prospection relies solely on autobiographical–episodic-memory networks, and argued that personal goals, which rely on personal abstract knowledge, provide an important framework for the overall organization of imagined events (Coolidge et al., 2016).

In work with Peter Gärdenfors on the evolution of causal cognition in humans (Gärdenfors & Lombard, 2018; Lombard & Gärdenfors, 2017), we suggest that causal network understanding or "causal grammar" enables hunters to draw together domain-specific knowledge into inter-domain networks of abstract causal understanding to facilitate speculative tracking. Such tracking is strongly associated with the Kalahari bow hunters of recent times (Liebenberg, 1990, 2013). Intimate knowledge of kin, non-kin, and animal behavior and their inanimate signs is incorporated with knowledge about the landscape (e.g., its geographic features, water sources, vegetation, etc.), abstract causal understanding, and the mental maps, thought processes, and social contexts of the tracker. Speculative tracking demonstrates how bow hunters create meaningful and imaginative causal network hypotheses, during which some of the activity happens abstractly, in the mind of the hunter—that is, prospection—enabling him or her to deal with complex, dynamic variables (Liebenberg, 1990; Gärdenfors & Lombard, 2018; Lombard & Gärdenfors, 2017). Such behavior therefore exemplifies a fully modern episodic-memory system capable of autobiographical-memory retrieval as well as episodic-memory simulations.

THE NEUROLOGY OF MODERN ARCHERS

Thus far, we have seen that each assessment aimed at eliminating bow hunting as proxy for aspects of cognitive evolution, potentially separating *H. sapiens* from other members in our family, has failed. Bow hunting, compared with spear hunting, consistently revealed more complex levels of cognition at work. If this is indeed the case, we could expect independent data to support our findings. Here I touch on only two independent neurological studies to demonstrate the potential of such work to help interpret aspects of cognitive evolution associated with techno-behaviors such as bow hunting, and how they can be used to augment and constrain such interpretations.

My first example is the work of Kim and colleagues (2011), who conducted a functional magnetic resonance imaging (fMRI) study in which expert archers and a non-archer control group were asked to watch a video of Western-style archery. Kim and colleagues found that expert archers showed stronger activation in the premotor cortex, inferior parietal cortex, superior temporal sulcus, dorsomedial prefrontal cortex, cingulate cortex, retrosplenial cortex, and parahippocampal gyrus while watching the video material—and none of the tested regions were significantly more activated in the non-archers compared with the expert archers. The increased activation of the dorsomedial prefrontal cortex of expert archers was interpreted as indicating that they spontaneously engaged in theory of mind (ToM) while watching others perform archery (Kim et al., 2011), possibly indicating an urge to "teach" or improve the skills of others (e.g., Lombard, 2015; Strauss, 2005). In addition, compared with non-archers, the expert archers displayed greater neural activity in the regions associated with episodic memory from familiar and meaningful information. These areas included the cingulate cortex, retrosplenial cortex, and parahippocampal gyrus (Kim et al., 2011). Kim and colleagues therefore suggested that expert archery stimulates brain activity in the mirror neuron system, as well as the neural networks associated with ToM (mind reading) and episodic memory.

The increased responses of the mirror neuron system in expert archers probably reflect motor familiarity with the activity (Kim et al., 2011). The mirror neuron system, composed of the premotor cortex and inferior parietal cortex, interacts with other neural systems to facilitate different types of social learning (e.g., Iriki, 2006). Interactions between the mirror neuron system and networks for ToM thus appear key to social mirroring and mind reading (the ability to form representations about the internal states of others; Gärdenfors, 2003, 2007). As noted by Kim and colleagues,

> The dorsomedial prefrontal cortex is selectively recruited for representing the triadic relationship between two minds and an object and the internal states of unfamiliar others. Therefore, the activation of the dorsomedial prefrontal cortex observed in expert archers probably occurred because they were required to represent the triadic relationship between two minds, a bow, and the internal state of an unfamiliar protagonist in the video. (Kim et al., 2011, p. 346)

(Also see Lombard, 2015, for similar discussion on the teaching of bow hunting in Kalahari hunter-gatherer groups.) Considering the working of a modern mind, it is also noteworthy that activation of the cingulate cortex and retrosplenial cortex is associated with imagination, navigation, and episodic recall from long-term memory maintained elsewhere in the brain (Kim et al., 2011; van Overwalle, 2009).

In a different fMRI study, explicitly aimed at assessing motor imagery, expert archers and non-archers were asked to mentally rehearse their archery shooting from a first-person perspective (Chang et al., 2011). In that study, it was shown that the premotor and supplementary motor areas, and the inferior frontal region, basal ganglia, and cerebellum, were more active in non-archers. Expert archers, by contrast, showed activation primarily in the supplementary motor areas bilaterally. These results were interpreted as indicating the increased participation of the cerebellum in non-archers when learning an unfamiliar archery task. The difference in cerebellar activation between archers and non-archers is considered evidence of the expertise effect in the mental rehearsal of archery, and the relative economy in the cortical processes of elite archers is thought to contribute to enhanced performance during archery. In this study, it was found that cortical regions, which show greater activation in non-archers, mainly involved the parietal cortex, including the precuneus, which might reflect a coupling between the visual imagery of object rotation and motor imagery (Chang et al., 2011). The data have also been interpreted to reflect the fact that non-archers have not yet mastered the complex task of accurate archery and thus strain to integrate the relevant sensory and cognitive information needed to plan a successful archery shot (Chang et al., 2011).

These two independent studies support our previous observations about conceptualization and the need for episodic memory associated with bow hunting as a techno-behavior (Coolidge et al., 2016). Additionally, a neurological link to ToM processing was highlighted (Kim et al., 2011). The study by Chang and colleagues (2011) revealed that the precuneus was more active in non-archers than in expert archers, which could indicate how techno-behaviors such as bow hunting might have contributed to the recent expansion of this brain region in *H. sapiens* (e.g., Bruner, 2010). These studies represent only a couple in a larger body of work pertaining to the neurology of modern archery. Future work will focus on a much more thorough analysis in an effort to start understanding possible pressures on brain regions and neurological networks exerted by the materiality of techno-behaviors that might have affected our cognitive evolution. Following Malafouris (2013) and Bruner and Lozano (2014), it will also explore ancient bow hunting as an example of the human mind expressed through extended and distributed embodiment, as opposed to being solely "brain-bound."

CONCLUSION

In the last few years it has become increasingly feasible to think about bow hunting at more than 60 Kya. During this time, colleagues and I have continued to test for aspects of cognition that can be associated with this uniquely human techno-behavior. The human mind of the past remains a difficult domain to interpret, and most inferences remain tentative. Yet, it is not unfeasible to explore, nor impossible to cautiously build, multiple strands of evidence that can provide increasingly robust insight into past human minds and their interaction with increasingly extended embodiment through technology. In this chapter I have introduced a body of work still in its infancy but which demonstrates that Stone Age/Paleolithic bow hunting can be associated with several cognitive traits considered to represent the "executive" and which are thus characteristic of the complex thinking associated with humans today. In addition, our pilot neuroarchaeological and independent neurological studies based on modern-day archery have revealed areas of the human brain stimulated during activities associated with bow hunting or archery. These studies demonstrate the inextricable interrelatedness between brain, body, mind, and using (or learning to use) technologies. Table 22.1 presents a summary of the cognitive and neurological aspects touched on in this chapter.

There are few, if any, other techno-behaviors, dating to more than 60 Kya for which such a direct and extensive list of cognitive and neurological associations currently exists. I am cautiously optimistic about the fact that the presently known functions of some of these brain regions correlate with aspects of our interpretations derived

Table 22.1. Current Cognitive and Neurological Associations with Bow Hunting

Cognition	*Neurology*
• Extended thought-and-action sequence or problem–solution distances • Cognitive flexibility or plasticity • Increased working-memory load • Active inhibition • Sustained internal attention • Praxis or mental rehearsal • Episodic/autonoetic- or autobiographical-memory processing • Imagination, mental imagery (prospection) • Theory of mind or mind reading	• Parietal cortex • Frontal midline cortex • Orbitofrontal cortex • Precuneus or superior medial parietal lobes • Premotor cortex • Inferior parietal cortex • Superior temporal sulcus • Dorsomedial prefrontal cortex • Cingulate cortex • Retrosplenial cortex • Parahippocampal gyrus • Inferior frontal region • Basal ganglia • Cerebellum

Note: Some of the neurological regions overlap. Table compiled by the author.

from problem–solution–distance analysis, cognition associated with the use of symbiotic technologies, and hypotheses about expert cognition and fully developed autobiographical memory. For example, for at least one of the listed brain regions, there exists evidence for recent, *H. sapiens*–expansion: the precuneus, located deep in the parietal cortex. Based on variation in the morphology in this region between different *Homo* groups, it has been suggested that the precuneus may have played a key role in our neurological evolution (Bruner, 2010). Documented functions associated with the precuneus include visuospatial integration, category recognition, praxis, numerical processing, and speech decoding. All of these are strongly associated with modern human ways of thinking, and Bruner's (2010) analysis shows that only *H. sapiens* displays a general enlargement of the complete parietal surface. What is more, the variation in morphological details of the region suggests neurofunctional differences in visuospatial integration, the "recognition and codification of the outer spatial environment and the associated integration between the outer frame and the inner perceptions" (Bruner, 2010, p. S77). In a previous section, I introduced the link between bow hunting and full autonoetic awareness (Coolidge et al., 2016), and Lou and colleagues (2004) have also suggested that a sense of self and self-representation possibly has its neurological foundations in the precuneus or superior medial parietal lobes.

Neandertals using their teeth as a "third hand" more so than contemporary humans could imply differences in the visuospatial integration between these two populations, with that of the Neandertals possibly being less extensive than that of *H. sapiens* (Bruner & Lozano, 2014). This hypothesis has been directly linked to between-species variation in parietal surface morphology, resulting in a lively "three hand" debate. A few topics pertaining to this chapter are as follows:

- A technological reorganization of the body's form and function is brought to mind by Malafouris' (2014) concept of "prosthesis" based on the "human-machine interface." In the case of Neandertals (and *H. heidelbergensis*) using their teeth as a "third hand," the emphasis is on the interface between body parts (Malafouris, 2014) as opposed to between body and machine.
- Wynn (2014, p. 291) suggests that the first evidence for hominin-specific visuospatial integration can be found in early Acheulean bifaces at about 1.8 Mya, with a second "major development" evidenced more than a million years later in the "fine 3-D symmetry" of late Acheulean bifaces. To him, these artifacts represented "the ability to conceive non-egocentric visual perspectives," and thus a "fully modern ability to conceptualize" through imagining (also see Garofoli, 2015; Langbroek, 2014, on Neandertal imagination).
- Levalloisian knapping (Rios-Garaizar, 2015; Spinapolice, 2015) and distance hunting (including distance guessing and good aiming for spear throwing; Rios-Garaizar, 2015) are brought to mind as good day-to-day examples of Neandertal visuospatial coordination.
- The further the reach of the tool, the more extensive the mapping of the body-schema and visuospatial integration to incorporate the technology in the human performance and experience. Thus, Overmann (2015, p. 165) argues that "the critical difference between the two species may not have been hand-eye coordination per se, . . . but a more pervasive prosthetic adaptation by *H. sapiens*."

- Variation in the sophistication of social network "wayfinding" (egocentric and allocentric), where that of *H. sapiens* is considered more developed than the Neandertal skill set, is based on symbolic expressions recorded for *H. sapiens* in Middle Stone Age Africa (Burke, 2015).
- Martín-Loeches (2014, p. 299) argues that although symbolic mediation seems to go largely unchallenged in archaeology as indicating the modern mind, "in cognitive science, the symbolic perspective is being replaced by alternative models of embodied cognition, according to which mental representations are directly grounded in sensorimotor experience." He suggests that from this perspective, symbolism should be replaced by evidence that reflects abstraction and integration because of their direct relation to "perceptual and action principles working in the brain according to neuroscientific traditions."
- *H. sapiens'* ability for "higher level abstractions, more complex mathematical and economic reasoning, and a more distinct sense of self and higher levels self-consciousness" is highlighted by Coolidge's contribution (2014, p. 297).

By its very nature, bow hunting and its associated tracking behaviors (Lombard, 2015; Lombard & Gärdenfors, 2017) require powerful visuospatial integration. I suggest that based on all the points raised here, the techno-behavior represents a profound prosthetic adaptation—the use of a machine that differs from the use of teeth as a hand, the knapping of stone tools (regardless of the complexity of technique), and the wielding of spears (even in the case of long-distance hunting). Techno-behaviors such as bow hunting, coupled with speculative tracking (Lombard & Gärdenfors, 2017), may well have contributed to exert the forces (on both minds and brains) that increased *H. sapiens'* capacity for integrating our "inner virtual space" with the visuospatial challenges and realities of our increasingly complex socioeconomic environments.

As a concluding thought, I want to highlight the last three topics listed under "Cognition" in Table 22.1. These are often omitted in discussions about what we can learn about human cognition from our techno-behaviors as reflected in the archaeological record. This is not surprising, because how does one excavate and analyze autobiographical-memory processing, imagination, or mind reading? Yet, the work presented in this chapter shows that these characteristics recur in association with archery. It is therefore not impossible to hypothesize about their presence in the past, especially when multiple, independent, and cross-disciplinary studies start to highlight similar trends that contribute to a uniquely human way of thinking. In this context, I want to draw attention to the work of Spreng and Grady (2010, p. 1112), who argue that "[r]emembering one's past (autobiographical memory), imagining one's future (prospection), and imagining the thoughts and feelings of others (ToM) are similar in that they all involve simulating an experience that is distinct from stimulus-driven behavior (Buckner & Carroll, 2007)." This study is relevant because all three cognitive processes have been associated with bow hunting, and they fall outside the realm of expert cognition, which relies on physical cues or stimuli for successful performance.

Spreng and Grady (2010) tested the hypothesis that adults who are engaged in autobiographical remembering, prospection, and ToM reasoning would activate the same set of brain regions across all three conditions. They applied partial least squares (PLS) analysis of fMRI data to examine whole-brain activity patterns associated with "thinking about the past, the future, and the thoughts and feelings of others" (Spreng & Grady, 2010, p. 1118). In sum, they found that autobiographical memory, prospection, and ToM shared a common pattern of brain activity, including midline structures in the frontal and parietal lobes associated with the default mode network (a highly interconnected core set of brain structures including the midline frontal and parietal structures, medial and lateral temporal lobes, and lateral parietal cortex), which is consistent with brain activity related to self- and other-referential thought (also see d'Argembeau et al., 2008; Johnson et al., 2006; Northoff & Bermpohl, 2004). They therefore suggested that the functional coordination of these heteromodal regions could be sufficient for producing self-relevant stimulus-independent thought. Iriki (2006) also suggested that humans have

> a genuinely voluntary intention system that can perform, or even just imagine, actions divorced from immediate context. This system not only enables us to mirror and understand the intentions of other people, it endows us with a fully developed theory of mind. The medial prefrontal cortex (MPFC), temporoparietal junction (TPJ, especially right hemisphere in humans) and temporal pole, together with a few additional structures, might carry out this function. Non-human primates show, to date, no sign of this kind of sense of the subjective self. (Iriki, 2006)

Tantalizingly, we once again see a correlation between uniquely human faculties and the inclusion of the parietal lobes—the precuneus—in the intricate neurological networks that facilitate them. I suggest that such neurological and thought-processing networks might have facilitated the necessary framework to imagine, innovate, apply, and modify symbiotic technologies such as the bow and arrow, something that cannot be accomplished through expert cognition alone and can no longer be perceived as not indicating cognitive variability compared to spear hunting. At the same time, applying pressure on these networks, as a result of increasingly complex techno-behaviors such as bow hunting, could have played a part in stimulating their expansion to facilitate our uniquely human way of thinking.

ACKNOWLEDGMENTS

I thank the editors for inviting me to participate in this volume and for its efficient administration. The constructive input from two anonymous reviewers is much appreciated, as is that of previous co-authors who worked with me on various aspects of bow hunting and cognition. My research is funded through an African Origins Platform grant (#98815) awarded by the National Research Foundation of South Africa, but opinions and mistakes remain my own.

REFERENCES

Aftanas, L. I., & Golocheikine, S. A. (2001). Human anterior and frontal midline theta and lower alpha reflect emotionally positive state and internalised attention: High-resolution EEG investigation of meditation. *Neuroscience Letters*, *310*(1), 57–60.

Allen, T. A., & Fortin, N. J. (2013). The evolution of episodic memory. *Proceedings of the National Academy of Sciences of the United States of America*, *110*(S2), 10379–10386.

Ayres, A. J. (1985). *Developmental dyspraxia and adult-onset apraxia*. Torrance, CA: Sensory Integration International.

Backwell, L., d'Errico, F., & Wadley, L. (2008). Middle Stone Age bone tools from the Howiesons Poort layers, Sibudu Cave, South Africa. *Journal of Archaeological Science, 35*(6), 1566–1580.

Barham, L. S. (2013). *From hand to handle: The first industrial revolution*. Oxford, UK: Oxford University Press.

Bland, B. H., & Oddie, S. D. (2001). Theta band oscillation and synchrony in the hippocampal formation and associated structures: The case for its role in sensorimotor integration. *Behavioural Brain Research, 127*(1), 119–136.

Bradfield, J., & Lombard, M. (2011). A macrofracture study of bone points used in experimental hunting with reference to the South African Middle Stone Age. *South African Archaeological Bulletin, 66*(193), 67–76.

Brooks, A. S., Nevell, L., Yellen, J. E., & Hartman, G. (2006). Projectile technologies of the African MSA: Implications for modern human origins. In E. Hovers & S. L. Kuhn (Eds.), *Transitions before the transition: Evolution and stability in the Middle Palaeolithic and Middle Stone Age* (pp. 233–255). New York: Springer.

Brown, K. S., Marean, C. W., Jacobs, Z., Schoville, B. J., Oestmo, S., Fisher, E. C., . . . Matthews, T. (2012). An early and enduring advanced technology originating 71,000 years ago in South Africa. *Nature, 491*(7425), 590–593.

Bruner, E. (2010). Morphological differences in the parietal lobes with the human genus: A neurofunctional perspective. *Current Anthropology, 51*(1), S77–S88.

Bruner, E., & Lozano, M. (2014). Extended mind and visuo-spatial integration: Three hands for the Neandertal lineage. *Journal of Anthropological Sciences, 92*, 273–280.

Buckner, R. L., & Carroll, D. C. (2007). Self-projection and the brain. *Trends in Cognitive Sciences, 11*(2), 49–57.

Burke, A. (2015). Visuo-spatial integration and spatial cognition: A comment on Bruner & Lozano (2014). *Journal of Anthropological Sciences, 93*, 177–180.

Carignani, G. (2016). On the origin of technologies: The invention and evolution of the bow-and-arrow. In F. Panebianco & E. Serrelli (Eds.), *Understanding cultural traits: A multidisciplinary perspective on cultural diversity* (pp. 315–339). Cham, Switzerland: Springer International Publishing.

Chang, Y., Lee, J., Seo, J.-H., Song, H.-J., Kim, Y., Lee, H. J., . . . Kim, J. G. (2011). Neural correlates of motor imagery for elite archers. *NMR in Biomedicine, 24*(4), 366–372.

Coolidge, F. L. (2014). The exaptation of the parietal lobes in *Homo sapiens*. *Journal of Anthropological Sciences, 92*, 295–298.

Coolidge, F. L., Haidle, M. N., Lombard, M., & Wynn, T. (2016). Bridging theory and bow hunting: Human cognitive evolution and archaeology. *Antiquity, 90*(349), 219–228.

Coolidge, F. L., & Wynn, T. (2009). *The rise of* Homo sapiens: *The evolution of modern thinking*. Oxford, UK: Wiley-Blackwell.

Coolidge, F. L., Wynn, T., & Overmann, K. A. (2012). The evolution of working memory. In T. P. Alloway & R. G. Alloway (Eds.), *Working memory: The connected intelligence* (pp. 37–60). New York: Psychology Press.

Cotterell, B., & Kamminga, J. (1990). *Mechanics of pre-industrial technology: An introduction to the mechanics of ancient and traditional materials culture*. Cambridge, UK: Cambridge University Press.

d'Argembeau, A., & Demblon, J. (2012). On the representational systems underlying prospection: Evidence from the event-cueing paradigm. *Cognition, 125*(2), 160–167.

d'Argembeau, A., Feyers, D., Majerus, S., Collette, F., van der Linden, M., Maquet, P., & Salmon, E. (2008). Self-reflection across time: Cortical midline structures differentiate between present and past selves. *Social Cognitive and Affective Neuroscience, 3*(3), 244–252.

Denny, M. (2007). *Ingenium: Five machines that changed the world.* Baltimore, MD: Johns Hopkins University Press.

d'Errico, F., Backwell, L., Villa, P., Degano, I., Lucejko, J. J., Bamford, M. K., . . . Beaumont, P. B. (2012). Early evidence of San material culture represented by organic artifacts from Border Cave, South Africa. *Proceedings of the National Academy of Sciences of the United States of America, 109*(33), 13214–13219.

di Paolo, E. A., Rohde, M., & De Jaegher, H. (2010). Horizons for the enactive mind: Values, social interaction, and play. In J. Stewart, O. Gapenne, & E. A. di Paolo (Eds.), *Enaction: Toward a new paradigm for cognitive science* (pp. 33–87). Cambridge, MA: MIT Press.

Donald, M. (1993). Precis of *Origins of the modern mind: Three stages in the evolution of culture and cognition. Behavioral and Brain Sciences, 16*(4), 737–748.

Dusseldorp, G. L., Lombard, M., & Wurz, S. (2013). Pleistocene *Homo* and the updated Stone Age sequence of South Africa. *South African Journal of Science, 109*(5/6), 1–7.

Fell, J., Klaver, P., Elfadil, H., Schaller, C., Elger, C. E., & Fernández, G. (2003). Rhinal-hippocampal theta coherence during declarative memory formation: Interaction with gamma synchronization? *European Journal of Neuroscience, 17*(5), 1082–1088.

Gärdenfors, P. (2003). *How* Homo *became* sapiens: *On the evolution of thinking.* Oxford, UK: Oxford University Press.

Gärdenfors, P. (2007). Evolutionary and developmental aspects of intersubjectivity. In H. Liljenström & P. Århem (Eds.), *Consciousness, transitions: Phylogenetic, ontogenetic and physiological aspects* (pp. 218–305). Amsterdam: Elsevier.

Gärdenfors, P., & Lombard, M. (2018). Causal cognition, force dynamics and early hunting technologies. *Frontiers in Psychology, 9*, article 87. doi: org/10.3389/fpsyg.2018.00087

Garofoli, D. (2015). Neandertal three hands: Epistemological foundations and a theory of visual impedance. *Journal of Anthropological Sciences, 93*, 185–190.

Geethanjali, B., Adalarasu, K., & Rajsekaran, R. (2012). Impact of music on brain function during mental task using electroencephalography. *World Academy of Science, Engineering and Technology, 66*, 883–887.

Gross, J., Schmitz, F., Schnitzler, I., Kessler, K., Shapiro, K., Hommel, B., & Schnitzler, A. (2004). Modulation of long-range neural synchrony reflects temporal limitations of visual attention in humans. *Proceedings of the National Academy of Sciences of the United States of America, 101*(35), 13050–13055.

Haidle, M. N. (2012). *How to think tools? A comparison of cognitive aspects in tool behavior of animals and during human evolution* (M. N. Haidle, Ed.). Tübingen, Germany: University of Tübingen.

Haidle, M. N. (2014). Building a bridge—An archeologist's perspective on the evolution of causal cognition. *Frontiers in Psychology, 5*(1472), 1–15.

Haidle, M. N., Bolus, M., Collard, M., Conard, N. J., Garofoli, D., Lombard, M., . . . Whiten, A. (2015). The nature of culture: An eight-grade model for the evolution and expansion of cultural capacities in hominins and other animals. *Journal of Anthropological Sciences, 93*, 43–70.

Hsieh, L.-T., & Ranganath, C. (2013). Frontal midline theta oscillations during working memory maintenance and episodic encoding and retrieval. *Neuroimage, 85*, 721–729.

Iovita, R., Schönekeß, H., Gaudzinski-Windheuser, S., & Jäger, F. (2016). Identifying weapon delivery systems using macrofracture analysis and fracture propagation velocity: A controlled experiment. In R. Iovita & K. Sano (Eds.), *Multidisciplinary approaches to the study of Stone Age weaponry* (pp. 13–27). Dordrecht, The Netherlands: Springer.

Iriki, A. (2006). The neural origins and implications of imitation, mirror neurons and tool use. *Current Opinion in Neurobiology, 16*(6), 660–667.

Jensen, O., Gelfand, J., Kounios, J., & Lisman, J. E. (2002). Oscillations in the alpha band (9–12 Hz) increase with memory load during retention in a short-term memory task. *Cerebral Cortex, 12*(8), 877–882.

Johnson, M. K., Raye, C. L., Mitchell, K. J., Touryan, S. R., Greene, E. J., & Nolen-Hoeksema, S. (2006). Dissociating medial frontal and posterior cingulate activity during self-reflection. *Social Cognitive and Affective Neuroscience, 1*(1), 56–64.

Kawasaki, M., & Yamaguchi, Y. (2012). Frontal theta and beta synchronisations for monetary reward increase visual working memory capacity. *Social Cognitive and Affective Neuroscience, 80*(5), 523–530.

Kim, Y., Seo, J.-H., Song, H.-J., Yoo, D.-S., Lee, H. J., Lee, J., ... Chang, Y. (2011). Neural correlates related to action observation in expert archers. *Behavioural Brain Research, 223*(2), 342–347.

Klimesch, W., Doppelmayr, M., Russegger, H., & Pachinger, T. (1996). Theta band power in the human scalp EEG and the encoding of new information. *Neuroreport, 7*(7), 1235–1240.

Klimesch, W., Doppelmayr, M., Stadler, W., Pöllhuber, D., Sauseng, P., & Roehm, D. (2001). Episodic retrieval is reflected by a process specific increase in human electroencephalographic theta activity. *Neuroscience Letters, 302*(1), 49–52.

Klimesch, W., Schack, B., & Sauseng, P. (2005). The functional significance of theta and upper alpha oscillations. *Experimental Psychology, 52*(2), 99–108.

Knyazev, G. G. (2012). EEG delta oscillations as a correlate of basic homeostatic and motivational processes. *Neuroscience and Biobehavioral Reviews, 36*(1), 677–695.

Kringelbach, M. L. (2005). The human orbitofrontal cortex: Linking reward to hedonic experience. *National Review of Neuroscience, 6*(9), 691–702.

Langbroek, M. (2014). Ice age mentalists: Debating neurological and behavioural perspectives on the Neandertal and modern mind. *Journal of Anthropological Sciences, 92*, 285–289.

Larquet, M., Coricelli, G., Opolczynski, G., & Thibaut, F. (2010). Impaired decision making in schizophrenia and orbitofrontal cortex lesion patients. *Schizophrenia Research, 116*(2), 266–273.

Liebenberg, L. (1990). *The art of tracking: The origin of science*. Claremont, South Africa: David Philip Publishers.

Liebenberg, L. (2013). *The origin of science: The evolutionary roots of scientific reasoning and its implications for citizen science*. Cape Town, South Africa: Cyber Tracker.

Lombard, M. (2008). Finding resolution for the Howiesons Poort through the microscope: Micro-residue analysis of segments from Sibudu Cave, South Africa. *Journal of Archaeological Science, 35*(1), 26–41.

Lombard, M. (2011). Quartz-tipped arrows older than 60 Ka: Further use-trace evidence from Sibudu, KwaZulu-Natal, South Africa. *Journal of Archaeological Science, 58*(8), 1918–1930.

Lombard, M. (2015). Hunting and hunting technologies as proxy for teaching and learning during the Stone Age of southern Africa. *Cambridge Archaeological Journal, 25*(4), 877–887.

Lombard, M. (2016). Mountaineering or ratcheting? Stone Age hunting weapons as proxy for the evolution of human technological, behavioral and cognitive flexibility. In M. N. Haidle, N. J. Conard, & M. Bolus (Eds.), *The nature of culture: Based on an interdisciplinary symposium "The nature of culture", Tübingen, Germany* (pp. 135–146). Dordrecht, The Netherlands: Springer Science + Business Media.

Lombard, M., & Gärdenfors, P. (2017). Tracking the evolution of causal cognition in humans. *Journal of Anthropological Sciences, 95*, 1–16.

Lombard, M., & Haidle, M. N. (2012). Thinking a bow-and-arrow set: Cognitive implications of Middle Stone Age bow and stone-tipped arrow technology. *Cambridge Archaeological Journal, 22*(2), 237–264.

Lombard, M., & Parsons, I. (2010). Fact or fiction? Behavioural and technological reversal after 60 Ka in southern Africa. *South African Archaeological Bulletin, 65*(192), 221–228.

Lombard, M., & Parsons, I. (2011). What happened to the human mind after the Howiesons Poort? *Antiquity, 85*(330), 1433–1443.

Lombard, M., & Phillipson, L. (2010). Indications of bow and stone-tipped arrow use 64,000 years ago in KwaZulu-Natal, South Africa. *Antiquity, 84*(325), 635–648.

Lombard, M., Schlebusch, C., & Soodyall, H. (2013). Bridging disciplines to better elucidate the evolution of early *Homo sapiens* in southern Africa. *South African Journal of Science, 109*(11/12), 27–34.

Lombard, M., & Wadley, L. (2016). Hunting technologies during the Howiesons Poort at Sibudu Cave: What they reveal about human cognition in KwaZulu-Natal, South Africa, between ~65 and 62 Ka. In R. Iovita & K. Sano (Eds.), *Multidisciplinary approaches to the study of Stone Age weaponry* (pp. 273–286). Dordrecht, The Netherlands: Springer.

Lou, H. C., Luber, B., Crupain, M., Keenan, J. P., Nowak, M., Kjaer, T. W., . . . Lisanby, S. H. (2004). Parietal cortex and representation of the mental self. *Proceedings of the National Academy of Sciences of the United States of America, 101*(17), 6827–6832.

Lu, H., Zuo, Y., Gu, H., Waltz, J. A., Zhan, W., Scholl, C. A., . . . Stein, E. A. (2007). Synchronized delta oscillations correlate with the resting-state functional MRI signal. *Proceedings of the National Academy of Sciences of the United States of America, 104*(46), 18265–18269.

Makeig, S., Delorme, A., Westerfield, M., Jung, T.-P., Townsend, J., Courchesne, E., & Sejnowski, T. J. (2004). Electroencephalographic brain dynamics following manually responded visual targets. *PLoS Biology, 2*(6), 747–762.

Malafouris, L. (2008). Beads for a plastic mind: The "blind man"s stick' (BMS) hypothesis and the active nature of material culture. *Cambridge Archaeological Journal, 18*(3), 401–414.

Malafouris, L. (2013). *How things shape the mind: A theory of material engagement*. Cambridge, MA: MIT Press.

Malafouris, L. (2014). Third hand prosthesis (comment for Bruner & Lozano). *Journal of Anthropological Sciences, 92*, 281–283.

Martín-Loeches, M. (2014). Extending the mind, embodying cognition: New light on old endeavours. *Journal of Anthropological Sciences, 92*, 299–301.

May-Benson, T. A., & Cermak, S. A. (2007). Development of an assessment for ideational praxis. *American Journal of Occupational Therapy, 61*(2), 148–153.

Nácher, V., Ledberg, A., Deco, G., & Romo, R. (2013). Coherent delta-band oscillations between cortical areas correlate with decision making. *Proceedings of the National Academy of Sciences of the United States of America, 110*(37), 15085–15090.

Northoff, G., & Bermpohl, F. (2004). Cortical midline structures and the self. *Trends in Cognitive Sciences, 8*(3), 102–107.

Obleser, J., Wöstmann, M., Hellbernd, N., Wilsch, A., & Maess, B. (2012). Adverse listening conditions and memory load drive a common alpha oscillatory network. *Journal of Neuroscience, 32*(36), 12376–12383.

Odar, B., & Maver, A. (2011). Archers at Potočka zijalka? *Arheološki Vestnik, 62*, 433–456.

Ostlund, S. B., & Balleine, B. W. (2007). The contribution of orbitofrontal cortex to action selection. *Annals of the New York Academy of Sciences, 1121*(1), 174–192.

Overmann, K. A. (2015). Teeth, tools and human becoming. *Journal of Anthropological Sciences, 93*, 163–167.

Palva, S., & Palva, J. M. (2007). New vistas for alpha-frequency band oscillations. *Trends in Neurosciences, 30*(4), 150–158.

Pargeter, J., Shea, J. J., & Utting, B. (2016). Quartz backed tools as arrowheads and hand-cast spearheads: Hunting experiments and macro-fracture analysis. *Journal of Archaeological Science, 73*, 145–157.

Riede, F. (2008). The Laacher See-eruption (12,920 BP) and material culture change at the end of the Allerød in Northern Europe. *Journal of Archaeological Science, 35*(3), 591–599.

Riley, M., Lambert, B., Qi, X.-L., & Constantinidis, C. (2013). Neural correlates of visual working memory capacity in the posterior parietal cortex. *Journal of Vision, 13*(9), 1167.

Rios-Garaizar, J. (2015). The visuo-spatial capacities of Neanderthals seen through material culture. *Journal of Anthropological Sciences, 93*, 173–176.

Robbins, L. H., Campbell, A. C., Brook, G. A., Murphy, M. L., & Hitchcock, R. K. (2012). The antiquity of the bow and arrow in the Kalahari Desert: Bone points from White Paintings Rockshelter, Botswana. *Journal of African Archaeology, 10*(1), 7–20.

Rolls, E. T., & Grabenhorst, F. (2008). The orbitofrontal cortex and beyond: From affect to decision-making. *Progress in Neurobiology, 86*(3), 216–244.

Rushworth, M. F. S., & Taylor, P. C. J. (2006). TMS in the parietal cortex: Updating representations for attention and action. *Neuropsychologia, 44*(13), 2700–2716.

Salazar, R. F., Dotson, N. M., Bressler, S. L., & Gray, C. M. (2012). Content-specific fronto-parietal synchronization during visual working memory. *Science, 338*(6110), 1097–1100.

Sauseng, P., Hoppe, J., Klimesch, W., Gerloff, C., & Hummel, F. C. (2007). Dissociation of sustained attention from central executive functions: Local activity and interregional connectivity in the theta range. *European Journal of Neuroscience, 25*(2), 587–593.

Schacter, D. L., & Addis, D. R. (2007). The cognitive neuroscience of constructive memory: Remembering the past and imagining the future. *Philosophical Transactions of the Royal Society of London. Series B, Biological Sciences, 362*(1481), 773–786.

Schlebusch, C. M., Malmström, H., Günther, T., Sjödin, P., Coutinho, A., Edlund, H., . . . Jakobsson, M. (2017). Southern African ancient genomes estimate modern human divergence to 350,000 to 260,000 years ago. *Science, 358*(6266), 652–655.

Schoenbaum, G., & Setlow, B. (2001). Integrating orbitofrontal cortex into prefrontal theory: Common processing themes across species and subdivision. *Learning and Memory, 8*(3), 134–147.

Shea, J. J. (2006). The origins of lithic projectile point technology: Evidence from Africa, the Levant, and Europe. *Journal of Archaeological Science, 33*(6), 823–846.

Shea, J. J., & Sisk, M. L. (2010). Complex projectile technology and *Homo sapiens* dispersal into Western Eurasia. *PaleoAnthropology*, 100–122.

Spinapolice, E. E. (2015). Third hand: A testable hypothesis. *Journal of Anthropological Sciences, 93*, 169–171.

Spreng, R. N., & Grady, C. L. (2010). Patterns of brain activity supporting autobiographical memory, prospection, and theory of mind, and their relationship to the default mode network. *Journal of Cognitive Neuroscience, 22*(6), 1112–1123.

Strauss, S. (2005). Teaching as a natural cognitive ability: Implications for classroom practice and teacher education. In D. B. Pillemer & S. H. White (Eds.), *Developmental psychology and social change: Research, history, and policy* (pp. 368–388). Cambridge, UK: Cambridge University Press.

Strydom, B., & Ferreira, J. T. (2010). The role of vision and visual skills in archery. *South African Optometry, 69*(1), 21–28.

Tesche, C. D., & Karhu, J. (2000). Theta oscillations index human hippocampal activation during a working memory task. *Proceedings of the National Academy of Sciences of the United States of America, 97*(2), 919–924.

Thut, G., Nietzel, A., Brandt, S. A., & Pascual-Leone, A. (2006). α-Band electroencephalographic activity over occipital cortex indexes visuo-spatial attention bias and predicts visual target detection. *Journal of Neuroscience, 26*(37), 9494–9502.

Tulving, E. (2002). Episodic memory: From mind to brain. *Annual Review of Psychology, 53*(1), 1–25.

Tulving, E., & Kim, A. (2007). The medium and the message of mental time travel. *Behavioral and Brain Sciences, 30*(3), 334–335.

van Overwalle, F. (2009). Social cognition and the brain: A meta-analysis. *Human Brain Mapping, 30*(3), 829–858.

Wadley, L. (2010). Were snares and traps used in the Middle Stone Age and does it matter? A review and a case study from Sibudu, South Africa. *Journal of Human Evolution, 58*(2), 179–192.

Wadley, L. (2013). Recognizing complex cognition through innovative technology in Stone Age and Palaeolithic sites. *Cambridge Archaeological Journal, 23*(2), 163–183.

Wadley, L., Hodgskiss, T., & Grant, M. (2009). Implications for complex cognition from the hafting of tools with compound adhesives in the Middle Stone Age, South Africa. *Proceedings of the National Academy of Sciences of the United States of America, 106*(24), 9590–9594.

Wadley, L., & Mohapi, M. (2008). A segment is not a monolith: Evidence from the Howiesons Poort of Sibudu, South Africa. *Journal of Archaeological Science, 35*(9), 2594–2605.

Walter, E., & Dassonville, P. (2008). Visuospatial contextual processing in the parietal cortex: An fMRI investigation of the induced Roelofs effect. *Neuroimage, 42*(4), 1686–1697.

Weiss, S., & Mueller, H. M. (2012). "Too many betas do not spoil the broth": The role of beta brain oscillations in language processing. *Frontiers in Psychology, 3*, 1–15.

Williams, V. M. E., Burke, A., & Lombard, M. (2014). Throwing spears and shooting arrows: Preliminary results of a pilot neuroarchaeological study. *South African Archaeological Bulletin, 69*(200), 199–207.

Worden, M. S., Foxe, J. J., Wang, N., & Simpson, G. V. (2000). Anticipatory biasing of visuospatial attention indexed by retinotopically specific α-band electroencephalography increases over occipital cortex. *Journal of Neuroscience, 20*(RC63), 1–6.

Wynn, T. (2014). Commentary on "Extended mind and visuo-spatial integration: Three hands for the Neandertal lineage." *Journal of Anthropological Sciences, 92*, 291–293.

Wynn, T., & Coolidge, F. L. (2004). The expert Neandertal mind. *Journal of Human Evolution, 46*(4), 467–487.

Wynn, T., & Coolidge, F. L. (2010). How Levallois reduction is similar to, and not similar to, playing chess. In A. Nowel & I. Davidson (Eds.), *Stone tools and the evolution of human cognition* (pp. 83–104). Boulder, CO: University of Colorado Press.

Wynn, T., Haidle, M. N., Lombard, M., & Coolidge, F. L. (2017). The expert cognition model in human evolutionary studies. In T. Wynn & F. L. Coolidge (Eds.), *Cognitive models in Palaeolithic archaeology* (pp. 21–43). Oxford, UK: Oxford University Press.

Xu, Y., & Chun, M. M. (2009). Selecting and perceiving multiple visual objects. *Trends in Neurosciences, 13*(4), 116–124.

23

EPILOGUE

SITUATING THE COGNITIVE IN COGNITIVE ARCHAEOLOGY

Thomas Wynn

Over the past 40 years, evolutionary cognitive archaeology (ECA) has established itself as an active player in the game of paleoanthropology (if perhaps second team . . .), and it has even begun to participate on the playing field of cognitive science itself (Coolidge & Wynn, 2016; Hecht, Gutman, Bradley, Preuss, & Stout, 2015; Stout, Passingham, Frith, Apel, & Chaminade, 2011). However, the degree of commitment to the ideas and methods of cognitive science varies dramatically, from passing references to quasi-cognitive features of the mind to full-scale investment in cognitive theory. It is thus appropriate to ask, "How cognitive is cognitive archaeology?"

First, however, it is necessary to address the bugaboos of language and symbolism. For many paleoanthropologists, language and symbolism continue to be proxies for hominin cognition, as if nothing else of importance ever occurred in hominin thinking. This has become especially evident in the increasingly bizarre debate about Neandertals and their putative similarities with and differences from "modern" humans (e.g., Villa & Roebroeks, 2014; also see the critique provided by Wynn, Overmann, & Coolidge, 2016). Advocates of indistinguishability suggest that because Neandertals made marks on cave walls (or wore feathers, or colored their bodies, etc.) that they were clearly using abstract symbols, had fully syntactical language, and were thus no different from contemporary *Homo sapiens* living in Africa. The poverty of this reasoning should be obvious, yet assertions like it appear again and again in academic publications, as well as in news media, where Neandertals hold a perennial fascination. Unfortunately, such simplistic assertions have had a negative impact on ECA. When casual readers encounter this kind of reasoning, it is no wonder they become skeptical of any attempt to document the prehistoric mind. Language and symbolism are undoubtedly important to modern life, and documenting their evolution should be one of the goals of paleoanthropology. But neither is an easily defined *cognitive* ability. They incorporate and rely on a variety of neurocognitive resources, but there is no single cognitive ability or brain region or neural network that is the nexus of symbolic communication.

ECA has made serious contributions to the study of language evolution, but these have had little lasting impact. For example, the most sophisticated archaeologically based argument for language evolution in general has been that of Iain Davidson and William Noble (Davidson & Noble, 1989; Noble & Davidson, 1996). It is solidly grounded in theory (i.e., Gibsonian perceptual psychology) and derives explicit predictions about the archaeological record. Yet it is widely ignored in paleoanthropological circles because its conclusions violate the ungrounded, common-sense expectations of the standard narrative of hominin evolution (which are that there was a long, gradual increase in brain size, and problem-solving with language led the way). This almost religious adherence to a narrative account is not how science is supposed to work. ECA studies of individual components of language have been more successful, as when Dietrich Stout and colleagues (Hecht, Gutman, Khreisheh, et al., 2015; Stout & Chaminade, 2012) described a potential common neural basis for object manipulation and syntax. The irony here is that Stout's research program is discovering a great deal of provocative information about tool use and technical cognition that is important to cognitive science in general, but it is the language-origin piece that draws the attention of paleoanthropologists. There was more to hominin cognitive evolution than the development of language and symbolism, yet pursuit of this elusive Holy Grail continues to divert attention and resources, to the detriment of ECA as a whole.

It is now common to encounter quasi-cognitive terms in archaeological discourse. These are terms that have no explicit grounding in a formal model of cognitive function but which refer to abilities that have potential bases in cognition. "Planning," "abstract thinking," "hierarchical complexity," and the fearsome "mental template" are now common topics even in accounts of hominin evolution that make little commitment to a cognitive stance (the justly influential McBrearty & Brooks, 2000, is an excellent example). Their use reflects the now-general tolerance of a cognitive perspective in paleoanthropology, and more specifically an understanding that prehistoric minds were not *tabulae rasae* molded by material context but active determinants of the archaeological record. Compared to 40 years ago, this is progress. A few quasi-cognitive approaches have been even more ambitious and border on archaeologically derived models of cognition. The most sophisticated of these is Miriam Haidle's cognigram approach (Haidle, 2009, 2014; Lombard & Haidle, 2012). It is grounded in her own system of cognitive components, including such qualities as "perception of need" and "problem-solution distance." Though not derived from a formal model of cognition, these components do seem to capture something of the complexity of technical cognition. Moreover, they appear amenable to reinterpretation from the perspective of more formal theory (Lombard & Haidle, 2012; Wynn, Haidle, Lombard, & Coolidge, 2017). Problem-solution distance in particular has shown considerable utility (Muller, Clarkson, & Shipton, 2017). Indeed, at the 2008 Wenner-Gren symposium on working memory (Wynn & Coolidge, 2010), the psychologists in the group expressed agreement that problem-solution distance had real potential as a measure of technical problem-solving. In a different vein, Mark Moore started his interpretation of early technology with the basic mechanical constraints of stone knapping, and borrowing loosely from Patricia Greenfield (1991), developed the notion of knapping as a "design space" that constrains "grammars of action" (Moore, 2011; Moore & Perston, 2016; also see Chapter 8 in this volume). The result is an understanding of stone knapping as a cognitive activity with its own internal logic that must be taken into account in any attempt to reconstruct cognition through analysis of lithic remains.

Quasi-cognitive approaches such as these may lack well-grounded links to established cognitive concepts, but they are unavoidable given the virtual absence of well-developed psychological models of technical cognition in natural settings. Unless and until psychology addresses technical cognition head on, archaeologists will inevitably be drawn to approaches in which their data sets are privileged, even if they must develop models of their own. In such approaches, the cognitive exists in the cognitive stance taken by the practitioners more than in terms and concepts taken from cognitive science directly. However, archaeologists can and do draw on cognitive science directly.

It is now not unusual to encounter interpretive concepts taken from cognitive science itself. Cognitive neuroscience is the most common source of understanding, which is not surprising given the tremendous success that this approach has experienced over the past three decades. **Cognitive control** (Hecht et al., 2013; Hecht, Gutman, Khreisheh, et al., 2015; Stout, Hecht, Khreisheh, Bradley, & Chaminade, 2015; Twomey, 2013; Vaesen, 2012), **visual attention** (Opitz, 2017), **cognitive fluidity** (Rossano, 2017), **working memory** (Lombard & Haidle, 2012; Wadley, Hodgskiss, & Grant, 2009; Wynn & Coolidge, 2010), **episodic memory** (De Smedt & de Cruz, 2011), **prospective memory** (Ambrose, 2010), **retrieval structures** (Wynn et al., 2017), **theory of mind** (Cole, 2014, 2017; Gärdenfors & Högberg, 2017; Henshilwood & Dubreuil, 2011; Shipton, 2010), **praxis** (Ruck, 2014), **skill** (Bril et al., 2012), and **spatial cognition** (Burke, 2012; Hodgson, 2010; Wynn, 2010) are just a few of the cognitive concepts that have appeared recently in archaeological discussions. In this way, ECA is beginning to recast the standard narrative in terms defined and tested by experimental psychology. The resulting picture will not only enhance paleoanthropology, it will enhance evolutionary psychology (EP) by providing the chronological and contextual information that EP typically eschews but desperately needs. Most of these studies have been *post hoc*. Archaeologists have taken archaeological data sets and patterns initially generated for other purposes (typically, materialist interpretations of one kind or another) and re-described or reinterpreted the results. Some of the most exciting examples have also relied on experimental replication studies (Lombard & Haidle, 2012; Stout et al., 2015; Wadley et al., 2009). But because the attributes and larger-scale patterns were not initially cognitive in intent, they are often not optimal for a cognitive analysis, as when Coolidge and Wynn initially struggled to recognize enhanced working memory in descriptions of Later Stone Age lithic technology (Coolidge & Wynn, 2001). If ECA aspires to contribute to cognitive science as a whole—and by no means is this a goal shared by everyone—then it will need to make a more serious commitment to the cognitive perspective.

The alternative to *post hoc* application of cognitive concepts is to ground the description and analysis in cognitive concepts from the very beginning. This is ECA at its most effective, and arguably the only avenue for ECA to contribute directly to cognitive science. Early examples include Wynn's use of a typology based on Piaget's scheme of spatial cognitive development (Wynn, 1979) and Davidson and Nobles' use of gesture and depiction as derived from Gibsonian perceptual psychology (Davidson & Noble, 1989). More recently, James Cole (2014, 2015, 2017; also see Chapter 17 in this volume) has constructed a typology of interpersonal interaction based on levels of intentionality and embodied cognition, and from it has derived expectations about patterns in the archaeological record. The result is an evolutionary sequence of a successively more powerful hominin theory of mind that both provides a pedigree for this important component of social cognition and enables a discussion of the actual context of development, not just a hypothetic environment of evolutionary adaptedness. Karenleigh Overmann (2013, 2016a, 2016b, 2017; also see Chapter 20 in this volume) initiated her analysis of numeracy with basic facts concerning numerical cognition, developed a hypothesis about how such thinking could have evolved through embodied and material resources, and then scoured the ethnographic and archaeological records for evidence. The result is a powerful model that challenges the standard (but underdetermined) "language first" model of numerical cognition. Interestingly, Overmann's model blithely erases the sacrosanct end-Pleistocene terminus for hominin evolution, and documents important evolutionary developments in cognition continuing into the Bronze Age. These are but two examples, but they underscore the potential of ECA. Not only does ECA enrich the standard narrative of hominin evolution, it also challenges the utility of the standard narrative itself as a framework for understanding the evolution of mind. ECA also provides an important alternative to evolutionary psychology, which largely ignores the paleoanthropological record in favor of a narrow Darwinian, adaptationist approach. But to participate on an equal footing with evolutionary psychology, ECA must commit to cognitive science, not just with *post hoc* recasting of archaeological interpretations, but with a complete package that starts with cognitive theory and concepts, generates predictions about archaeologically visible behavior, and examines the archaeological record for evidence.

ECA practitioners, including those with a serious commitment to cognitive theory, often run into a nagging problem—the experimental targets of cognitive neuroscience rarely include the kinds of activities and problem-solving that archaeologists regularly encounter in their data. Stone knapping, gathering plants, setting traps for small mammals, and finding ochre for coloring are just not the kinds of things that psychologists ask their undergraduate participants to do. To be fair, psychologists have begun to study "ecologically valid" activities through experimental protocols that have real-world relevance (unlike, say, the Stroop task, in which color words are presented to participants in different colors—something hardly anyone ever encounters in the real world), but thus far most of these studies focus on activities that are invisible to the archaeologist (e.g., Bethlehem et al., 2017). Ecological validity is one of the primary concerns of EP, and not surprisingly, the studies focus on themes that have always resonated for EP: mate selection, altruism, sex differences, and so on. There is, of course, no reason to expect evolutionary psychologists to study activities of relevance to archaeologists, though this would certainly be nice. Archaeologists will have to do it for themselves, and this is the largest missing piece in the methodological tessellation that is ECA.

Archaeologists have begun to do their own experimental work. As mentioned earlier, archaeologists such Lyn Wadley (e.g., Wadley et al., 2009) have turned to actualistic replication of prehistoric technical activities. Methodologically, this approach shares more with ethnography and cognitive anthropology than cognitive neuroscience and often requires *post hoc* invocations of cognitive mechanisms. Archaeologists have also begun to use experimental protocols based in cognitive neuroscience. Morgan and colleagues (Morgan et al., 2015) and Putt and colleagues (Putt, Woods, & Franciscus, 2014) have recently tested the efficacy of different learning and teaching strategies in stone knapping, and Stout and colleagues (Hecht, Gutman, Khreisheh, et al., 2015; Stout et al., 2015) have used a longitudinal study in which novice participants learned stone knapping and were tested at regular points in the process. This is the direction that ECA must go if it hopes not only to enrich the standard narrative of hominin evolution but also to make contributions to cognitive neuroscience as a whole. True, it might require a revision of the archaeological *Weltanschauung* wherein researchers see themselves more as Indiana Jones than B. F. Skinner, but the pay-off would be a much more effective cognitive archaeology.

Of course, ECA need not ally itself with cognitive neuroscience. There are alternative understandings of the nature of mind that might, in fact, be more amenable to archaeological data. Cognitive neuroscience takes a decidedly Cartesian stance on the nature of mind: Thinking occurs in the brain, which is firmly situated in the head. But in the last 25 years, non-Cartesian understandings of mind have become more and more influential. "Embodied" and "extended" approaches to cognition emphasize the active, and indissociable, roles that bodies and material objects play in mental life. The way we think is enabled by what our bodies can do and what our artifacts can do. The advantage of such a perspective for the archaeologist should be obvious. Archaeologists find things. And if prehistoric things played roles in prehistoric thought, then archaeologists are finding actual components of thinking. The most sophisticated archaeological perspective on this approach is Lambros Malafouris' material engagement theory (Malafouris, 2013), which has begun to make substantive contributions to the understanding of the evolutionary record (Malafouris, 2008; Overmann, 2016a; Roberts, 2016). There are others. Recently, Bril and colleagues have made productive use of Gibsonian perceptual psychology (Bril et al., 2012; Bril, Parry, & Dietrich, 2015), a perspective first used in ECA by Davidson and Noble (1989) to investigate tool use in general and the origin of stone knapping in particular.

There is a final move that ECA practitioners could take that would enhance their value to cognitive science in general, and perhaps establish the approach as a viable scientific discipline in its own right: abandon the standard paleoanthropological narrative. Too much of ECA has been aimed at elucidating the minds of Neandertals, or *Homo erectus*, or even *Australopithecus*. Such reconstructions are entertaining, perhaps, but in the long run not very useful. A complete picture of the mind of an early hominin is beyond reach, even for a hominin as well known as Neandertals. But individual components of cognition are accessible, and of significant interest to cognitive science in general. Knowing that enhancements in cognitive control occurred at several points in hominin evolution, for example, enables science to identify the contexts in which they occurred. It also lends support to arguments for cognitive control's importance in human mental life. It is in this way that ECA can make serious contributions to the scientific understanding of the mind.

Prognostication is a perilous task, especially when it concerns a methodological approach that has yet to achieve maturity. It seems likely that ECA will continue as a minority perspective in paleoanthropology, adding much-needed cognitive content to the standard narrative, but making only occasional contributions to cognitive science in general. However, if ECA practitioners aspire to make significant contributions to the understanding of the mind itself, they will need to make a much more serious commitment to cognitive theory and experimental methods.

REFERENCES

Ambrose, S. H. (2010). Coevolution of composite-tool technology, constructive memory, and language: Implications for the evolution of modern human behavior. *Current Anthropology, 51*(S1), S135–S147.

Bethlehem, R. A. I., Allison, C., van Andel, E. M., Coles, A. I., Neil, K., & Baron-Cohen, S. (2017). Does empathy predict altruism in the wild? *Social Neuroscience, 12*(6), 743–750.

Bril, B., Parry, R., & Dietrich, G. (2015). How similar are nut-cracking and stone-flaking? A functional approach to percussive technology. *Philosophical Transactions of the Royal Society of London. Series B, Biological Sciences, 370*(1682), 1–13.

Bril, B., Smaers, J., Steele, J., Rein, R., Nonaka, T., Dietrich, G., . . . Roux, V. (2012). Functional mastery of percussive technology in nut-cracking and stone-flaking actions: Experimental comparison and implications for the evolution of the human brain. *Philosophical Transactions of the Royal Society of London. Series B, Biological Sciences, 367*(1585), 59–74.

Burke, A. (2012). Spatial abilities, cognition and the pattern of Neanderthal and modern human dispersals. *Quaternary International, 247*, 230–235.

Cole, J. (2014). The identity model: A theory to access visual display and hominin cognition within the Palaeolithic. In R. I. M. Dunbar, C. Gamble, & J. A. J. Gowlett (Eds.), *Lucy to language: The benchmark papers* (pp. 90–107). Oxford, UK: Oxford University Press.

Cole, J. (2015). Handaxe symmetry in the Lower and Middle Palaeolithic: Implications for the Acheulean gaze. In F. Coward, R. Hosfield, M. I. Pope, & F. Wenban-Smith (Eds.), *Settlement, society and cognition in human evolution: Landscapes in mind* (pp. 234–257). Cambridge, UK: Cambridge University Press.

Cole, J. (2017). Accessing hominin cognition: Language and social signalling in the Lower to Middle Paleolithic. In T. Wynn & F. L. Coolidge (Eds.), *Cognitive models in Palaeolithic archaeology* (pp. 157–195). New York: Oxford University Press.

Coolidge, F. L., & Wynn, T. (2001). Executive functions of the frontal lobes and the evolutionary ascendancy of *Homo sapiens*. *Cambridge Archaeological Journal, 11*(3), 255–260.

Coolidge, F. L., & Wynn, T. (2016). An introduction to cognitive archaeology. *Current Directions in Psychological Science, 25*(6), 386–392.

Davidson, I., & Noble, W. (1989). The archaeology of perception: Traces of depiction and language. *Current Anthropology, 30*(2), 125–155.

De Smedt, J., & de Cruz, H. (2011). The role of material culture in human time representation: Calendrical systems as extensions of mental time travel. *Adaptive Behavior, 19*(1), 63–76.

Gärdenfors, P., & Högberg, A. (2017). The archaeology of teaching and the evolution of *Homo docens*. *Current Anthropology, 58*(2), 188–208.

Greenfield, P. M. (1991). Language, tools and brain: The ontogeny and phylogeny of hierarchically organized sequential behavior. *Behavioral and Brain Sciences, 14*(4), 531–595.

Haidle, M. N. (2009). How to think a simple spear. In S. A. de Beaune, F. L. Coolidge, & T. Wynn (Eds.), *Cognitive archaeology and human evolution* (pp. 57–74). Cambridge, UK: Cambridge University Press.

Haidle, M. N. (2014). Building a bridge—An archeologist's perspective on the evolution of causal cognition. *Frontiers in Psychology, 5*(1472), 1–15.

Hecht, E. E., Gutman, D. A., Bradley, B. A., Preuss, T. M., & Stout, D. (2015). Virtual dissection and comparative connectivity of the superior longitudinal fasciculus in chimpanzees and humans. *NeuroImage, 108*, 124–137.

Hecht, E. E., Gutman, D. A., Khreisheh, N., Taylor, S. V, Kilner, J., Faisal, A. A., . . . Stout, D. (2015). Acquisition of Paleolithic toolmaking abilities involves structural remodeling to inferior frontoparietal regions. *Brain Structure and Function, 220*(4), 2315–2331.

Hecht, E. E., Murphy, L. E., Gutman, D. A., Votaw, J. R., Schuster, D. M., Preuss, T. M., . . . Parr, L. A. (2013). Differences in neural activation for object-directed grasping in chimpanzees and humans. *Journal of Neuroscience, 33*(35), 14117–14134.

Henshilwood, C. S., & Dubreuil, B. (2011). The Still Bay and Howiesons Poort, 77-59 ka: Symbolic culture and the evolution of the mind during the African Middle Stone Age. *Current Anthropology, 52*(3), 361–400.

Hodgson, D. (2010). Another side of symmetry: The Acheulean biface debate. *Antiquity, 84*, 325.

Lombard, M., & Haidle, M. N. (2012). Thinking a bow-and-arrow set: Cognitive implications of Middle Stone Age bow and stone-tipped arrow technology. *Cambridge Archaeological Journal, 22*(2), 237–264.

Malafouris, L. (2008). Beads for a plastic mind: The "blind man"s stick' (BMS) hypothesis and the active nature of material culture. *Cambridge Archaeological Journal, 18*(3), 401–414.

Malafouris, L. (2013). *How things shape the mind: A theory of material engagement.* Cambridge, MA: MIT Press.

McBrearty, S., & Brooks, A. S. (2000). The revolution that wasn't: A new interpretation of the origin of modern human behavior. *Journal of Human Evolution, 39*(5), 453–563.

Moore, M. W. (2011). The design space of stone flaking: Implications for cognitive evolution. *World Archaeology, 43*(4), 702–715.

Moore, M. W., & Perston, Y. (2016). Experimental insights into the cognitive significance of early stone tools. *PLoS One, 11*(7), 1–37.

Morgan, T. J. H., Uomini, N. T., Rendell, L. E., Chouinard-Thuly, L., Street, S. E., Lewis, H. M., . . . Laland, K. N. (2015). Experimental evidence for the co-evolution of hominin toolmaking teaching and language. *Nature Communications, 6*(6029), 1–18.

Muller, A., Clarkson, C., & Shipton, C. (2017). Measuring behavioural and cognitive complexity in lithic technology throughout human evolution. *Journal of Anthropological Archaeology, 48*, 166–180.

Noble, W., & Davidson, I. (1996). *Human evolution, language, and mind: A psychological and archaeological inquiry*. Cambridge, UK: Cambridge University Press.

Opitz, R. (2017). An experiment in using visual attention metrics to think about experience and design choices in past places. *Journal of Archaeological Method and Theory, 24*(4), 1203–1226.

Overmann, K. A. (2013). Material scaffolds in numbers and time. *Cambridge Archaeological Journal, 23*(1), 19–39.

Overmann, K. A. (2016a). Beyond writing: The development of literacy in the Ancient Near East. *Cambridge Archaeological Journal, 26*(2), 285–303.

Overmann, K. A. (2016b). The role of materiality in numerical cognition. *Quaternary International, 405*, 42–51.

Overmann, K. A. (2017). Materiality and numerical cognition: A material engagement theory perspective. In T. Wynn & F. L. Coolidge (Eds.), *Cognitive models in Palaeolithic archaeology* (pp. 89–112). Oxford, UK: Oxford University Press.

Putt, S. S., Woods, A. D., & Franciscus, R. G. (2014). The role of verbal interaction during experimental bifacial stone tool manufacture. *Lithic Technology, 39*(2), 96–112.

Roberts, P. (2016). "We have never been behaviourally modern": The implications of material engagement theory and metaplasticity for understanding the Late Pleistocene record of human behaviour. *Quaternary International, 405*, 8–20.

Rossano, M. (2017). Cognitive fluidity and Acheulean over-imitation. *Cambridge Archaeological Journal, 27*(3), 495–509.

Ruck, L. (2014). Manual praxis in stone tool manufacture: Implications for language evolution. *Brain and Language, 139*, 68–83.

Shipton, C. (2010). Imitation and shared intentionality in the Acheulean. *Cambridge Archaeological Journal, 20*(2), 197–210.

Stout, D., & Chaminade, T. (2012). Stone tools, language and the brain in human evolution. *Philosophical Transactions of the Royal Society of London. Series B, Biological Sciences, 367*(1585), 75–87.

Stout, D., Hecht, E. E., Khreisheh, N., Bradley, B. A., & Chaminade, T. (2015). Cognitive demands of Lower Paleolithic toolmaking. *PLoS One, 10*(4), 1–18.

Stout, D., Passingham, R. E., Frith, C. D., Apel, J., & Chaminade, T. (2011). Technology, expertise and social cognition in human evolution. *European Journal of Neuroscience, 33*(7), 1328–1338.

Twomey, T. (2013). The cognitive implications of controlled fire use by early humans. *Cambridge Archaeological Journal, 23*(1), 113–128.

Vaesen, K. (2012). Cooperative feeding and breeding, and the evolution of executive control. *Biology and Philosophy, 27*(1), 115–124.

Villa, P., & Roebroeks, W. (2014). Neandertal demise: An archaeological analysis of the modern human superiority complex. *PLoS One, 9*(4), 1–10.

Wadley, L., Hodgskiss, T., & Grant, M. (2009). Implications for complex cognition from the hafting of tools with compound adhesives in the Middle Stone Age, South Africa. *Proceedings of the National Academy of Sciences of the United States of America, 106*(24), 9590–9594.

Wynn, T. (1979). The intelligence of later Acheulean hominids. *Man, 14*, 371–391.

Wynn, T. (2010). The evolution of human spatial cognition. In F. L. Dolins & R. W. Mitchell (Eds.), *Spatial cognition, spatial perception: Mapping the self and space* (pp. 213–236). Cambridge, UK: Cambridge University Press.

Wynn, T., & Coolidge, F. L. (2010). Beyond symbolism and language: An introduction to Supplement 1, Working Memory. *Current Anthropology, 51*(S1), S5–S16.

Wynn, T., Haidle, M. N., Lombard, M., & Coolidge, F. L. (2017). The expert cognition model in human evolutionary studies. In T. Wynn & F. L. Coolidge (Eds.), *Cognitive models in Palaeolithic archaeology* (pp. 21–43). Oxford, UK: Oxford University Press.

Wynn, T., Overmann, K. A., & Coolidge, F. L. (2016). The false dichotomy: A refutation of the Neandertal indistinguishability claim. *Journal of Anthropological Sciences, 94*, 1–21.

INDEX